AF400713

Geometry: The Line and the Circle

AMS/MAA | TEXTBOOKS

VOL **44**

Geometry: The Line and the Circle

Maureen T. Carroll

Elyn Rykken

Providence, Rhode Island

2010 *Mathematics Subject Classification.* Primary 51-01.

For additional information and updates on this book, visit
www.ams.org/bookpages/text-44

Library of Congress Cataloging-in-Publication Data

Names: Carroll, Maureen T., 1966– author. | Rykken, Elyn, 1967– author.
Title: Geometry: The line and the circle / Maureen T. Carroll, Elyn Rykken.
Description: Providence, Rhode Island: MAA Press, an imprint of the American Mathematical Society, [2018] | Series: AMS/MAA textbooks; volume 44 | Designed for an upper-level college geometry course. | Includes bibliographical references and indexes.
Identifiers: LCCN 2018034790 | ISBN 9781470448431 (alk. paper)
Subjects: LCSH: Geometry–Textbooks. | Geometry–Study and teaching (Higher)
Classification: LCC QA445 .C2985 2018 | DDC 516–dc23
LC record available at https://lccn.loc.gov/2018034790

To our parents for their love, support and encouragement

Contents

Contents ix

Note to the Instructor

This book is an introduction to Euclidean and non-Euclidean geometry designed for an upper-level college geometry course. Its content and narrative grew out of our experience teaching junior/senior level advanced geometry and history of mathematics courses for over twenty years. We have learned that, independent of ability and mathematical maturity, most students have only faint memories of the synthetic Euclidean geometry that they studied in high school. This, coupled with the fact that mathematics students rarely have occasion to read primary sources in their major, led us to take full advantage of the unique opportunity available in geometry and introduce our readers to the bible of mathematics, Euclid's *Elements*. With Euclid as a guide, the reader begins by travelling along the same path as millions of geometry students spanning multiple millennia, continents and languages.

Before beginning our journey with Euclid, we introduce the two most important and familiar geometric objects as our main characters, *the line* and *the circle*. These characters coincide with the Euclidean tools provided by the axioms at the start of the *Elements*. For us, they serve as a narrative touchstone as we periodically check in on them as we make our way through the *Elements*, noting that, while some books emphasize the line, others highlight the circle. In Book I, for example, the line has the starring role in the proposition statements, but the circle is doing a lot of the heavy lifting behind the scenes within the proofs. When we move beyond Euclidean geometry, we identify the behavior of our main characters in each of their new environments in order to keep track of any changes. Comparing and contrasting the nature of these fundamental figures in other geometries has the added benefit of challenging the reader's preconceived notions of straightness and roundness, forcing a re-examination of basic geometry concepts. In particular, we encourage the reader to contemplate provocative questions such as: What is a straight line? What does parallel mean? What is distance? What is area?

Euclid's *Elements* is a mathematical achievement of historical significance. No other mathematics text has been published as many times or read by as many people as the *Elements*. Its longevity is due to its clarity, rigor and, most importantly, its superior organization and development of geometry as an axiomatic system. It is the proper gateway to the study of axiomatic systems. Of course, the only way to fully appreciate and understand Euclidean geometry is to step outside of it in order to gain perspective. We do this by exploring Spherical, Taxicab, Hyperbolic, and finite affine and projective geometries. In fact, we take a detour halfway through Euclid's first book in order to consider two of these geometries. This change of perspective at an early stage of our exploration of the *Elements* provides a natural way to expose hidden flaws in Euclid's reasoning. It also sheds light on the importance of the axiomatic development of mathematics, and creates an avenue to discuss the difference between axiomatic systems and their models, as well as the desirable properties of such systems.

The history of geometry comprises the majority of the history of mathematics. Fittingly, this text includes discussions of many important historical figures and developments in geometry, including the controversy surrounding Euclid's fifth postulate, the impossible constructions of Greek antiquity, and the development of non-Euclidean, projective and finite geometries. It is important for students to understand that new mathematics does not arrive fully formed on the page, but rather, evolves as it is discovered and created by individuals, sometimes spanning centuries.

Outline of the book

This book covers traditional Euclidean geometry with an axiomatic approach through the lens provided by the *Elements*. After introducing our two main characters in Chapter 1, we discuss Euclid's definitions, Postulates, and Common Notions in Chapter 2. Chapter 3 presents Neutral geometry, covering the first 28 propositions of Book I along with I.31. We take a detour from the *Elements* to explore Spherical and Taxicab geometries in Chapters 4 and 5. These geometries force us to reconsider our preconceived ideas about straightness, distance and area. They also reveal the gaps in Euclid's original set of axioms. Opening the door to other geometries early in the book allows us to consider axiomatic systems in general, and to introduce Hilbert's axioms for plane geometry in Chapter 6 as a way to shore up the gaps in Euclid's foundation. Here, we also detail the fundamental and desirable properties of axiomatic systems and the mathematicians who were the first to successfully navigate these metamathematical waters.

After discussing Neutral, Spherical and Taxicab geometries, we head back to Euclid's Book I in Chapter 7. By including Book I in its entirety, readers experience the beauty of Euclid's reasoning and his modular approach to the systematic development of propositions. He meticulously avoids any use of the Parallel Postulate for as long as possible and builds new tools as allowed. It is only at the very end of Book I, after carefully working his way through triangle congruence, parallelism, area and quadrature, that we discover how these pieces fit together to achieve his ultimate goal, the proof of the Pythagorean Theorem and its converse.

Geometric algebra is the topic of Book II in Chapter 8, where readers should experience a newfound appreciation for algebra as it greatly simplifies the propositions of this book. This chapter is particularly helpful for secondary education students since it includes the Law of Cosines. Euclid's book ends with the quadrature of any polygon, but the chapter ends with a return to Spherical geometry as we consider which figures can be "squared" on a sphere. Chapter 9 briefly covers the topic of similarity (found in Book VI) and ends with a generalization of the Pythagorean Theorem. Chapter 10 explores Euclid's third book where the focus is squarely on the circle, and our oft-neglected main character gets some long overdue "me time." Chapter 11 covers Book IV which concerns concurrency points of a triangle, constructions of regular polygons, and results about inscribed and circumscribed circles.

In Chapter 12 we shift to Hyperbolic geometry where we continue to re-examine the concepts of straightness, parallelism, distance and area. Understanding the usefulness of models, we present three models for Hyperbolic geometry, ultimately focusing on the Poincaré Half-Plane model. In Chapter 13, we give an axiomatic development of Hyperbolic geometry and prove the following surprising results: parallel lines are not everywhere equidistant, the angle sum of a triangle is less than two right angles, rectangles and squares do not exist, AAA is a congruence scheme, and area is a function of angle sums rather than lengths.

We follow this in Chapter 14 with an axiomatic development of finite affine and projective planes, including a lengthy discussion of the history of the development of projective geometry. Chapter 15 covers isometries of the Euclidean and hyperbolic planes. We finish the book with a return to Euclidean geometry to tell the story of four classic construction problems of Greek antiquity and how the nineteenth-century solutions to these problems unambiguously marked the limitations of the Euclidean tools while simultaneously opening new paths ripe for mathematical exploration.

Designing a course using this text

With a healthy amount of both Euclidean and non-Euclidean geometry, instructors are free to choose their focus based on the needs, abilities and mathematical maturity of their students. We outline an option for a course with a majority emphasis on Euclidean topics below, but this can easily be revised to form a minority Euclidean course. Regarding the Euclidean content of the book, rather than simply picking and choosing highlights of the first six books of Euclid's *Elements*, we have included all of the propositions from the first two books, over half of the propositions from the third book, and nearly all of the fourth book. We do not, however, intend for the instructor to show all of these propositions in class. Our courses are designed to be more interactive than a standard lecture format course, as we find that it helps our students take ownership of the material. We recommend a plan where students rewrite Euclid's propositions using modern notation and present their updated version to the class. Through this process, students gain skill in both writing and presenting proofs since they must carefully read the proofs in order to separate assumptions from conclusions, and to determine the key definitions, postulates and previous propositions needed in each argument. All of our students, not simply those seeking a degree in secondary education, benefit from these oral presentations (as recommended by Mathematical Association of America [MAA] guidelines). For the construction propositions, we suggest that students reproduce these results using dynamic geometry software. This methodical approach to Euclidean geometry highlights the strength of an axiomatic development and has the added benefit of clearly distinguishing Neutral from non-Neutral propositions. Certainly, some of these propositions can be skipped in the classroom altogether, and they are available in the book as a resource.

While there is more content than can be covered in a semester, this book is, nevertheless, designed to be used for a one-semester course in geometry. Its intended audience is junior/senior undergraduate mathematics majors (be they seeking certification in secondary education or not). One *could* spend the entire semester on the first eleven chapters, but by carefully choosing topics, we expect that an instructor can cover these chapters in roughly eight or nine weeks. To get a sense of how to pace a course, we have included a breakdown of time that we typically spend on each of these chapters. We have taught the course three days a week with 50-minute periods, and two days a week with 75-minute periods.

We fully expect students to read this book. In that spirit, we assign the reading of Chapter 1 before the first class, and we discuss Chapter 1, as well as Chapter 2, on the first day. We then cover the first three propositions of Book I and introduce dynamic geometry software such as Geometer's Sketchpad®or GeoGebra in class. We have found that a basic familiarity with geometric terms facilitates a smooth transition from understanding Heath's translation of Euclid's propositions and proofs to updating the proofs with revised notation. Consequently, for roughly a week we have the students present their updated versions of the propositions from Chapter 3, skipping the proofs of the construction propositions which are included in a lab assignment utilizing the software.

For Chapter 4, we cover the basic ideas of Spherical geometry and then have the students explore which propositions of Neutral geometry still hold in Spherical geometry. Since this chapters lends itself to hands-on exploration, we have our students work in groups to explore the nature of lines, circles and triangles in this strange new universe by utilizing strings, markers, tape and inexpensive plastic balls. The sections on Spherical trigonometry and uniquely Spherical constructions are optional. Likewise, for Chapter 5, we discuss only the basic ideas of Taxicab geometry before ceding the remainder of the class to student exploration to determine which propositions of Neutral geometry still hold in Taxicab geometry. We do not cover all of the proofs in Chapter 6 in the classroom, instead choosing to highlight desirable properties of axiomatic systems as well as Hilbert's axioms. We spend over a week in Chapter 7, once again having students re-write and present select propositions up to I.46. We take the helm to present the Pythagorean Theorem and its converse. Chapters 8 and 9 take about a week, with "Quadrature on the Sphere" and "A Generalized Pythagorean Theorem" as optional sections. Students present select propositions from Chapter 10 for roughly a week, and we use dynamic software to discuss much of Chapter 11.

- Chapters 1–3 (roughly two weeks)
- Chapters 4–6 (roughly two weeks)
- Chapters 7–9 (roughly two weeks)
- Chapters 10–11 (roughly two weeks)

After Chapter 11, there is considerable flexibility for an instructor to choose a subset of the remaining five chapters according to his or her own interests. One of the authors typically spends the rest of the semester on Chapter 16 (one week), followed by Chapters 12 and 13 (roughly four weeks—skipping some of the proofs from Chapter 13), and finishing with Chapter 15 (roughly two weeks—omitting inversions and Hyperbolic isometries). The other author replaces Chapter 15 with Chapter 14. It is also possible to reserve more weeks for the last five chapters of the book by choosing only the highlights of Chapters 8, 9, 10 and 11.

Common Core State Standards for Mathematics [CCSS]. The following recommendations are taken from the *Geometry Course report* of the Geometry Study Group [GSG]. The group was charged by the MAA's Committee on the Undergraduate Program in Mathematics [CUPM] with making recommendations about geometry in the undergraduate mathematics curriculum. Their report is part of the 2015 CUPM Curriculum Guide to Majors in the Mathematical Sciences. Below, we briefly describe how this book addresses each recommended topic.

GSG writes: "To be prepared to teach a geometry course based on CCSS, future teachers should take a college geometry course in which definitions and proof are emphasized. In addition, the course they take should include coverage of the following topics:"

- **Proof**
  We emphasize reading, writing and presenting proofs throughout the book.

- **Transformations**
  In Chapter 15, we prove that any isometry in Neutral geometry can be written as the composition of three or fewer reflections. We then study reflections, rotations, translations and glide reflections in the Euclidean plane. We also consider Euclidean inversions and their role as reflections in the Poincaré Half-plane model of the hyperbolic plane.

- **Parallel Postulate**
  By separating Book I into Chapters 3 and 7, our book is clear on which results in Euclidean geometry depend on the Parallel Postulate. By presenting Spherical, Hyperbolic and projective geometries, we provide multiple two-dimensional geometries in which the Parallel Postulate does not hold.

- **Pythagorean Theorem**
  In addition to Euclid's proof, we include seven other well-known proofs of this famous theorem in our exercises, including proofs by Bhāskara, Leonardo da Vinci, U.S. President James A. Garfield and Thābit ibn Qurra. We also consider a generalized version of this theorem.

- **Dynamic geometry software**
  We encourage the use of dynamic geometric software throughout the textbook as it provides valuable insight into geometric relationships. We routinely use Geometer's Sketchpad® or GeoGebra in our courses as well as Spherical Easel and Non-Euclid.

- **Historical perspectives**
  We incorporate historical context throughout the book, particularly regarding the resolution of the controversy surrounding Euclid's fifth postulate, the development of Hyperbolic, affine and projective geometries, the classic impossible constructions of Greek antiquity, and the development of the mathematics required to prove the impossibility of these constructions.

- **Real-life applications**
  Chapters 11 and 14 include connections between art, architecture and geometry.

Based on these recommendations, an appropriate course could include the following:

- Chapters 1–3
- Chapter 4, sections 1–5
- Chapters 5–7
- Chapter 8, sections 1–2 (while algebraic in nature, helpful for future teachers)
- Chapter 9, sections 1–2
- Chapter 10
- Chapter 11 (optional)
- Chapters 12–13

- Chapter 14 (optional)
- Chapter 15, sections 1–3
- Chapter 16 (optional)

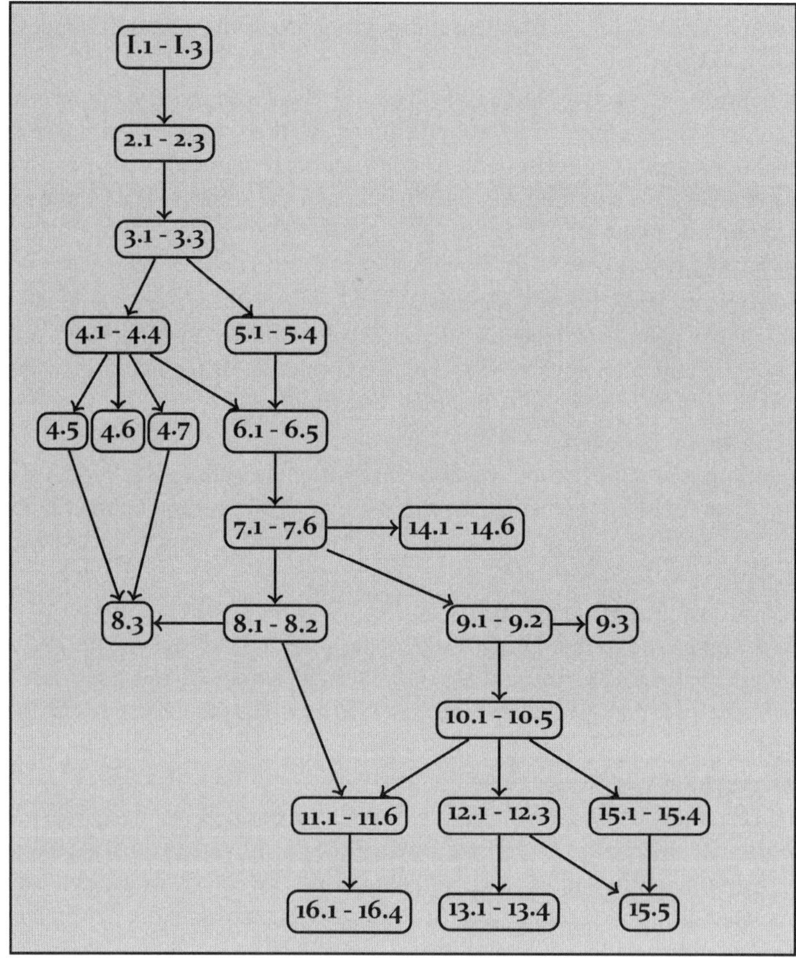

Figure 0.1. Section dependency chart

Note to the Reader

Good stories have conflict and resolution. The story of geometry is no exception. The characters in this story are geometric objects known since childhood: lines, circles, triangles and squares, to name a few. What conflict could these characters possibly generate? Are we not confident in our deeply ingrained understanding of these fundamental figures? Perhaps you recall a few core facts about triangles—say, the Pythagorean Theorem for right triangles or the 180-degree angle sum of any triangle. While we will revisit these and other well-known results, we will also visit geometric lands where these bedrocks no longer hold, where in some worlds lines are circles, and in others, circles are squares. How is this possible? To paraphrase Walt Whitman, geometry is large; it contains multitudes.

Mathematics, by its very nature, is logical and systematic and, yet, it can still produce results that astonish. One famous case involves Georg Cantor (1845–1918) who, while exploring the nature of infinite sets, documented his incredulity upon discovering that intuition had led him astray, writing to a friend, "Je le vois, mais je ne le crois pas!" ("I see it, but I don't believe it!") Cantor was surprised by the conflict that arose when his findings contradicted his expectation that the infinite would play by the same rules as the finite, and yet, he was delighted by the resolution his mathematical reasoning provided. In the same spirit, we aim to present you with a few surprises in geometry that run counter to your intuition.

As Cantor's story illustrates, surprise requires expectation, and expectation comes from experience. Thus, we need to build our experience and examine our pre-existing assumptions. We start with this fundamental question: How many geometries are there? If you think there is only one then you are in fine company. For 2000 years, there was only one geometry to study, and an ancient Greek mathematician named Euclid was its primary expositor. His book, the *Elements*, is the most famous and most published mathematics book of all time. Translated into many languages, it was standard reading for students through the centuries and, fittingly, we have chosen it as the starting point for our explorations. It was only relatively recently in the history of mathematics that Euclid's geometry was found to be just one of many interesting and equally valid geometric worlds. What triggered this revolution? While our two-dimensional figures appear unambiguous, their properties and very nature are more elusive than suggested by first glance. As we will see, a quest to resolve basic questions about the nature of parallel lines was responsible for this seismic paradigm shift.

As we embark on this trip, we first take a closer look at Euclid's geometry, its origins and its axioms. The trip is all-inclusive; though we have prompted you here for your geometric recollections, we will provide all of the definitions, theorems and proofs necessary for the journey (even the Pythagorean Theorem). At the start, the words of

Euclid's propositions and proofs will be familiar, but the style will be unlike others you have encountered. They are verbose, lacking most of the symbolic language and notation in use today. To better understand a Euclidean proposition and its proof, we suggest that you rewrite it in your own words using standard symbols and notation. Mathematics is a language, and the act of translating this language is a good way to learn how to read it and write it. Austrian Stefan Zweig (1881–1942), the author upon whom *The Grand Budapest Hotel* is based, shared this view. As a young writer he spent several years translating the works of French masters as an improvised apprenticeship in the literary arts. In doing so, he learned the structure of a good book without the pressure of creating the characters, plot or narrative. We echo Zweig's advice to young writers to translate a seasoned author's work into your language, as this is a reliable method of learning Euclid's geometry and the art of writing proofs. As a translator, you do not have to create the mathematics, but you will come to understand the logical structure necessary to write clear, correct proofs.

Finally, to state the obvious, we wrote this book to be read by you. To that end, we have included a considerable amount of commentary, history and explanation to help guide you through the story of geometry. Even with the additional narrative, reading a mathematics book is neither an easy nor a passive endeavor. To understand the mathematics you will need to read and then reread the axioms, definitions, theorems and proofs. We find it best to be an active reader with pencil and paper at the ready. Most importantly, as we journey to other strange worlds, keep an open and agile mind and be prepared to abandon preconceived notions as we reconsider and revise our assumptions about geometry and its most familiar objects.

Acknowledgments

Before we extend our thanks to the organizations and individuals whose efforts were directly related to the preparation of this book, we would like to acknowledge the ripple effect created by two professional programs sponsored by the Mathematical Association of America over two decades ago. We thank Christine Stevens and Jim Leitzel for starting Project NExT, and Victor Katz and V. Frederick Rickey for creating the Institute in the History of Mathematics and Its Use in Teaching. Both programs helped shape our early careers, and most importantly, we owe our meeting to the former.

As mathematicians the world is our office. With that in mind, we thank the constructors and supporters of the 165-mile long Pennsylvania Delaware & Lehigh Trail for providing the beautiful path where we logged thousands of miles bicycling in quiet contemplation or lively discussion about geometry and all book-related matters.

To Muhlenberg College and the University of Scranton we extend our sincere thanks for sabbaticals, awards, grants and funding in support of our scholarship. We are grateful to our students who have carefully read various drafts of the manuscript over the years: Elyn's Advanced Geometry students from 2013, 2015 and 2017, and Maureen's Geometry students from 2014, 2016 and 2018. Special thanks go to Myles Dworkin and Eric Jovinelly for their insights and suggestions, and to Danny Clark, Jessica Hollister and Rob McCloskey for combing the manuscript for typographical errors. Finally, thanks to the reviewers and editors at the MAA, in particular Steve Kennedy and Stan Seltzer, and to the American Mathematical Society production staff, in particular Christine Thivierge, Peter Sykes and Becky Rivard for helping us to make our vision of this book a reality.

1

The Line and the Circle

Figure 1.1. *Circles in a Circle* by Vasily Kandinsky (1923) courtesy of the Philadelphia Museum of Art

1.1 Introduction

Every great story has great characters. *Charlotte's Web* has Wilbur and Charlotte, *Pride and Prejudice* has Elizabeth Bennet and Mr. Darcy, and *The Hound of the Baskervilles* has Sherlock Holmes and Dr. Watson. The story of geometry is no exception. We start this book by introducing our two main characters: *the line* and *the circle*. Under their undeniable milquetoast veneer, we will reveal them to be interesting and complex characters as they take our narrative in unexpected directions. We certainly hold no claim to primacy here. In his 1884 novel *Flatland: A Romance of Many Dimensions*, author Edwin Abbott imbues these objects with personalities and voices (the straight lines are women, the circles are priests). In his 1963 book *The Dot and the Line: A Romance in Lower Mathematics*, Norton Juster tells the tale of a straight line who falls in love with a dot and then attempts to woo her away from a slothful squiggle. It begins:

> Once upon a time there was a sensible straight line who was hopelessly in love with a dot. 'You're the beginning and the end, the hub, the core and the quintessence,' he told her tenderly, but the frivolous dot wasn't a bit interested, for she only had eyes for a wild and unkempt squiggle who never seemed to have anything on his mind at all. [76]

Both of these stories have been turned into films. The ten-minute animated short, *The Dot and the Line,* directed by Chuck Jones and Maurice Noble, won the 1965 Oscar for Best Animated Short Film. Both the thirty-four minute *Flatland: The Movie* (2007) and the ninety-eight minute *Flatland: The Film* (2007) are based on Abbott's book. While our main characters are not fictional and will not be given voices, we will learn about them in the same way we come to know all great literary and film characters, largely by observing how they behave.

1.2 Which came first?

The actions of any fictional character are viewed through the cultural and historical lens of the reader. Though our main protagonists are geometric figures, different cultures have historically identified one of these shapes as more fundamental or basic than the other. In particular, certain cultures observe the world and see lines, others find circles. In her book, *Ethnomathematics: A Multicultural View of Mathematical Ideas*, Marcia Ascher provides us with two such examples [6]. Espousing a rectilinear view of the world, the first is an excerpt from *The Stretched String,* an essay from the book *The Mathematical Experience* written by Philip J. Davis and Reuben Hersh in 1981.

"The Stretched String" from *The Mathematical Experience*

> In some primitive cultures there are no number words except one, two, and many. But in every human culture that we will ever discover, it is important to go from one place to another, to fetch water or dig roots. Thus human beings were forced to discover – not once, but over and over again, in each new human life – the concept of the straight line, the shortest path from here to there, the activity of going directly towards something.
>
> In raw nature, untouched by human activity, one sees straight lines in primitive form. The blades of grass or stalks of corn stand erect, the rock falls down straight, objects along a common line of sight are located rectilinearly. But nearly all the straight lines we see around us are human artifacts put there by human labor. The ceiling meets the wall in a straight line, the doors and windowpanes and

tabletops are all bounded by straight lines. Out the window one sees rooftops whose gables and corners meet in straight lines, whose shingles are layered in rows and rows, all straight.

The world, so it would seem, has compelled us to create the straight line so as to optimize our activity, not only by the problem of getting from here to there as quickly and easily as possible, but by other problems as well. For example, when one goes to build a house of adobe blocks, one finds quickly enough that if they are to fit together nicely, their sides must be straight. Thus the idea of a straight line is intuitively rooted in the kinesthetic and the visual imaginations. We feel in our muscles what it is to go straight toward our goal, we can see with our eyes whether someone else is going straight. The interplay of these two sense intuitions gives the notion of straight line a solidity that enables us to handle it mentally as if it were a real physical object that we handle by hand.

By the time a child has grown up to become a philosopher, the concept of a straight line has become so intrinsic and fundamental a part of his thinking that he may imagine it as an Eternal Form, part of the Heavenly Host of Ideals which he recalls from before birth. Or, if his name be not Plato but Aristotle, he imagines that the straight line is an aspect of Nature, an abstraction of a common quality he has observed in the world of physical objects. [**30**]

Providing an alternative philosophy from a Native American viewpoint, our second excerpt is from the book *Black Elk Speaks*, the story of a holy man of the Sioux tribe as told through John G. Neihardt.

Excerpt from *Black Elk Speaks*

I came to live here where I am now between Wounded Knee Creek and Grass Creek. Others came too, and we made these little gray houses of logs that you see, and they are square. It is a bad way to live, for there can be no power in a square.

You have noticed that everything an Indian does is in a circle, and that is because the Power of the World always works in circles, and everything tries to be round. In the old days when we were a strong and happy people, all our power came to us from the sacred hoop of the nation, and so long as the hoop was unbroken, the people flourished. The flowering tree was the living center of the hoop, and the circle of the four quarters nourished it. The east gave peace and light, the south gave warmth, the west gave rain, and the north with its cold and mighty wind gave strength and endurance. This knowledge came to us from the outer world with our religion. Everything the Power of the World does is done in a circle. The sky is round, and I have heard that the earth is round like a ball, and so are all the stars. The wind, in its greatest power, whirls. Birds make their nests in circles, for theirs is the same religion as ours. The sun comes forth and goes down again in a circle. The moon does the same, and both are round. Even the seasons form a great circle in their changing, and always come back again to where they were. The life of a man is a circle from childhood to childhood, and so it is in everything where power moves. Our tepees were round like the nests of birds, and these were always set in a circle, the nation's hoop, a nest of many nests, where the Great Spirit meant for us to hatch our children.

But the Wasichus (whitemen) have put us in these square boxes. Our power is gone and we are dying, for the power is not in us any more. You can look at our boys and see how it is with us. When we were living by the power of the circle in the way we should, boys were men at twelve or thirteen years of age. But now it takes them very much longer to mature.

Well, it is as it is. We are prisoners of war while we are waiting here. But there is another world. [**90**]

As we reflect on Davis and Hersh's claim that the sides of a building must be straight, we may wish to think about the National Museum of the American Indian in Washington, DC. Douglas Joseph Cardinal, the architect of the building, describes his creation as a "a majestic curvilinear form that represents the nurturing female forms of Mother Earth" [**36**]. Though it is surrounded by buildings that reflect a rectilinear model of physical structure, this building has no straight sides. As an exercise, the reader will compare and contrast the opposing viewpoints represented in these readings.

Figure 1.2. National Museum of the American Indian, Washington, D.C.

1.3 What *is* a straight line, anyways?

At the beginning of his book *Experiencing Geometry,* David Henderson challenges his readers to consider the meaning of the word "straight." In a geometry book whose central characters are the line and the circle, reflecting upon the answer to this question is an excellent way to begin our journey. The following thought exercise is adapted from his book.

When do you call a line straight?

The central focus of this thought exercise is to guide the reader in cobbling together a reasonable definition for a straight line, or, at a minimum, to appreciate the difficulty of defining this familiar concept. As a first attempt at an answer, you may try to fit your definition to personal experience. It may help to consider some related practical concerns that arise when constructing lines. For example, suppose you had to mark the first and third base lines on a baseball field, or mark the lines on a volleyball or tennis court. Can you determine a method to produce these straight lines? Once completed, is there a way to check if the lines are straight? Alternatively, suppose you need to construct some object out of 2×4 lumber. Assuming you would like to purchase non-warped lumber, how can you determine if a 2×4 is straight? More generally, how can you determine if any physical object is straight?

As is true with most concepts, it is often helpful to consider what distinguishes a straight line from its opposite, a non-straight line. If you find yourself relying on the use of a ruler to make this distinction then you must ask yourself how you determined the straightness of the ruler. You may also find yourself relying on the instinctual animal calculation at the heart of the idiom "as the crow flies": *A straight line is the shortest path between two points.* Technically, we can only employ this idea if we were to measure all possible paths between two points and then take the shortest. It appears to be a hopeless task and reminiscent of the claims touted by many a merchant. There are only finitely many food markets in any city, or the world for that matter, but claiming to be the world's best would surely raise some eyebrows. The incredulity factor increases when

defining a straight line by the "crow's method" since there are infinitely many paths between any two points. How can anyone claim to find the shortest path out of *infinitely* many? Another related question to ponder: If the shortest path between two points is a straight line, then is a straight line between two points always the shortest path?

Oftentimes in this book we will find ourselves considering the symmetries of geometric objects. Saving all formal explanation of symmetry for a later chapter, can you informally describe any symmetries related to a line, and does this help us to define "straight?" One immediate benefit of introducing symmetry is the realization that we can use it to produce a straight line with ease. How? Simply fold a piece of paper.

Lastly, since demonstrative mathematics began with the work of Greek mathematicians, we would be remiss if we did not consider the view of Greek scholars on this matter. To be clear, most high school geometry courses consist entirely of theorems originating in Greece over two millennia ago. Furthermore, the word *mathematics* has its roots in the Greek language. The Greek philosopher Plato (427–347 BCE) believed that a person could not be considered educated without learning mathematics, specifically, the systematic deductions of geometry. There is a well-known story that at the entryway to his famous Academy stood a sign reading, "Let no one ignorant of geometry enter here." For Plato, geometric understanding implied an understanding of logic and, hence, the ability to study philosophy. In *Parmenides* he writes: "the round is that of which all the extreme points are equidistant from the centre," and "the straight is that of which the centre intercepts the view of the extremes." The modern reader may be satisfied with his description of *round*, but may find that of *straight* lacking. The Greek mathematician Euclid of Alexandria (ca. 325–ca. 265 BCE) wrote the *Elements*, a geometry text that we begin to explore in the following chapter. In this book he writes: "A straight line is a line which lies evenly with the points on itself." Do these descriptions help us in our quest to determine the meaning of "straight" or to give a definition of "straight line?" Our readers are asked to provide their own description in the exercises.

Exercises 1.3

1. Write a short essay to answer the question: " What is a straight line?"

2. Write a short essay (1–2 pages) comparing and contrasting the two readings: "The Stretched String" and the excerpt from *Black Elk Speaks*.

3. Watch *Flatland: The Movie*. Carefully explain how the three-dimensional sphere appears to two-dimensional objects, and hence, how a four-dimensional being would appear to us. What strange abilities would four-dimensional beings have?

2

Euclid's *Elements:* Definitions and Axioms

Figure 2.1. CALVIN AND HOBBES ©1991 Watterson. Reprinted with permission of UNIVERSAL Uclick. All rights reserved.

2.1 The *Elements*

The *Elements* is a geometry book written by Euclid of Alexandria (ca. 325–ca. 265 BCE) over two thousand years ago. In its original ancient Greek, the title $\Sigma\tau o\iota\chi\varepsilon\hat{\iota}\alpha$, or Stoicheia, is translated as *Elements* which refers to the rudimentary principles or primary results presented in the text. Though there is not much known about the author of the book, there are few books as widely translated, published, read and referenced as his *Elements.* It "has appeared in more editions than any work other than the *Bible*. It has been translated into countless languages and has been continuously in print in one country or another nearly since the beginning of printing" [77]. With these credentials, we rightly consider it to be the "bible of mathematics."

The *Elements* is a collection of 465 geometric results organized into thirteen books. These books are essentially chapters, though they are called books just as The Bible consists of books. Instead of giving each book a name, Euclid assigns numbers. He also numbers the geometric results, called propositions. Over time, Roman and Arabic numerals were adopted for clarity, and a reference to Proposition III.2 means the second proposition in the third book. It is now customary for mathematicians to follow the same structure when numbering their theorems, though Arabic numerals typically replace the Roman. Euclid's book has stood the test of time for over two millennia due to an organizational genius that goes much deeper than the numbering of books and propositions. He set the standard for all authors of mathematics texts by giving a systematic development of geometric results from first principles.

While it is known that Euclid was a teacher at Alexandria in Egypt, much of his life is a mystery. The most reliable information about him is found in the commentaries of the Greek philosopher Proclus (412–485) who lived a distant 700 years after Euclid. Proclus writes that Euclid lived in the time between Plato (427–347 BCE) and Archimedes (287–212 BCE), and that he organized and perfected the mathematical results of his time. It is generally assumed that Euclid gathered results of both plane and solid geometry, and also number theory, based on the works of predecessors such as Thales of Miletus (ca. 624–ca. 547 BCE), Pythagoras of Samos (ca. 569–ca. 475 BCE), Hippocrates of Chios (ca. 470–ca. 410 BCE), Plato, and Eudoxus of Cnidus (ca. 408–ca. 355 BCE). According to Proclus, Hippocrates was the first to write a book on the "elements" of geometry. The fact that there are no surviving copies of Hippocrates' text is, perhaps, indicative of the superior logical structure found in Euclid's text.

> Euclid lived during the reign of Ptolemy I (366–282 BCE). Proclus wrote that Ptolemy, like many a weary geometry student before and after his time, "asked Euclid if there was not a shorter road to geometry than through the *Elements*, and Euclid replied that there was no royal road to geometry" [**98**].

Euclid begins Book I by defining his geometric terms. (The exercise of defining a "straight line" in Section 1.3 should help the reader appreciate the inherent difficulty of this task.) He then specifies his *axioms*. An axiom is a statement that the reader must assume to be true in order to proceed with the text. As we will see, these axioms are not giant leaps of faith, but rather, have a self-evident nature. Once he has these preliminaries out of the way he states his first proposition, I.1, and furnishes its proof using his definitions, axioms and the standard rules of logic, convincing the reader of its truth. With the completion of the first proof he states his second proposition. In the proof of I.2, Euclid is now free to employ the statement of Proposition I.1 since it has already been proven. Book I proceeds as a careful building of results, the proof of every proposition resting solely on the standard rules of logic, definitions, axioms and earlier propositions. This structure is described as the *axiomatic method*, and Euclid's *Elements* is the oldest surviving text demonstrating the axiomatic development of a field of mathematics.

For the 2100 years that followed the introduction of the *Elements*, all geometry was Euclidean geometry. As we will see in Chapter 6, it took 2200 years before geometers improved upon his logical structure. It's now 2300 years later and nearly all of the geometry that we learn in high school can be found in these books. Euclid is, without question, the master of the geometry textbook, and our journey to follow our main characters, the line and the circle, must begin with him. As we read the translation of this ancient text, our goal is to understand the structure chosen by Euclid and to appreciate the benefits of this axiomatic approach to our study of geometry.

> The two oldest surviving copies of the *Elements* date to the ninth century, one housed at Oxford University and the other at the Vatican. Only fragments of the book have been found that predate these copies. The manuscript housed at Oxford was copied on parchment in 888 AD by Stephen the Clerk for Arethas of Patras in Constantinople. With the help of the Clay Mathematics Institute, this manuscript has been digitized and can be viewed in high resolution with corresponding Greek and English text at the Institute's website, *www.claymath.org/euclids-elements*.

2.2 Definitions

The first book of the *Elements* begins with twenty-three definitions. The first few do not sound very mathematical, but by the time we reach the definition of perpendicular lines, they become much more familiar. Sir Thomas Heath (1861–1940) is credited with authoring the definitive English translation of the *Elements*. Throughout this book we rely on Heath's translation, and any portion of the *Elements* reprinted from this translation is done so with the permission of Dover Publications [**40**].

(1) A *point* is that which has no part.

(2) A *line* is breadthless length.

(3) The extremities of a line are points.

(4) A *straight line* is a line which lies evenly with the points on itself.

(5) A *surface* is that which has length and breadth only.

(6) The extremities of a surface are lines.

(7) A *plane surface* is a surface which lies evenly with the straight lines on itself.

(8) A *plane angle* is the inclination to one another of two lines in a plane which meet one another and do not lie in a straight line.

(9) And when the lines containing the angle are straight, the angle is called *rectilineal*.

(10) When a straight line set up on a straight line makes the adjacent angles equal to one another, each of the equal angles is *right*, and the straight line standing on the other is called a *perpendicular* to that on which it stands.

(11) An *obtuse angle* is an angle greater than a right angle.

(12) An *acute angle* is an angle less than a right angle.

(13) A *boundary* is that which is an extremity of anything.

(14) A *figure* is that which is contained by any boundary or boundaries.

(15) A *circle* is a plane figure contained by one line such that all the straight lines falling
upon it from one point among those lying within the figure are equal to one another;

(16) And the point is called the *centre* of the circle.

(17) A *diameter* of the circle is any straight line drawn through the centre and terminated in both directions by the circumference of the circle, and such a straight line
also bisects the circle.

(18) A *semicircle* is the figure contained by the diameter and the circumference cut off
by it. And the centre of the semicircle is the same as that of the circle.

(19) *Rectilineal figures* are those which are contained by straight lines, *trilateral* figures
being those contained by three, *quadrilateral* those contained by four, and *multilateral* those contained by more than four straight lines.

(20) Of trilateral figures, an *equilateral triangle* is that which has its three sides equal, an
isosceles triangle that which has two of its sides alone equal, and a *scalene triangle*
that which has its three sides unequal.

(21) Further, of trilateral figures, a *right-angled triangle* is that which has a right an-
gle, an *obtuse-angled triangle* that which has an obtuse angle, and an *acute-angled
triangle* that which has its three angles acute.

(22) Of quadrilateral figures, a *square* is that which is both equilateral and right-angled;
an *oblong* that which is right-angled but not equilateral; a *rhombus* that which is
equilateral but not right-angled; and a *rhomboid* that which has its opposite sides
and angles equal to one another but is neither equilateral nor right-angled. And
let quadrilaterals other than these be called *trapezia*.

(23) *Parallel* straight lines are straight lines which, being in the same plane and being
produced indefinitely in both directions, do not meet one another in either direction.

A few comments about these definitions are in order. As we will see, it is not possible to define every term in a geometry (or any branch of mathematics for that matter).
To try to do so inevitably invites circular reasoning, which mathematicians strive to
avoid at all costs. For example, the definition of *point* requires us to define *part*. Our
definition for *part* would require more terms that would also require definition. Where
would the definitions end? The accepted point of view acknowledges that any logical
system must begin with some undefined terms. Here, the words *point* and *straight line*
will be two of our undefined terms. This does not mean, however, that Euclid's definitions are not useful in helping us visualize these two concepts. Nor does it mean
that you wasted your time in the previous chapter trying to describe straight lines. Just
because a term remains undefined does not mean that we cannot understand it. Even
if we were able to read these words without preconceived notions, we could still come
to understand them and visualize them by learning their properties as we did in elementary school.

Note that Euclid uses *line* much in the same way that we would use the term **curve**.
In particular, for Euclid a line *can* be straight but it does not have to be, and it can also

extend infinitely in one or both directions, or it can be finite, which we will refer to as a **line segment**. We will refer to a line extending infinitely in only one direction as a **ray**. In general, the context will clarify which of these objects is under consideration. For this book, we will use the common notations $\overleftrightarrow{AB}$ for a straight line extending infinitely in both directions and $\overrightarrow{AB}$ for a ray starting at point A. A straight line segment will be denoted by $\overline{AB}$ or AB. Finally, we drop the adjective straight when referring to a line, ray or line segment.

Euclid gives definitions that he never uses (e.g., oblong) and definitions that are more general than he requires (e.g., angle). Some historians take these as evidence that Euclid drew from earlier works that no longer exist. Others theorize that some definitions were added after Euclid's time, or perhaps, were controversial. In Euclid's definition of an angle, he allows two curves to intersect at a point, but in 464 of his 465 theorems he only considers angles formed by lines. The lone exception occurs in the third book, and, as we will see, was the subject of much controversy amongst Greek geometers. His definition is also restrictive in that it does not allow an angle to be formed by three points which lie on a line. As for angle notation, we use $\angle ABC$ to represent the angle at B formed by the rays $\overrightarrow{BA}$ and $\overrightarrow{BC}$. In this case, we call point B the **vertex** of the angle. As demonstrated in Figure 2.2, these rays create two angles at vertex B. In the next chapter, we will see that, as expected, these two angles sum to four right angles. The measure of $\angle ABC$ is defined to be the smaller of the two angles, with a maximum, yet unobtainable, measure of two right angles. Euclid never measures his angles in either degrees or radians. The solitary unit of angle measurement in the *Elements* is the right angle. For example, we will have two right angles rather than 180° or π radians. We will use $\angle ABC$ to refer to both the angle and the angle's measure, allowing the context to clarify the intention.

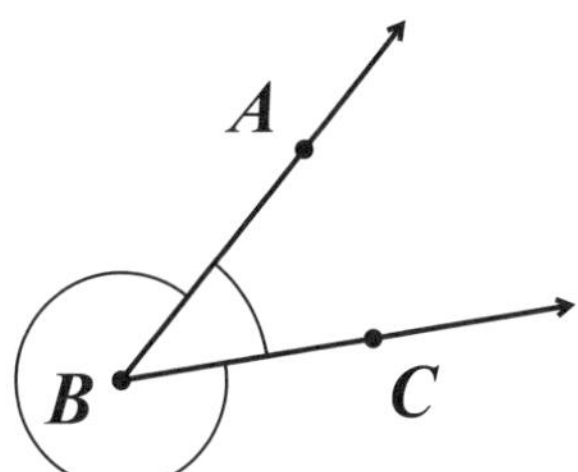

Figure 2.2. $\angle ABC$ is defined as the smaller angle

While many of Euclid's descriptions of rectilineal figures in definitions 19 through 22 are familiar to us, there are some subtle differences. First, since we are more familiar with the term **polygon** we will make use of this substitution for rectilineal figure. Likewise, though Euclid never uses a trapezium (singular form of trapezia) in the *Elements*, we will substitute the more familiar **trapezoid**. Technically, Euclid's trapezium as given here includes any quadrilateral with *at most* one pair of parallel sides. In another of Euclid's works, *On Divisions of Figures*, Euclid's trapezium *must* have a pair of parallel sides, thus agreeing with our standard definition for trapezoid. Such discrepancies are suggestive of the many versions and editions of the *Elements* before, during and after Euclid's time.

Of the quadrilaterals defined in Book I, the only one Euclid uses is the square. Though not given a definition here, Euclid does refer to *parallelogram* and *parallelogramic area* in Propositions I.35 and I.34, respectively. It is clear from the proof of Proposition I.34 that a **parallelogram** is a quadrilateral whose opposite sides are parallel. Similarly, Euclid does not specifically define an object called a rectangle, though he refers to one by name in Book II. Consistent with Euclid's use of this term, we define a **rectangle** to be any right-angled quadrilateral. Contemporary definitions allow equilateral triangles to be a subset of isosceles triangles. Euclid, however, partitions triangles into one of three distinct categories: equilateral, isosceles, or scalene. In a similar fashion, we often consider squares to be a subset of rectangles which are, in turn, a subset of parallelograms. Euclid again partitions his quadrilaterals into nonoverlapping categories. When defining these polygons, we prefer the present-day convention of nesting them and, since it will not cause any confusion, will take this meaning of these words in the book. While Euclid's rhomboid may appear to be the same as a parallelogram, technically it is not since there is no requirement that the opposite sides be parallel. (In Proposition I.34 Euclid proves that any parallelogram that is not a square, rhombus or rectangle must be a rhomboid. He never proves that a rhomboid is a parallelogram, though it would be easy to show. Also of note, he never uses the defined terms rhombus or rhomboid in the *Elements*.) Finally, the reader should be careful to note that parallel lines are simply lines in the plane that do not intersect. This definition takes on greater importance when we shift our gaze from Euclidean to other geometries.

We add a final definition for polygons that Euclid did not include. In planar geometry, a polygon is called **convex** if the line segment joining any two points of the figure lies entirely in the figure. In Figure 2.3, *ABCDE* is convex but *FGHIJ* is not, as demonstrated by its dashed line.

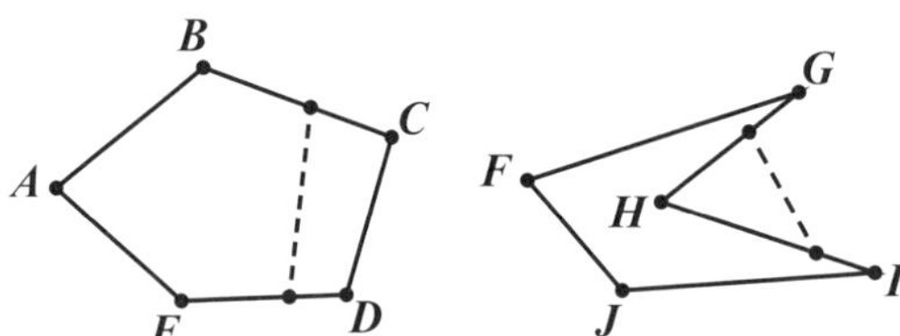

Figure 2.3. Convexity in the plane

2.3 Postulates and common notions

After introducing our main characters, the line and the circle, as well as the concepts of points, triangles and polygons, Euclid then presents his postulates and common notions. Postulates and common notions are the axioms for our geometric system. Axioms are statements to be taken on faith with the understanding that they are so obvious that it would be difficult not to accept them, that is, unless you are the precocious and playful comic strip character at the start of this chapter. For the most part, this is the case with Euclid's axioms. However, in the statement of Euclid's fifth postulate, we find the beginnings of a story that will take mathematicians nearly two millennia to finish. It is for this reason that we will delay a careful consideration of the fifth postulate for several chapters and spend our time concentrating on the first four postulates. The geometry resulting from the common notions along with the first four postulates is so important that it has its own name: **Neutral geometry.**

POSTULATES

Let the following be postulated:

(1) To draw a straight line from any point to any point.

(2) To produce a finite straight line continuously in a straight line.

(3) To describe a circle with any center and distance.

(4) That all right angles are equal to one another.

(5) That, if a straight line falling on two straight lines make the interior angles on the same side less than two right angles, the two straight lines, if produced indefinitely, meet on that side on which are the angles less than the two right angles.

COMMON NOTIONS

(1) Things which are equal to the same thing are also equal to one another.

(2) If equals be added to equals, the wholes are equal.

(3) If equals be subtracted from equals, the remainders are equal.

(4) Things which coincide with one another are equal to one another.

(5) The whole is greater than the part.

Euclid's axioms come in two flavors; the postulates are geometric while the common notions, dealing with magnitudes of lengths, angles and areas, are algebraic to a contemporary reader. The first common notion is known as the transitive property, namely, *If $a = b$ and $b = c$, then $a = c$.* The second and third common notions can be translated as: *If $a = b$, then $a + c = b + c$, and $a - c = b - c$*, respectively. While the third may seem unnecessary to someone schooled in algebra, we must remember that lengths, angles and areas are always positive quantities for Euclid. To understand the fourth common notion, we need to explain Euclid's notion of "to coincide." Loosely speaking, we will say that two objects coincide when one object can be superposed on the other in such a way that they completely match. In this common notion, Euclid makes a distinction between the ability to superpose two objects versus their equal measurement. Here, Euclid means that if two line segments coincide, then they have the same length; if two angles coincide, they have the same measure; and finally, if two triangles coincide, then they have the same area. The converse of this common notion, namely, that things with equal measure must coincide, is not true, in general. In the exercises, the reader is asked to provide an example to demonstrate this.

Though not obvious, the fifth common notion ensures that geometric figures, however measured, must have nonzero size. Thus, for example, a line segment cannot have zero length and a triangle cannot have zero area. To explain why this common notion prevents the existence of figures of zero measure, we can explore what happens if we assume the existence of such a figure. For example, suppose A, B and C are distinct points on a line, segment AB has zero length, and segment AC contains AB. Since the length of the whole segment, AC, equals that of its part, BC, this would contradict the fifth common notion, an impossibility since it is an axiom.

The first three postulates concern our book's main characters: the line and the circle. The first and second postulates give us the geometric tool known as an **unmarked straightedge**. We can visualize this tool as a ruler, but there are no markings to keep track of distances. Notice that while Euclid's first postulate allows us to draw a line segment between any two points, he does not explicitly state that this segment must be unique. In Proposition I.4, however, we will see that Euclid assumes there can only be one segment joining any two points.

The third postulate allows us to construct circles given a specified center and a distance called the *radius*. This postulate gives us the geometric tool known as the **collapsible compass**. To be clear, if we are given a point A and a radius AB, we can then construct the circle with center A and radius AB. We cannot, however, use this postulate to construct a circle with center A and radius given by a segment BC. If we were to place the ends of the compass on the points B and C and then lift the compass off of the page to transfer this length to A, the compass would collapse.

While the first three postulates are abstract statements, the unmarked straightedge and collapsible compass *do* exist in the real world (and for us, the electronic world!) and Euclid is clearly interested in determining which geometric objects can be created with these tools. To construct a geometric figure using only these two tools is called a *Euclidean construction*. We will not have to wait long for our first example as Euclid uses these tools in the first proposition of Book I to construct an equilateral triangle from a given line segment. After constructing the equilateral triangle in I.1, we will see that Euclid goes on to construct other *regular polygons* (equal sides, equal angles). In Chapter 16, we will see that it is not possible to construct a regular polygon of every possible number of sides. The quest to determine exactly which regular polygons can be constructed using only these two tools took two millennia, eventually completed by Carl Friedrich Gauss (1777–1855) at the end of the eighteenth century. As we will see, other questions of constructibility took even longer.

Most readers are familiar with the compass used in high school geometry classes. This compass is usually a *rigid compass*, meaning that given a point A and a length BC, we can transfer the length BC so that we may construct a circle with center A and radius BC. Fortunately, Euclid proves that a collapsible compass and a rigid compass are mathematically equivalent as the second proposition of Book I. That is, any object constructible with one of these tools is constructible with the other. Once we have proven this proposition, we will typically assume that our compass is rigid. It is interesting to note that it is the compass, not the straightedge, that allows us to transfer lengths in the plane.

The fourth postulate appears fairly self-evident, though necessary given Euclid's definition of a right angle: "When a straight line set up on a straight line makes the adjacent angles equal to one another, each of the equal angles is *right*." Since Euclid chooses the right angle as his basis of measuring all other angles, he needs to assert that all right angles are, in fact, equal to each other, no matter where they lie in the plane.

The fifth postulate is commonly referred to as the *Parallel Postulate* in spite of the fact that it gives a condition for determining when two lines will intersect. It is so called because, if we assume all of Neutral geometry, then the fifth postulate is equivalent to a much more familiar statement known as Playfair's Axiom: *Through a point not on a given straight line, there exists at most one straight line that is parallel to the given line.* We explore the equivalence of these statements in Chapter 7.

Now that we have familiarized ourselves with his definitions, postulates and common notions, we are ready to delve into the propositions and their proofs. Euclid organized the thirteen books of the *Elements* so that each proposition's proof relied only on definitions, postulates, common notions, and previously proven propositions. Thus, he builds his geometry from scratch. By assuming only what we have presented in this chapter, he develops nearly all of the familiar results of Euclidean geometry in the plane. In the exercises, readers are asked to recall a few of these results from their high school geometry course.

Exercises 2.3

1. Draw a picture of a plane angle where the sides are not straight lines. Speculate as to how we could measure such an angle. Explain.

2. What do we usually call Euclid's *oblong*?

3. The converse of Common Notion 4 is the statement: *Things which are equal to one another must coincide.* To demonstrate that this statement is false, give an example of two geometric figures of equal measure that do not coincide.

4. Find a straightedge and a compass and use them to draw lines and circles on a page.

5. List ten theorems/results from your high school geometry course. When stating each theorem/result, be precise.

3

Book I of Euclid's *Elements:* Neutral Geometry

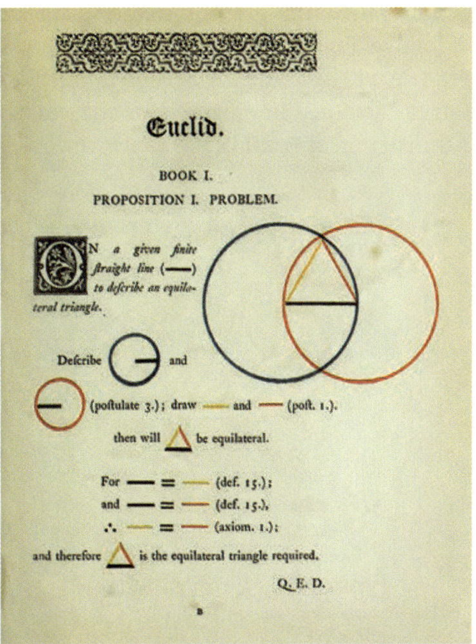

Figure 3.1. Proposition I.1 by Oliver Byrne [**19**]

If we imagine the *Elements* as a play where the first book is Act I, then lines, and the angles, triangles and quadrilaterals they create, play the starring role. Circles are akin to the set crew in this production. They toil in the background and do much of the heavy lifting in Euclid's proofs, but they are nowhere to be seen on stage in the proposition statements. In general, the statements of Book I are of two types. There are Euclidean constructions, that is, statements that assert the constructibility of a certain geometric object, for example, an angle bisector or a square. Alternatively, there are statements which explain a relationship between geometric objects, for example, the exterior angle in any triangle must be greater than either opposite interior angle. Often, but not always, the proof of a construction ends with *Q.E.F.*, which is an abbreviation of the Latin phrase *quod erat faciendum*, meaning "that which was to have been done" or "precisely what was required to be done" [**39**]. The others end with *Q.E.D.*, which stands for *quod erat demonstrandum*, meaning "that which was to have been demonstrated" or "precisely what was required to be proved." [**39**] We will use a small square to indicate the end of a proof.

In addition to learning geometric terms and theorems, one of the goals of this chapter is to learn the axiomatic method from its first expositor. Euclid's book has survived the millennia for good reason: his structure, logic and presentation is hard to beat for this foundational material. He is, quite simply, a master. As we walk through the propositions in his first book, we begin our journey to learn the fundamentals of Euclidean geometry with a focus on developing proof-reading and proof-writing skills.

3.1 Propositions I.1 through I.8

The first three propositions are constructions. We give two different proofs for each of the first two propositions. The first proof is Heath's translation of Euclid's proof, and the second version follows Euclid's steps but has updated language and notation. Throughout this chapter, you will be asked to read and update Euclid's proofs with your own language and notation. As noted in the previous chapter, propositions and proofs reprinted from Heath's translation of Euclid's *Elements* are done so with the permission of Dover Publications.

Proposition I.1. *On a given finite straight line to construct an equilateral triangle.*

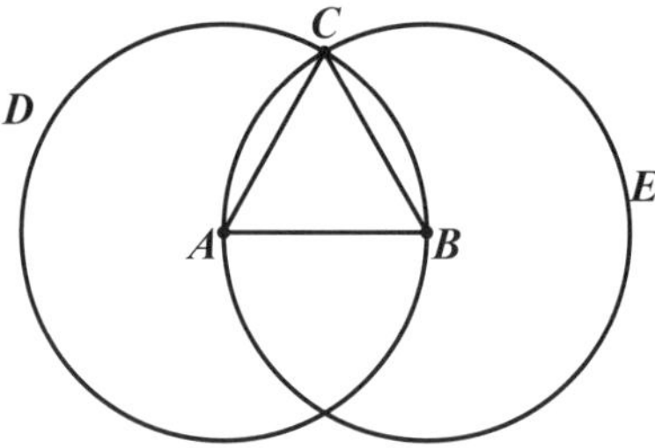

Figure 3.2. Proposition I.1

Proof.   Let *AB* be the given finite straight line.

Thus it is required to construct an equilateral triangle on the straight line *AB*.

With centre *A* and distance *AB* let the circle *BCD* be described; [Post. 3] again, with centre *B* and distance *BA* let the circle *ACE* be described; [Post. 3] and from the point *C*, in which the circles cut one another, to the points *A*, *B* let the straight lines *CA*, *CB* be joined. [Post. 1]

Now, since the point *A* is the centre of the circle *CDB*, *AC* is equal to *AB*. [Def. 15]

Again, since the point *B* is the centre of the circle *CAE*, *BC* is equal to *BA*. [Def. 15]

But CA was also proved equal to AB; therefore each of the straight lines CA, CB is equal to AB.

And things which are equal to the same thing are also equal to one another; [C.N. 1] therefore CA is also equal to CB.

Therefore the three straight lines CA, AB, BC are equal to one another.

Therefore the triangle ABC is equilateral; and it has been constructed on the given finite straight line AB.

Being what it was required to do.   □

Before we give an updated proof, it is important to understand that, in all of his proofs, Euclid starts by stating the givens and then proceeds to explicitly state what he must prove. He then shows what is required, and finally, ends by restating what he has accomplished. It may take a few propositions to grow accustomed to this style. Secondly, Euclid is careful to explain which postulates, common notions and definitions justify each of his steps. Notice, however, that Euclid does not explain why the two circles he constructs will necessarily intersect. His accompanying diagram certainly suggests that such an intersection is inevitable, but Euclid did not intend for his proofs to rely on such drawings. The sketches are only meant as a useful tool to help the reader visualize the proof. Eventually we will have to address this particular oversight by Euclid, and as we proceed, be alert for other logical gaps where an assumption is made without full justification.

In his commentary on the *Elements*, Proclus details the critiques, and sometimes disparages the critics, of these proofs. Though it seems unimaginable, the relatively straightforward proof of Proposition I.1 generated a full 18 pages of commentary! We will note some of Euclid's unstated assumptions as we proceed, but we will not be taking a Proclus-sized microscope to these proofs. Instead, in Chapter 4 we highlight the quagmire resulting from unstated assumptions by exploring a different geometry, and in Chapter 6 we detail the eventual modifications made to Euclid's axioms in the late nineteenth century.

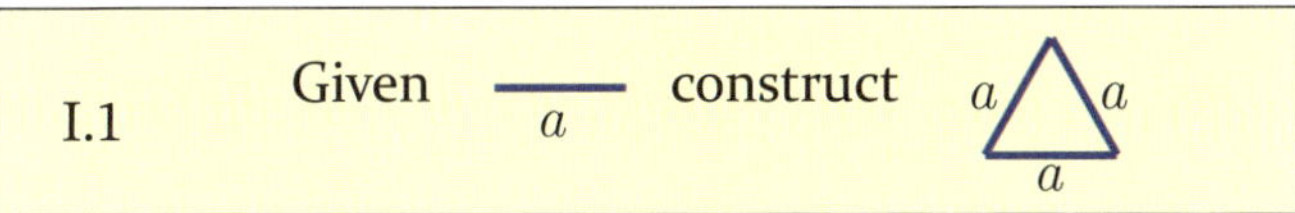

We will follow each Euclidean proof with a box containing the statement of the proposition represented pictorially in miniature, as shown above. These shorthand versions of the propositions are presented in the style of Oliver Byrne (1810-1880), author of the most spectacularly colorful reinvention of Euclid's work, *The First Six Books of The Elements of Euclid in which Coloured Diagrams and Symbols Are Used Instead of Letters for the Greater Ease of Learners*. As demonstrated by the proof of I.1 shown in Figure 3.1, this 1847 book is a translation of Euclid into a hybrid language where shapes of vibrant yellow, red, blue and black substitute for geometric objects such as lines, circles, triangles and rectangles, and symbols such as $+$, $\perp$, $\therefore$ and $=$ take the place of words. (This minimalist aesthetic clearly did not extend to the titling of his work as a simple word count proves the title to be more verbose than his proof of the first proposition.)

Byrne intended to revolutionize the teaching of the standard curriculum of his day with this visual approach to geometry, making it easier for every student to learn Euclid. In his preface, he claims that with his "enticing mode of communicating knowledge, that the Elements of Euclid can be acquired in less than one third the time usually employed, and the retention by the memory is much more permanent." His use of a primary color palette was entirely in support this goal, for at a glance, one can easily see that two segments or two angles are equal in measure if they share the same color. Sadly, none of his mathematics books were commercially or critically successful during his lifetime, and in particular, this work, now his most famous, was described as a mere "curiosity" by mathematics historian Florian Cajori in 1928 [20].

> Though it did not produce a pedagogical revolution, Oliver's Byrne's edition of Euclid was lauded in the latter half of the twentieth century for its beauty and artistry in typographic design and printing. That the praise came from outside the mathematical community is not too surprising, for at a distance, the colorful arrangement of woodblock print geometric shapes on many a page gives one the impression of gazing at a work of abstract art in the MoMA or the Pompidou. In particular, Byrne's style bears an unmistakable likeness to the paintings of Piet Mondrian (1872-1944) and the designs of Frank Lloyd Wright (1867-1959), both artist and architect working over a half century after the appearance of Byrne's book. Of recent note, the publication of a facsimile reproduction in 2010 has once again revived interest in both the book and the Irish born-and-educated Byrne [19]. For a well-researched account of Byrne's life we recommend the 2015 article by Hawes and Kolpas, and for a recent revival of Byrne's style applied to the work of Omar Khayyam (1048–1131), see Kent and Muraki's 2016 article [68][79].

The colorful miniature propositions given here, our homage to Byrne, utilize a similar set of shapes, colors and algebraic symbols. While the miniature for I.1 requires just one color, propositions that relate two geometric objects, for example I.4, require a larger color palette. We provide miniatures for all Book I propositions and a subset of those from the other books. For ease of reference, we include all of these miniatures in Appendix C. While we do not make any Byrne-like claims as to their time-saving benefits, you may find yourself, like the authors, relying on this Appendix as an at-a-glance visual reference to the Euclidean propositions *"for the Greater Ease of Learners"* who wish to apply earlier results in later proofs.

We are now ready to provide an updated proof of Proposition I.1.

Proposition I.1. *On a given finite straight line to construct an equilateral triangle.*

Proof 2.   (Use the diagram given in Figure 3.2 since it is unchanged.)

Let AB be the given line segment.

Using Postulate 3, construct two circles, one centered at A, the other at B, both with radius AB.

Let C be one of the two intersections of these two circles.

Using Postulate 1, draw line segments AC and BC.

Since AC and AB are radii of the same circle, by Definition 15, they are equal.

By similar reasoning, $AB = BC$.

By Common Notion 1, we have $AC = AB = BC$, and hence, we have constructed an equilateral triangle $\triangle ABC$ as desired. $\square$

Take note that our main characters both play a role here. While the first proposition is fundamentally a result about lines, its proof cannot be established without circles. The circles in the proof and represented in the diagram come by way of Postulate 3, giving us the ability to construct a circle at a given center with a given radius, and essentially, providing every geometer with a theoretical collapsible compass. As mentioned in the previous chapter, compasses are either *collapsible* or *rigid*. Clearly, anything that can be constructed with a collapsible compass can also be constructed with a rigid one. This next proposition proves the converse, that is, anything that can be constructed with a rigid compass can also be constructed with a collapsible compass. Thus, the two tools are mathematically equivalent. You may wonder why Euclid did not choose to make his third postulate stronger, and hence, avoid the need to prove this second proposition. In general, mathematicians agree with Aristotle (384–322 BCE) that "other things being equal, that proof is the better which proceeds from the fewer postulates." [15] More generally, an economy of assumptions makes for a stronger theory. After this proof and for the remainder of the book (including when we explore other geometries), we will assume that we have a rigid compass.

Proposition I.2. *To place at a given point [as an extremity]*[1] *a straight line equal to a given straight line.*

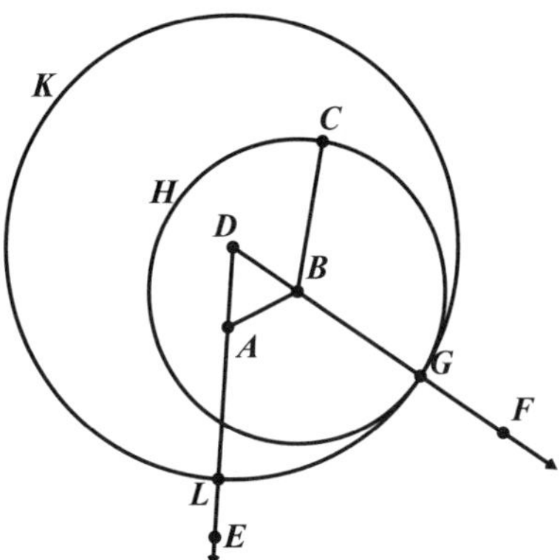

Figure 3.3. Proposition I.2

Proof. Let A be the given point, and BC the given straight line.

Thus it is required to place at the point A [as an extremity] a straight line equal to the given straight line BC.

From the point A to the point B let the straight line AB be joined; [Post. 1] and on it let the equilateral triangle DAB be constructed. [I.1]

Let the straight lines AE, BF be produced in a straight line with DA, DB; [Post. 2] with centre B and distance BC let the circle CGH be described; [Post. 3] and again, with centre D and distance DG let the circle GKL be described. [Post. 3]

---

[1]The square brackets within the statement of a proposition indicate material added by Heath to clarify the Greek text.

Then, since the point B is the centre of the circle CGH, BC is equal to BG.

Again, since the point D is the centre of the circle GKL, DL is equal to DG.

And in these DA is equal to DB; therefore the remainder AL is equal to the remainder BG. [C.N. 3]

But BC was also proved equal to BG; therefore each of the straight lines AL, BC is equal to BG.

And things which are equal to the same thing are also equal to one another; [C.N. 1] therefore AL is also equal to BC.

Therefore at the given point A the straight line AL is placed equal to the given straight line BC.

Being what it was required to do.   □

Proof 2.   (Use the diagram given in Figure 3.3 since it is unchanged.)

Let A be the given point and BC be the given line segment.

Using Postulate 1, draw segment AB.

With Proposition 1, construct an equilateral triangle using AB as a side, let the other vertex be labeled D.

Draw rays $\overrightarrow{DA}$ and $\overrightarrow{DB}$ with Postulate 2.

With Postulate 3, construct a circle with center B and radius BC.

Label its intersection with $\overrightarrow{DB}$ as G.

Construct a second circle with center D and radius DG.

Label the intersection of this circle with $\overrightarrow{DA}$ as L.

We claim that $AL = BC$ as follows:

Since DL and DG are both radii of the same circle, they must be equal. [Def. 15]

By construction, $DA = DB$, thus by Common Notion 3, we have $AL = BG$.

But BG and BC are radii of the same circle so $BG = BC$. [Def. 15]

Hence by Common Notion 1, we have $AL = BC$ as desired.   □

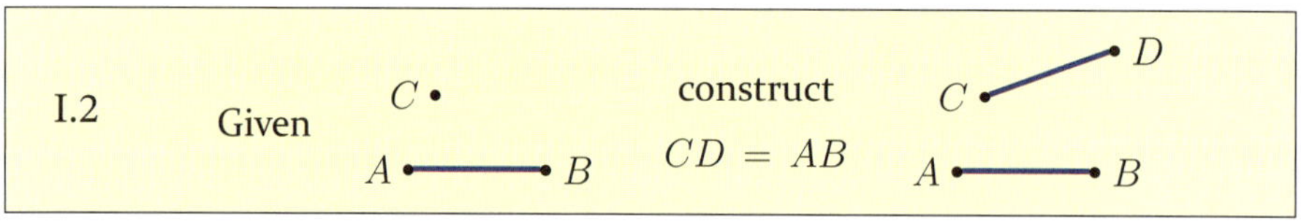

I.2   Given C •   construct $CD = AB$ C •——• D
                        A •——• B A •——• B

With a rigid compass comes the ability to transfer a length in the plane, which in turn produces the ability to add one segment to another, or subtract a shorter segment from a longer as we see in the next proposition. This is addition and subtraction, but not with numbers, plus signs or minus signs. In fact, the first plus sign in this chapter does not appear until page 38, and it is only the second equation sporting the symbol in the first 38 pages of the book. Arithmetic and algebraic symbols are our mathematical language, but they would be Greek to Euclid, so to speak. For the simple fact that the mathematical notation we use today developed over the millennium and a half after his lifetime, Euclid used words to convey his ideas. We favor algebraic symbols. So, when Euclid writes the lengthy, "Let AB, CD be the two given unequal straight lines, and let AB be the greater.", we substitute "Let $AB > CD$." We cannot help but include our mathematical language when we give an updated translation of a Euclidean proof. Keep this in mind, but don't lose sight of the fact that geometry is the scope through which Euclid views arithmetic and algebra throughout the *Elements*.

Proposition I.3. *Given two unequal straight lines, to cut off from the greater a straight line equal to the less.*

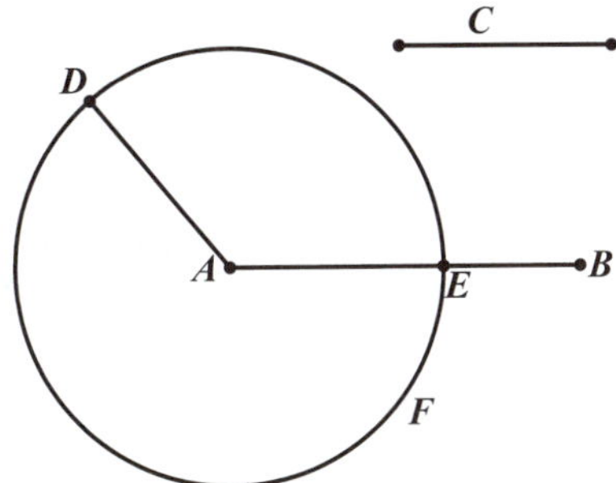

Figure 3.4. Proposition I.3

Proof. Let AB, C be the two given unequal straight lines, and let AB be the greater of them.

Thus it is required to cut off from AB the greater a straight line equal to C the less.

At the point A let AD be placed equal to the straight line C; [I.2] and with centre A and distance AD let the circle DEF be described. [Post. 3]

Now, since the point A is the centre of the circle DEF, AE is equal to AD. [Def. 15] But C is also equal to AD.

Therefore each of the straight lines AE, C is equal to AD; so that AE is also equal to C. [C.N. 1]

Therefore, given the two straight lines AB, C, from AB the greater AE has been cut off equal to C the less.

Being what it was required to do. $\square$

The third proposition reinforces the fact that it is the compass, not the straightedge, that allows us to transfer lengths in the plane. As directed in the exercises, the reader should write their own version of the proof using the previous examples as a guide.

As noted earlier, Euclid's propositions come in two flavors, some are constructions and others relate geometric figures. The fourth proposition is our first example where a relationship between geometric figures is described. Specifically, Euclid shows that two triangles with two corresponding equal sides and two equal angles formed by those sides, must coincide with each other. Over time, Euclid's term *coincide* was replaced with the term congruent, defined here.

Definition 3.1. Two triangles $\triangle ABC$ and $\triangle DEF$ are **congruent**, denoted $\triangle ABC \cong \triangle DEF$, if both their corresponding sides and their corresponding angles are equal, that is, $AB = DE, BC = EF, AC = DF$ and $\angle ABC = \angle DEF, \angle BCA = \angle EFD, \angle BAC = \angle EDF$.

It is important to note that the order of the vertices in each triangle must correspond to the congruence. That is, if $\triangle ABC \cong \triangle GHI$, then we know $AB = GH, BC = HI, AC = GI$ and $\angle A = \angle G, \angle B = \angle H, \angle C = \angle I$. For notational ease, we will abbreviate $\angle ABC$ as $\angle B$ whenever there is only one angle to be found at B, and hence, no possibility for confusion. For general polygons, two polygons are said to be congruent if their corresponding sides and angles are all equal.

Proposition I.4 is commonly known as the *SAS (side-angle-side) congruence scheme* since Euclid proves that the equivalence of *two* sides and their included angles is enough to guarantee the congruence of two triangles. Before the end of this chapter we will encounter three other congruence schemes for triangles, *SSS* [I.8] *(side-side-side)*, *ASA* [I.26] *(angle-side-angle)* and *AAS* [I.26] *(angle-angle-side)*.

Proposition I.4 [SAS]. *If two triangles have the two sides equal to two sides respectively, and have the angles contained by the equal straight lines equal, they will also have the base equal to the base, the triangle will be equal to the triangle, and the remaining angles will be equal to the remaining angles respectively, namely those which the equal sides subtend.*

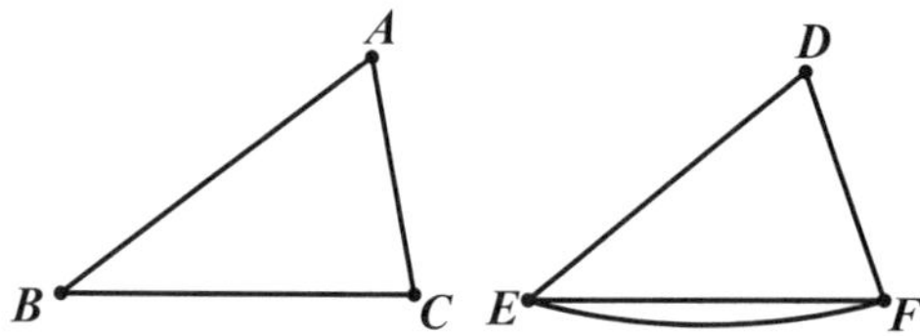

Figure 3.5. Proposition I.4 SAS

Proof.   Let ABC, DEF be two triangles having the two sides AB, AC equal to the two sides DE, DF respectively, namely AB to DE and AC to DF, and the angle BAC equal to the angle EDF.

I say that the base BC is also equal to the base EF, the triangle ABC will be equal to the triangle DEF, and the remaining angles will be equal to the remaining angles respectively, namely those which the equal sides subtend, that is, the angle ABC to the angle DEF, and the angle ACB to the angle DFE.

For, if the triangle ABC be applied to the triangle DEF, and if the point A be placed on the point D and the straight line AB on DE, then the point B will also coincide with E, because AB is equal to DE.

Again, AB coinciding with DE, the straight line AC will also coincide with DF, because the angle BAC is equal to the angle EDF; hence the point C will also coincide with the point F, because AC is again equal to DF.

But B also coincided with E; hence the base BC will coincide with the base EF, and will be equal to it. [C.N. 4]

Thus the whole triangle ABC will coincide with the whole triangle DEF, and will be equal to it. [C.N. 4]

And the remaining angles will also coincide with the remaining angles and will be equal to them, the angle ABC to the angle DEF, and the angle ACB to the angle DFE. [C.N. 4]

Therefore etc. Q.E.D.   □

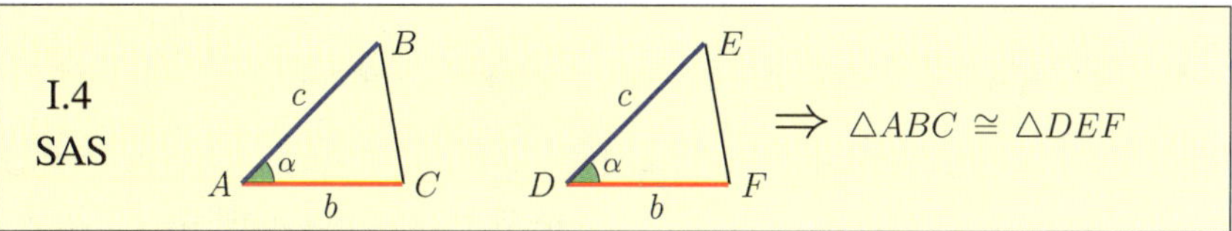

There are a few items of interest in the proof of Proposition I.4. First, it is within the body of the proof that Euclid claims that there is exactly one line segment joining any two points. Therefore, he assumes that a line between two points, as given by Postulate 1, is unique. This is another Euclidean omission to be addressed in a later chapter. More importantly, Euclid employs a proof technique that only appears twice in Book I, and it raises some questions. Essentially, Euclid picks up $\triangle ABC$ and "applies," or superposes, it on top of $\triangle DEF$. He then argues that the triangles will precisely match up, and thus, they coincide. This is called a **proof by superposition,** and we will see this technique again in Proposition I.8 [SSS]. One of Euclid's unstated assumptions here is that moving these triangles in the plane does not deform them like a funhouse mirror would. Although related to Common Notion 4, none of Euclid's postulates or common notions allows him to "apply" one triangle to another. This is another example of Euclid's omissions that we will need to address in a subsequent chapter. Lastly, it is in this proposition that Euclid calls one of the segments forming the trilateral its **base**, and the remaining segments its **sides**, as a convenient naming scheme.

The most difficult of the early propositions is the fifth, which subsequently came to be known as *pons asinorum*, a Latin phrase translated as *bridge of fools* or *ass's bridge*. The reason for its nickname is not entirely clear, but it is speculated that the proof was less of a bridge and more of dead-end for poor geometry students. We will include two proofs, the first from the *Elements*.

Proposition I.5. *In isosceles triangles the angles at the base are equal to one another, and, if the equal straight lines be produced further, the angles under the base will be equal to one another.*

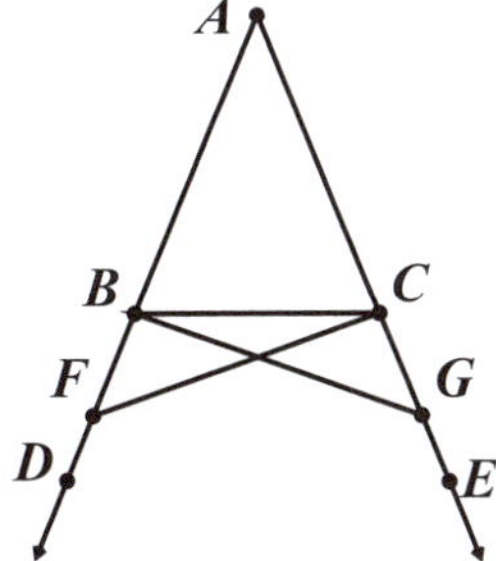

Figure 3.6. Proposition I.5

Proof.   Let ABC be an isosceles triangle having the side AB equal to the side AC; and let the straight lines BD, CE be produced further in a straight line with AB, AC. [Post. 2]

I say that the angle ABC is equal to the angle ACB, and the angle CBD to the angle BCE.

Let a point F be taken at random on BD; from AE the greater let AG be cut off equal to AF the less; [I.3] and let the straight lines FC, GB be joined. [Post. 1]

Then, since AF is equal to AG and AB to AC, the two sides FA, AC are equal to the two sides GA, AB, respectively; and they contain a common angle, the angle FAG.

Therefore the base FC is equal to the base GB, and the triangle AFC is equal to the triangle AGB, and the remaining angles will be equal to the remaining angles respectively, namely those which the equal sides subtend, that is, the angle ACF to the angle ABG, and the angle AFC to the angle AGB. [I.4]

And, since the whole AF is equal to the whole AG, and in these AB is equal to AC, the remainder BF is equal to the remainder CG.

But FC was also proved equal to GB; therefore the two sides BF, FC are equal to the two sides CG, GB respectively; and the angle BFC is equal to the angle CGB, while the base BC is common to them; therefore the triangle BFC is also equal to the triangle CGB, and the remaining angles will be equal to the remaining angles respectively, namely those which the equal sides subtend; therefore the angle FBC is equal to the angle GCB, and the angle BCF to the angle CBG.

Accordingly, since the whole angle ABG was proved equal to the angle ACF, and in these the angle CBG is equal to the angle BCF, the remaining angle ABC is equal to the remaining angle ACB; and they are at the base of the triangle ABC.

But the angle FBC was also proved equal to the angle GCB; and they are under the base.

Therefore etc. Q.E.D.   □

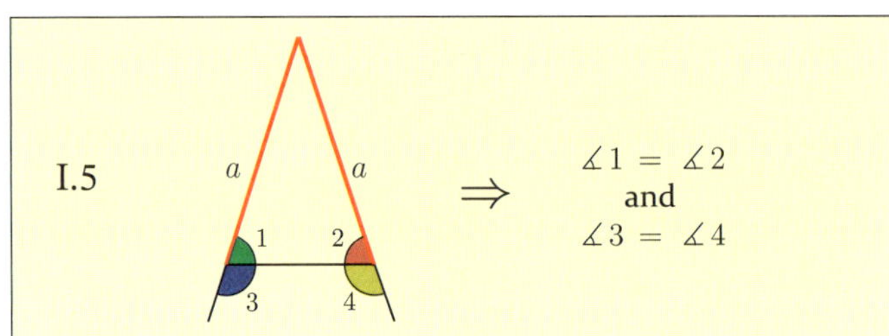

Notice that Euclid writes that two triangles are "equal" when we would more precisely describe them as "congruent." This becomes particularly important later in Book I when Euclid uses the term "equal" to mean that the triangles have the same area (but are not necessarily congruent). In general, it is clear from Euclid's context which meaning is correct. Also, when writing an updated version of the proof the reader may be tempted to use Common Notion 3 in order to claim that angles $\angle FBC$ and $\angle GCB$ are equal. Since we are only allowed to rely on the axioms and any earlier propositions, such a claim is invalid since it rests on a geometric result that has not been shown yet. It is not until Proposition I.13 that Euclid proves that a straight line standing upon a straight line makes either two right angles or angles whose sum equals two right angles.

The second proof offers an alternative approach to the first part of the theorem and is written with updated notation. Proclus attributes this proof to Pappus of Alexandria (ca. 290–ca. 350 BCE).

Proof 2.   Let $\triangle ABC$ be an isosceles triangle with $AB = AC$.

We will think of this triangle in two different ways, namely as both $\triangle ABC$ and $\triangle ACB$.

Since $AB = AC$, we clearly have $AC = AB$.

We also have $\angle BAC = \angle CAB$.

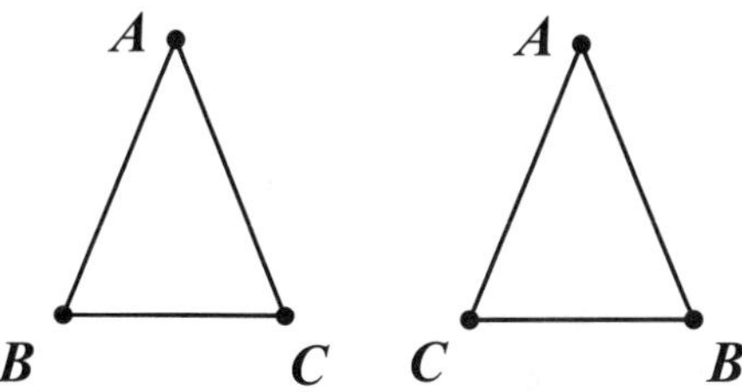

Figure 3.7. Proposition I.5

Thus by I.4 [SAS], we have $\triangle ABC \cong \triangle ACB$ and hence $\angle ABC = \angle ACB$ as desired. $\square$

The next proposition is the converse of the previous. A statement of the form *"If P, then Q."* is called an **implication**, and its **converse** is the statement *"If Q, then P."* In general, the validity of an implication does not imply the validity of its converse. For example, consider the statement: *If it is raining, then I carry an umbrella.* This statement merely implies that I am well-prepared for rain. Its converse is: *If I carry an umbrella, then it is raining.* The validity of this statement, on the other hand, would imply that I am, in fact, a rain god. (See Exercise 2.3.3 for another example.) When both an implication and its converse are valid, Euclid often follows the proof of the implication with its converse.

The proof of Proposition I.6 is our first formal example of the proof technique known as a **proof by contradiction**. (We say "formal" since we rather informally used this technique in our discussion of Common Notion 5 in the previous chapter.) When proving the implication *"If P, then Q."* with this technique, we begin by assuming P and the negation of Q, and show that this leads to an absurdity, or contradiction. This technique, also called *reductio ad absurdum*, offers an alternative to a **direct proof** of the implication *"If P, then Q."* where P is assumed to be true and Q must be shown to be true. In the proof of Proposition I.6, Euclid assumes that two angles in a triangle are equal (P) and the sides subtending the angles are not equal (negation of Q), and then finds a contradiction resulting from these assumptions. More specifically, he assumes $\angle ABC = \angle ACB$ and $AB \neq AC$, and then derives a contradiction.

Proposition I.6. *If in a triangle two angles be equal to one another, the sides which subtend the equal angles will also be equal to one another.*

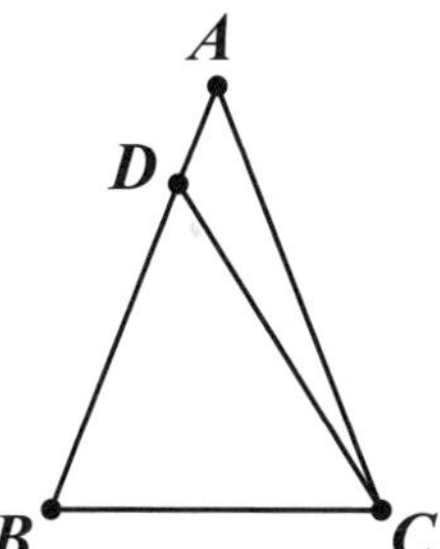

Figure 3.8. Proposition I.6

Proof. Let ABC be a triangle having the angle ABC equal to the angle ACB; I say that the side AB is also equal to the side AC.

For, if AB is unequal to AC, one of them is greater.

Let AB be greater; and from AB the greater let DB be cut off equal to AC the less; let DC be joined.

Then, since DB is equal to AC, and BC is common, the two sides DB, BC are equal to the two sides AC, CB respectively; and the angle DBC is equal to the angle ACB; therefore the base DC is equal to the base AB, and the triangle DBC will be equal to the triangle ACB, the less to the greater: which is absurd.

Therefore AB is not unequal to AC; it is therefore equal to it.

Therefore etc. Q.E.D.   □

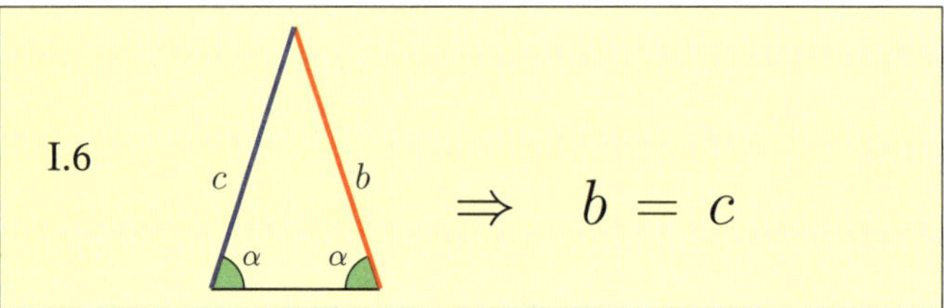

When working with an isosceles triangle and trying to recall the distinction between Propositions I.5 and I.6 in order to justify some step in a proof, it may help to know that Euclid never uses Proposition I.6 in any proof for the remainder of Book I.

The following proposition gives the result Euclid needs to prove the congruence scheme SSS of Proposition I.8. We call Proposition I.7 a **lemma** as it is not particularly interesting in its own right, but *is* helpful when proving a more substantial theorem. As with the previous proposition, Euclid employs a proof by contradiction.

Proposition I.7. *Given two straight lines constructed on a straight line [from its extremities] and meeting in a point, there cannot be constructed on the same straight line [from its extremities], and on the same side of it, two other straight lines meeting in another point and equal to the former two respectively, namely each to that which has the same extremity with it.*

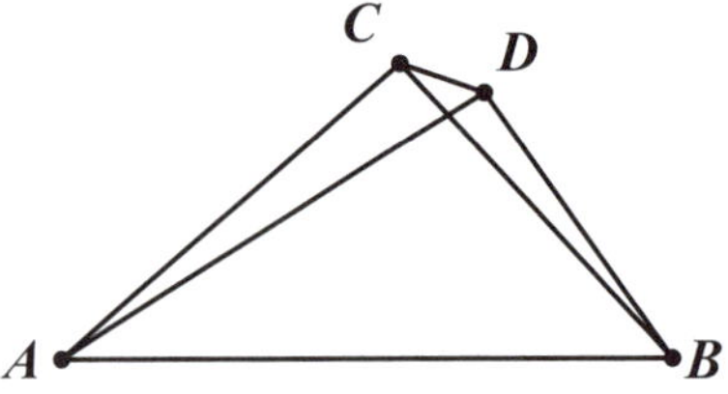

Figure 3.9. Proposition I.7

Proof. For, if possible, given two straight lines AC, CB constructed on the straight line AB and meeting at the point C, let two other straight lines AD, DB be constructed on the same straight line AB, on the same side of it, meeting in another point D and equal to the former two respectively, namely each to that which has the same extremity with it, so that CA is equal to DA which has the same extremity A with it, and CB to DB which has the same extremity B with it; and let CD be joined.

Then, since AC is equal to AD, the angle ACD is also equal to the angle ADC; [I.5] therefore the angle ADC is greater than the angle DCB; therefore the angle CDB is much greater than the angle DCB.

Again, since CB is equal to DB, the angle CDB is also equal to the angle DCB. But it was also proved much greater than it: which is impossible.

Therefore etc. Q.E.D.   □

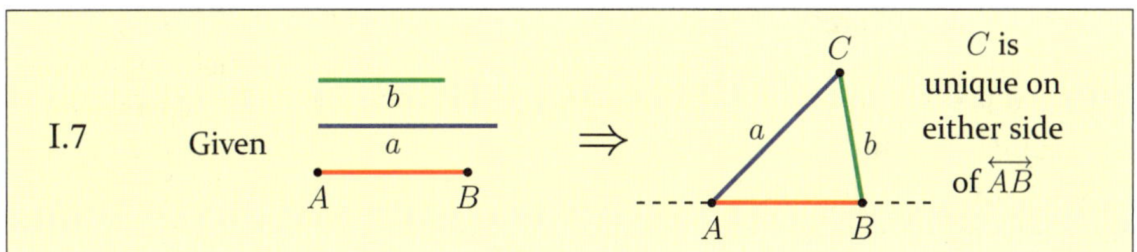

Euclid's proof of I.7 depends upon the configuration of points C and D. He assumes that D lies outside triangle $\triangle ABC$, but we must consider the alternative case where D lies inside the triangle, as illustrated in Figure 3.10. We leave it to the reader as Exercise

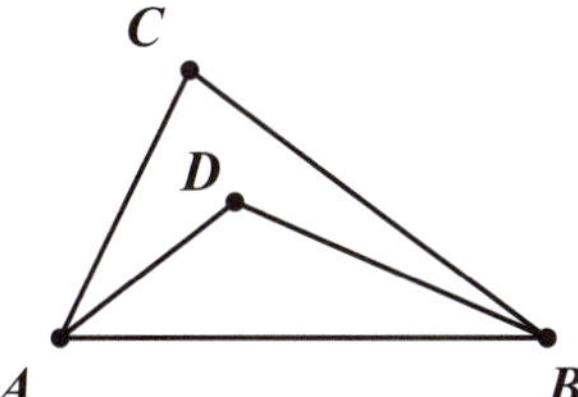

Figure 3.10. Alternative configuration for Proposition I.7

3.1.5 to find the contradiction in this case, and take this opportunity to warn the reader about the tendency to overlook the subtleties in a proof when relying on diagrams. We must be careful to read these proofs with a critical eye since Euclid does not always provide a proof for every possible case. It was the tradition of the Greek geometers to give one case, usually the most difficult, and to leave any other cases to the reader.

Proposition I.8 [SSS]. *If two triangles have the two sides equal to two sides respectively, and have also the base equal to the base, they will also have the angles equal which are contained by the equal straight lines.*

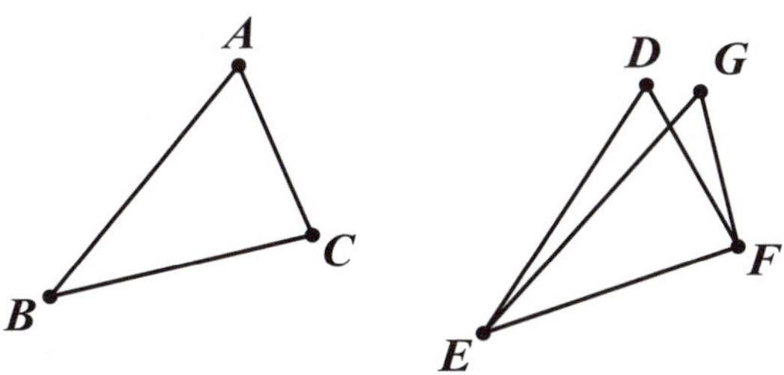

Figure 3.11. Proposition I.8 SSS

Proof.   Let ABC and DEF be two triangles having the two sides AB, AC equal to the two sides DE, DF respectively, namely AB to DE, and AC to DF; and let them have the base BC equal to the base EF.

I say that the angle BAC is also equal to the angle EDF.

For, if the triangle ABC is applied to the triangle DEF, and if the point B be placed on the point E and the straight line BC on EF, the point C will also coincide with F, because BC equals EF.

Then, BC coinciding with EF, BA, AC will also coincide with ED, DF; for, if the base BC coincides with the base EF, and the sides BA, AC do not coincide with ED, DF but fall beside them as EG, GF, then given two straight lines constructed on a straight line [from its extremities] and meeting in a point, there will have been constructed on the same straight line [from its extremities], and on the same side of it, two other straight lines meeting in another point and equal to the former two respectively, namely each to that which has the extremity with it.

But they cannot be so constructed. [I.7]

Therefore it is not possible that, if the base BC is applied to the base EF, the sides BA, AC should not coincide with ED, DF; they will therefore coincide, so that the angle BAC will also coincide with the angle EDF, and will be equal to it.

If therefore etc. Q.E.D.   □

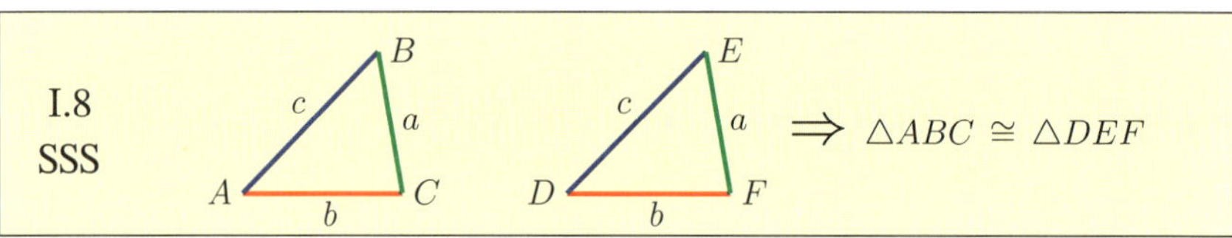

A careful reader may notice that Euclid does not specifically state that points D and G are on the same side of the plane determined by line EF. Since this requirement is part of the hypotheses of Proposition I.7, Euclid clearly means that the points meet this condition (as shown in his diagram for I.8), else he could not use I.7.

Before we begin the exercises, we introduce the following two definitions.

Definition 3.2. The point M of the line segment AB such that $AM = MB$ is called the **midpoint** of AB.

Definition 3.3. The line segment joining a vertex of a triangle to the midpoint of the opposite side is called a **median**.

A subtle omission in Euclid's common notions can be found in the proof of Proposition I.6. If $AB \neq AC$, then how can AB and AC be related to each other? Euclid assumes that either $AB > AC$ or $AB < AC$. This is the **law of trichotomy** for lines, and Euclid uses it often in Book I. (The law of trichotomy for real numbers says that any real number is either positive, negative or zero.) Not unlike Plato

> and Aristotle's classic laws of thought, this law has an air of common sense to it, but what are axioms if not things that are self-evident?

Exercises 3.1

1. Following the examples of Proposition I.1 and Proposition I.2, give an updated version of Euclid's proof of Proposition I.5. Be sure to justify each step, substitute mathematical symbols where appropriate, and include helpful diagrams as needed. Compare Euclid's proof with that of Pappus. Which do you prefer? Why? Why do you think Euclid chose his own proof instead of the shorter version?

2. Give an updated version of Euclid's proof of Proposition I.6. Be sure to justify each step, substitute mathematical symbols where appropriate, and include helpful diagrams as needed. When writing a proof by contradiction it is helpful to the reader to declare your intention at the start. This can be accomplished by writing *Proof (by contradiction):* at the start, or announcing your intended direction with the first sentence of the proof.

3. Give an updated version of Euclid's proof of Proposition I.7. Be sure to justify each step, substitute mathematical symbols where appropriate, and include helpful diagrams as needed.

4. Euclid's proof of Proposition I.7 uses the law of trichotomy as well as two other properties of inequality which are not given as common notions. One of them is the transitivity of inequality and another is a closely related property of inequality. [*Hint for the latter: If $a = b$ and $b < c$ ....*] Use your proof in Exercise 3 to give these missing properties.

5. For the proof of Proposition I.7, consider the alternative configuration where D lies inside $\triangle ABC$. We will reproduce Proclus' contradiction for this case by extending AC and AD to points E and F, respectively, as illustrated in Figure 3.10. Show that angle $\angle BDC$ is both greater than $\angle BCD$, and equal to angle $\angle BCD$, which is clearly not possible.

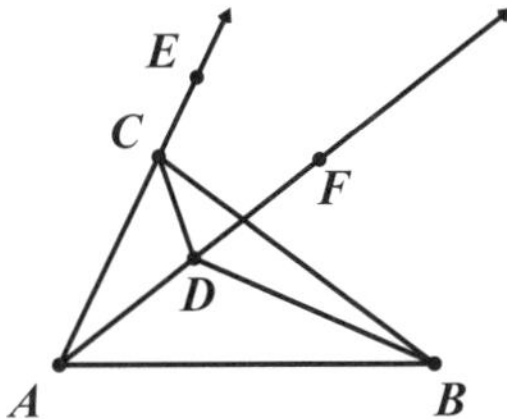

Figure 3.12. Exercise 3.1.5: Alternative configuration for Proposition I.7

6. Follow the instructions in Exercise 3 for Proposition I.8.

7. A **kite** is a convex quadrilateral with two pairs of congruent adjacent sides. The convex quadrilateral $ABCD$ shown in Figure 3.13 is a kite since $AB = BC$ and $AD = CD$. Prove that the diagonals of a kite intersect at right angles.

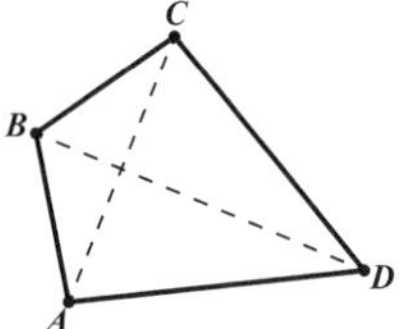

Figure 3.13. Exercise 3.1.7: Kite $ABCD$ where $AB = BC$ and $AD = CD$

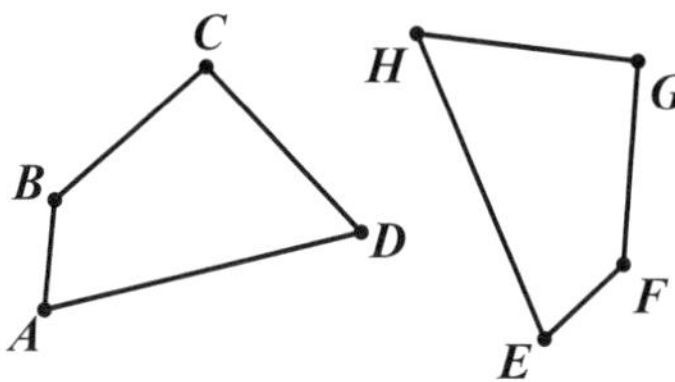

Figure 3.14. Exercise 3.1.8: Congruent quadrilaterals $ABCD$ and $EFGH$

8. Two quadrilaterals $ABCD$ and $EFGH$ are **congruent** if both their corresponding sides and their corresponding angles are equal, that is, $AB = EF$, $BC = FG$, $CD = GH$ and $AD = EH$, and $\angle A = \angle E$, $\angle B = \angle F$, $\angle C = \angle G$ and $\angle D = \angle H$, as illustrated in Figure 3.14.

   Though Euclid does not give any congruence schemes for quadilaterals, we add congruence scheme SASAS for convex quadrilaterals here. Prove congruence scheme SASAS for convex quadrilaterals. In other words, assuming that $AB = EF$, $BC = FG$, $CD = GH$, $\angle B = \angle F$ and $\angle C = \angle G$, prove that $\angle A = \angle E$, $\angle D = \angle H$ and $AD = EH$.

   Note: for the next two exercises, we need the fact that if $x = y$, then $\frac{1}{2}x = \frac{1}{2}y$. In the proof of Proposition I.37, Euclid claims, but does not prove, the following: "But the halves of equal things are equal to one another." We will wait until Chapter 7 to formally prove this.

9. Prove that the medians to the equal sides of an isosceles triangle are equal to each other.

10. Prove that the triangle formed by joining the midpoints of the three sides of an isosceles triangle is also isosceles.

11. Consider the isosceles triangle $\triangle ABC$ with $AB = AC$. Choose points D and E on AB and AC, respectively, such that $AD = AE$. Draw a corresponding picture and then prove that $CD = BE$.

12. Consider the isosceles triangle $\triangle ABC$ with $AB = AC$. Choose points D and E on side BC such that $BD = CE$. Draw a corresponding picture and then prove that $AD = AE$.

3.2 Propositions I.9 through I.15

The next four propositions are all constructions. Proposition I.9 gives us the ability to bisect any given angle while Proposition I.10 gives us the ability to bisect any given segment. In Propositions I.11 and I.12 we construct perpendicular lines to a given line segment, the first from a point on the segment, the second from a point not on it. These four along with Proposition I.23 (which allows us to copy a given angle onto a segment) and Proposition I.31 (in which we construct a line parallel to a given line through a point not on the given line), provide a basic toolbox for Euclid. It may surprise the reader that while bisecting any angle is a straightforward task, **trisecting** it, or dividing it into three equal angles, is not. It would take until 1837 before Pierre Wantzel (1814–1848) proved that it is, in fact, impossible to trisect every angle using only a straightedge and compass. Finally, while these next four propositions are constructions, Euclid's reliance on previous propositions within the body of each proof makes the resulting steps inefficient and impractical as a construction algorithm. In the exercises, the reader is asked to provide a more succinct set of instructions for each construction and provide the corresponding proof.

Proposition I.9. *To bisect a given rectilineal angle.*

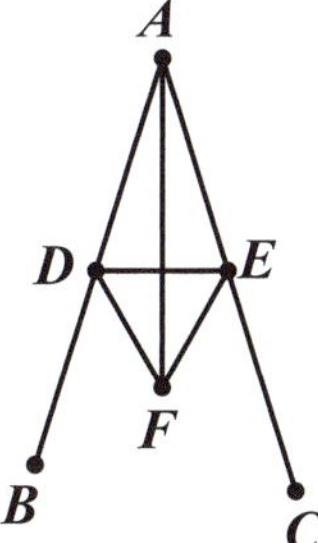

Figure 3.15. Proposition I.9

Proof. Let the angle BAC be the given rectilineal angle.

Thus it is required to bisect it.

Let a point D be taken at random on AB; let AE be cut off from AC equal to AD; [I.3] let DE be joined, and on DE let the equilateral triangle DEF be constructed; let AF be joined.

I say that the angle BAC has been bisected by the straight line AF.

For, since AD is equal to AE, and AF is common, the two sides DA, AF are equal to the two sides EA, AF respectively.

And the base DF is equal to the base EF; therefore the angle DAF is equal to the angle EAF. [I.8]

Therefore the given rectilineal angle BAC has been bisected by the straight line AF.

Q.E.F.   □

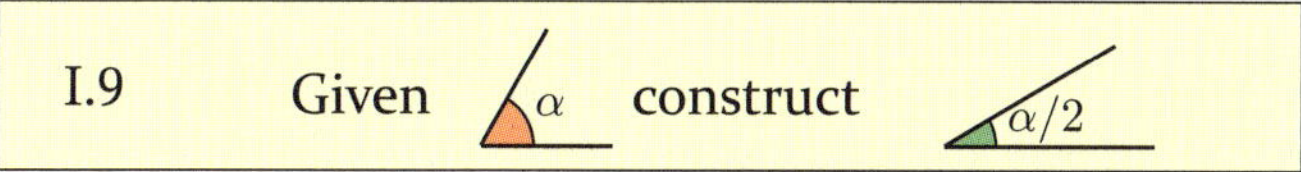

The line AF in Proposition 9 is called the **angle bisector** of angle $\angle BAC$.

Proposition I.10. *To bisect a given finite straight line.*

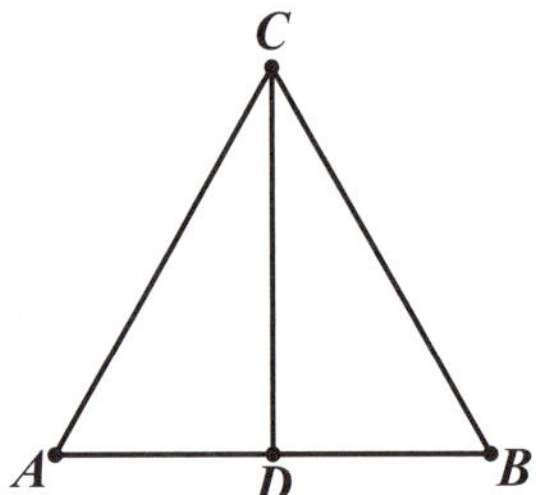

Figure 3.16. Proposition I.10

Proof.   Let AB be the given finite straight line.

Thus it is required to bisect the finite straight line AB.

Let the equilateral triangle ABC be constructed on it, [I.1] and let the angle ACB be bisected by the straight line CD; [I.9]

I say that the straight line AB has been bisected at the point D.

For, since AC is equal to CB, and CD is common, the two sides AC, CD are equal to the two sides BC, CD respectively; and the angle ACD is equal to the angle BCD; therefore the base AD is equal to the base BD. [I.4]

Therefore the given finite straight line AB has been bisected at D.

Q.E.F.   □

> I.10   Given $A \bullet\!\!-\!\!\underset{a}{}\!\!-\!\!\bullet B$ construct $A \bullet\!\!\underset{\frac{a}{2}}{}\!\!\bullet\overset{M}{}\!\!\underset{\frac{a}{2}}{}\!\!\bullet B$

While it is a subtle point, Euclid does not justify why the angle bisector for $\angle ACB$ intersects side AB at a point D. We will add this to our list of unstated assumptions that will eventually need to be addressed. The line CD in the proof is called the *perpendicular bisector* of AB, which we define in general as follows. Additionally, using updated notation, we write $CD \perp AB$ to indicate that CD is perpendicular to AB.

Definition 3.4. Given a line segment AB, the straight line through its midpoint that is also perpendicular to AB is called its **perpendicular bisector**.

The following theorem is not found in the *Elements* and is our first example of a **biconditional statement**, namely, a statement of form *"P if and only if Q."* which is abbreviated as *"P iff Q."* This type of statement includes both the implication *"If P, then Q."* and its converse *"If Q, then P."* The proof of a biconditional statement typically consists of separate proofs of each implication. Alternatively, a proof which utilizes a sequence of other logically equivalent biconditional statements can be given. This is often the case with proofs from high-school algebra. We leave the proof of this theorem as two exercises, one for each implication. Also, two extra notes about *distance* are in order. While we will discuss the notion of distance more formally in later chapters, for now please take the distance between two points as the length of the line segment between them. Lastly, **equidistant** means *equally distant*, or *of the same distance*.

Theorem 3.5. *Given a line segment AB, a point C is equidistant from A and B (that is, $CA = CB$), if and only if C lies on the perpendicular bisector of AB.*

Proposition I.11. *To draw a straight line at right angles to a given straight line from a given point on it.*

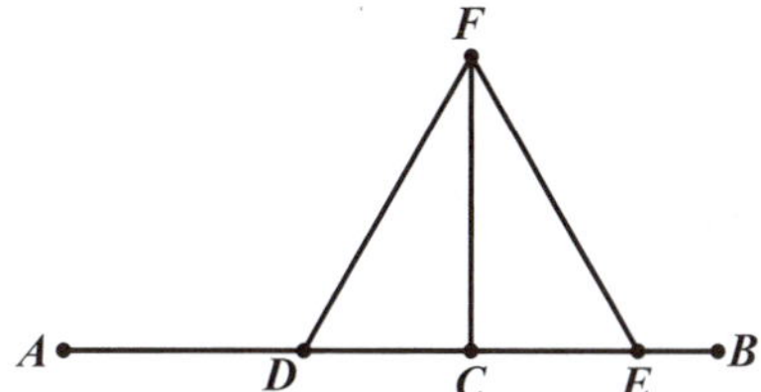

Figure 3.17. Proposition I.11

Proof.   Let AB be the given straight line, and C the given point on it.

Thus it is required to draw from the point C a straight line at right angles to the straight line AB.

Let a point D be taken at random on AC; let CE be made equal to CD; [I.3] on DE let the equilateral triangle FDE be constructed, [I.1] and let FC be joined;

I say that the straight line FC has been drawn at right angles to the given straight line AB from C the given point on it.

For, since DC is equal to CE, and CF is common, the two sides DC, CF are equal to the two sides EC, CF respectively; and the base DF is equal to the base FE; therefore the angle DCF is equal to the angle ECF; [I.8] and they are adjacent angles.

But, when a straight line set up on a straight line makes the adjacent angles equal to one another, each of the equal angles is right; [Def. 10] therefore each of the angles DCF, FCE is right.

Therefore the straight line CF has been drawn at right angles to the given straight line AB from the given point C on it.

Q.E.F.   □

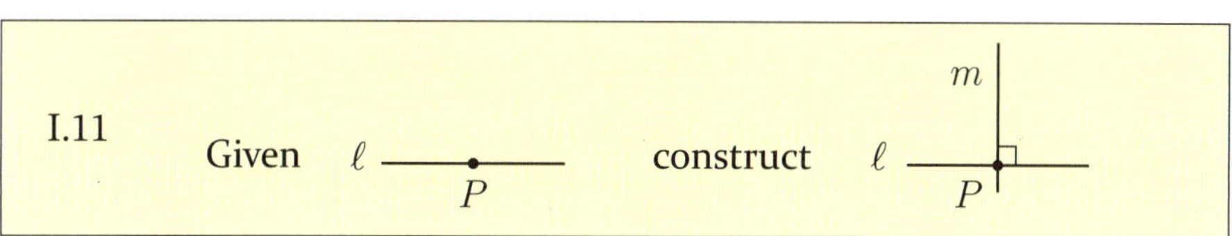

Proposition I.12. *To a given infinite straight line, from a given point which is not on it, to draw a perpendicular straight line.*

Proof.   Let AB be the given infinite straight line, and C the given point which is not on it; thus it is required to draw to the given infinite straight line AB, from the given point C which is not on it, a perpendicular straight line.

For let a point D be taken at random on the other side of the straight line AB, and with centre C and distance CD let the circle EFG be described; [Post. 3] let the straight line EG be bisected at H, [I.10] and let the straight lines CG, CH, CE be joined. [Post. 1]

I say that CH has been drawn perpendicular to the given infinite straight line AB from the given point C which is not on it.

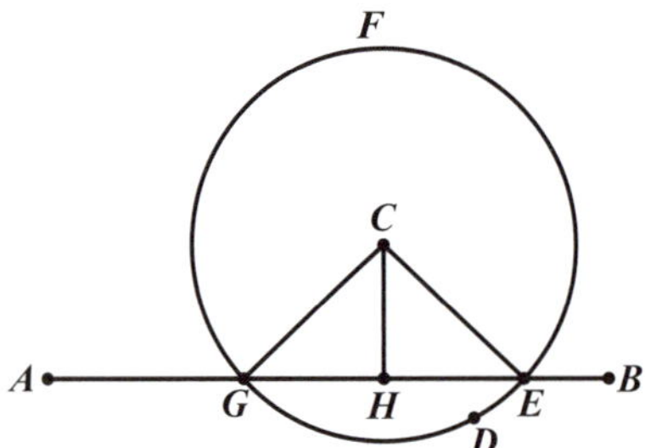

Figure 3.18. Proposition I.12

For, since GH is equal to HE, and HC is common, the two sides GH, HC are equal to the two sides EH, HC respectively; and the base CG is equal to the base CE; therefore the angle CHG is equal to the angle EHC. [I.8]

And they are adjacent angles.

But, when a straight line set up on a straight line makes the adjacent angles equal to one another, each of the equal angles is right, and the straight line standing on the other is called a perpendicular to that on which it stands. [Def. 10]

Therefore CH has been drawn perpendicular to the given infinite straight line AB from the given point C which is not on it.

Q.E.F.   □

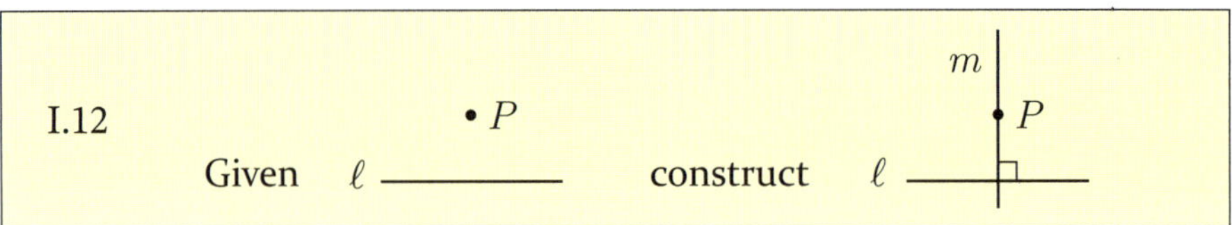

When constructing a perpendicular line and trying to recall the distinction between Propositions I.11 and I.12 in order to justify some step in a proof, it may help to know that Euclid never uses Proposition I.12 in any proof for the remainder of Book I.

The next few propositions concern the angles created by intersecting lines. Here are a few relevant definitions giving updated terminology.

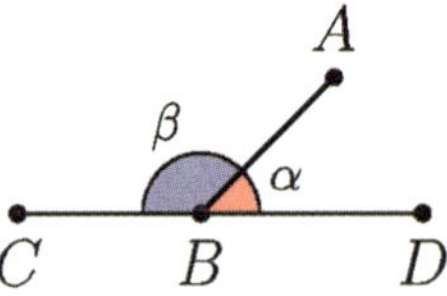

Figure 3.19. Supplementary angles α and β

Definition 3.6. If A is a point not on the straight line CD and B is a point on CD between C and D, then $\angle ABD$ and $\angle ABC$ are said to be **supplementary angles**.

Definition 3.7. If AB and CD are straight lines intersecting at point E, then angles $\angle AEC$ and $\angle BED$ are said to be **vertical angles**.

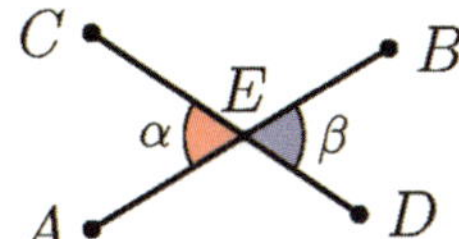

Figure 3.20. Vertical angles α and β

Proposition I.13 states that supplementary angles sum to two right angles and Proposition I.14 is its converse. While these *could* be combined to form an iff statement, Euclid rarely utilizes biconditional statements in the *Elements*, the first appearing in Book III. Though these two propositions may seem fairly obvious, the more interesting, I.15, is often referred to as the "Vertical Angle Theorem" as it asserts the equivalence of vertical angles.

Proposition I.13. *If a straight line set up on a straight line make angles, it will make either two right angles or angles equal to two right angles.*

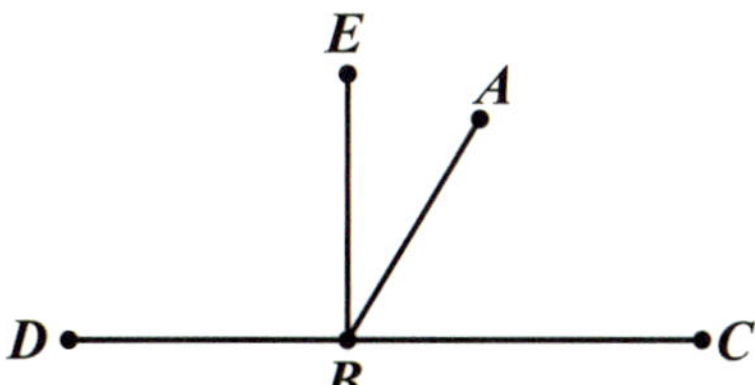

Figure 3.21. Proposition I.13

Proof. For let any straight line AB set up on the straight line CD make the angles CBA, ABD;

I say that the angles CBA, ABD are either two right angles or equal to two right angles.

Now, if the angle CBA is equal to the angle ABD, they are two right angles. [Def. 10]

But, if not, let BE be drawn from the point B at right angles to CD; [I.11] therefore the angles CBE, EBD are two right angles.

Then, since the angle CBE is equal to the two angles CBA, ABE, let the angle EBD be added to each; therefore the angles CBE, EBD are equal to the three angles CBA, ABE, EBD. [C.N. 2]

Again, since the angle DBA is equal to the two angles DBE, EBA, let the angle ABC be added to each; therefore the angles DBA, ABC are equal to the three angles DBE, EBA, ABC. [C.N. 2]

But the angles CBE, EBD were also proved equal to the same three angles; and things which are equal to the same thing are also equal to one another; [C.N. 1] therefore the angles CBE, EBD are also equal to the angles DBA, ABC.

But the angles CBE, EBD are two right angles; therefore the angles DBA, ABC are also equal to two right angles.

Therefore etc. Q.E.D. □

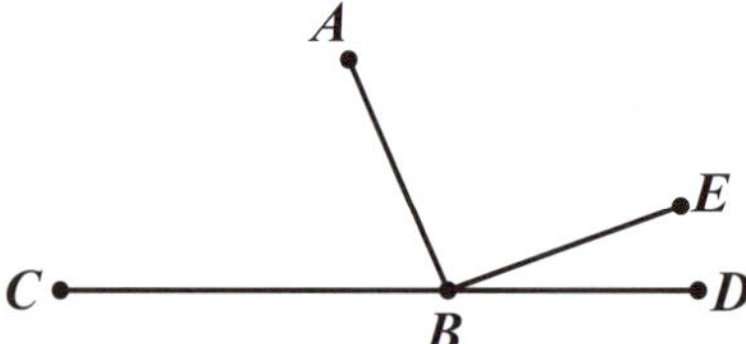

Notice that in this proof, Euclid assumes that $\angle CBA$ is acute and hence $\angle CBE = \angle CBA + \angle ABE$. Certainly this matches his diagram, but from the start he could have chosen $\angle ABD$ to be acute and drawn the resulting diagram. Upon reflection, it is clear that either choice would have led to the same conclusion and only the letters and diagram would be different. To acknowledge that there is another choice that can be made and note the resulting logical irrelevance of the choice, within the body of the proof we preface the choice with the phrase "*Without loss of generality, assume...*" which is abbreviated with **WLOG**. In an updated version of this particular proof we could write, "*WLOG, assume $\angle CBA$ is acute.*" One last note for I.13 concerns the new symbol found in its miniature representation shown in the box. Here, the symbol ∟ is used to denote a right angle.

Proposition I.14. *If with any straight line, and at a point on it, two straight lines not lying on the same side make the adjacent angles equal to two right angles, the two straight lines will be in a straight line with one another.*

A

E

C   B   D

Figure 3.22. Proposition I.14

Proof.   For with any straight line AB, and at the point B on it, let the two straight lines BC, BD not lying on the same side make the adjacent angles ABC, ABD equal to two right angles;

I say that BD is in a straight line with CB.

For, if BD is not in a straight line with BC, let BE be in a straight line with CB.

Then, since the straight line AB stands on the straight line CBE, the angles ABC, ABE are equal to two right angles. [I.13]

But the angles ABC, ABD are also equal to two right angles; therefore the angles CBA, ABE are equal to the angles CBA, ABD. [Post. 4 and C.N. 1]

Let the angle CBA be subtracted from each; therefore the remaining angle ABE is equal to the remaining angle ABD, [C.N. 3] the less to the greater: which is impossible. Therefore BE is not in a straight line with CB.

Similarly we can prove that neither is any other straight line except BD.

Therefore CB is in a straight line with BD.

Therefore etc. Q.E.D.   □

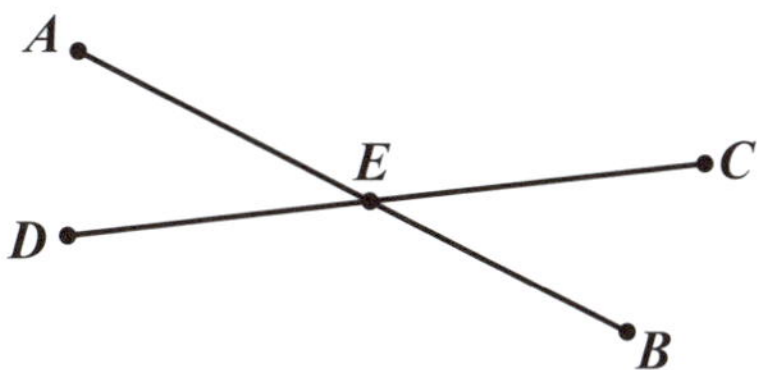

It is a bit odd that Euclid bothers to write "Similarly we can prove that neither is any other straight line except BD." For when he assumes that BD is not in a straight line with CB, Postulate 2 tells us that we can extend CB. Thus, there must exist an E such that BE is in a straight line with CB. Since the proof works regardless of whether or not E lies within $\angle ABD$, perhaps this is an acknowledgment by Euclid that his diagram does not represent all possible cases.

Proposition I.15 [Vertical Angle Theorem]. *If two straight lines cut one another, they make the vertical angles equal to one another.*

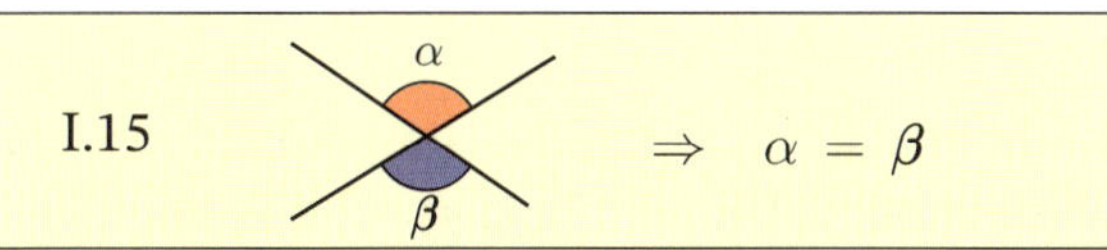

Figure 3.23. Proposition I.15

Proof.   For let the straight lines AB, CD cut one another at the point E;
   I say that the angle AEC is equal to the angle DEB, and the angle CEB to the angle AED.
   For, since the straight line AE stands on the straight line CD, making the angles CEA, AED, the angles CEA, AED are equal to two right angles. [I.13]
   Again, since the straight line DE stands on the straight line AB, making the angles AED, DEB, the angles AED, DEB are equal to two right angles. [I.13]
   But the angles CEA, AED were also proved equal to two right angles; therefore the angles CEA, AED are equal to the angles AED, DEB. [Post. 4 and C.N. 1]
   Let the angle AED be subtracted from each; therefore the remaining angle CEA is equal to the remaining angle BED. [C.N. 3]
   Similarly it can be proved that the angles CEB, DEA are also equal.
   Therefore etc. Q.E.D.   □

The updated proof of the Vertical Angle Theorem given below illustrates how the numbering of angles in a diagram can make a proof much easier to follow, especially when there are a good number of angles of interest in a proof.

Proof 2.   Let AB and CD intersect at a point E and label the angles as shown in the figure.
   Since CE stands on AB, by proposition 13, we have $\angle 1 + \angle 2 = 2$ right angles.

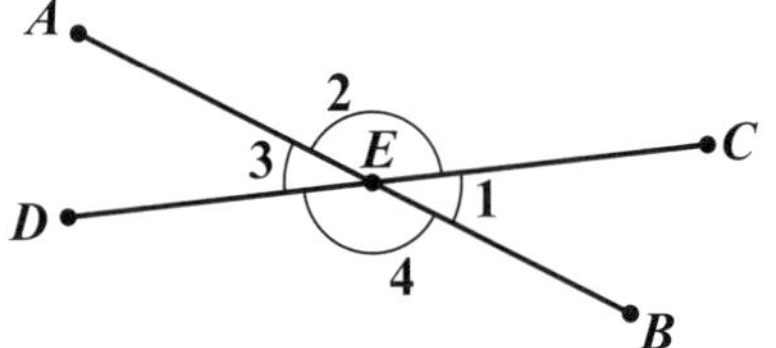

Figure 3.24. Proposition I.15

Similarly, since AE stands on CD, we have $\angle 2 + \angle 3 = 2$ right angles.

By common notion 1, $\angle 1 + \angle 2 = \angle 2 + \angle 3$. Subtracting $\angle 2$ from both sides, we get $\angle 1 = \angle 3$ as desired.

Similarly, it can be shown that $\angle 2 = \angle 4$.   □

Notice that the sum of all four angles meeting at point E in Proposition I.15 is four right angles. This is true in general, and Proclus stated it as a corollary to this proposition. It is included as an exercise for the reader. Before we begin the exercises, we introduce the following definition.

Definition 3.8. In any triangle, a line segment starting from a vertex and meeting the line defined by its opposite side perpendicularly is an **altitude** of the triangle.

Exercises 3.2

1. Give a construction and corresponding proof for each of the listed propositions. Unlike Euclid's versions for these propositions, make sure that each proof only relies on Propositions I.1–I.8. Be sure to justify each step and include helpful diagrams as needed.

   (a) Proposition I.9   (c) Proposition I.11

   (b) Proposition I.10   (d) Proposition I.12

2. Give an updated version of Euclid's proof of each of the listed propositions. Be sure to justify each step, substitute mathematical symbols where appropriate, and include helpful diagrams as needed.

   (a) Proposition I.13   (b) Proposition I.14

3. Prove Theorem 3.5 by completing each direction of the biconditional as given separately in the following two parts.

   (a) Prove that every point on the perpendicular bisector of a line segment AB is equidistant from the segment's endpoints.

   (b) Prove that every point that is equidistant from the endpoints of a line segment AB lies on its perpendicular bisector.

4. Prove that the median to the base of an isosceles triangle is perpendicular to the base and bisects the opposite angle.

5. Let $\triangle ABC$ be isosceles where $AB = AC$. Prove that the bisector of $\angle CAB$ is perpendicular to BC and bisects BC.

6. Prove that if one of the altitudes of a triangle is also a median, then the triangle is isosceles.

7. Give a proof of the corollary Proclus gave to Proposition I.15: *If any number of straight lines cut one another at a single point, the angles that result will be equal in sum to four right angles.*

8. Prove that if a line bisects one of the angles in a pair of vertical angles, then it also bisects the other. Specifically, let $\overleftrightarrow{AB}$ and $\overleftrightarrow{CD}$ intersect at a point E. Prove that if $\overleftrightarrow{EF}$ bisects $\angle AED$, then it bisects the vertically opposite angle, $\angle BEC$, as well. Include a diagram.

9. Prove that the angle bisectors for a pair of vertical angles lie in the same straight line. Specifically, let $\overleftrightarrow{AB}$ and $\overleftrightarrow{CD}$ intersect at a point E. Suppose $\overrightarrow{EF}$ bisects $\angle AED$, and $\overrightarrow{EG}$ bisects the vertically opposite angle $\angle BEC$. Prove that E, F and G lie in the same straight line. [*Hint: You may use Exercise 7, Proclus' corollary to Proposition I.15.*]

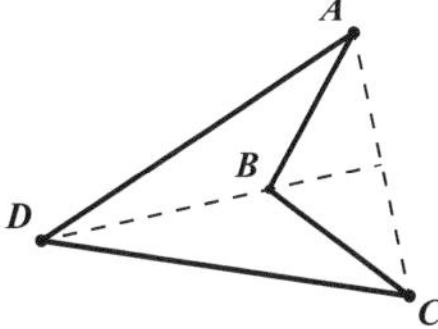

Figure 3.25. Exercise 3.2.10: Dart $ABCD$ where $AB = BC$ and $AD = CD$

10. A **dart** is a nonconvex quadrilateral with two pairs of congruent adjacent sides. The nonconvex quadrilateral $ABCD$ shown in Figure 3.25 is a dart since $AB = BC$ and $AD = CD$. Prove that the lines containing the diagonals of a dart intersect at right angles.

3.3 Propositions I.16 through I.28 and I.31

The next six propositions are all inequalities relating angle measures or segment lengths. Proposition I.16 is often called the "Exterior Angle Theorem" as it compares interior angles of a triangle with an angle exterior to the triangle. Proposition I.17 asserts that the sum of any two interior angles in a triangle is less than two right angles. This proposition will seem rather weak after Proposition I.32, the proposition every high school geometry student remembers: the sum of the angles in any triangle is equal to two right angles. Unlike Proposition I.32, Proposition I.17 belongs to Neutral geometry since its proof does not rely on the fifth postulate. Propositions I.18 and I.19 relate angle magnitude with side length and are converses of each other.

Proposition I.16 [Exterior Angle Theorem]. *In any triangle, if one of the sides be produced, the exterior angle is greater than either of the interior and opposite angles.*

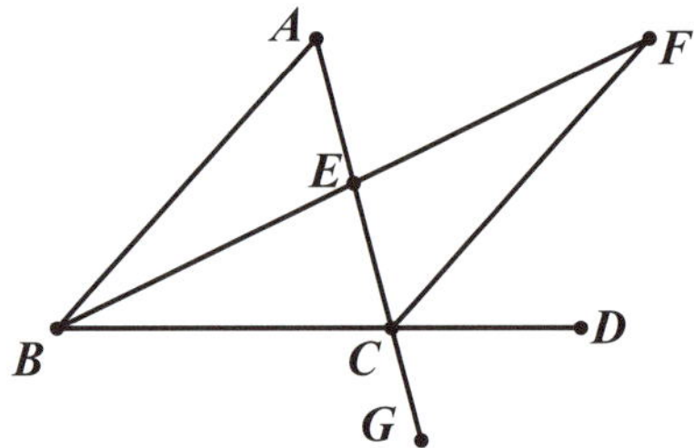

Figure 3.26. Proposition I.16

Proof.   Let *ABC* be a triangle, and let one side of it *BC* be produced to *D*;

I say that the exterior angle *ACD* is greater than either of the interior and opposite angles *CBA, BAC*.

Let *AC* be bisected at *E* [I.10], and let *BE* be joined and produced in a straight line to *F*;

let *EF* be made equal to *BE* [I.3], let *FC* be joined [Post. 1], and let *AC* be drawn through to *G* [Post. 2].

Then, since *AE* is equal to *EC*, and *BE* to *EF*, the two sides *AE, EB* are equal to the two sides *CE, EF* respectively; and the angle *AEB* is equal to the angle *FEC*, for they are vertical angles. [I.15]

Therefore the base *AB* is equal to the base *FC*, and the triangle *ABE* is equal to the triangle *CFE*, and the remaining angles are equal to the remaining angles respectively, namely those which the equal sides subtend; [I.4]

therefore the angle *BAE* is equal to the angle *ECF*.

But the angle *ECD* is greater than the angle *ECF*; [C.N. 5] therefore the angle *ACD* is greater than the angle *BAE*.

Similarly also, if *BC* be bisected, the angle *BCG*, that is, the angle *ACD* [I.15], can be proved greater than the angle *ABC* as well.

Therefore etc. Q.E.D.   □

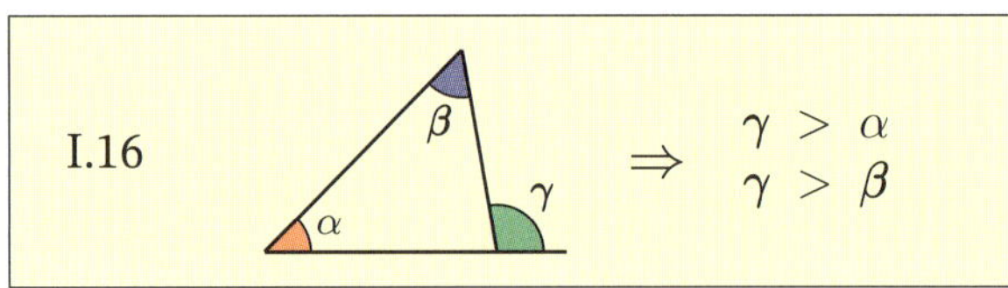

The proof of the following related theorem, which uses Proposition I.16 to establish the uniqueness of a perpendicular through a point, is left as an exercise.

Theorem 3.9. *Let C be a point not on the line $\overleftrightarrow{AB}$. There is exactly one line through C that is perpendicular to $\overleftrightarrow{AB}$.*

Proposition I.17. *In any triangle two angles taken together in any manner are less than two right angles.*

Proof.   Let *ABC* be a triangle; I say that two angles of the triangle *ABC* taken together in any manner are less than two right angles. For let *BC* be produced to *D*. [Post. 2]

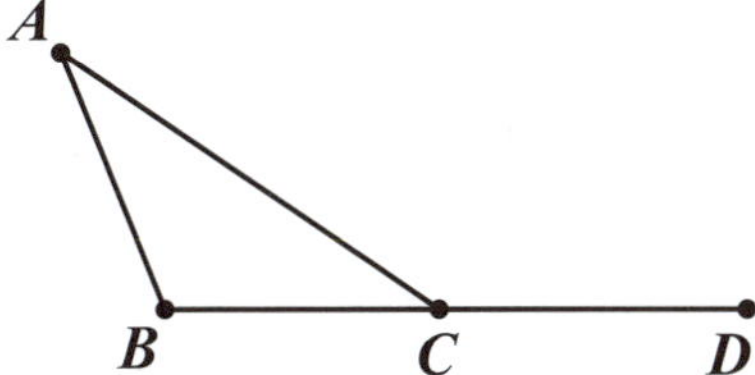

Figure 3.27. Proposition I.17

Then, since the angle ACD is an exterior angle of the triangle ABC, it is greater than the interior and opposite angle ABC. [I.16]

Let the angle ACB be added to each; therefore the angles ACD, ACB are greater than the angles ABC, BCA.

But the angles ACD, ACB are equal to two right angles. [I.13]

Therefore the angles ABC, BCA are less than two right angles.

Similarly we can prove that the angles BAC, ACB are also less than two right angles, and so are the angles CAB, ABC as well.

Therefore etc. Q.E.D. $\square$

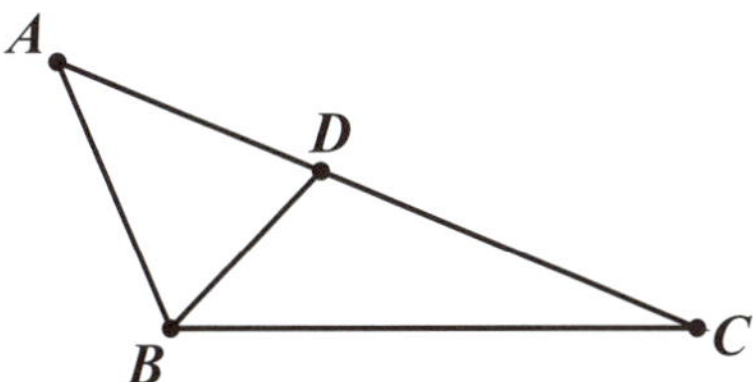

Proposition I.18. *In any triangle the greater side subtends the greater angle.*

Figure 3.28. Proposition I.18

Proof. For let ABC be a triangle having the side AC greater than AB;

I say that the angle ABC is also greater than the angle BCA.

For, since AC is greater than AB, let AD be made equal to AB [I.3], and let BD be joined.

Then, since the angle ADB is an exterior angle of the triangle BCD, it is greater than the interior and opposite angle DCB. [I.16]

But the angle ADB is equal to the angle ABD, since the side AB is equal to AD;

therefore the angle ABD is also greater than the angle ACB;

therefore the angle ABC is much greater than the angle ACB.

Therefore etc. Q.E.D. $\square$

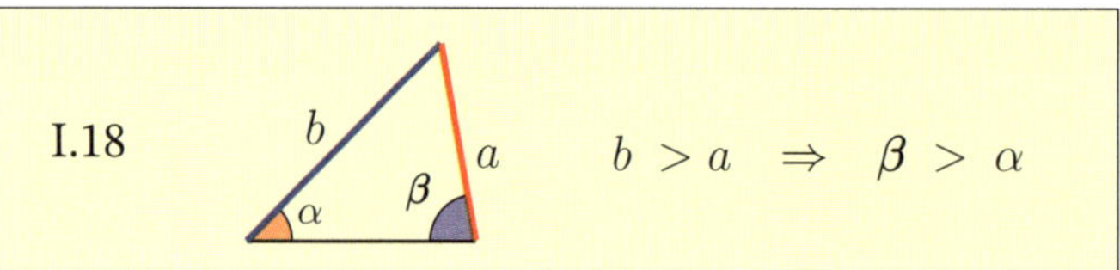

Proposition I.19. *In any triangle the greater angle is subtended by the greater side.*

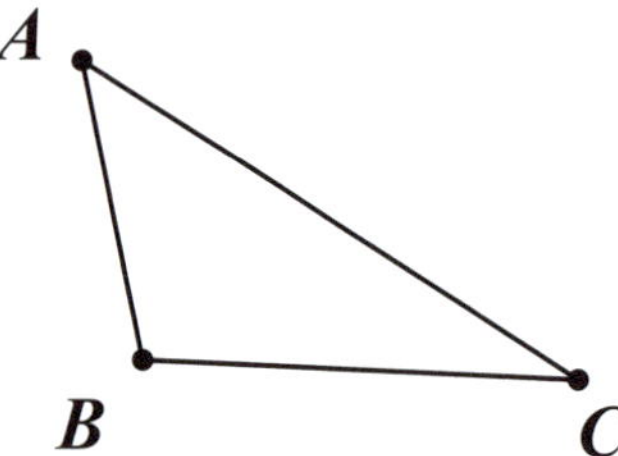

Figure 3.29. Proposition I.19

Proof.   Let ABC be a triangle having the angle ABC greater than the angle BCA;
I say that the side AC is also greater than the side AB.

For, if not, AC is either equal to AB or less.

Now AC is not equal to AB; for then the angle ABC would also have been equal to the angle ACB; [I.5] but it is not; therefore AC is not equal to AB.

Neither is AC less than AB, for then the angle ABC would also have been less than the angle ACB; [I.18] but it is not; therefore AC is not less than AB.

And it was proved that it is not equal either.

Therefore AC is greater than AB.

Therefore etc. Q.E.D.   □

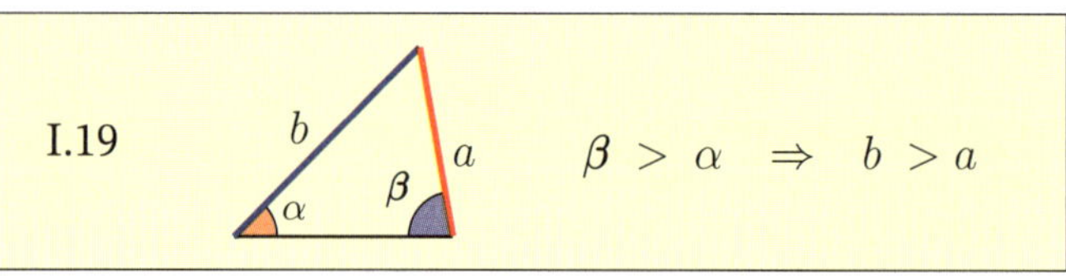

Definition 3.10. The **distance from a point A to a line** $\overleftrightarrow{BC}$ is defined to be the length of the shortest segment AD where D is any point on $\overleftrightarrow{BC}$.

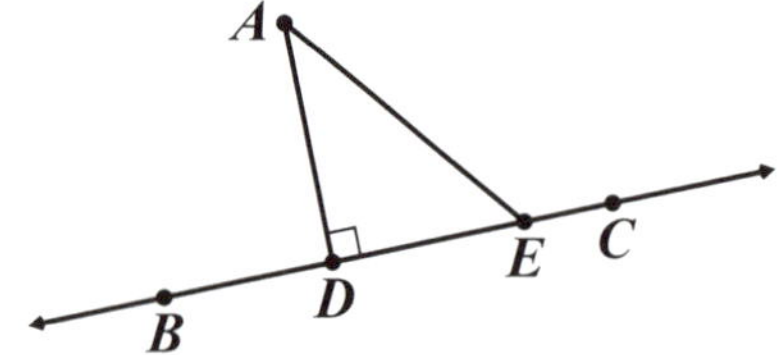

Figure 3.30. Distance from a point to a line

We leave the proof of the following theorem as Exercise 3.3.7.

Theorem 3.11. *Of all line segments joining a point not on a given line to the line, the unique shortest segment is perpendicular to the given line.*

Known as the Triangle Inequality Theorem, Proposition I.20 asserts that the sum of the lengths of two sides of a triangle is greater than the third side. According to Proclus, the Epicureans of Greece ridiculed the inclusion of this proposition in the *Elements* since it is "evident even to an ass and requiring no proof" [40]. They claimed that if you were to place a donkey at one vertex of a triangle and his food at another, he would never traverse the two sides over the one to get to it. Proclus' response was, "that a mere perception of the truth of a theorem is a different thing from a proof of it and a knowledge of why it is true" [40]. We add to Proclus' wisdom with the observation that it is often the most obvious that is most difficult to prove.

Proposition I.20 [Triangle Inequality Theorem]. *In any triangle two sides taken together in any manner are greater than the remaining one.*

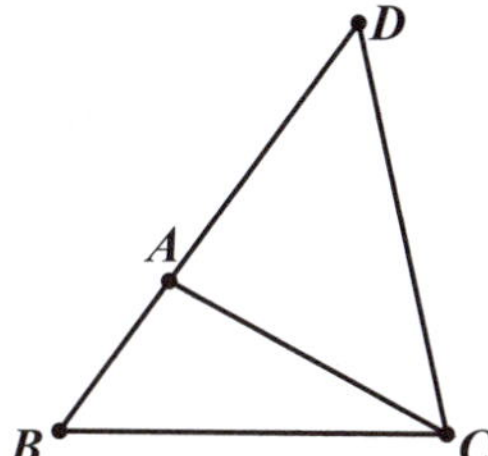

Figure 3.31. Proposition I.20

Proof. For let ABC be a triangle; I say that in the triangle ABC two sides taken together in any manner are greater than the remaining one, namely BA, AC greater than BC, AB, BC greater than AC, BC, CA greater than AB.

For let BA be drawn through to the point D, let DA be made equal to CA, and let DC be joined.

Then, since DA is equal to AC, the angle ADC is also equal to the angle ACD; [I.5] therefore the angle BCD is greater than the angle ADC. [C.N. 5]

And, since DCB is a triangle having the angle BCD greater than the angle BDC, and the greater angle is subtended by the greater side, [I.19] therefore DB is greater than BC.

But DA is equal to AC; therefore BA, AC are greater than BC.

Similarly we can prove that AB, BC are also greater than CA, and BC, CA than AB.

Therefore etc. Q.E.D. $\square$

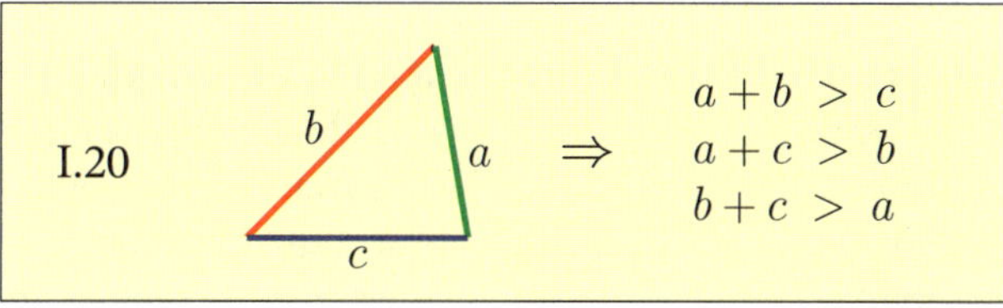

Proposition I.21. *If on one of the sides of a triangle, from its extremities, there be constructed two straight lines meeting within the triangle, the straight lines so constructed will be less than the remaining two sides of the triangle, but will contain a greater angle.*

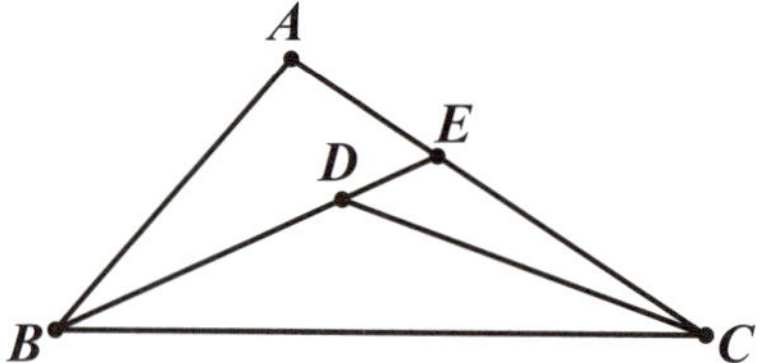

Figure 3.32. Proposition I.21

Proof.   On BC, one of the sides of the triangle ABC, from its extremities B, C, let the two straight lines BD, DC be constructed meeting within the triangle;

I say that BD, DC are less than the remaining two sides of the triangle BA, AC, but contain an angle BDC greater than the angle BAC.

For let BD be drawn through to E.

Then, since in any triangle two sides are greater than the remaining one, [I.20] therefore, in the triangle ABE, the two sides AB, AE are greater than BE.

Let EC be added to each; therefore BA, AC are greater than BE, EC.

Again, since, in the triangle CED, the two sides CE, ED are greater than CD, let DB be added to each; therefore CE, EB are greater than CD, DB.

But BA, AC were proved greater than BE, EC; therefore BA, AC are much greater than BD, DC.

Again, since in any triangle the exterior angle is greater than the interior and opposite angle, [I.16] therefore, in the triangle CDE, the exterior angle BDC is greater than the angle CED.

For the same reason, moreover, in the triangle ABE also, the exterior angle CEB is greater than the angle BAC. But the angle BDC was proved greater than the angle CEB; therefore the angle BDC is much greater than the angle BAC.

Therefore etc. Q.E.D.   □

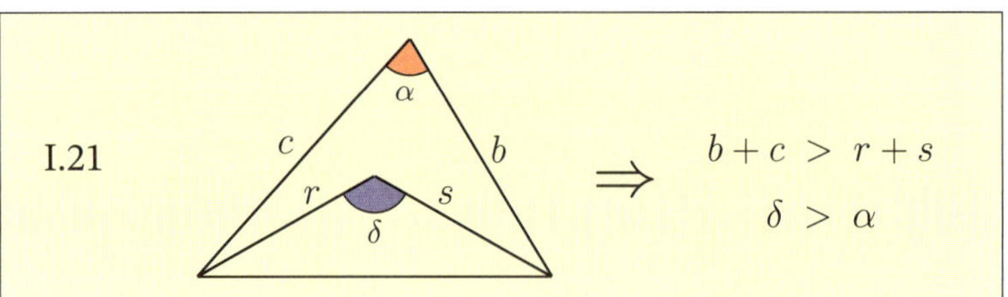

In Proposition I.22, Euclid constructs a triangle from three segments where the sum of any two is greater than the third. Euclid's statement of I.22 incorporates his conclusion from I.20, which may cause confusion. In the "thus it is necessary that..." clause of this proposition, Euclid includes the requirement from I.20 that the sum of any two sides of a triangle must be greater than the third. In other words, if we were given three line segments that fail to meet this condition, say segments of length 1, 1 and 3, then it would be impossible to construct a triangle with these lengths. To be clear, Euclid's secondary clause in the proposition could be folded into the hypothesis of the statement as follows: *Given three straight lines where two of the straight lines taken together in any manner should be greater than the remaining one, to construct a triangle.* We take this opportunity to note that there are many equivalent ways of rephrasing the conditional statement "If P, then Q." These include "Q, if P.; "P only if Q.; "P is a sufficient condition for Q."; and "Q is a necessary condition for P." So, while in I.20 Euclid shows that "the sum of any two sides greater than the third" is a necessary condition for a triangle, here he shows that it is also a sufficient condition. This in turn allows Euclid to copy any triangle and hence, as we will see in Proposition I.23, any angle.

Proposition I.22. *Out of three straight lines, which are equal to three given straight lines, to construct a triangle: thus it is necessary that two of the straight lines taken together in any manner should be greater than the remaining one. [I.20]*

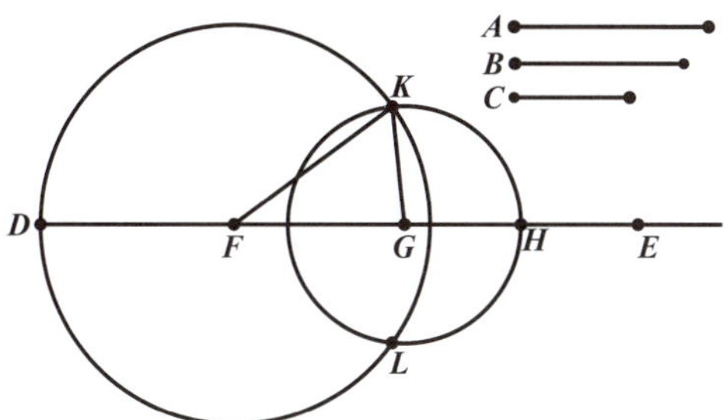

Figure 3.33. Proposition I.22

Proof. Let the three given straight lines be A, B, C, and of these let two taken together in any manner be greater than the remaining one, namely A, B greater than C, A, C greater than B, and B, C greater than A; thus it is required to construct a triangle out of straight lines equal to A, B, C.

Let there be set out a straight line DE, terminated at D but of infinite length in the direction of E, and let DF be made equal to A, FG equal to B, and GH equal to C. [I.3]

With centre F and distance FD let the circle DKL be described; again, with centre G and distance GH let the circle KLH be described; and let KF, KG be joined;

I say that the triangle KFG has been constructed out of three straight lines equal to A, B, C.

For, since the point F is the centre of the circle DKL, FD is equal to FK.

But FD is equal to A; therefore KF is also equal to A.

Again, since the point G is the centre of the circle LKH, GH is equal to GK.

But GH is equal to C; therefore KG is also equal to C. And FG is also equal to B; therefore the three straight lines KF, FG, GK are equal to the three straight lines A, B, C.

Therefore out of the three straight lines KF, FG, GK, which are equal to the three given straight lines A, B, C, the triangle KFG has been constructed.

Q.E.F.   ☐

I.22	Given $\begin{array}{c} a \\ c \\ b \end{array}$	where $\begin{array}{c} a+b > c \\ a+c > b \\ b+c > a \end{array}$	construct

A careful reader will notice that Euclid does not justify the intersection of the two constructed circles. This is the same unstated assumption we found in the proof of Proposition I.1.

Proposition I.23 is the theorem that allows Euclid to copy an angle. However, Euclid's reliance on Proposition I.22 makes his proof particularly unhelpful when viewed as a set of instructions. As an exercise, the reader is asked to give a more succinct set of instructions for this construction and provide the corresponding proof.

Proposition I.23. *On a given straight line and at a point on it to construct a rectilineal angle equal to a given rectilineal angle.*

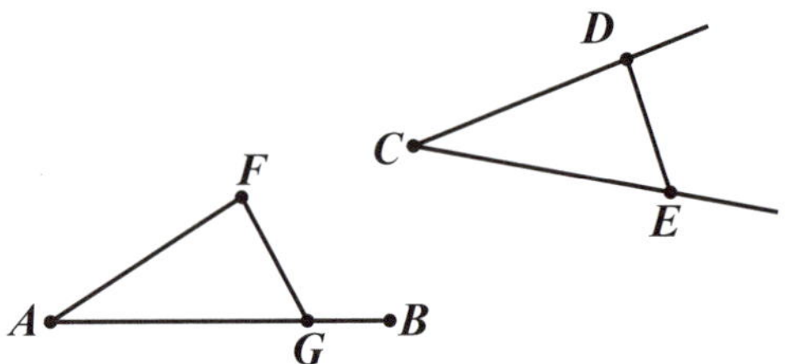

Figure 3.34. Proposition I.23

Proof. Let AB be the given straight line, A the point on it, and the angle DCE the given rectilineal angle;

thus it is required to construct on the given straight line AB, and at the point A on it, a rectilineal angle equal to the given rectilineal angle DCE.

On the straight lines CD, CE respectively let the points D, E be taken at random; let DE be joined, and out of three straight lines which are equal to the three straight lines CD, DE, CE let the triangle AFG be constructed in such a way that CD is equal to AF, CE to AG, and further DE to FG.

Then, since the two sides DC, CE are equal to the two sides FA, AG respectively, and the base DE is equal to the base FG, the angle DCE is equal to the angle FAG. [I.8]

Therefore on the given straight line AB, and at the point A on it, the rectilineal angle FAG has been constructed equal to the given rectilineal angle DCE.

Q.E.F.   □

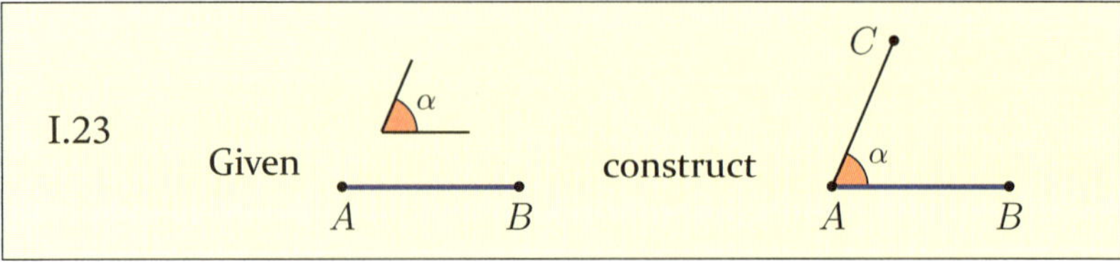

These next two propositions provide inequalities relating angle magnitude with side length. They are similar in nature to Propositions I.18 and I.19, but while the earlier propositions considered only one triangle, these compare two.

Proposition I.24. *If two triangles have the two sides equal to two sides respectively, but have the one of the angles contained by the equal straight lines greater than the other, they will also have the base greater than the base.*

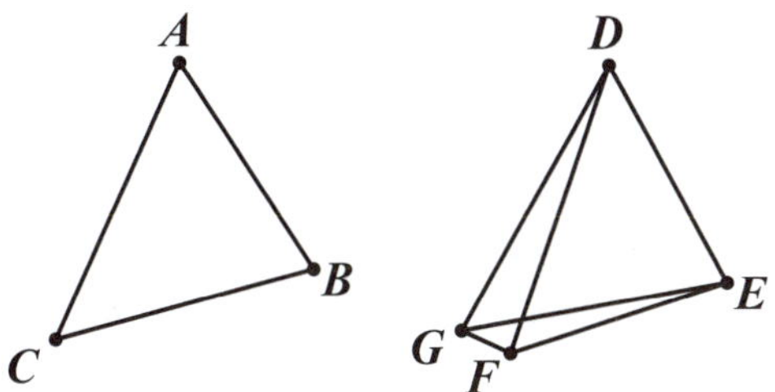

Figure 3.35. Proposition I.24

Proof. Let ABC, DEF be two triangles having the two sides AB, AC equal to the two sides DE, DF respectively, namely AB to DE, and AC to DF, and let the angle at A be greater than the angle at D;

I say that the base BC is also greater than the base EF.

For, since the angle BAC is greater than the angle EDF, let there be constructed, on the straight line DE, and at the point D on it, the angle EDG equal to the angle BAC; [I.23] let DG be made equal to either of the two straight lines AC, DF, and let EG, FG be joined.

Then, since AB is equal to DE, and AC to DG, the two sides BA, AC are equal to the two sides ED, DG, respectively; and the angle BAC is equal to the angle EDG; therefore the base BC is equal to the base EG. [I.4]

Again, since DF is equal to DG, the angle DGF is also equal to the angle DFG; [I.5] therefore the angle DFG is greater than the angle EGF.

Therefore the angle EFG is much greater than the angle EGF.

And, since EFG is a triangle having the angle EFG greater than the angle EGF, and the greater angle is subtended by the greater side, [I.19] the side EG is also greater than EF.

But EG is equal to BC. Therefore BC is also greater than EF.

Therefore etc. Q.E.D. □

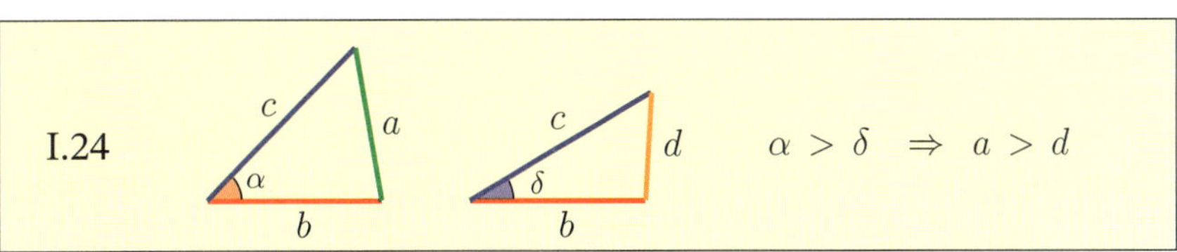

Proposition I.25. *If two triangles have the two sides equal to two sides respectively, but have the base greater than the base, they will also have the one of the angles contained by the equal straight lines greater than the other.*

Proof. Let ABC, DEF be two triangles having the two sides AB, AC equal to the two sides DE, DF respectively, namely AB to DE, and AC to DF; and let the base BC be greater than the base EF;

I say that the angle BAC is also greater than the angle EDF.

For, if not, it is either equal to it or less.

Now the angle BAC is not equal to the angle EDF; for then the base BC would also have been equal to the base EF, [I.4] but it is not; therefore the angle BAC is not equal to the angle EDF.

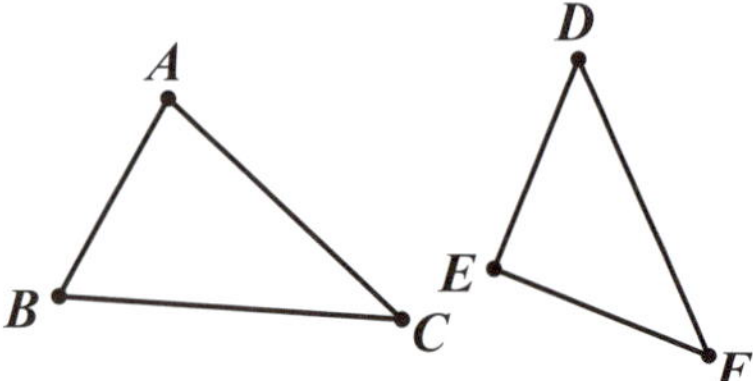

Figure 3.36. Proposition I.25

Neither again is the angle *BAC* less than the angle *EDF*; for then the base *BC* would also have been less than the base *EF*, [I.24] but it is not; therefore the angle *BAC* is not less than the angle *EDF*.

But it was proved that it is not equal either; therefore the angle *BAC* is greater than the angle *EDF*.

Therefore etc. Q.E.D.   □

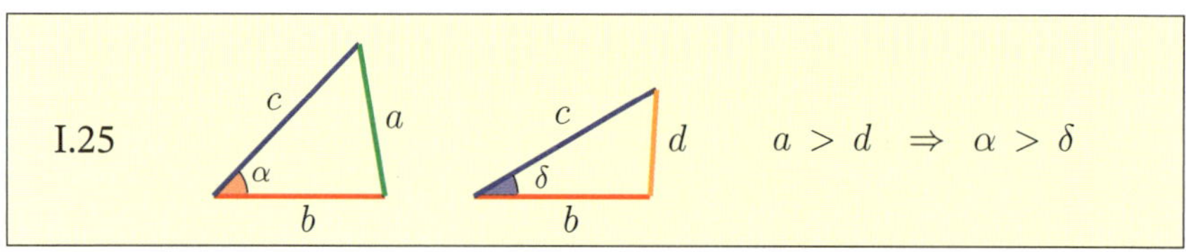

With the completion of the following proposition we will have proven the four congruence schemes for triangles: SAS, SSS, ASA and AAS. In terms of the possible different arrangements of letters when dealing with the three angles (A) or three sides (S) of a triangle, the only two missing possibilities are side-side-angle and angle-angle-angle. These are not, however, valid congruence schemes. It is easy to see that angle-angle-angle does not guarantee congruence by imaging a triangle along with a "pinch to zoom" enlargement of the same triangle. The smaller and larger versions share the same angles but clearly have different side lengths and are thus, not congruent. We ask the reader to produce a counterexample to side-side-angle as an exercise.

Proposition I.26 [ASA], [AAS]. *If two triangles have the two angles equal to two angles respectively, and one side equal to one side, namely, either the side adjoining the equal angles, or that subtending one of the equal angles, they will also have the remaining sides equal to the remaining sides and the remaining angle to the remaining angle.*

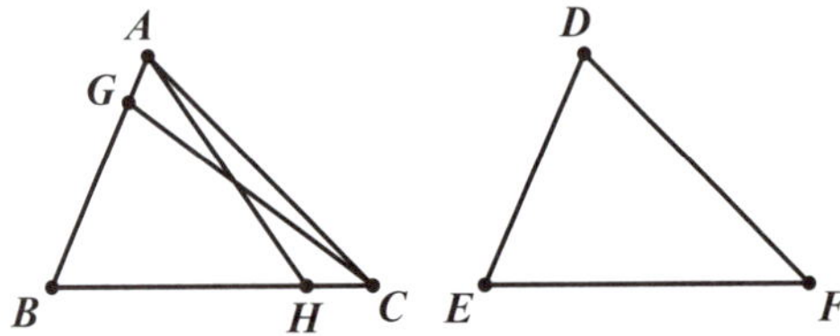

Figure 3.37. Proposition I.26

Proof. Let *ABC*, *DEF* be two triangles having the two angles *ABC*, *BCA* equal to the two angles *DEF*, *EFD* respectively, namely the angle *ABC* to the angle *DEF*, and the angle BCA to the angle EFD; and let them also have one side equal to one side, first that adjoining the equal angles, namely BC to EF;

I say that they will also have the remaining sides equal to the remaining sides respectively, namely AB to DE and AC to DF, and the remaining angle to the remaining angle, namely the angle BAC to the angle EDF.

For, if AB is unequal to DE, one of them is greater.

Let AB be greater, and let BG be made equal to DE; and let GC be joined.

Then, since BG is equal to DE, and BC to EF, the two sides GB, BC are equal to the two sides DE, EF respectively; and the angle GBC is equal to the angle DEF; therefore the base GC is equal to the base DF, and the triangle GBC is equal to the triangle DEF, and the remaining angles will be equal to the remaining angles, namely those which the equal sides subtend; [I.4] therefore the angle GCB is equal to the angle DFE.

But the angle DFE is by hypothesis equal to the angle BCA; therefore the angle BCG is equal to the angle BCA, the less to the greater: which is impossible.

Therefore AB is not unequal to DE, and is therefore equal to it.

But BC is also equal to EF;

therefore the two sides AB, BC are equal to the two sides DE, EF respectively, and the angle ABC is equal to the angle DEF; therefore the base AC is equal to the base DF, and the remaining angle BAC is equal to the remaining angle EDF. [I.4]

Again, let sides subtending equal angles be equal, as AB to DE;

I say again that the remaining sides will be equal to the remaining sides, namely AC to DF and BC to EF, and further the remaining angle BAC is equal to the remaining angle EDF.

For, if BC is unequal to EF, one of them is greater.

Let BC be greater, if possible, and let BH be made equal to EF; let AH be joined.

Then, since BH is equal to EF, and AB to DE, the two sides AB, BH are equal to the two sides DE, EF respectively, and they contain equal angles; therefore the base AH is equal to the base DF, and the triangle ABH is equal to the triangle DEF, and the remaining angles will be equal to the remaining angles, namely those which the equal sides subtend; [I.4] therefore the angle BHA is equal to the angle EFD.

But the angle EFD is equal to the angle BCA; therefore, in the triangle AHC, the exterior angle BHA is equal to the interior and opposite angle BCA: which is impossible. [I.16]

Therefore BC is not unequal to EF, and is therefore equal to it.

But AB is also equal to DE; therefore the two sides AB, BC are equal to the two sides DE, EF respectively, and they contain equal angles; therefore the base AC is equal to the base DF, the triangle ABC equal to the triangle DEF, and the remaining angle BAC equal to the remaining angle EDF. [I.4]

Therefore etc. Q.E.D.   □

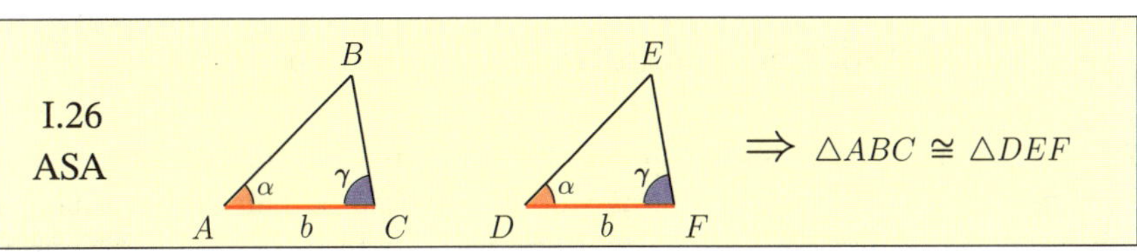

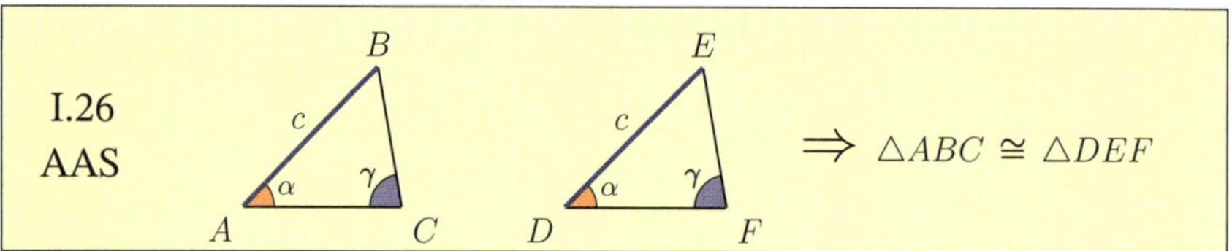

While side-side-angle is not a valid congruence scheme for triangles in general, there *are* two special cases where it holds: when the congruent angles are either right or obtuse. In a right triangle, the sides adjacent to the right angle are commonly referred to as the **legs**, while the side opposite to the right angle is called the **hypotenuse**. Consequently, when the congruent angles of two triangles are right, the scheme is more commonly referred to as hypotenuse-leg, abbreviated HL. We state the theorem and begin the proof, but leave the remainder to the reader. We also leave it as an exercise for the reader to provide a proof for SSA in the case where the congruent angles are obtuse.

Theorem 3.12 [HL]. *If the hypotenuse and a leg of one right triangle are congruent to the hypotenuse and a leg of a second right triangle, then the triangles are congruent.*

Proof. Consider right triangles $\triangle ABC$ and $\triangle DEF$ with right angles $\angle B$ and $\angle E$, respectively. Assume that $AC = DF$ and $AB = DE$. Extend the ray $\overrightarrow{CB}$ to a point G such that $GB = EF$. Join AG... $\square$

Recalling from Theorem 3.11 that the distance between a line and a point not on it is given by the length of the unique perpendicular segment from the point to the line, we have the following theorem. The proof is left to the reader as Exercise 3.3.11.

Theorem 3.13. *A point interior to an angle $\angle BAC$ lies on the angle's bisector if and only if it is equidistant from the rays $\overrightarrow{AB}$ and $\overrightarrow{AC}$.*

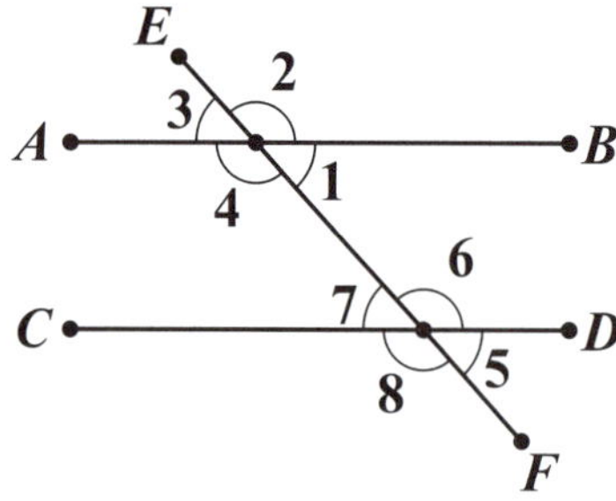

Figure 3.38. Angles created by a transversal to parallel lines

Propositions I.27 and I.28 are noteworthy in that they are the first two to mention parallel lines. Before we state the propositions, it would be helpful to explain some new terminology. When a pair of lines is crossed by another line, the crossing line is called a **transversal**. This arrangement of lines creates many angles of interest, of which several pairs are given identifying names. We will refer to Figure 3.38, which shows lines AB and CD crossed by transversal EF, to define our new terms. Angles $\angle 1$, $\angle 4$, $\angle 6$ and $\angle 7$ are called **interior angles**. The pair of angles $\angle 1$ and $\angle 7$, or the pair $\angle 4$ and $\angle 6$, are called **alternate interior angles**. Angles $\angle 1$ and $\angle 6$, or angles $\angle 4$ and $\angle 7$, are interior and on the same side of the transversal. Angles $\angle 2$, $\angle 3$, $\angle 5$ and $\angle 8$ are called **exterior angles**. The pair of angles $\angle 2$ and $\angle 8$, or the pair $\angle 3$ and $\angle 5$, are called **alternate exterior angles**. Angles $\angle 2$ and $\angle 5$, or angles $\angle 3$ and $\angle 8$, are exterior and on the same side of the transversal. Angle $\angle 6$ is interior and **opposite** to the exterior angle $\angle 2$. The pair of angles $\angle 1$ and $\angle 2$, or the pair $\angle 2$ and $\angle 3$, are examples of supplementary, or **adjacent angles**. The pair of angles $\angle 1$ and $\angle 3$, or the pair $\angle 2$ and $\angle 4$, are examples of vertical angles.

Propositions I.27 and I.28 give conditions on these angles which guarantee that AB and CD are parallel. We write $AB \parallel CD$ when these lines are parallel.

Proposition I.27. *If a straight line falling on two straight lines make the alternate angles equal to one another, the straight lines will be parallel to one another.*

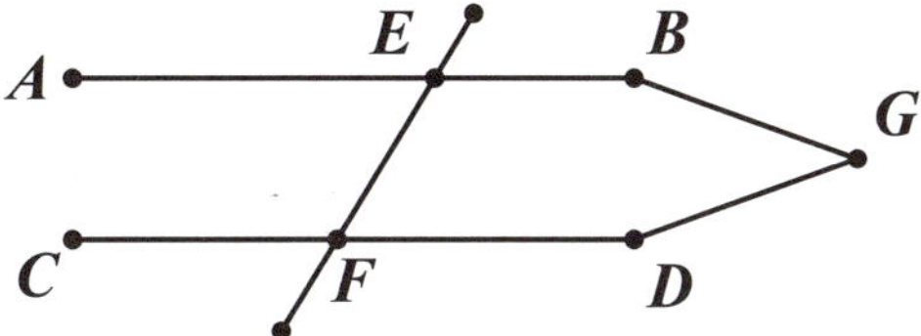

Figure 3.39. Proposition I.27

Proof. For let the straight line EF falling on the two straight lines AB, CD make the alternate angles AEF, EFD equal to one another;

I say that AB is parallel to CD.

For, if not, AB, CD when produced will meet either in the direction of B, D or towards A, C.

Let them be produced and meet, in the direction of B, D, at G.

Then, in the triangle GEF, the exterior angle AEF is equal to the interior and opposite angle EFG: which is impossible. [I.16]

Therefore AB, CD when produced will not meet in the direction of B, D.

Similarly it can be proved that neither will they meet towards A, C.

But straight lines which do not meet in either direction are parallel; [Def. 23] therefore AB is parallel to CD.

Therefore etc. Q.E.D.   □

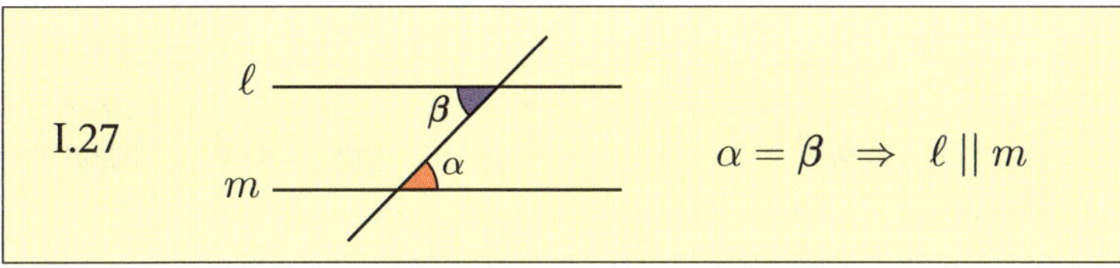

Referring back to the numbered angles in Figure 3.38, Proposition I.27 states that if either $\angle 1 = \angle 7$ or $\angle 4 = \angle 6$, then $AB \parallel CD$. In the next proposition, we will show that if $\angle 2 = \angle 6$, $\angle 3 = \angle 7$, $\angle 4 = \angle 8$, or $\angle 1 = \angle 5$, then $AB \parallel CD$. Moreover, if $\angle 1 + \angle 6$ or $\angle 4 + \angle 7$ equals two right angles, then $AB \parallel CD$.

Proposition I.28. *If a straight line falling on two straight lines make the exterior angle equal to the interior and opposite angle on the same side, or the interior angles on the same side equal to two right angles, the straight lines will be parallel to one another.*

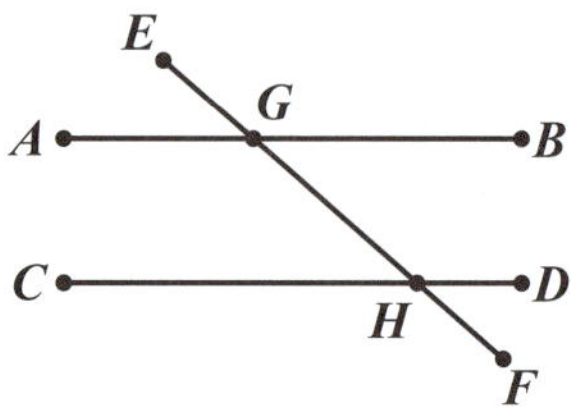

Figure 3.40. Proposition I.28

Proof. For let the straight line EF falling on the two straight lines AB, CD make the exterior angle EGB equal to the interior and opposite angle GHD, or the interior angles on the same side, namely BGH, GHD, equal to two right angles;

I say that AB is parallel to CD.

For, since the angle EGB is equal to the angle GHD, while the angle EGB is equal to the angle AGH, [I.15] the angle AGH is also equal to the angle GHD; and they are alternate; therefore AB is parallel to CD. [I.27]

Again, since the angles BGH, GHD are equal to two right angles, and the angles AGH, BGH are also equal to two right angles, [I.13] the angles AGH, BGH are equal to the angles BGH, GHD.

Let the angle BGH be subtracted from each; therefore the remaining angle AGH is equal to the remaining angle GHD; and they are alternate; therefore AB is parallel to CD. [I.27]

Therefore etc. Q.E.D.   □

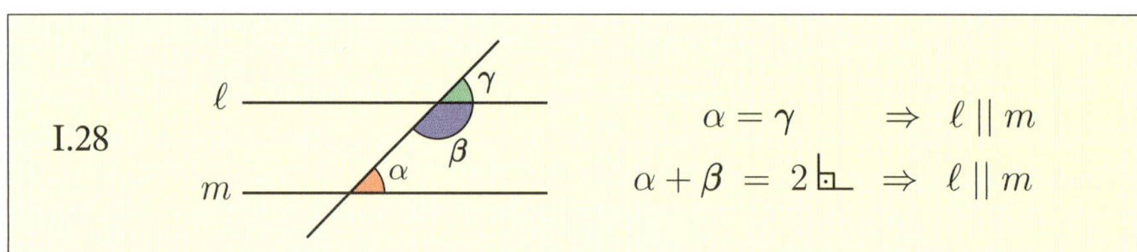

While the previous two propositions had our first references to parallel lines, neither of the proofs relied on Postulate 5, the Parallel Postulate. As such, these propositions belong to Neutral geometry. Finally, though it is out of numerical order, we include Proposition I.31 since it, too, belongs to Neutral geometry even though it concerns the construction of a parallel line. Once again, because of its reliance on Proposition I.23, the proof is not useful as a set of instructions for constructing the parallel line. As an exercise, you will be asked to give a more succinct set of steps and provide the corresponding proof.

Proposition I.31. *Through a given point to draw a straight line parallel to a given straight line.*

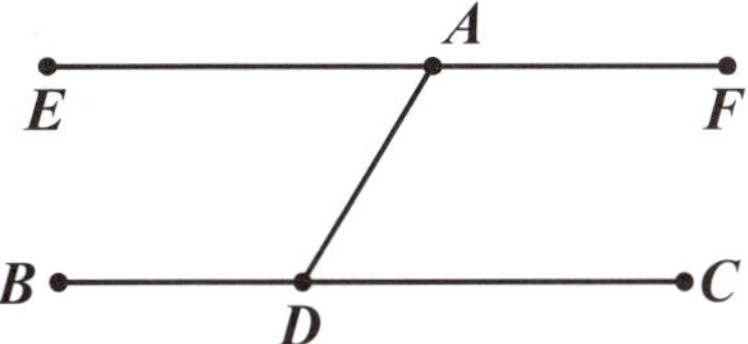

Figure 3.41. Proposition I.31

Proof. Let A be the given point, and BC the given straight line; thus it is required to draw through the point A a straight line parallel to the straight line BC.

Let a point D be taken at random on BC, and let AD be joined; on the straight line DA, and at the point A on it, let the angle DAE be constructed equal to the angle ADC [I.23]; and let the straight line AF be produced in a straight line with EA.

Then, since the straight line AD falling on the two straight lines BC, EF has made the alternate angles EAD, ADC equal to one another, therefore EAF is parallel to BC. [I.27]

Therefore through the given point A the straight line EAF has been drawn parallel to the given straight line BC.

Q.E.F.   □

I.31	Given	A • ℓ ______	construct $m \parallel \ell$ through A	m ———A•——— ℓ —————

With the completion of this proof, we have finished the 29 propositions of Book I that do not rely on the Parallel Postulate. We previously classified these results under two general headings, either constructions or relationships between geometric objects. Now we can refine this classification. Just over a third of these propositions are constructions: triangles in 1 and 22; lines in 11, 12 and 31; bisections in 9 and 10; and copying in 2, 3 and 23. Six propositions establish the congruence of two objects: triangles in 4, 8 and 26; and angles or sides in 5, 6 and 15. [Note: Proposition I.7 is a lemma for I.8.] There are eight propositions comparing relative magnitudes of angles or sides: angles in 16 and 17; sides in 20; angles and sides in 18, 19, 21, 24 and 25. Two propositions establish when, and if, an object is a straight line: 13 and 14. Lastly, two propositions establish when lines are parallel: 27 and 28.

Exercises 3.3

1. Give an updated version of Euclid's proof of each of the listed propositions. Be sure to justify each step, substitute mathematical symbols where appropriate, and include helpful diagrams as needed.

(a) Proposition I.16	(g) Proposition I.22
(b) Proposition I.17	(h) Proposition I.24
(c) Proposition I.18	(i) Proposition I.25
(d) Proposition I.19	(j) Proposition I.26
(e) Proposition I.20	(k) Proposition I.27
(f) Proposition I.21	(l) Proposition I.28

2. Whereupon $\angle ACB$ is "added to each," Euclid's proof of Proposition I.17 utilizes a nonexistent property of magnitudes. Give the missing property using inequality notation to justify this step of the proof.

3. Give a construction and corresponding proof of Proposition I.23 that only relies on Propositions I.1–I.8.

4. Give a construction and corresponding proof of Proposition I.31 that only relies on Propositions I.1–I.8 and Proposition I.27.

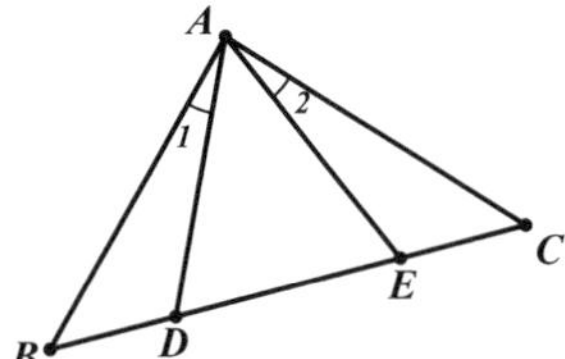

Figure 3.42. Exercise 3.3.5

5. Consider the isosceles triangle $\triangle ABC$ as shown in Figure 3.42. Assume that $AB = AC$ and $\angle 1 = \angle 2$. Prove that $\triangle ADE$ is also an isosceles triangle.

6. Prove Theorem 3.9: Through a given point C not on line $\overleftrightarrow{AB}$, there is only one straight line perpendicular to $\overleftrightarrow{AB}$.

7. Prove Theorem 3.11. To do this, first use the uniqueness of the perpendicular as in Exercise 6 (Theorem 3.9), then prove that the perpendicular is of shortest length. To do this, let C be a point not on line $\overleftrightarrow{AB}$, and let D be constructed on $\overleftrightarrow{AB}$ so that $CD \perp \overleftrightarrow{AB}$. Prove that if E is any other point on $\overleftrightarrow{AB}$, then $CE > CD$.

8. Provide a counterexample to demonstrate why side-side-angle (SSA) is not a valid congruence scheme for triangles.

9. Finish the proof of Theorem 3.12 (HL). Include a diagram.

10. Prove the congruence scheme SSA in the case when the congruent angles are obtuse. That is, consider triangles $\triangle ABC$ and $\triangle DEF$, where $AB = DE, BC = EF, \angle ACB = \angle DFE$, and $\angle ACB$ is obtuse. Prove that $\triangle ABC \cong \triangle DEF$. Include a diagram. [*Hint: Using a proof by contradiction, WLOG, assume AC > DF. Construct G on AC such that GC = DF. Prove $\triangle BGC \cong \triangle EDF$, and then consider isosceles triangle $\triangle ABG$.*]

11. Prove Theorem 3.13 by completing each direction of the biconditional as given separately in the following two parts.

   (a) Prove that every point on an angle's bisector is equidistant to that angle's sides. That is, let $\overrightarrow{AD}$ be the angle bisector of $\angle BAC$. Pick E on $\overrightarrow{AD}$. Construct perpendiculars EF and EG to $\overrightarrow{AB}$ and $\overrightarrow{AC}$, respectively, where F lies on $\overrightarrow{AB}$, and G lies on $\overrightarrow{AC}$. Prove that $EF = EG$.

(b) Prove that every point interior to an angle and equidistant to its sides lies on the angle's bisector. That is, let D lie on the interior of $\angle BAC$. Construct perpendiculars DE and DF to $\overrightarrow{AB}$ and $\overrightarrow{AC}$, respectively, where E lies on $\overrightarrow{AB}$, and F lies on $\overrightarrow{AC}$. If $DE = DF$, prove that $\overrightarrow{AD}$ bisects $\angle BAC$.

12. Prove that if the diagonals of a convex quadrilateral bisect each other, then the quadrilateral is a parallelogram.

13. Prove that a rectangle is a parallelogram.

14. As given in Euclid's twenty-second definition of Book I, a rhombus is an equilateral quadrilateral. Give a proof for each of the following statements related to a rhombus.

  (a) Prove that a rhombus is a parallelogram.
  (b) Prove that the diagonals of a rhombus intersect at right angles and bisect each other.
  (c) Prove that if the diagonals of a quadrilateral intersect at right angles and bisect each other, then the quadrilateral is a rhombus.

15. In $\triangle ABC$, assume $AB < AC$ and let D be the intersection of the angle bisectors at B and C. Prove that $DB < DC$.

16. In $\triangle ABC$, assume $AB < BC$ and let E be the midpoint of AC. Prove that $\angle CBE < \angle ABE$. [*Hint: Extend BE to a point F such that $BE = EF$.*]

17. In $\triangle ABC$, assume $AB < BC$ and let the angle bisector of $\angle ABC$ intersect AC at D. Prove that $AD < CD$. [*Hint: Use Exercise 16.*]

18. Consider triangle $\triangle ABC$. Extend sides AB and AC to H and J, respectively. Let the angle bisectors of exterior angles $\angle CBH$ and $\angle BCJ$ intersect at a point D. From D, construct perpendicular DE to $\overrightarrow{AH}$, as illustrated in Figure 3.43. Prove that the length of AE is half the perimeter of $\triangle ABC$. [*Hint: From D, construct perpendiculars DF and DG to $\overrightarrow{AJ}$ and BC, respectively.*]

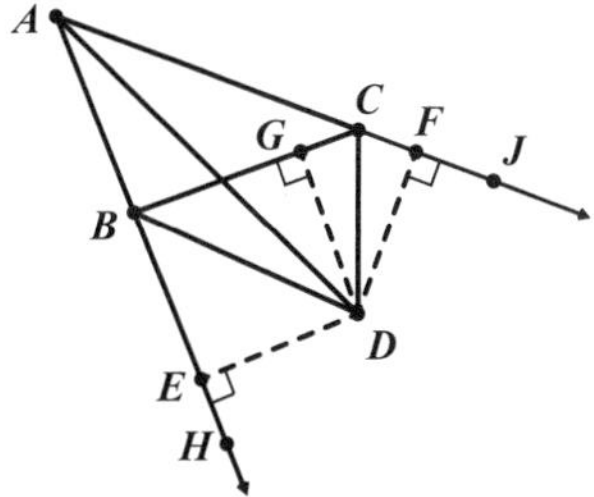

Figure 3.43. Exercise 3.3.18: AE is half the perimeter of $\triangle ABC$

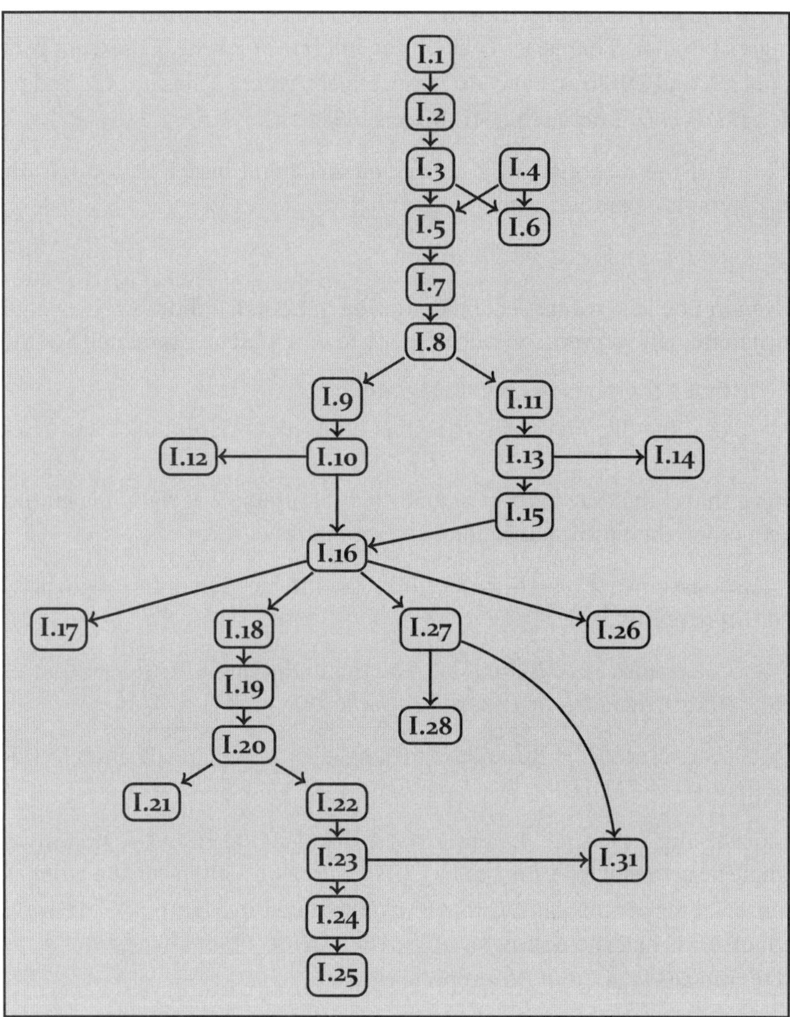

Figure 3.44. Partial dependency graph for Neutral geometry propositions

4

Spherical Geometry

Figure 4.1. Photograph from NASA's Earth Observatory

Now that we have introduced Euclid's definitions, axioms and common notions, and have spent some time exploring propositions from the first book of the *Elements,* let's take a step back and look at a completely different geometry, *Spherical geometry*. In doing so, we shift our gaze from the familiar flat Euclidean world to a round, three-dimensional world ripe for exploration. While Spherical geometry may present a new mathematical world, it is in many ways as natural as Euclidean geometry. After all, as the NASA photograph reminds us, we live on a giant sphere (albeit, imperfect).

As a first task, we will determine how our main characters behave on this new surface. To do this, we must ask ourselves, "What do a line and a circle look like on a sphere?" To help visualize and formulate an answer to this question, we encourage the reader to have a sphere in hand before reading any further. This is similar to using a piece of paper as a model for a Euclidean plane. There are many readily available models of spheres which can be used. A baseball or an orange will work in a pinch, but these are a bit too small. We prefer a Lénárt Sphere™, but realize it is pricey and hard to find. A volleyball or basketball has the proper dimensions, but since we will be writing on the sphere, an inexpensive plastic ball of roughly the same size is best as it works well with dry or wet-erase markers. It is also helpful to have a handful of rubber bands that fit around the sphere, a piece of string able to stretch halfway around the sphere, and some chalk, pencils, or markers to write (and hopefully, erase) on your spherical model.

4.1 What *is* a straight line, anyways? - Part 2

With sphere in hand, let's try to build our intuition for working within Spherical geometry. The first thing we will do is determine the behavior of our main characters, with an initial focus on the circle since it is more natural to visualize this figure on a sphere. As in life, it's often good to start with what you know, in this case, Euclidean geometry. Since we are quite skilled at constructing a circle on a plane, perhaps the same idea will translate to the sphere. Our definition of a circle is the set of all points equidistant from a given center. So, first mark a point on your spherical model which will be the center of our circle. Take a moment to think about how we could mark all of the points on the sphere that are the same distance from your chosen center. On a plane we can use a compass to form a circle, but if we did not have that tool handy, we could also use a pin, a piece of string, and a marker. With the pin placed at the center, the string tied to the pin, and the marker tied to the other end of the string at the desired length, a careful sweep of the marker over the plane, keeping the string taut, will produce a circle on the plane. This construction on the plane should be enough to convince you that we can produce a circle in a similar manner on the sphere.

Let's construct a circle on our spherical model. Since a pin might puncture the sphere, we will use a pencil as a substitute, and a shorter pencil will be easier to manage. Tie one end of the string around the pencil in a tight knot on the metal band near the eraser, then loop the other end of the string around the marker in a loose knot so we can adjust the length of taut string between them. If your sphere is roughly the size of a soccer ball, then three inches is a good distance between the pencil and marker for an initial attempt. Place the eraser on the point marked as the center of the circle. Depending upon your level of dexterity, the next step might work better with two people. Have one person hold the pencil in place while the other sweeps out the circle with the marker, ensuring that the string remains taut at its fixed length. The resulting figure on the sphere is a circle. See Figure 4.2 for a representation. (Figure 4.2 and several other figures were made using *Spherical Easel*, a program written by David Austin and Will Dickinson for creating interactive diagrams in Spherical geometry. It can be downloaded at *merganser.math.gvsu.edu/easel*.) Take some time to draw a few circles of various length on your sphere. A compass, more precisely, a spherical compass, makes this procedure much easier, but this tool is quite specialized and probably not in your desk drawer. As you draw some circles, try to imagine the shape of this time-saving tool.

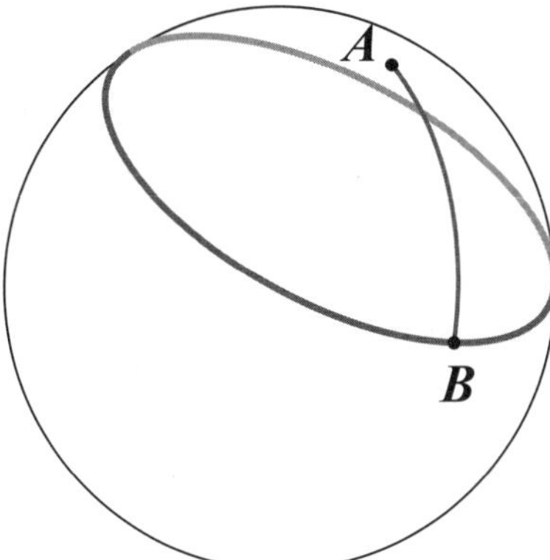

Figure 4.2. Circle with center A and radius AB.

Let's look at some similarities and differences between circles in Euclidean geometry and circles on the sphere. The definition and construction method are essentially the same. When you hold the sphere up for inspection, you may notice that your circle looks a lot like a circle in the plane. There is, however, a major difference in the size of circles that can be produced in one geometry versus the other. On a plane there is no theoretical limit to the size of a circle. On a sphere, the largest possible circle is one that splits the sphere precisely into two equal halves, or hemispheres. If we imagine the earth as a perfect sphere, the equator (90° latitude) is such a circle. A circle of maximal size is called a **great circle**, and we define it to be the intersection of the sphere with a plane that passes through the center of the sphere. On the earth, in addition to the equator, the most well-known examples of great circles are the longitude lines. Notice that, with the exception of the equator, no latitude is a great circle. For example, when we imagine a plane cutting through the 66.5° S latitude line (roughly the Antarctic circle), we lop off the bottom of a globe, but certainly do not go through its center.

On a sphere, two points are said to be **antipodal** if they are diametrically opposed, that is, situated at opposite sides of a diameter of the sphere, like the north and south pole. Each point A on the sphere, has a unique antipodal point A'. Furthermore, the antipodal point of A' is A. Notice that great circles contain infinitely many pairs of antipodal points since a plane cutting through the center of a sphere contains infinitely many diameters of the sphere.

Let's go back to our spherical model again. Suppose we have a plastic ball for which every great circle is 72 centimeters in circumference. If we mark a point N for the north pole, we could construct a great circle at the "equator" by making our string 18 cm (one-quarter the circumference of a great circle) and sweeping out the circle with the marker after centering the pencil at N. If we make the string 36 cm in length (half the circumference of a great circle) and center the pencil at N, we find that it is impossible to sweep out a circle and we can only mark the point S at the south pole. If we make the string 20 cm in length and center the pencil at N, then we produce a circle in the southern hemisphere. Notice that this same circle could have been produced by centering the pencil at the antipodal point, S, and setting the string at 16 cm in length. This ability to express the same circle using two different centers separates spherical circles from Euclidean circles. By utilizing antipodal points in this way, a string with a length of one-quarter the circumference of a great circle is sufficient to produce any circle on a sphere.

Now let's consider lines in Spherical geometry. First take a moment to review both the Euclidean definition of line and your answer to the question about the meaning of "straight line" in Section 1.3. Do either of these offer any help in determining the meaning of a straight line on a sphere? Following our Euclidean intuition about circles and their construction has led us to their spherical counterparts, the former produced by a compass and the latter by a spherical compass. Is there a spherical ruler we can use to construct a line on our spherical model with ease? It's more than likely that neither Euclid's definition nor your answers to the "What is a straight line?" thought experiment are offering much help here.

In order to get moving in the proper direction, we need to make a distinction between being an internal or an external viewer of a geometric world.[1] (If you watched *Flatland: The Movie* as directed in Exercise 1.3.3, then you have already spent some time thinking about this distinction. There is a sequel, *Flatland 2: Sphereland*, that may also prove helpful while learning to visualize this geometry [53].) Imagine a small creature, let's say a robotic ant, living in a two-dimensional Euclidean world. The ant is only aware of the plane on which it crawls, and sees no third dimension. The ant can crawl forward by moving its left and right legs forward in a highly precise robotic manner. There is no up or down. An external viewer lives outside this plane, say in a three-dimensional world, and can watch the ant crawl around on its plane. Let's take an internal view; suppose you *are* the ant and the plane is solid white with no markings. How can you know when you are travelling straight ahead? Take a few minutes to think about what "straight" means for you as the ant. Remember, since you are a robotic ant your movement is precisely programmed. It may help to consider what happens when the ant is turning, that is, *not* travelling in a line. In a manner similar to a marching band holding formation through a turn, to proceed in a rightwards direction, the ant's left legs travel farther than its right legs. If the right legs travel farther than the left legs, then the ant is turning left. If the right and left legs travel the same distance, then the ant is headed straight ahead.

Now let's suppose the ant lives on a sphere roughly the size of our spherical model. Analogously, here the ant is only aware of the surface of the sphere on which it crawls, and there are no markings on the sphere. How can the ant know when it is travelling in a straight line? Suppose, for example, an external viewer estimates that the ant is walking along the $75°S$ latitude line of the sphere, which is close to the south pole. In order to maintain that path, the ant has to move the legs which are closer to the equator farther than the legs which are closer to the south pole. Alternatively, if the ant were crawling on the equator then both sets of legs travel the same distance. More generally, unless the ant *is* crawling on a great circle, one set of legs is travelling farther than the other. Thus, by the same reasoning we applied to the plane, this means that a great circle is a straight line on a sphere. Could that be correct? A line on a sphere is a circle!?

For further convincing, here are three experiments in increasing order of complexity and difficulty. First, try placing an appropriately sized rubber band on your spherical model. Where does it have to be placed so that it rests without shooting off

---

[1] As in Chapter 1, we take inspiration for this line of questioning from David Henderson [70].

and taking out somebody's eye? Second, after locating or creating a shallow puddle of water, roll a tennis ball very quickly through the puddle. When the ball travels through the puddle at a good speed then its path on the ground will be a straight line. What is the shape of the water mark transferred from the puddle to the tennis ball? Finally, only those of you who are circus performers should try this last experiment. In a pool, stand on a floating sphere. To cause the least harm to yourself and others when you fall (and you, most certainly, *will* fall), maintain a safe distance of at least twenty feet from any other person or object. Start walking on the sphere. If your path is a great circle and you have perfect balance then you could be up there impressing your audience for quite some time.

Here we have it: **On a sphere, straight lines are the great circles.** Perhaps you still find this troubling since you cannot help but see that a great circle is curvy, not straight. This is true, but this is the observation of an external viewer who sees the sphere but does not live on its surface. To the surface dwellers, great circles are intrinsically straight. So does this mean our two main characters have been pared down to a single, multiple-personality headliner in Spherical geometry? No. While every line *is* a circle, every circle is *not* a line. That said, the circle clearly has the leading role in Spherical geometry.

Now that we have an intuitive understanding for lines on a sphere, let's take a look at how they behave as compared to the their counterparts in Euclidean geometry. Suppose we have two antipodal points, let's say N and S, thinking of the north and south poles. By simply considering the great circles corresponding to longitudinal lines, it's easy to see that there are infinitely many lines through N and S. Likewise, there are infinitely many lines through *any* pair of antipodal points. Suppose N and A are not antipodal. Just as every city on the planet is on exactly one longitudinal line, there is exactly one great circle which passes through N and A. By this reasoning, there is exactly one line through any pair of distinct nonantipodal points.

How will we define a **spherical line segment**? Informally, in Euclidean geometry, we think of a line segment as a finite part of a line with exactly two distinct endpoints. Given any great circle on a sphere and distinct points A and B that lie on it, how do we define a line segment joining them? As illustrated by Figure 4.3, the Euclidean line segment notation, AB, is ambiguous here since there are two arcs that connect these points. Between nonantipodal points A and B, there is a shorter arc through C called the **minor arc**, and a longer arc through D called the **major arc**. We will distinguish these arcs by including identifying points. In the figure, we see that the minor arc is $\overparen{ACB}$ and the major arc is $\overparen{ADB}$. In general, we write $\overparen{EFG}$ to denote the arc of the great circle between points E and G that passes through F. At first glance, both arcs seem to satisfy our informal notion of what a line segment is. Thus, arcs $\overparen{ACB}$ and $\overparen{ADB}$ prompt us to discuss the difference between a straight line joining two points and the shortest path between those two points. As mentioned in Chapter 1, the following statement is often taken as a definition in high school geometry: *A straight line is the shortest path between two points.* This is still true on a sphere, which is why airplanes travel great circle routes around the earth. In Spherical geometry, the shortest path between nonantipodal points A and B is the minor arc, namely $\overparen{ACB}$ in Figure 4.3. The major arc, $\overparen{ADB}$, illustrates that not every arc between A and B is the shortest path. To ensure unambiguous notation, when A and B are nonantipodal we let AB denote the unique minor arc that joins them. When A and B are antipodal we will not use this

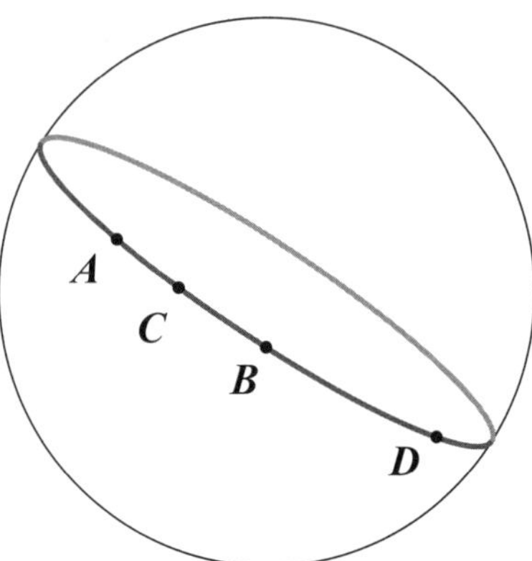

Figure 4.3. Arcs $\overarc{ACB}$ and $\overarc{ADB}$ joining A and B

notation since there are infinitely many great circles joining them. Instead, we will be careful to specify a particular line, or arc joining the points by giving a third point on the arc, when necessary.

Let's go back to our understanding of circles armed with this new notation. Recall that given two nonantipodal points A and B, if A is the center of a circle and B is a point on the circle, then we will use the minor arc connecting them, AB, as our radius, and we will refer to this circle as the circle with center A and radius AB. Note also that this circle can be expressed as having center A' and radius $A'B$, where A' is the unique antipodal point for A.

Finally, we end this section by briefly discussing the difference between *distance* and *length*.[2] Distance is measured between two objects, say points, and length is a measurement of one object, say an arc. A minor arc, for example, has a length less than half the circumference of a great circle. The distance between any two nonantipodal points on the sphere is the length of the minor arc joining them, and hence, is less than half the circumference of a great circle. The distance between antipodal points is exactly half the circumference of a great circle. So, the distance between any two points on the sphere is at most half the circumference of a great circle. A major arc has *length* greater than half the circumference of a great circle but less than the full circumference. Lastly, a great circle has a length equal to its circumference.

A careful reader will notice that we've paused just short of defining a line segment in this section. We prefer to wait until after we've turned our attention to one of the key characters in Euclid's first book, the triangle.

Exercises 4.1

1. Use your spherical model to help visualize the shape of a tool that would allow you to draw lines and line segments on your model, namely, a spherical ruler. Describe this construction tool. A diagram corresponding to your description would be helpful.

2. Use your spherical model to help visualize the shape of a tool that would allow you to draw circles on your model, namely, a spherical compass. The radius adjustment should be easier and faster than untying and retying knots in your string. Describe

---

[2]In Chapter 5 we give a rigorous definition of distance, but for now, we informally say that the distance between two points is the length of the shortest path joining them.

this construction tool. A diagram corresponding to your description would be helpful.

3. Review your answer to Exercise 1.3.1. Does your essay make sense within the context of Spherical geometry?

4.2 Triangles in Spherical geometry

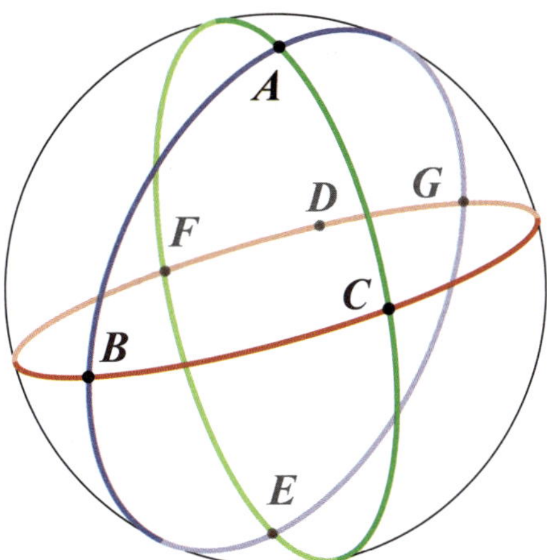

Figure 4.4. Trilateral figures on a sphere

Before we can examine the axioms and propositions of Book I within Spherical
geometry, we need to carefully define a triangle on a sphere. Euclid himself does not
specifically define a "triangle," but rather classifies trilateral figures as either equilateral, isosceles or scalene. The idea that a triangle is simply a three-sided figure turns
out to be too broad a classification on the sphere. In order to be able to prove things
about these figures, we would like triangles on a sphere to resemble their counterparts
in the Euclidean plane. In particular, we would like to preserve as many congruence
schemes as possible for triangles. This means that we must explore the sides and angles
of trilateral figures in Spherical geometry.

First, we define angles the same way that we do in the plane; namely, angles are
formed when two lines intersect. For example, as we will see in Section 4.4, the intersection of any longitude line with the equator creates four right angles. Next, due to
the nature of the sphere, we must ensure that the sides of any triangle only intersect
at its vertices. For example, consider the figure created by arcs $\overarc{ABG}$, $\overarc{ACF}$ and $\overarc{FDG}$
in Figure 4.4. To make the figure a bit easier to identify, we have flattened it, giving a
two-dimensional representation in Figure 4.5. Even though the figure has three sides,
we do not want to call this trilateral figure a triangle since major arcs $\overarc{ABG}$ and $\overarc{ACF}$
intersect at E, which is not a vertex of this three-sided figure. So, we need to place
restrictions on arcs specifying which are allowable as sides of a triangle. For now, we
will just note that not every arc can be the side of a triangle.

Given the importance of SAS in Euclidean geometry, ideally, we would like it to
hold in Spherical geometry. Let's explore this on our spherical model. Make a point on

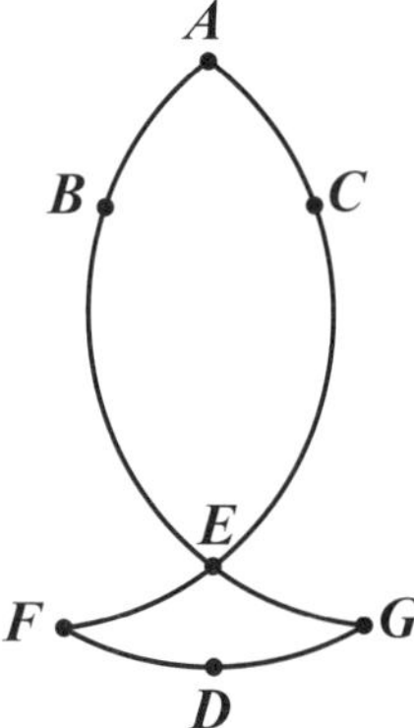

Figure 4.5. Self-intersecting trilateral figure

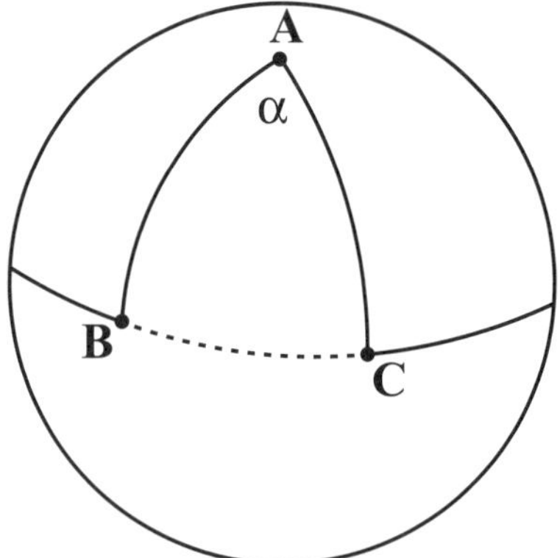

Figure 4.6. What could go wrong with SAS?

the sphere using a marker. We'll refer to this point as the north pole, A. Starting at the north pole, draw a minor arc that ends in the northern hemisphere, that is, it *does not* extend past the equator. Call the point at the end of this arc B. Draw another minor arc starting at the north pole and ending in the northern hemisphere at a point, C, that is not antipodal to B. Consider the figure contained by sides AB and AC, and angle α, as shown in Figure 4.6. We seem to have two choices for the third side, namely, either BC or major arc $\widehat{BDC}$, where D is some point on the great circle through B and C that is obscured from our view by the sphere. The figure created by choosing arc BC resembles a triangle. Though odd to imagine, the trilateral figure created by choosing $\widehat{BDC}$ includes the lower hemisphere, as well as the region resembling a triangle, since it must contain angle α. So, the figure created by choosing BC will lie completely inside the figure created by choosing $\widehat{BDC}$. If we were to call both trilateral regions "triangles" then the congruence scheme SAS would clearly not hold. To avoid the problem created by allowing $\widehat{BDC}$ to be side the of a triangle, we restrict the length of a side to be less than half of the circumference of a great circle. A pleasant consequence is that this restriction also resolves the issue of the self-intersecting trilateral.

Another new concern when working on the sphere is determining the interior of a figure. In the Euclidean plane, a triangle divides the plane into two regions, one with finite and one with infinite area.[3] We choose the **interior** of the triangle to be the region of finite area, and the **exterior** to be the region with infinite area. With a surface area of $4\pi R^2$, there is no such choice on a sphere of radius R since both regions have finite area. Let's revisit the minor arcs AB, AC and BC from our previous example. These three arcs form a trilateral figure that actually resembles a triangle in the plane, though it's a bit bloated. See Figure 4.7 for a representative diagram. Shade this figure using a marker. Notice that sides AB, AC and BC create two regions, the shaded and

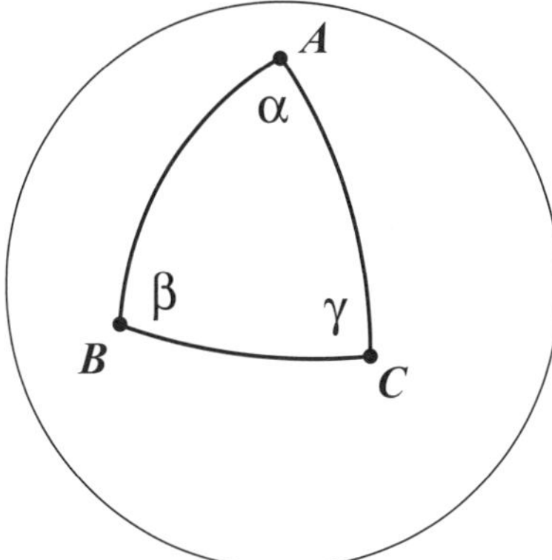

Figure 4.7. Trilateral ABC: What region is the interior?

the unshaded, both of which are trilateral with finite area. Only the smaller region, however, resembles a triangle. If we were to call both regions "triangles" then the congruence scheme SSS clearly would not hold as both trilaterals share the same sides, but one has angles α, β, and γ radians, while the other has angles $2\pi - \alpha$, $2\pi - \beta$, and $2\pi - \gamma$ radians, respectively. Since we'd like SSS to hold, we cannot allow both the shaded and unshaded trilateral regions to be considered triangles. So, we need to place restrictions on the allowable angles for a spherical triangle. As is the case for Euclidean triangles, we require that each angle be less than π.

We specify these two restrictions in the following definition of a spherical triangle.

Definition 4.1. A **spherical triangle** is a trilateral figure in which

- Any angle must be less than two right angles.
- Any side must be less than half of the circumference of a great circle.

Such figures are often referred to as "small triangles," but when we use the term *triangle* in Spherical geometry we will mean these figures. Notice that the first restriction determines the interior of a trilateral figure. The second condition ensures that the sides joining any two vertices are well-defined minor arcs. You may wonder why we don't allow sides that are exactly half the circumference of a great circle. In Exercise 4.2.1, the reader is asked to consider whether this is even possible in a trilateral figure.

We are now finally ready to define a line segment.

---

[3]In Chapter 7 we give an axiomatic treatment of area. Here, we appeal to our informal notion of area.

Definition 4.2. In Spherical geometry, when A and B are nonantipodal points, the **spherical line segment** AB is the shortest path, or minor arc, joining the points. If A and A' are antipodal points lying on great circle, ℓ, then each arc joining A and A' is a line segment.

Thus, while major arc $\overparen{ADB}$ is part of the line that joins A and B, we will not call it a line segment.

Now that we have an understanding of the behavior of our main characters in this new geometry, we turn our attention to interpreting Euclid's axioms on a sphere.

Exercises 4.2

1. Let's examine what happens if we allow the definition of a spherical triangle to include a side of length one-half the circumference of a great circle. Let N, S and E be distinct points where N and S are antipodal. Let A be any point not on segment $\overparen{NES}$. Construct segments AN and AS. Do AN, AS and $\overparen{NES}$ form a trilateral figure?

4.3 Euclid's axioms viewed in Spherical geometry

Starting with five algebraic Common Notions and five geometric Postulates, Euclid proves one proposition after another using well-established logical   constructs. Having interpreted and defined the basic terms of *circles*, *lines*, *angles* and *triangles* on a sphere, we are now ready to consider Euclid's axioms within our new spherical world. Since the Common Notions are independent of the shape of our world, we can see that these five axioms are still valid in our new geometry. The five Postulates, however, concern our main characters, and their behavior has changed in Spherical geometry. Are these geometric axioms still valid? Using our working understanding of lines and circles on spheres, let's revisit Euclid's five postulates to carefully determine whether each postulate holds as it is stated. As the first two postulates have the most nuanced interpretations, for these we also consider how Euclid uses these postulates in the proofs, but we limit the scope of our analysis to Neutral geometry since this will be sufficient for our investigations. Before reading further, take a few minutes to carefully consider each postulate to determine whether or not there is an interpretation which admits its validity in Spherical geometry for our analysis of the propositions of Neutral geometry.

Postulate 1: To draw a straight line from any point to any point.

Given two points A and B, this postulate is used to create a line segment, AB, that joins them. If A and B are nonantipodal points, then there is a unique great circle passing through them and the line segment AB is the unique minor arc on this circle between them. If A and B are antipodal points, there are infinitely many great circles that contain them, thus the segment between them is not uniquely defined. In either case, given any two points A and B, there is at least one line segment that joins them. Hence, with a strict interpretation, this postulate holds in Spherical geometry. If this matches your interpretation of Postulate 1, then skip ahead to Postulate 2.

If you are reading this paragraph then either you are simply curious or our interpretation does not match yours, and we are certain that uniqueness is at issue. How do we know this? There is often a tacit understanding that, while not explicitly stated, the line segment is unique. If Euclid's postulate is interpreted to include uniqueness then this postulate fails in Spherical geometry. We concede that this alternative is a valid interpretation, but to paraphrase Calvin in the comic strip at the start of Chapter 2, we do not want you to be excused from further study. Instead, in the case of antipodal points, we ask whether the lack of uniqueness of the segment will present any problems for our analysis of the Neutral geometry propositions on a sphere? Upon examination of Euclid's use of this postulate within his proofs, it is common for him to use this postulate to join two points with a line segment, but Proposition I.4 is the only proof where he assumes that the line will be unique. What exactly does he assume in I.4? In the proof, Euclid "applies" one triangle to another and, while he does not cite any postulate, he claims that given two vertices of a triangle, E and F, there is only one line segment connecting them. In this case, EF is a side of a triangle. In Spherical geometry, vertices of a triangle are necessarily nonantipodal, and thus, there is a unique line segment joining E and F. Therefore, in spite of the fact that the line between any two points is not necessarily unique in Spherical geometry, it is reasonable to argue that Postulate 1 still holds, *and*, this interpretation is sufficient for analyzing the Neutral geometry propositions.

Postulate 2: To produce a finite straight line continuously in a straight line.

Euclid uses this postulate to extend a given line segment. It is clear that given a line segment joining two points, we can always extend this segment into a great circle that contains the two points. Moreover, this great circle is unique. The word that may give us pause is "continuously." Here it is helpful to distinguish between the notions of infinite and boundaryless.[4] While a great circle is certainly finite in length, it is also free of any boundaries, in that we can continuously produce the line, never reaching an end. Let's appeal to the distinction between the extrinsic and intrinsic views for greater clarity. As sailors quickly discovered, it is not possible to fall off the "edge of the world." No such boundaries exist for the surface dwellers of a sphere. With this interpretation, it is reasonable to argue that this postulate holds in Spherical geometry. If this matches your interpretation of Postulate 2, then skip ahead to Postulate 3.

If you are reading this paragraph then our interpretation does not match yours, and we are certain that the finiteness of a great circle is at issue. If Euclid's second postulate is interpreted to mean that a line has infinite length, then this postulate fails in Spherical geometry. We admit that there is one particular reference to this in Neutral geometry, specifically in the proof of Proposition I.22, which supports this view: "let there be set out a straight line DE ... of infinite length...". We concede that this alternative is a valid interpretation, but hear us out one more time. We ask again whether our interpretation of "continuously" will present any problems for our analysis of the validity of the Neutral geometry propositions on a sphere. Upon examination of Euclid's use of this postulate within his proofs, it is clear that he intended for there to

---

[4]In his book *A Survey of Geometry* [**44**], Howard Eves writes: "Although Postulate 2 asserts that a straight line may be produced indefinitely, it does not necessarily imply that a straight line is infinite in extent, but merely that it is endless, or boundless. The arc of a great circle joining two points on a sphere may be produced indefinitely along the great circle, making the prolonged arc endless, but certainly it is not infinite in extent. Now it is conceivable that a straight line may behave similarly, that after a finite prolongation it, too, may return on itself. It was Bernhard Riemann who in his famous dissertation, *Über die Hypothesen welche der Geometrie zu Grunde liegen,* of 1854, distinguished between the boundlessness and the infinitude of straight lines."

be only one way to extend a line segment, and in this regard, our interpretation is on solid ground. Only occasionally does Euclid require that a given line be extended by a particular length. On the sphere, all but two of these cases translate to extending a minor arc of a line by at most a segment equal to a given side of a triangle.[5] Euclid's language in I.22 spoke of the infinite but the word went far beyond his need. With our restriction on the length of a side of a spherical triangle, we will still be able to extend our lines as needed on a sphere. Therefore, in spite of the fact that producing a line continuously on a sphere means going around and around the same great circle, it is reasonable to argue that Postulate 2 holds for our purposes, that is, this interpretation is sufficient for analyzing the Neutral geometry propositions on a sphere.

Postulate 3: To describe a circle with any center and distance.

The word "distance" may at first seem troubling, but as discussed in Section 4.1, the distance between two points is the length of the shortest path joining them. For nonantipodal points, this is the length of the unique minor arc between them, and for antipodal points this is half the circumference of a great circle. So, distance can be at most πR on a sphere of radius R. Given point A and segment AB, using our compass we can construct a circle with the given center and radius since AB refers to a minor arc of a great circle. In the maximal case where we are given point C and segment $\overset{\frown}{CDC'}$ where C' is antipodal to C, the circle with center C and radius $\overset{\frown}{CDC'}$ will be a point, namely C'. Therefore, we can describe a circle with any center and any *possible* distance. Thus, it is reasonable to argue that Postulate 3 holds.

Postulate 4: That all right angles are equal to one another.

By the symmetry of the sphere, Postulate 4 holds.

Postulate 5: That, if a straight line falling on two straight lines make the interior angles on the same side less than two right angles, the two straight lines, if produced indefinitely, meet on that side on which are the angles less than two right angles.

To consider whether this postulate holds, we must first consider two straight lines in Spherical geometry. Without loss of generality, suppose that one of the lines is the equator, e. Now imagine any other great circle, ℓ, on the sphere and you will see why any two distinct lines in Spherical geometry must intersect at two points. In fact, they will intersect at antipodal points, say A and A'. (Try this on your spherical model!) In Euclidean geometry, a line that crosses a pair of lines (but not at an intersection point of the lines) is called a transversal. Every line cuts the plane into two sides, and the Parallel Postulate specifies on which side of the transversal the pair of lines must meet. In Spherical geometry, every line cuts the sphere in two hemispheres, and the Parallel Postulate specifies in which hemisphere (relative to the transversal) the pair of lines must meet. Suppose t is a transversal to lines e and m. (Note that t cannot meet e and ℓ at their antipodal points of intersection, A and A', for the same reason a transversal in Euclidean geometry cannot meet the pair of lines where the lines intersect.) Since e and ℓ meet at antipodal points and t divides the sphere into two hemispheres, A will be on one side and A' on the other. Therefore, e and ℓ meet on both sides of the transversal. Since t is an arbitrary transversal, the implication is always true and Postulate 5 holds.

---

[5]The lone exceptions occur in Euclid's proof of I.2, which we handle carefully in the next section by adding it to our Spherical axioms, and in his proof of I.22, which requires adding the lengths of all three sides of a triangle, a sum we show to be less than the circumference of a great circle.

With the given interpretations, all of Euclid's axioms as they are written and used in Neutral geometry can be considered valid in Spherical geometry. In the next section we consider the validity of the propositions of Neutral geometry on the sphere.

4.4 Neutral geometry on the sphere

In Chapter 3 we proved the 29 propositions of Neutral geometry from Book I of the *Elements* assuming ten axioms (Well actually, we only need nine for Neutral geometry!) and utilizing well-established logical constructs. Since these ten axioms are valid in Spherical geometry under the interpretations given in the previous section, the same 29 propositions of Book I *should* be true on a sphere. Let's take another look at the statements and proofs of these Neutral geometry propositions.

We'll start with Proposition I.1: *On a given finite straight line to construct an equilateral triangle.* At first glance, this appears to hold exactly as it does in Euclidean geometry. However, if we pick a segment of length between one-third and one-half of the circumference of a great circle, something surprising happens. Try it for yourself on your spherical model, but as Figure 4.8 illustrates, the two circles we construct in the proof of Proposition I.1 will not intersect. As we mentioned in Chapter 3 after

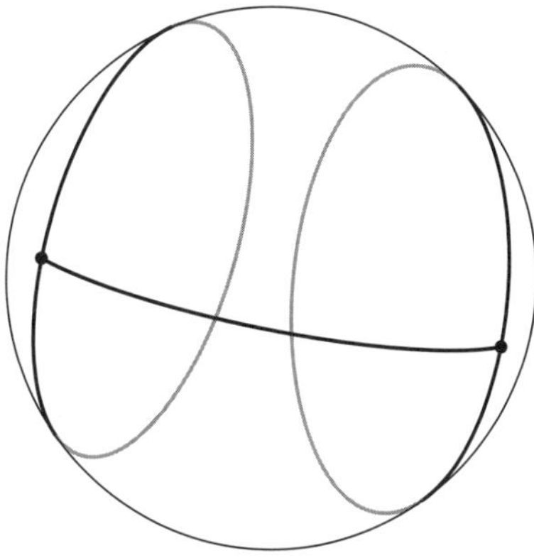

Figure 4.8. Two nonintersecting circles

the proof of I.1, none of Euclid's Common Notions or Postulates guarantees that the constructed circles must intersect. Therefore, there must be an unstated assumption at work that Euclid is utilizing in order to claim that these circles meet. While not our first encounter with this "missing Euclidean postulate," the sphere has given us a *new* way to identify this omission by Euclid, and perhaps, it will reveal others. For now, we will delay discussion of further implications of this observation in favor of merely noting that, in Spherical geometry, Proposition I.1 holds only if the segment is small enough, specifically, less than one-third of the circumference of a great circle. This restriction places a strict limit on the size of equilateral triangles that can be constructed on the sphere. We replace Euclid's proposition with Proposition I.1*S* given below with the modification in boldface type, but we will delay its justification until the end of Section 4.6.

Proposition I.1*S*. *On a given finite straight line* **less than one-third of the circumference of a great circle**, *to construct an equilateral triangle.*

Let's consider Proposition I.2: *To place at a given point [as an extremity] a straight line equal to a given straight line.* This is the proposition that allows us to assume that we have a rigid compass. Since the proof of this proposition requires the construction of an equilateral triangle on a segment of arbitrary length, we are in trouble. (The partial dependency tree for the propositions of Neutral geometry given in Figure 3.44 at the end of Chapter 3 will be helpful here.) Given the failure of I.1, clearly we cannot do this for segments of length at least one-third the circumference of a great circle. Euclid chose a collapsible compass as a construction tool in his postulates because he could fairly easily prove its equivalence to a rigid compass with his second proposition. We do not have this luxury in Spherical geometry, and thus, we have no choice but to revise our postulates to assume that our compass is rigid from the start.

Spherical Postulate 6. In Spherical geometry, the compass is rigid.

With the adoption of a rigid compass, both the statement and proof of Proposition I.3 (*Given two unequal straight lines, to cut off from the greater a straight line equal to the less.*) still hold. Given our definition of spherical triangles, Proposition I.4 (*If two triangles have the two sides equal to two sides respectively, and have the angles contained by the equal straight lines equal, they will also have the base equal to the base, the triangle will be equal to the triangle, and the remaining angles will be equal to the remaining angles respectively, namely those which the equal sides subtend.*) holds as well, and we have SAS for spherical triangles.

We will leave the investigation of the statements and proofs of Propositions I.5 through I.8 as exercises and will jump to Proposition I.9: *To bisect a given rectilineal angle.* In the proof of this proposition, it is crucial to notice that we have control over the length of segment DE. By picking D close enough to A, we can ensure that equilateral triangle $\triangle DEF$ can be constructed. Thus, the proof holds with only one minor modification. While we are typically only interested in bisecting angles less than π, we leave it to the reader to show that, as is the case in Euclidean geometry, the constructed line AF will also bisect the angle that is greater than π. Modifying the proof of Proposition I.10 (*To bisect a given finite straight line.*) proves to be a bit trickier. While we claim that the proposition still holds, Euclid's proof will only work for segments that are less than one-third of the circumference of a great circle. We will defer a new, more general proof of this proposition until after our consideration of Proposition I.11 (*To draw a straight line at right angles to a given straight line from a given point on it.*) and Proposition I.13 (*If a straight line set up on a straight line make angles, it will make either two right angles or angles equal to two right angles.*). Since the proofs of Proposition I.11 and Proposition I.13 do not use Proposition I.10, we are free to reorder them without fear of circular reasoning. We claim that Proposition I.11 and its proof still hold because, as in Proposition I.9, we are free to pick D close to C so that the equilateral triangle $\triangle FDE$ can be constructed. Proposition I.13 and its proof, which relies on Proposition I.11, will also still hold. Before we proceed with the proof of Proposition I.10, we define the **polar points**, or **poles**, of a line.

Definition 4.3. Let A and B be any two nonantipodal points on an arbitrary line in Spherical geometry. Construct perpendicular lines to AB at points A and B, and label the intersection points of these lines C and D. Points C and D are called **polar points**, or **poles**, of the line containing AB.

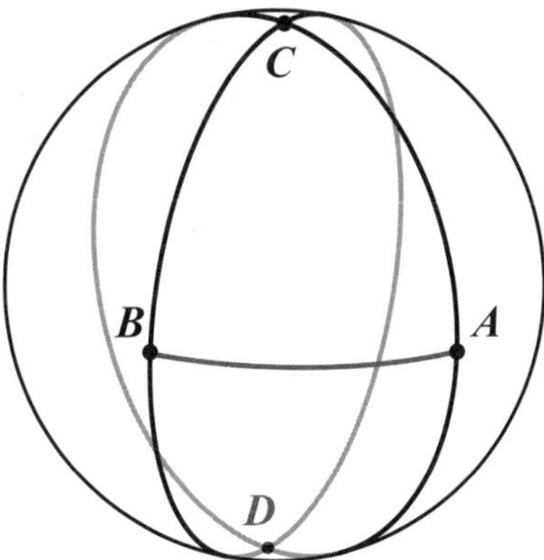

Figure 4.9. Polar points

We leave the proof of the following lemma to the reader, but note that it requires the congruence scheme ASA, which is proved as the first part of Proposition I.26. Since the proof of this part of I.26 does not require any proposition after Proposition I.4, it still holds and can be assumed prior to proving Proposition I.10.

Lemma 4.4. *The polar points for line AB are antipodal points located a distance of one-quarter the circumference of the sphere from all points on the line AB.*

We are now ready to give a new proof for Proposition I.10. We leave it to the reader to provide a diagram for each case.

Proposition I.10. *To bisect a given finite straight line.*

Proof. There are two cases depending upon whether or not the given segment is exactly half the circumference of a great circle in length.

Case 1. Suppose the endpoints of the given finite line are not antipodal. Let AB be the line segment to be bisected. Construct a polar point C of line AB. Join AC and BC. By Lemma 4.4, we have $AC = BC$. Using I.9, bisect $\angle ACB$ and extend the line to point E on AB. Then by Proposition I.4 (SAS), we have $\triangle ACE \cong \triangle BCE$. Thus, $AE = BE$ and we have bisected AB at E.

Case 2. Suppose the endpoints of the given finite line are antipodal. Let $\widehat{ABA'}$ be the given finite line where B is some point between antipodal points A and A'. Construct a polar point C for segment AB. Join AC. Using I.11, construct a perpendicular to AC at C. Let the intersection of this perpendicular with $\widehat{ABA'}$ be E. We claim that E bisects the given $\widehat{ABA'}$. Since C is a polar point to AB, and B is on line $\widehat{ABA'}$, by Lemma 4.4 we have $AC = A'C$. Furthermore, AC can be extended to a line, and the antipodal point of A is on that line. Therefore, $\widehat{ACA'}$ is a straight line. As $\angle ACE$ is right by construction, then by I.13, angles $\angle ACE$ and $\angle A'CE$ are *both* right. Thus, by I.4 (SAS), $\triangle ACE \cong \triangle A'CE$, and hence $AE = A'E$ as desired. $\square$

We leave it to the reader to analyze Propositions I.12, I.14 and I.15 in a similar manner.

We now turn our attention to Proposition I.16 [Exterior Angle Theorem]: *In any triangle, if one of the sides be produced, the exterior angle is greater than either of the interior and opposite angles.* As illustrated in Figure 4.10, we construct the spherical triangle $\triangle NBC$ by placing one vertex, N, at the North Pole and placing the two other vertices, B and C, on the equator. We can see that the exterior angle $\angle NCD$ as well as the two interior angles $\angle NBC$ and $\angle NCB$ are all right angles. Thus the exterior angle $\angle NCD$ is equal to its opposite interior angle $\angle NBC$. Furthermore, we can choose $\angle BNC$ to be larger than a right angle. Thus, is it spectacularly clear that Proposition I.16 does not hold in Spherical geometry. We leave it as an exercise to investigate why the proof fails. We also leave the reader to analyze the validity of Proposition I.17 as we take a look at Propositions I.18 through I.20.

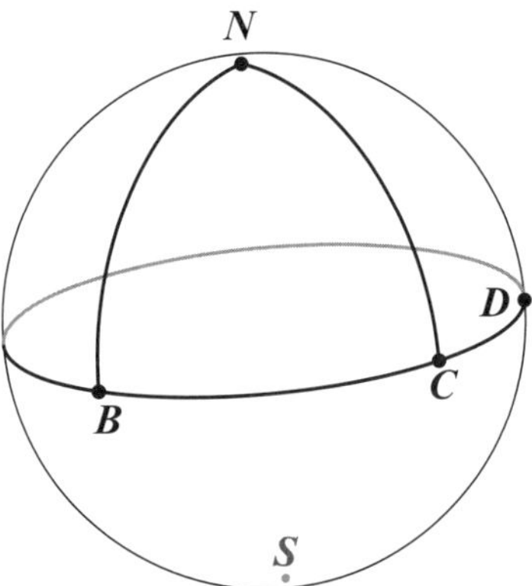

Figure 4.10. A counterexample for Proposition I.16

Euclid's proof of Proposition I.18 (*In any triangle the greater side subtends the greater angle.*) relies on I.16. Continuing down this path, his proof of Proposition I.19 (*In any triangle the greater angle is subtended by the greater side.*) relies on Proposition I.18 and his proof of Proposition I.20 (*In any triangle two sides taken together in any manner are greater than the remaining one.*) relies on I.19. At first thought, this may lead us to question the validity of all three of these propositions in Spherical geometry since each relies on the failed I.16. Contrary to this initial speculation, on the sphere these propositions can be proven *without* calling on Proposition I.16.

As you may recall from Chapter 3, the Epicureans thought that the validity of Proposition I.20, also known as the Triangle Inequality, was obvious (even to an ass). This Epicurean observation holds in Spherical geometry as I.20 is valid, but we will not present the proof here since it relies on results from three-dimensional Euclidean geometry [**112, 117**]. The important thing to note about the proof is that it relies on neither I.18 nor I.19. This means that we are free to use Proposition I.20 to prove either I.18 or I.19. We will prove Proposition I.19, but since the proof requires the copying of an angle we must once again venture a bit further ahead in the propositions.

We leave it to the reader to analyze the two parts of Proposition I.21. In Proposition I.22, a triangle is constructed from three segments where the sum of any two lengths is greater than the third. Unfortunately, not all such triangles are constructible on the sphere since it is possible to give three lengths for which the circles in the proof do not intersect. For example, $a = b = c = \frac{3\pi}{4}$ meets the condition of I.22, but a triangle with these sides is not constructible by I.1S. If, however, we also assume that the sum of the sides is less than the circumference of a great circle, then the triangle *is* constructible and the proof will hold. We replace Euclid's proposition with Proposition I.22S given below with the modification in boldface type. We justify this modification with the following lemma.

Lemma 4.5. *The total length of the sides of any spherical triangle on a sphere of radius R is less than the circumference of a great circle, that is, $2\pi R$.*

Proof. Consider a spherical triangle $\triangle ABC$. Extend AB and AC so that they meet at A', the antipodal point of A. Then $\overparen{ABA'} + \overparen{ACA'} = 2\pi R$. Consider the spherical triangle $\triangle A'BC$. By I.20, we have $BC < BA' + CA'$. Thus, $AB + AC + BC < AB + AC + BA' + CA' = \overparen{ABA'} + \overparen{ACA'} = 2\pi R$. $\qquad\square$

Proposition I.22S. *Given three straight lines whose* **sum is less than the circumference of a great circle,** *and such that the sum of any two is greater than the third, to construct a triangle.*

This is good news for the proof of Proposition I.23 on the sphere since we can control the length of the segment required in the proof. Therefore, Proposition I.23 holds, and we can copy angles in Spherical geometry.

Now we are ready to go back to Proposition I.19 and present the proof from the 1871 text, *Spherical Trigonometry for the Use of Colleges and Schools*, by Isaac Todhunter (1820–1884) [**117**].

Proposition I.19. *In any triangle the greater angle is subtended by the greater side.*

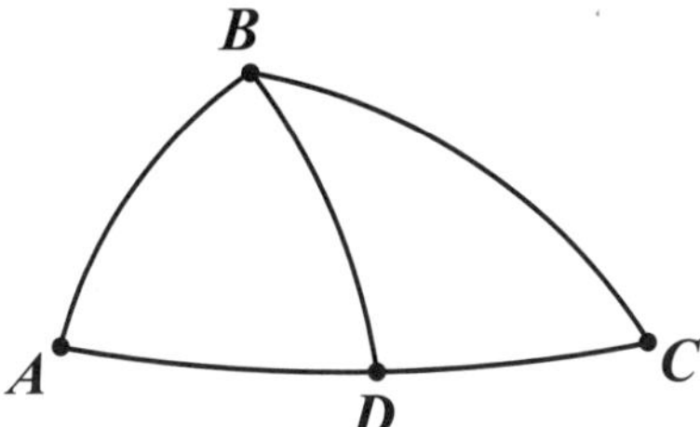

Figure 4.11. Proposition I.19

Proof. Let $\triangle ABC$ be a spherical triangle, and let angle $\angle ABC$ be greater than angle $\angle BAC$. We wish to prove that side AC will be greater than side BC. Let D be the point on AC such that angle $\angle ABD$ is equal to angle $\angle BAD$ [Proposition I.23]; then BD is equal to AD [Proposition I.6 on $\triangle ABD$], and $BD + DC$ is greater than BC [Proposition I.20 on $\triangle BDC$]; therefore $AD + DC$ is greater than BC; that is, AC is greater than BC. $\qquad\square$

We leave it to the reader to prove Proposition I.18 using Proposition I.19. We also leave the analysis of Propositions I.24 and I.25 as an exercise. This brings us to the final Euclidean triangle congruence scheme, AAS, which is found in the second half of Proposition I.26. The scheme AAS is not valid in Spherical geometry and we leave it to the reader to find a counterexample. Lastly, we leave it to the reader to analyze Propositions I.27, I.28 and I.31.

Let's recap our findings and discuss the consequences. We established an interpretation for the validity of Euclid's ten axioms for investigating the propositions of Neutral geometry on the sphere. Based on our work in Chapter 3 and our understanding of axiomatic systems, this means that these propositions should be true in both Euclidean *and* Spherical geometries since they start with the same axioms. However, we have clearly shown in this section that this is not the case since some of these propositions fail in Spherical geometry. So, where is the error in the reasoning? While one could argue that there are other possible interpretations of the postulates within Spherical geometry, we must acknowledge that the omissions we have highlighted, that is, assumptions made by Euclid that were not based on his axioms or previous propositions, are cause for concern. In other words, we know that Euclid's set of ten axioms is *incomplete* and we will have to add some more. We are certainly missing axioms that distinguish Euclidean geometry from Spherical geometry, but we will postpone our augmentation of the axioms until after we consider another type of geometry in Chapter 5. For the remainder of this chapter, we investigate area, trigonometry and constructions in this strange, and yet familiar, new world.

> **Elliptic and Double Elliptic geometries**
>
> Spherical geometry is sometimes referred to as *Double Elliptic geometry*. The "double" refers to the fact that any two lines intersect in two distinct points. Felix Klein (1849-1925) created the closely related *Elliptic geometry* by identifying antipodal points on the sphere. That is, every antipodal pair, $\{A, A'\}$, becomes one point. As a consequence, any two lines in Elliptic geometry intersect exactly once. Somewhat surprisingly, a line no longer separates the plane into two halves in this geometry. However, as is the case in Spherical Geometry, the Exterior Angle Theorem (I.16) does not hold.

Exercises 4.4

1. For Propositions I.5 through I.8 consider the following:

   - Does the proposition hold on a sphere?
   - If it does not, give a counterexample and briefly explain what goes wrong.
   - If it does hold, does Euclid's proof work? If it does not, briefly explain what goes wrong. Note: you do not need to provide a valid proof.

2. Prove that the line AF constructed in the proof of Proposition I.9 will also bisect the angle that is greater than π.

3. Prove Lemma 4.4: Show that the polar points for line AB are antipodal points located a distance of one-quarter the circumference of the sphere from all points on the line containing AB.

4. Give a diagram to illustrate each case in the proof of Proposition I.10 in Spherical geometry.

5. Follow the directions given in Exercise 1 to analyze the validity of Propositions I.12, I.14 and I.15 in Spherical geometry.

6. Proposition I.16 does not hold in Spherical geometry. Consider the following triangle: place one vertex at the North Pole and place the two other vertices on the equator. Make the angle at the North Pole a right angle (or bigger). Carefully explain what goes wrong in Euclid's **proof** of Proposition I.16.

7. Follow the directions given in Exercise 1 to analyze the validity of Proposition I.17 in Spherical geometry.

8. Consider a spherical triangle $\triangle ABC$. Show that $AB + BC \neq AC$ as if it were Proposition I.15$\frac{1}{2}$. In your proof, be careful to use only the propositions that are valid in Spherical geometry.

9. Give a proof of Proposition I.18 that relies only on Proposition I.19 and propositions between I.4 and I.15 that are valid in Spherical geometry.

10. Follow the directions given in Exercise 1 to analyze the validity of Proposition I.21 in Spherical geometry. Be sure to analyze each of the two pieces separately.

11. Follow the directions given in Exercise 1 to analyze the validity of Propositions I.24 and I.25 in Spherical geometry.

12. Find a counterexample for AAS in Spherical geometry. That is, find two triangles $\triangle ABC$ and $\triangle DEF$ that have two pairs of congruent angles and a pair of congruent sides, but are not congruent triangles.

13. Follow the directions given in Exercise 1 to analyze the validity of Propositions I.27, I.28 and I.31 in Spherical geometry.

14. Give a proof of the following problem from a nineteenth century spherical trigonometry book: *"If one angle of a triangle be equal to the sum of the other two, the greatest side is double of the distance of its middle point from the opposite angle"* [**117**].

4.5 Area in Spherical geometry

While we will postpone the strict axiomatic treatment of area until Chapter 7, we appeal to an informal notion of area to consider a very beautiful and somewhat unexpected result about the area of a spherical triangle. In order to prove this result, we must first introduce the spherical *lune*, a two-sided figure that does not exist in Euclidean geometry. As you read the definition, it may help to visualize the rind of a wedge of an orange, or to place two rubber bands on your spherical model.

Definition 4.6. Two distinct lines on a sphere divide the sphere into four regions, each of which is called a **lune**. Each lune has an angle, α, which is given by the angle of intersection of the two lines.

Notice that if one of the lunes has angle α radians, the lune directly across the sphere from it will also have angle α. The other pair of lunes will each have the supplementary angle $\pi - \alpha$ radians.

Lemma 4.7. *On a sphere of radius R, the area of a lune with angle α radians is given by* $Area = 2\alpha R^2$.

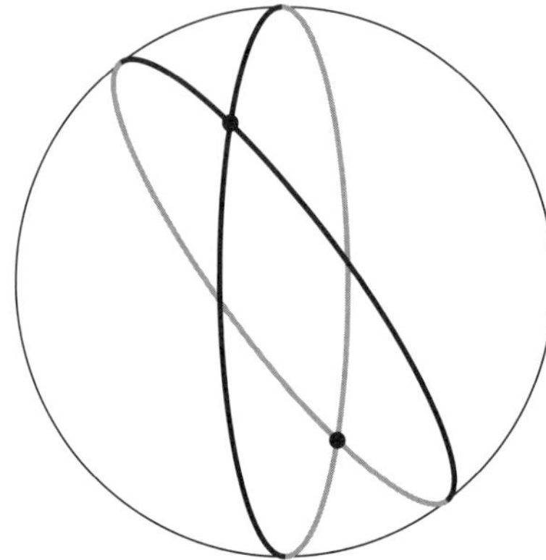

Figure 4.12. Lunes created by two lines

Proof.   A sphere of radius R has surface area $4\pi R^2$. A lune of angle $\frac{\pi}{2}$ has one-quarter of the sphere's total surface area, or πR^2, since it is a proportion of $\frac{\frac{\pi}{2}}{2\pi} = \frac{1}{4}$ of the total surface area. Thus, a lune of angle α must have a proportion of $\frac{\alpha}{2\pi}$ of the total surface area, or $\left(\frac{\alpha}{2\pi}\right)4\pi R^2 = 2\alpha R^2$. $\square$

We are now ready for our area result.

Theorem 4.8. *On a sphere of radius R, the area of spherical triangle $\triangle ABC$ with angles α, β and γ (measured in radians) is given by $Area(\triangle ABC) = (\alpha + \beta + \gamma - \pi)R^2$.*

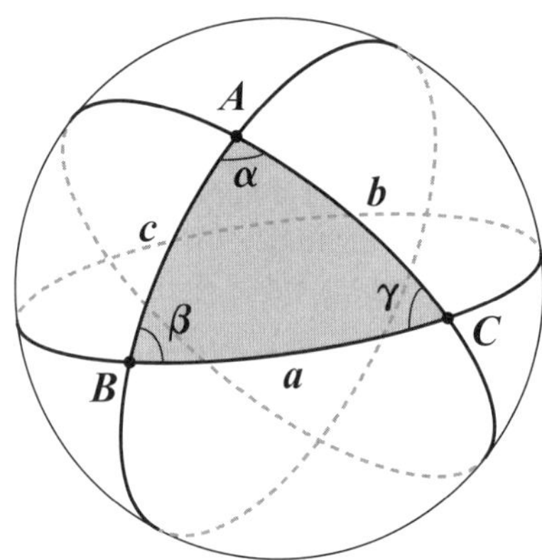

Figure 4.13. Area of a spherical triangle

Proof.   The sides of $\triangle ABC$ create lunes of angles α, β and γ, with areas of $2\alpha R^2$, $2\beta R^2$ and $2\gamma R^2$, respectively. Together, these three lunes cover exactly half of the area of the sphere, or $2\pi R^2$. (Use your spherical model to convince yourself of this by shading the three lunes.) Since each lune contains the triangle $\triangle ABC$, when we sum the area of the three lunes we count the area of the triangle $\triangle ABC$ three times. Adjusting for this overcounting, we have

$$\text{sum of lunar areas} - 2Area(\triangle ABC) = \frac{1}{2}(\text{surface area of sphere}).$$

Thus, we have

$$2\alpha R^2 + 2\beta R^2 + 2\gamma R^2 - 2Area(\triangle ABC) = 2\pi R^2.$$

Solving for the area of $\triangle ABC$ gives $Area(\triangle ABC) = (\alpha + \beta + \gamma - \pi)R^2$, as desired.

$\square$

The quantity $(\alpha + \beta + \gamma - \pi)$ is referred to as the **spherical excess** of the triangle. Notice that, in order for the area of a triangle to be a positive number, the sum of the angles in a spherical triangle must be greater than π radians. Also, recall that by definition, every angle of a spherical triangle is less than two right angles. This means that the sum of the angles in a spherical triangle must be less than six right angles, or 3π radians. Combining these upper and lower bounds on the angle sum gives the following corollary to the previous theorem.

Corollary 4.9. *The angle sum for every spherical triangle lies strictly between π and 3π radians.*

Let's extend our definition of a triangle to a general polygon and see what we can determine about its area. We define a **spherical polygon**, or **spherical *n*-gon**, as an *n*-sided figure (closed and nonself-intersecting) where $n \geq 4$ and the length of each side is less than half of the circumference of a great circle. For example, a *spherical quadrilateral* is a spherical 4-gon. An example of a spherical 8-gon, or octagon, is shown in Figure 4.14. As is the case in Euclidean geometry, angles in an *n*-gon can be less than or greater than π, but not equal to π.

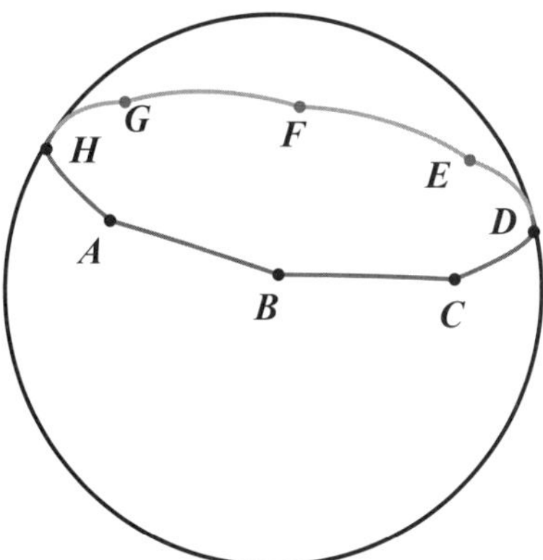

Figure 4.14. Spherical octagon

Recall, in planar geometry, a polygon is called **convex** if the line segment joining any two points of the figure lies entirely in the figure. In Figure 4.15, $ABCDE$ is convex but $FGHIJ$ is not, as demonstrated by its dashed line. Is the octagon shown in Figure 4.14 convex in Spherical geometry? As with the spherical triangle in Section 4.2, we encounter the problem of identifying the interior of an *n*-gon on a sphere since it creates

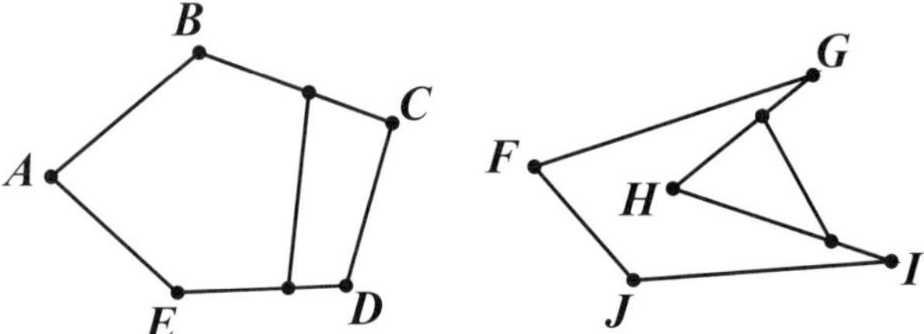

Figure 4.15. Convexity in the plane

two trapped regions. In Figure 4.14, for example, we see both a smaller and a larger region trapped by the 8-gon. The smaller region is convex since the line segments (the minor arcs) between any two points on the octagon will stay within the smaller region. This is not true for the larger region since AC, for example, does not live in the larger region. Next, let's consider the vertex angles. Notice that for the smaller octagonal region the vertex angles are all less than π radians, whereas, in the larger octagonal region the vertex angles are all greater than π. Also, the smaller region lives strictly within a hemisphere (all points live in the same hemisphere, as defined by some great circle), whereas the larger region does not. Figure 4.16 shows a 6-gon for which neither the smaller nor the larger trapped region is convex. While the smaller region lives strictly within a hemisphere, its vertex angles are a mix of angles both less than π and greater than π. In general, it can be shown that the polygonal region trapped by a spherical polygon is convex if and only if every vertex angle is less than π radians. Furthermore, any convex polygon lives strictly within a hemisphere, with at most one side lying on its boundary. As a consequence, no two points lying either on the boundary or within the figure can be antipodal. We finish this section by calculating the area of a

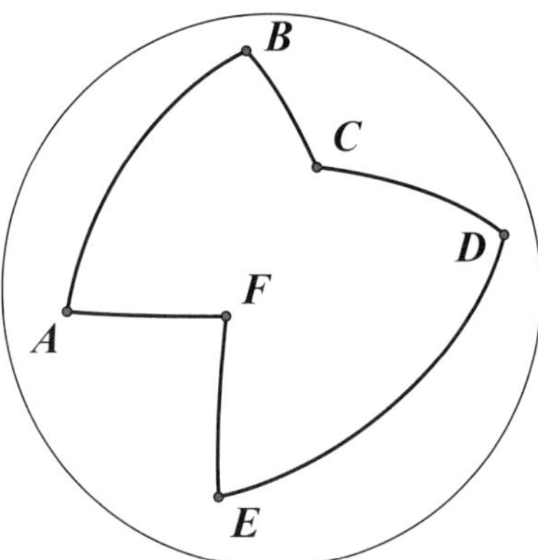

Figure 4.16. Neither polygonal region trapped by the 6-gon is convex

polygonal region meeting this convexity condition, then consider the area of its larger complementary region.

Although we have not formally discussed the axioms of area, we note that one way to calculate the area of a figure is to break it up into smaller figures of known area, then sum the areas of these component pieces. (We will revisit this idea more formally in

Chapter 7.) In the following proof, we decompose a given quadrilateral region into two spherical triangles in order to determine its area.

Theorem 4.10. *On a sphere of radius $R = 1$, the area of the spherical quadrilateral region, Q, formed by quadrilateral ABCD with vertex angles α, β, γ and δ, where $0 < \alpha, \beta, \gamma, \delta < \pi$, is given by*

$$Area(Q) = (\alpha + \beta + \gamma + \delta) - 2\pi.$$

Proof. Let quadrilateral region, Q, formed by $ABCD$, be given as specified, and join AC. Since all vertex angles are less than π, Q is convex. Thus, AC lies entirely inside our quadrilateral region and, since $R = 1$, is less than π in length. Consider the two trilateral regions, ABC and ADC: each side is less than π by hypothesis and convexity, and every angle is less than π radians by hypothesis and Common Notion 5. Thus, these regions are spherical triangles, $\triangle ABC$ and $\triangle ADC$. By Theorem 4.8, $Area(\triangle ABC) = (\angle BAC + \beta + \angle BCA) - \pi$ and $Area(\triangle ADC) = (\angle CAD + \gamma + \angle ACD) - \pi$. Adding these to find the area of quadrilateral region Q, we have $Area(Q) = Area(\triangle ABC) + Area(\triangle ADC) = (\angle BAC + \angle CAD + \beta + \angle BCA + \angle CAD + \gamma) - 2\pi = (\alpha + \beta + \gamma + \delta) - 2\pi$, as desired. $\square$

We can generalize Theorem 4.10 to the following result, the proof of which we leave to the reader.

Theorem 4.11. *On a sphere of radius $R = 1$, the area of the spherical polygonal region, P, formed by n-gon $A_1 A_2 \ldots A_n$ with vertex angles α_1, α_2, $\ldots$, α_n, where each $\alpha_i < \pi$, is given by*

$$Area(P) = \left(\sum_{i=1}^{n} \alpha_i\right) - (n-2)\pi.$$

When P is a convex polygonal region as specified in the previous theorem, then the sum of the angles must be less than $n\pi$. Thus, by Theorem 4.11, the area of P is less than 2π. Since the area of a sphere of radius $R = 1$ is 4π, we can determine the area of the complement of this region, denoted by $\overline{P}$, by subtracting the area of P from the sphere's area. We also note that $\overline{P}$ is the polygonal region given by the same vertices, $A_1, A_2, \ldots, A_n$, but with angles $\beta_i = 2\pi - \alpha_i > \pi$, since each $\alpha_i < \pi$. The area of $\overline{P}$ is calculated as follows:

$$Area(\overline{P}) = 4\pi - area(P)$$

$$= 4\pi - \left[\left(\sum_{i=1}^{n} \alpha_i\right) - (n-2)\pi\right]$$

$$= (n+2)\pi - \sum_{i=1}^{n} \alpha_i.$$

Since each angle $\alpha_i < \pi$, the area of $\overline{P}$ is greater than 2π. Of course, we already knew this since these regions are complements of each other with areas that sum to 4π. Notice further, that we can rewrite this area equation as follows:

$$area(\overline{P}) = \left(2n\pi - \sum_{i=1}^{n} \alpha_i\right) - (n-2)\pi$$

$$= \left(\sum_{i=1}^{n}(2\pi - \alpha_i)\right) - (n-2)\pi$$

$$= \left(\sum_{i=1}^{n} \beta_i\right) - (n-2)\pi.$$

Though $\overline{P}$ is not convex, the formula for its area matches the area formula for a convex polygonal region as given in Theorem 4.11! So, the area equation holds regardless of whether all vertex angles are less than π, or all are greater than π. Thus, we have our final result for spherical area.

Theorem 4.12. *On a sphere of radius $R = 1$, the area of the spherical polygonal region, P, formed by n-gon $A_1 A_2 \ldots A_n$ with vertex angles $\theta_1, \theta_2, \ldots, \theta_n$, where either $\theta_i < \pi$ for all $1 \le i \le n$, or $\theta_i > \pi$ for all $1 \le i \le n$, is given by*

$$Area(P) = \left(\sum_{i=1}^{n} \theta_i\right) - (n-2)\pi.$$

Exercises 4.5

1. Explain why, on a sphere with $R = 1$, the sum of the angles in a triangle that contains at least one right angle must be less than 2π.

2. Use the results of Section 4.5 to explain why, given our definition, there are no rectangles in Spherical geometry. Be very specific.

3. Prove Theorem 4.11. [*Hint: Decompose the polygonal region by joining one vertex to all nonadjacent vertices.*]

4. A **regular spherical polygon** is an equilateral, equiangular spherical polygon.

   (a) Determine the lower bound for the interior angle of a regular spherical n-gon whose interior angle is less than π. [For example, in the case of of a regular spherical pentagon, the angle α must satisfy the inequality $\frac{3\pi}{5} < \alpha < \pi$.]

   (b) Determine the interior angle of a regular *Euclidean* n-gon.

   (c) How does the lower bound for the interior angle of a regular spherical n-gon in part (a) relate to the interior angle of a regular Euclidean n-gon in part (b)?

4.6 Trigonometry for spherical triangles

This next section is a bit of a departure from our geometric discussion as we take a short trip through spherical trigonometry to present three well-known trigonometric formulas that hold for triangles on a sphere. Known as the First Spherical Law of Cosines

$[C_1]$, the Second Spherical Law of Cosines $[C_2]$, and the Spherical Law of Sines $[S]$, they are given below. Since we will be measuring lengths of segments on a sphere, we need to know the radius, R, of the sphere. It is most convenient to assume that $R = 1$, although the formulas can be appropriately adjusted if this is not the case. For ease, we assume that we are working on a sphere of radius one unit for the remainder of this chapter unless explicitly stated otherwise.

Let $\triangle ABC$ be a spherical triangle with sides a, b, c and interior angles α, β, γ as shown in the figure. The laws are given by:

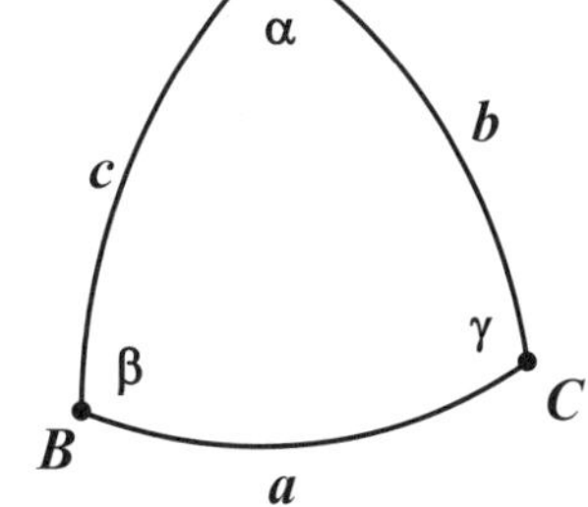

$$[C_1] \qquad \cos a = \cos b \cos c + \cos \alpha \sin b \sin c$$

$$[C_2] \qquad \cos \alpha = \cos a \sin \beta \sin \gamma - \cos \beta \cos \gamma$$

$$[S] \qquad \frac{\sin \alpha}{\sin a} = \frac{\sin \beta}{\sin b} = \frac{\sin \gamma}{\sin c}$$

In the special case where α is a right angle, $[C_1]$ simplifies to

$$\cos a = \cos b \cos c.$$

This particular formula is often referred to as the **Spherical Pythagorean Theorem**. Since it is helpful to have the First and Second Spherical Laws of Cosine solved for both the cosine of an angle and a side, here are their alternative versions:

$$[C_1^*] \qquad \cos \alpha = \frac{\cos a - \cos b \cos c}{\sin b \sin c},$$

$$[C_2^*] \qquad \cos a = \frac{\cos \alpha + \cos \beta \cos \gamma}{\sin \beta \sin \gamma}.$$

Because lengths are not measured in degrees, it is best to work in radians for the angles in these formulas. If angle measurement is given in degrees, use the conversion factor π radians $= 180°$ to convert to radians before applying the laws. Since we started with the assumption that the radius of the sphere is one unit, these formulas also only work when $R = 1$. When $R \neq 1$, simply divide the given lengths, a, b and c, by R before applying the laws. When solving, rewrite these formulas (just as we would the Euclidean Law of Cosines) depending upon which pieces of information are given and which are unknown. For example, an alternative version of the First Spherical Law of Cosines, $[C_1^*]$, is

$$\cos \beta = \frac{\cos b - \cos a \cos c}{\sin a \sin c}.$$

When the measures of certain angles or sides are known, we can use the formulas to find the unknowns. If we have a spherical triangle and are given two sides and the included angle, for example b, c and α, then we can solve for the third side, a, by applying law $[C_1]$. Then, by rewriting $[C_1^*]$ we can solve for β and γ, thereby completely determining the triangle, as the following example illustrates.

Example 4.13. Suppose we are given angle α, and two sides b and c, of $\triangle ABC$ as $\alpha = \frac{\pi}{2}, b = \frac{2\pi}{3}$ and $c = \frac{\pi}{2}$. To find the remaining angles and side of the triangle, using

$[C_1]$ we have $\cos a = -\frac{1}{2} \cdot 0 + 0$, which gives $a = \frac{\pi}{2}$. Using the alternative version of

$[C_1^*]$ as above gives $\cos \beta = \dfrac{-\frac{1}{2} - 0}{1 \cdot 1} = -\frac{1}{2}$. So, $\beta = \frac{2\pi}{3}$. Finally, $\cos \gamma = 0$ which gives

$\gamma = \frac{\pi}{2}$. (If, for example, we start with $\triangle A'B'C'$ on a sphere of radius $R = 2$, where

$\alpha' = \frac{\pi}{2}, b' = \frac{4\pi}{3}$ and $c' = \pi$, we first find its corresponding triangle on the unit sphere

in order to utilize the formulas given above. To do this, we convert lengths but leave

the angles unchanged. Specifically, here we convert b' and c' to $b = \dfrac{b'}{R} = \dfrac{2\pi}{3}$ and

$c = \dfrac{c'}{R} = \dfrac{\pi}{2}$, respectively. Notice that the triangle on the unit sphere corresponding to

$\triangle A'B'C'$ is identical to the triangle given at the start of this example, $\triangle ABC$. Hence,

$a' = Ra = \pi, \beta' = \beta = \frac{2\pi}{3}$ and $\gamma' = \gamma = \frac{\pi}{2}$.)

With this example, it is evident that these formulas provide another way to see that
the congruence scheme SAS must hold in Spherical geometry, that is, there is only one
such triangle given two sides and their included angle. Alternatively, if we are given
three sides of a spherical triangle, a, b and c, then we can find each of the three angles,
α, β and γ, after rewriting Law $[C_1^*]$ for each angle. Thus, these formulas give a different
justification for congruence scheme SSS in Spherical geometry.

Before discussing other possible scenarios, a word of caution regarding inverse
trigonometric functions is in order. Keep in mind that each angle of a spherical triangle
lives in the interval $(0, \pi)$, and while the range of the inverse cosine function is $[0, \pi]$,
the range of the inverse sine function is $[-\frac{\pi}{2}, \frac{\pi}{2}]$. Therefore, as we shall see in the next
example, solutions obtained by employing the arcsine function must be verified since
this function never produces an obtuse angle.

Example 4.14 (Example 4.13 revisited). Suppose we are given three sides of spherical
$\triangle ABC$ as $a = \frac{\pi}{2}, b = \frac{2\pi}{3}$ and $c = \frac{\pi}{2}$. To find the angles, using $[C_1^*]$ we have $\cos \alpha = 0$
which gives $\alpha = \frac{\pi}{2}$. We *could* continue to use the Law of Cosines to find the remaining
unknowns as we did in the previous example. If instead, we use the Law of Sines to

find β, then we start with the equation $\dfrac{1}{1} = \dfrac{\sin \beta}{\frac{\sqrt{3}}{2}}$. So, $\sin \beta = \dfrac{\sqrt{3}}{2}$, and the arcsine

function returns angle $\frac{\pi}{3}$ for β. However, by Proposition I.18, the greater side subtends
the greater angle. In our given triangle, since $b > a$, we know $\beta > \alpha = \frac{\pi}{2}$. Therefore,
our initial solution, $\frac{\pi}{3}$, cannot be angle β. Where is the problem in our reasoning?
As noted above, since arcsine only returns angles between $-\frac{\pi}{2}$ and $\frac{\pi}{2}$, we will never
obtain an obtuse solution. In the given $\triangle ABC$, β must be obtuse, and the obtuse
angle satisfying $\sin \beta = \dfrac{\sqrt{3}}{2}$ is $\beta = \frac{2\pi}{3}$. The lesson: When θ is a solution to the Law
of Sines equation, either θ or $\pi - \theta$ will be found in the given spherical triangle, but
more work is needed to determine which is correct. Since the range of the arccosine
function is $[0, \pi]$, no such additional step is necessary when working with the Law of
Cosines.

Continuing our exploration of congruence schemes, if we are given two angles
and the included side, for example β, γ and a, then we can find the third angle, α, by

applying Law $[C_2]$. Then, by using $[C_2^*]$, we can solve for sides b and c. Thus, ASA must hold. This line of reasoning leads us to a fourth congruence scheme, one that is not valid in Euclidean geometry. If we know α, β and γ, then by rewriting and using $[C_2^*]$ we can solve for sides a, b and c. Thus, in Spherical geometry, AAA is a valid congruence scheme. Exercise 4.6.1 demonstrates this congruence scheme as there is a unique triangle associated with any given set of three angles. We will state this as a theorem and give its proof in Section 4.7.

With these laws providing justification to four triangle congruence schemes on the sphere, the temptation to extend this line of reasoning to the unexplored AAS and SSA schemes is strong. How about AAS? In Section 4.4, we noted that this scheme does not hold in Spherical geometry and asked the reader to provide a counterexample. Does the trigonometry support our earlier work? Suppose we are given $\alpha = \frac{\pi}{2}, \beta = \frac{\pi}{2}$ and $a = \frac{\pi}{2}$ for $\triangle ABC$. The Law of Sines gives $b = \frac{\pi}{2}$, and $[C_1]$ gives $\cos\gamma = \cos c$. Thus, while we know $\gamma = c$, we do not have enough information to solve for this last unknown. Try $c = \frac{\pi}{3}$ and $c = \frac{2\pi}{3}$ for just two of the infinitely many possible values of c that satisfy these equations. Thus, unlike Euclidean geometry, two angles and an unincluded side is not enough information to determine a spherical triangle. The exploration of the possibility of SSA as a triangle congruence scheme on the sphere is left as an exercise.

For our final investigation, we revisit the restriction placed on constructible equilateral triangles in Proposition I.1S. First, since the total angle sum for spherical triangles ranges strictly between π and 3π, it is clear that the interior angle of a spherical equilateral triangle ranges strictly between $\frac{\pi}{3}$ and π by Corollary 4.9. Its Euclidean counterpart, however, is fixed at $\frac{\pi}{3}$. The Law of Cosines tells us how the side of an equilateral triangle is related to its angle. Assuming that the triangle has side of length a and angle α, Law $[C_2^*]$ gives

$$\cos a = \frac{\cos\alpha + \cos^2\alpha}{\sin^2\alpha} = \frac{\cos\alpha(1 + \cos\alpha)}{1 - \cos^2\alpha} = \frac{\cos\alpha}{1 - \cos\alpha}.$$

As shown in Figure 4.17, the graph of function $a = \arccos\left(\dfrac{\cos\alpha}{1 - \cos\alpha}\right)$ over the domain of $\frac{\pi}{3} < \alpha < \pi$ has a range of $0 < a < \frac{2\pi}{3}$. Thus, the side length is necessarily less than

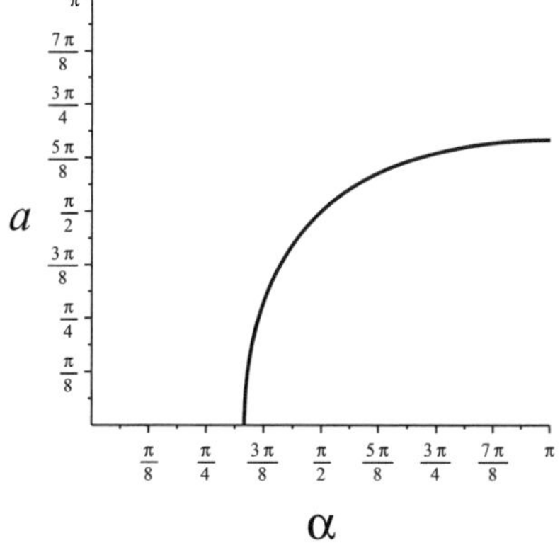

Figure 4.17. $a = \arccos\left(\dfrac{\cos\alpha}{1 - \cos\alpha}\right)$

$\frac{2\pi}{3}$, or one-third of the circumference of a great circle. This explains why the circles in the proof of Proposition I.1 do not intersect when the given segment has length at least one-third of the circumference of a great circle, as clearly, no such spherical equilateral triangle exists. Hence, we have justified the additional restriction placed in Proposition I.1S.

Exercises 4.6

1. On a sphere of radius $R = 1$, solve the spherical triangle with angles:

    (a) $\alpha = 60°, \beta = 60°, \gamma = 90°$

    (b) $\alpha = 60°, \beta = 90°, \gamma = 90°$

    (c) $\alpha = 90°, \beta = 90°, \gamma = 90°$

    (d) $\alpha = 60°, \beta = 90°, \gamma = 120°$

2. On a sphere of radius $R = 1$, solve the spherical triangle with sides:

    (a) $a = \dfrac{\pi}{2}, b = \dfrac{\pi}{2}, c = \dfrac{\pi}{2}$

    (b) $a = 1, b = 1, c = 1$

    (c) $a = \dfrac{\pi}{6}, b = \dfrac{\pi}{4}, c = \dfrac{\pi}{4}$

    (d) $a = \dfrac{\pi}{4}, b = \dfrac{\pi}{3}, c = \dfrac{\pi}{2}$

3. On a sphere of radius $R = 1$, solve the spherical triangle with:

    (a) $a = \dfrac{\pi}{2}, \beta = 60°, \gamma = 60°$

    (b) $b = \dfrac{\pi}{4}, \alpha = 90°, \gamma = 45°$

    (c) $c = \dfrac{\pi}{2}, \alpha = 60°, \beta = 120°$

    (d) $a = \dfrac{\pi}{2}, \beta = 45°, \gamma = 135°$

4. On a sphere of radius $R = 1$, solve the spherical triangle with:

    (a) $a = \dfrac{\pi}{2}, b = \dfrac{\pi}{2}, \gamma = 60°$

    (b) $a = \dfrac{\pi}{4}, b = \dfrac{\pi}{2}, \gamma = 90°$

    (c) $b = \dfrac{\pi}{4}, c = \dfrac{3\pi}{4}, \alpha = 60°$

5. On a sphere of radius R, solve the spherical triangle:

    (a) with sides R, R, R

    (b) with $b = 2R, \alpha = \dfrac{\pi}{3}, \gamma = \dfrac{\pi}{2}$

6. Calculate the area of each spherical triangle given in the first five exerciese.

7. Explore the possibility of an SSA triangle congruence scheme on the sphere: Suppose $a = \dfrac{\pi}{2}, b = \dfrac{\pi}{2}$ and $\alpha = \dfrac{\pi}{2}$ for $\triangle ABC$.

    (a) Use the Laws of Cosines and Sines to determine if this information is sufficient to determine triangle $\triangle ABC$. If the answer is yes, then prove it. If the answer is no, then provide two different triangles with the given a, b and α.

(b) Is SSA a valid congruence scheme in Spherical geometry?

8. Given an equilateral triangle with angle α and side length a, give a formula for $\cos \alpha$ in terms of $\cos a$.

9. Prove the following proposition: *"If two angles of a spherical triangle are supplements of each other, their opposite sides are also supplements of each other"* [112].

10. *"Find the angles and sides of an equilateral spherical triangle whose area is one-fourth of that of the sphere on which it is described"* [117].[6] Note: Assume we are working on a sphere of radius $R = 1$.

11. In an equilateral spherical triangle with angle α and side a, show that

$$\sec \alpha = 1 + \sec a.$$

12. In a spherical triangle with sides a, b and c and opposite angles α, β and γ, respectively, if $a = \alpha$, show that either $b = \beta$ or $b = \pi - \beta$, and either $c = \gamma$ or $c = \pi - \gamma$. That is, b and β are equal or supplemental, as are c and γ.

13. *"If two angles of a spherical triangle be respectively equal to the sides opposite them, show that the remaining side is the supplement of the remaining angle; or else that the triangle has two quadrants (sides with lengths equal to one-quarter of the circumference) and two right angles, and then the remaining side is equal to the remaining angle"* [117]. In other words, in a spherical triangle with sides a, b and c and opposite angles α, β and γ, respectively, assume that $a = \alpha$ and $b = \beta$. Show that if $a = b = \pi/2$, then $c = \gamma$, otherwise $c = \pi - \gamma$.

14. In a spherical triangle with sides a, b and c and opposite angles α, β and γ, respectively, if $b + c = \pi$, show that $\sin 2\beta + \sin 2\gamma = 0$.

15. Consider the spherical triangle $\triangle ABC$ with $a = BC, b = AC$ and $c = AB$. Let D be the midpoint of AB, and let $d = CD$, as shown in Figure 4.18. Prove that

$$\cos a + \cos b = 2 \cos \frac{c}{2} \cos d.$$

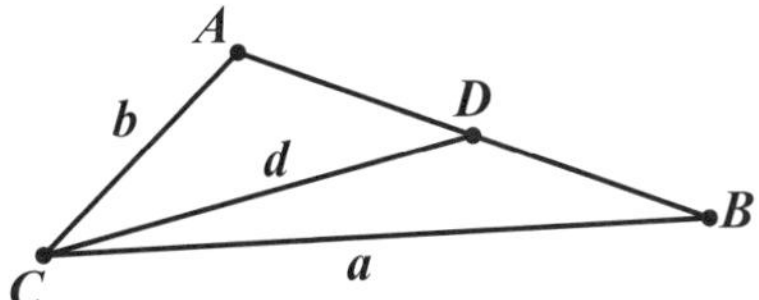

Figure 4.18. Exercise 4.6.15

16. *"A triangle has the sum of two sides equal to a semicircumference: find the arc joining the vertex with the middle of the base"* [117]. In other words, consider the spherical triangle $\triangle ABC$ with $a = BC$, $b = AC$ and $c = AB$. Let D be the midpoint of AB, and let $d = CD$, as shown in Figure 4.18. Since the semicircumference is π when

---

[6]Exercises 4.6.10 through 4.6.17 are adapted from Todhunter's nineteenth century spherical trigonometry book, [117].

$R = 1$, we assume $a + b = \pi$, and we must find d. [*Hint: Use the formula in Exercise 15.*]

17. Consider the spherical triangle $\triangle ABC$. Let D and E be the midpoints of AB and AC, respectively. Join DE and let F be a polar point for DE. Join FB, FD, FE, and FC. Show that $\angle BFC = 2\angle DFE$. [*Hint: Join AF and then show that $\angle AFD = \angle BFD$ and $\angle AFE = \angle CFE$.*]

4.7 Uniquely spherical constructions

In Chapter 3, we encountered the ten constructions of Neutral geometry, representing just over one-third of its propositions. In Section 4.4, our analysis of these constructions on the sphere showed that six of these are valid, namely Propositions I.3, I.9, I.10, I.11, I.12 and I.23, though some require modification to Euclid's proof; two are not valid, namely I.1 and I.22, unless a restriction is placed upon the length of at least one of the given segments; two are invalid, namely I.2 (which was added to the postulates) and I.31 (since parallel lines do not exist on the sphere). In this final section of the chapter, we explore uniquely Spherical constructions resulting from the close relationship that exists between lengths and angles on the sphere. Don't try these in Euclidean geometry! Here, we continue to assume that we are working on a unit sphere ($R = 1$) and our angle measure is in radians. We start with two somewhat surprising results that allow us to convert a given length to an angle, or vice versa.

Theorem 4.15. *To construct a length equal to a given angle α, where $\alpha < \pi$.*

Proof. Let $\angle BAC = \alpha$. Extend AB and AC until they intersect at A' which is antipodal to A. Bisect $\overset{\frown}{ABA'}$ and $\overset{\frown}{ACA'}$. Label these points D and E, respectively, and construct segment DE. We claim that $DE = \alpha$. Consider the line containing DE. By definition, A is a polar point of DE. Note that the total angle surrounding A is 2π. Similarly, if we extend segment DE to form great circle DE, it too has a length of 2π. Thus

$$\frac{\alpha}{2\pi} = \frac{DE}{2\pi},$$

and $\alpha = DE$, as desired. $\qquad\qquad\square$

Theorem 4.16. *To construct an angle equal to a given length less than half of the circumference.*

We leave the proof of the second theorem, which is similar to the previous, as an exercise. Thus, we can convert angles to lengths, and vice versa.

Continuing down this path of exploration, we would like to construct a new triangle from a given triangle in such a way that the angles of the new triangle are based on the sides of the given triangle, and vice versa. This construction produces a new object called a *polar triangle* which was introduced by Abu Nasr Mansur ibn Iraq (ca. 970–ca. 1036), an astronomer and mathematician who worked in Ghazna in present-day Afghanistan [**121**].

Definition 4.17. Let $\triangle ABC$ be a spherical triangle. Let C' be the polar point of AB that lies in the same hemisphere as C. In a similar manner, construct points A' and B'.

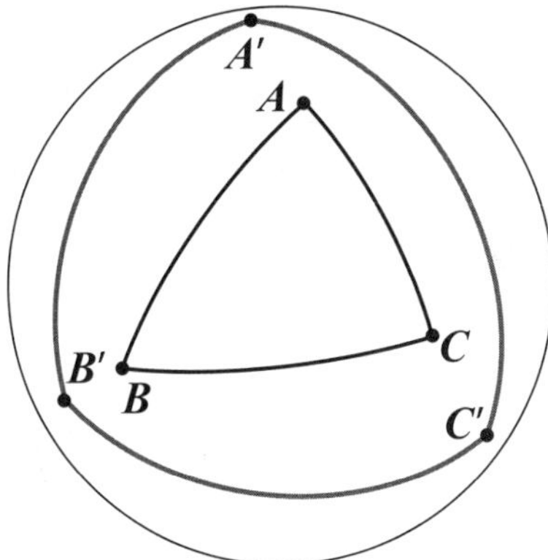

Figure 4.19. The polar triangle of $\triangle ABC$ is $\triangle A'B'C'$

Points A', B' and C' are polar points of sides BC, AC and AB, respectively. We will call $\triangle A'B'C'$ the **polar triangle** of $\triangle ABC$.

Although it is far from obvious, there is a very close relationship between a triangle and its associated polar triangle. This relationship is best described by two well-known theorems, the first of which shows that the process of constructing a polar triangle is an **involution**, that is, it is its own inverse.

Theorem 4.18 [Polar Triangle Involution Theorem]. *The polar triangle of a polar triangle is the original triangle.*

Proof. Let $\triangle ABC$ be a given spherical triangle and $\triangle A'B'C'$ be its polar triangle. Since C' is a pole of AB and A' is a pole of BC, both A' and C' are a length of $\frac{\pi}{2}$ from B. Thus B is a pole of $A'C'$.

We will now show that B is on the same side of $A'C'$ as B'. Since B' is a pole of AC on the same side as B, the distance between B and B' must be less than one-quarter of the circumference. If we assume that B is not on the same side of $A'C'$ as B', then the antipodal point of B, which we will call B'', must be on the same side of $A'C'$ as B'. But this would imply that the distance between B' and B'' is also less than one-quarter of the circumference of a great circle, which is not possible as the distance between antipodal points B and B'' is exactly half of the circumference, and B' lies on a unique line between them. Therefore B and B' must be on the same side of $A'C'$.

By similar reasoning, A and C are the appropriate polar points for $B'C'$ and $A'B'$, respectively. $\qquad\square$

Theorem 4.19 [Polar Duality Theorem]. *The sides of a polar triangle are the supplements of the angles of the original triangle, and the angles of a polar triangle are the supplements of the sides of the original. [Note that the supplement of side a is $\pi - a$.]*

Proof. Let $\triangle ABC$ be a given spherical triangle with angles α, β and γ at vertices A, B and C, respectively. Consider its polar triangle $\triangle A'B'C'$. We will start by showing that $B'C'$ has length $\pi - \alpha$. As demonstrated in Figure 4.20, extend rays $\overrightarrow{AB}$ and $\overrightarrow{AC}$ so that the first intersection with the line containing $B'C'$ are points D and E, respectively.

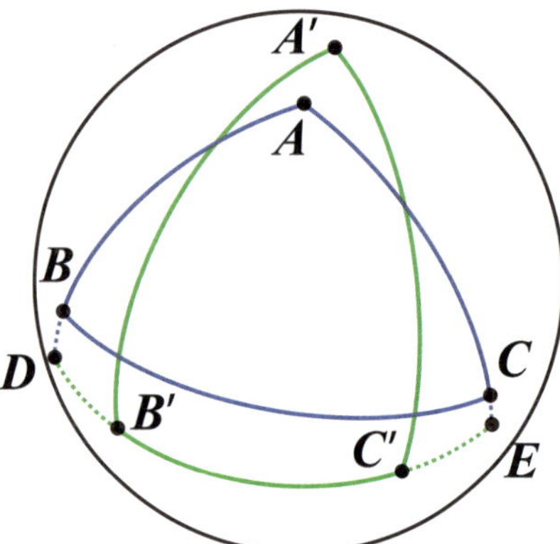

Figure 4.20. Polar Duality Theorem

Notice that $\angle DAE = \alpha$. Since A is a polar point for the line $B'C'$, by Theorem 4.15 we have $DE = \alpha$. Furthermore, since C' is a polar point for the line AB, and D lies on AB, we have $DC' = \pi/2$. By similar reasoning, $B'E = \pi/2$. We leave it to the reader to show that DE cannot straddle $B'C'$, leaving only two possible cases.

Case 1. $B'C'$ contains segment DE. In this case, we have the four collinear points as specified by $\overset{\frown}{B'DEC'}$. Thus $B'C' = B'E + DC' - DE = \pi/2 + \pi/2 - \alpha = \pi - \alpha$, as desired.

Case 2. DE contains segment $B'C'$, as illustrated in Figure 4.20. In this case, we have collinear points as specified by $\overset{\frown}{DB'C'E}$. Thus, once again, $B'C' = DC' + B'E - DE = \pi/2 + \pi/2 - \alpha = \pi - \alpha$.

By similar reasoning, we have $A'B' = \pi - \gamma$ and $A'C' = \pi - \beta$. To show that the angles are supplements of the sides of the original, we can apply the results we have just proven along with Theorem 4.18 to polar triangle $\triangle A'B'C'$.   □

Notice that while our name for this useful trilateral figure corresponding to a spherical triangle is polar *triangle*, we are only now guaranteed that a polar triangle *is* a spherical triangle. By the Polar Duality Theorem, when $\triangle ABC$ has sides $a = \pi/6$, $b = \pi/4$ and $c = \pi/3$, for example, its polar triangle has angles $\alpha' = 5\pi/6$, $\beta' = 3\pi/4$ and $\gamma' = 2\pi/3$.

You may recall that Lemma 4.5 guarantees that the sum of the sides of a spherical triangle is less than the circumference of a great circle, or, in the case of a unit sphere, 2π. The Polar Duality Theorem provides an alternative proof for this lemma as follows. Given $\triangle ABC$ with sides a, b and c, consider its polar triangle $\triangle A'B'C'$. By the Polar Duality Theorem, $\triangle A'B'C'$ has angles $\alpha' = \pi - a$, $\beta' = \pi - b$ and $\gamma' = \pi - c$. Adding the angles of $\triangle A'B'C'$ gives

$$\alpha' + \beta' + \gamma' = 3\pi - (a + b + c).$$

Thus, we have

$$a + b + c = 3\pi - (\alpha' + \beta' + \gamma').$$

By Corollary 4.9, the sum of the angles of $\triangle A'B'C'$ is larger than π. Therefore, $a + b + c < 2\pi$, and the total length of the sides for any spherical triangle is less than 2π.

When combined with the Triangle Inequality, the duality theorem also determines a condition on the angles of a spherical triangle. Specifically, since any two sides taken together must be greater than the third, by the Polar Duality Theorem we have that the sum of any two angles must be less than the remaining angle augmented by π, for example, $\alpha + \beta < \pi + \gamma$.

Corollary 4.20. *In a spherical triangle, the sum of any two angles is less than the remaining angle augmented by π.*

With this corollary, it is easy to see that while there are fewer restrictions on the angles of a spherical triangle than there are in Euclidean geometry, there are still sets of angles whose sum lies between π and 3π for which there is no corresponding spherical triangle. For example, if we let $\alpha = \frac{3\pi}{4}$, $\beta = \frac{3\pi}{4}$ and $\gamma = \frac{\pi}{4}$, then $\alpha + \beta \not< \pi + \gamma$. Therefore, a spherical triangle with these angles does not exist. The Law of Cosines, $[C_2^*]$, further supports this, giving

$$\cos a = \frac{-\frac{\sqrt{2}}{2} + \left(-\frac{\sqrt{2}}{2}\right)\left(\frac{\sqrt{2}}{2}\right)}{\frac{\sqrt{2}}{2} \cdot \frac{\sqrt{2}}{2}} = -\frac{\sqrt{2}}{2} - 1 < -1.$$

Clearly, there is no side a which satisfies this equation.

When a triangle with a given set of angles *does* exist, we can use the Polar Duality Theorem in conjunction with the close relationship between lengths and angles established in this section to construct a triangle with the given angles. We give this construction as the following theorem.

Theorem 4.21. *Given three angles between 0 and π whose sum is greater than π, and such that the sum of any two angles is less than the remaining angle augmented by π, to construct a triangle.*

Proof. Let α, β and γ be the given angles. By assumption, $\alpha, \beta, \gamma < \pi$. By Theorem 4.15, construct a segment with length α and label its endpoints A and B. Construct A', the antipodal point to A. Segment $A'B$ has length $a' = \pi - \alpha$, which is less than half the circumference of a great circle. Similarly, construct segments of length $b' = \pi - \beta$ and $c' = \pi - \gamma$. Because the sum of any two angles is less than the remaining angle augmented by π, the lengths a', b' and c' satisfy the Triangle Inequality, namely, that any two sides taken together will be greater than the third. Moreover, since $\alpha + \beta + \gamma > \pi$, we have $a' + b' + c' < 2\pi$. Using Proposition I.22S, construct triangle $\triangle CDE$ with sides of these lengths. By the Polar Duality Theorem, the polar triangle of $\triangle CDE$ is the desired triangle since it has the given angles α, β and γ. $\qquad\square$

The final congruence scheme for spherical triangles, AAA_S, follows directly from the theorem. We will refer to this theorem as AAA_S, with the subscript **S** as a reminder that it is a congruence scheme for **S**pherical geometry.

Corollary 4.22 [AAA_S]. *Two spherical triangles with equal angles are congruent.*

Before leaving this world, we take one last look at equilateral triangles. Note that the polar triangle of an equilateral triangle is equilateral. As discussed in Section 4.6, the angle, α, of an equilateral triangle on a sphere lies between $\pi/3$ and π. Given such an α, we can construct the equiangular triangle with this angle using Theorem 4.21. Thus, by polar duality, an equilateral triangle must have side length strictly between 0 and $2\pi/3$, giving yet another justification for Proposition I.1S.

We are now ready to leave the spherical world, where angles behave in unexpected ways, to explore a very different world where distances behave in curious ways.

Exercises 4.7

1. Prove Theorem 4.16: *To construct an angle equal to a given length less than half of the circumference.*

2. Prove that in the proof of Theorem 4.19 the only possible cases are that either $B'C'$ contains DE or DE contains $B'C'$.

3. Determine when a triangle and its polar triangle coincide.

4. Consider the triangle $\triangle ABC$, where $a = b = \pi/2$ and $c = \pi/4$. Determine the angles and sides of its polar triangle.

5

Taxicab Geometry

Before we head back to Euclidean geometry, we'll take a look at another geometry, *Taxicab geometry*. In the history of mathematics, this is a relatively new geometry as its name was coined by the Austrian-American mathematician Karl Menger (1902–1985) in 1952 for a geometry exhibit at the Museum of Science and Industry of Chicago. Its unusual measure for distance is due to the work of German mathematician Hermann Minkowski (1864–1909). We take inspiration for this chapter from Eugene Krause's book, *Taxicab Geometry: An Adventure in Non-Euclidean Geometry*. While this geometry evokes an image of the streets that create the rectangular grid of a city, we will only be in the city for a short visit as we quickly leave the asphalt behind and venture back to Euclid's realm in the next chapter.

Rectangular grids are essentially coordinate systems, and as such, we must work on the Cartesian plane in Taxicab geometry. Working with an x-axis, y-axis, ordered pairs and algebraic equations is certainly not how Euclid approached his geometry. In fact, it would have been impossible as it was nearly two thousand years before René Descartes (1596–1650) linked algebra and geometry into our familiar coordinate system where a curve has an algebraic and a geometric representation. As Euclid reasons through one proposition after another in Book I of the *Elements*, the associated diagrams are certainly helpful but not logically necessary, and they live in a coordinate-free plane. This approach to geometry is referred to as **synthetic**, or axiomatic, geometry. In order to define the objects in Taxicab geometry, we head to the land of **analytic geometry,** that is, the land of coordinate systems and equations. Though it would not have contributed to our pedagogical goal of studying axiomatic systems, we certainly could have ventured into the Cartesian plane earlier in the book when we introduced our main characters. If we had, then lines and circles would have been described algebraically as the set of points satisfying $ax + by = c$ and $(x - h)^2 + (y - k)^2 = r^2$, respectively, where a, b, c, h, k and r are real numbers. This is Euclidean coordinate geometry, familiar to every algebra, trigonometry, and calculus student, but not the focus of our studies. In this chapter, however, we will bring our considerable analytic geometry skills to bear on Taxicab geometry. As with each new geometry we consider, we start by determining how our main characters behave.

5.1 Points, lines, angles, distances and circles

In the previous chapter, the nature of the sphere necessitated a careful consideration of the behavior of lines, in particular, the notions of straight and line segment. In Taxicab geometry, you may be either disappointed or relieved to learn that the points, lines and angles are exactly the same as in Euclidean coordinate geometry. Points are ordered pairs of real numbers, such as $P = (-1, 1)$ or $Q = (2, 5)$. A line is the set of all points that satisfy an equation of the form $ax + by = c$, where a, b, and c are real numbers. An angle is measured in the same manner as in Euclidean coordinate geometry. What, then, is different? The answer lies in how we measure distance.

While we have informally discussed the concept of distance in the previous two chapters, we are now ready to give a precise definition of a distance function. Let's start by noting some of the obvious properties that we expect of a function that gives the distance between points A and B. The distance should be positive unless $A=B$, when it is zero. The starting point for our measurement, A or B, should make no difference. Lastly, the distance from A to B over two separate legs of travel, say through some intermediate point, C, will be the same as, or longer than, the distance between A and B. The following definition incorporates these ideas.

Definition 5.1. A **metric**, or **distance**, function on a set of points, $\mathscr{P}$, is a function $d : \mathscr{P} \times \mathscr{P} \to \mathbb{R}$ such that, for any points $A, B, C \in \mathscr{P}$, the following properties hold:

(1) $d(A, B) \geq 0.$

(2) $d(A, B) = 0$ if and only if $A = B$.

(3) $d(A, B) = d(B, A).$

(4) $d(A, B) + d(B, C) \geq d(A, C).$

Note: The fourth property is the metric function equivalent of the Triangle Inequality (I.20).

Every high school algebra student learns that the Euclidean distance between two points $P = (x_1, y_1)$ and $Q = (x_2, y_2)$ is given by the formula

$$d_E(P, Q) = \sqrt{(x_1 - x_2)^2 + (y_1 - y_2)^2}.$$

We leave it to the reader to check that this function satisfies the four properties of a metric. In Euclidean coordinate geometry we use this distance formula to calculate the length of the line segment PQ. As Figure 5.1 shows, the well-known Pythagorean

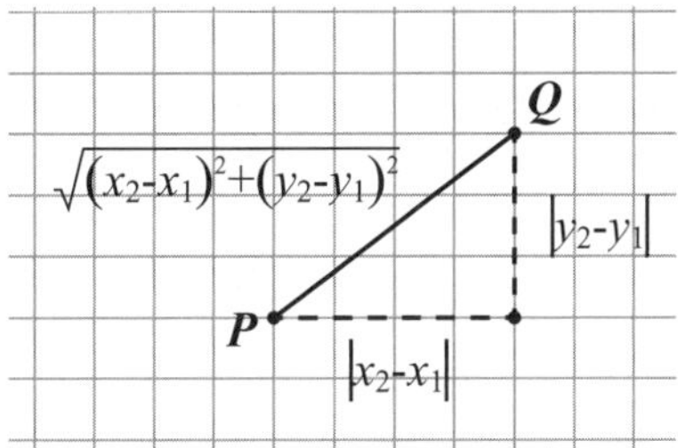

Figure 5.1. Euclidean distance

Theorem justifies both the formula and our use of it to find the length of segment PQ.

In Taxicab geometry, we use a different metric to find the distance between two points, specifically

$$d_T(P,Q) = |x_1 - x_2| + |y_1 - y_2|.$$

Once again, we leave it to the reader to check that the four properties of a metric

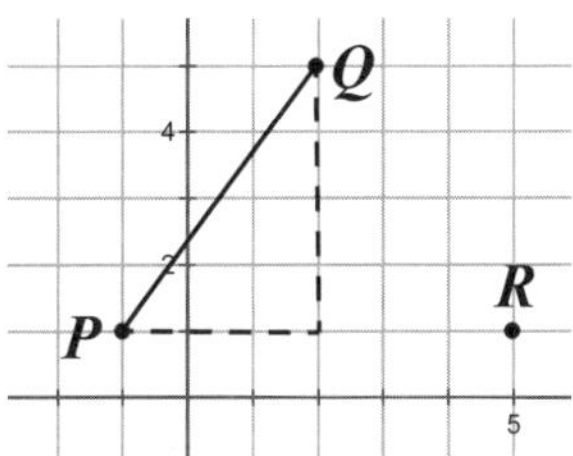

Figure 5.2. Distances in Taxicab geometry

hold for d_T. While we no longer have the Pythagorean Theorem to guide us, we still use the horizontal and vertical displacements between points P and Q to define the length of line segment PQ in Taxicab geometry. Thus, if $P = (-1, 1)$ and $Q = (2, 5)$ as shown in Figure 5.2, then the Euclidean distance between P and Q is $d_E(P,Q) = 5$, but the Taxicab distance is $d_T(P,Q) = 7$. This may seem strange, so we'll reiterate: in Euclidean geometry the line segment PQ of Figure 5.2 has length 5, but in Taxicab geometry its length is 7, and thus, the Pythagorean Theorem is not valid for this right-angled triangle.

The name of the geometry evokes the appropriate analogy. Imagine being in a city with the streets arranged in a perfect grid. In order to travel from P to Q you cannot go through buildings or travel "as the crow flies." It'll take you at least seven blocks of city travel to get from P to Q. It's also interesting to note that, in Taxicab geometry, point R (as shown in Figure 5.2) is closer to point P than Q, though the opposite is true in Euclidean geometry. Here, we have $d_T(P,R) = 6$ and $d_T(Q,R) = 7$.

With a new way to calculate distance, we must be careful to differentiate between a straight line joining two points and a path of shortest length between the points. As we mentioned in Chapters 1 and 4, the following statement is often taken as a definition in high school geometry: *A straight line is the shortest path between two points.* While it is true that the line segment joining $A = (x_1, y_1)$ and $B = (x_2, y_2)$ is the shortest path between A and B, it is not unique. For example, in Figure 5.2, there are many street paths of length 7 blocks that you can travel in our imagined city to get from P to Q. As you experiment with different paths between these two points, be careful not to take the city grid metaphor literally: in Taxicab geometry there are no "streets," so we do not have to limit ourselves to travelling on them. You may travel in any direction, producing any path, and the length of the path is the sum of distances traveled in each of the four compass directions: east, west, north and south. Thus, there are infinitely many paths of shortest length between any two points unless the points are directly east/west or north/south of one another when the shortest path is unique and horizontal or vertical, respectively. Use Figure 5.3 to consider the paths from A to B of shortest

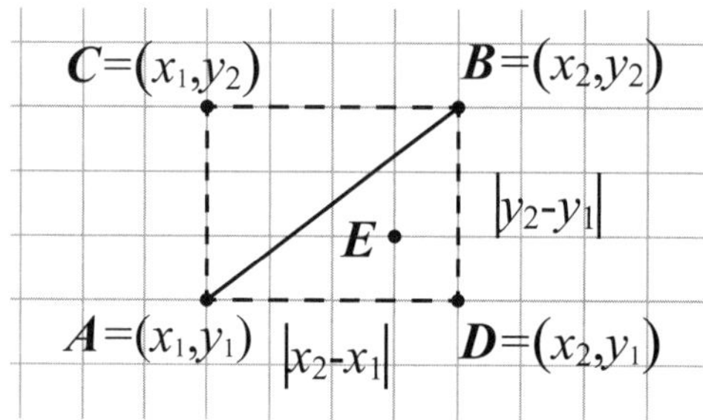

Figure 5.3. Multiple shortest paths

length, $|x_1 - x_2| + |y_1 - y_2|$. To create two such paths, consider points $C = (x_1, y_2)$ and $D = (x_2, y_1)$. The piecewise linear path given by travelling from A to C to B, and the path given by travelling from A to D to B both have length $|x_1 - x_2| + |y_1 - y_2|$. What other paths share this length? How about A to E to B? If it helps to use specific points, then try $A = (1, 3)$, $B = (5, 6)$, $C = (1, 6)$, $D = (5, 3)$ and $E = (4, 4)$.

How will leaving angles the same while changing how we calculate lengths affect results from Euclidean geometry? Let's discuss two theorems involving perpendicular lines where this creates some differences. In Theorem 3.5, we proved that the set of points equidistant from two given points is exactly the same as the set of points that lie on the perpendicular bisector of the line segment joining the two points. Let's examine

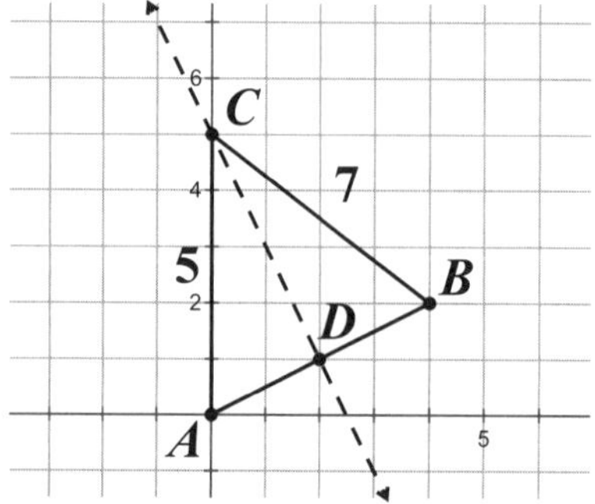

Figure 5.4. Perpendicular bisector for AB

this theorem for a particular pair of points in Taxicab geometry. First, we note that the slope of any line in Taxicab geometry remains the same as in Euclidean geometry since it is defined as "rise over run," with "rise" as a vertical displacement and "run" as a horizontal. A vertical or a horizontal distance is unchanged in this new geometry. As shown in Figure 5.4, when $A = (0,0)$ and $B = (4,2)$, the perpendicular bisector of AB is given by equation $y = -2x + 5$, shown as the dashed line $\overleftrightarrow{CD}$. Notice that line $\overleftrightarrow{CD}$ is the bisector of AB since AD and BD are both 3 units in length, and it is perpendicular to AB since it has negative, reciprocal slope to AB. Notice that $C = (0,5)$ lies on this line, but $d_T(C,A) = 5$ while $d_T(C,B) = 7$. Thus, in Taxicab geometry, the set of points equidistant from A and B is different from the set of points on the perpendicular bisector of AB, and the theorem does not hold in Taxicab geometry. If the set of points equidistant from A and B is not a line, is it some other recognizable figure? We leave it as an exercise to describe this set of points, namely all P such that $d_T(P,A) = d_T(P,B)$.

Next, we recall from Definition 3.10 that the *distance from a point C to a line $\overleftrightarrow{AB}$* is defined to be the length of the shortest segment CD where D is any point on $\overleftrightarrow{AB}$. In Euclidean geometry, by Theorem 3.11 we know this occurs when $CD \perp \overleftrightarrow{AB}$. Notice that in Figure 5.4, $d_T(C,D) = 6$, but $d_T(C,A) = 5$. Clearly, we must reevaluate our rule for finding the distance from a point to a line in this new geometry. We leave it as an exercise to determine the procedure for finding this distance.

Since lines have been hogging the spotlight, we have not set eyes on our other main character in this strange new world. So, let's turn our attention to circles. In analytic geometry, a Euclidean circle with center C and radius r can be described as the set of points equidistant to C, that is, $\{P \mid d_E(P,C) = r\}$. Thus, a Taxicab circle with center C and radius r consists of the set $\{P \mid d_T(P,C) = r\}$. This all seems rather routine, but things are about to take a weird turn. For example, Figure 5.5 shows a Taxicab circle with center $C = (1,2)$ and radius $r = 3$. Take a few minutes to verify this

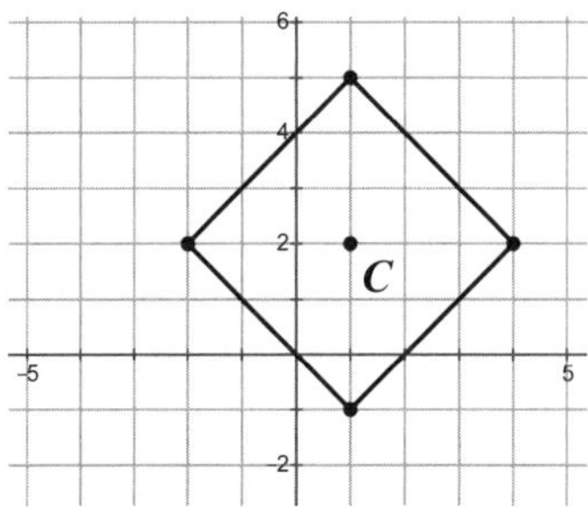

Figure 5.5. A circle in Taxicab geometry

example and marvel at its strangeness. While in Spherical geometry, lines are great circles, here each circle is composed of four line segments! This delighted Menger, who wrote "Square circles or round squares have haunted many diverse philosophical writers as the archetype of the impossible and the absurd; they were assigned a place near - or rather below - golden mountains, unicorns and mermaids" [60]. We will revisit this idea of the impossibility of "square circles" in Chapter 16. For now, let's go down the rabbit hole to see where these odd Taxicab circles lead us. The following is, perhaps, the most shocking result in this strange new world. In Euclidean geometry, π is defined to be the ratio of the circumference of a circle to its diameter. The diameter of the Taxicab circle in Figure 5.5 is 6, and its circumference is its perimeter, $4(6) = 24$. Using this definition, we have $\pi_T = 4$ in Taxicab geometry.

You may be curious as to whether the formulas for the area of a triangle or rectangle in Euclidean geometry might also work in Taxicab geometry. When we discuss the concept of area in Chapter 7, we shall see that this is not the case. In fact, to this date no one has developed a way to define an area function in this geometry.

> Initially, we wrote the equation $\pi_T = 4$ without the subscript, then we remembered a tiny footnote in history and thought better of it. In 1897, the Indiana General Assembly introduced Bill #246 at the behest of Edwin Goodwin, a physician and amateur mathematician. In it, he claimed that the area of a circle is equal to the area of a square with sides of length one-fourth of the circumference of the circle. This would imply that for a circle of radius, r, Goodwin's area formula is $\dfrac{\pi^2 r^2}{4}$ instead of the correct area, πr^2. Since Goodwin claimed that his formula gave the area of a circle, setting these two equations equal and solving for π gives 4. When this story is retold, it is often said that Indiana tried to pass a law proclaiming that 4 is the value of π. Fortunately, Clarence Waldo (1852–1926), a mathematics professor at Purdue University, stepped in and convinced the legislature not to pass the bill, saving Indiana much embarrassment.

Exercises 5.1

1. On a sheet of grid paper, plot each pair of points and then find both the Euclidean and Taxicab distances between them.

   (a) $A = (1, 6), B = (6, 1)$
   (b) $C = (-3, 2), D = (4, 1)$
   (c) $E = (-2, -4), F = (3, -1)$
   (d) $G = (-1, 3), H = (-1, 5)$

2. Give an example of four distinct points A, B, C and D that satisfies each condition.

   (a) $d_T(A, B) = d_T(C, D)$ but $d_E(A, B) \neq d_E(C, D)$.
   (b) $d_E(A, B) = d_E(C, D)$ but $d_T(A, B) \neq d_T(C, D)$.
   (c) $d_E(A, B) = d_E(C, D)$ and $d_T(A, B) = d_T(C, D)$.

3. What conditions on points A and B must hold to ensure that $d_T(A, B) = d_E(A, B)$? Prove this.

4. Prove that for all points A and B, $d_T(A, B) \geq d_E(A, B)$.

5. Show that the Euclidean distance, $d_E(P, Q) = \sqrt{(x_1 - x_2)^2 + (y_1 - y_2)^2}$, satisfies the four properties of a metric.

6. Show that the Taxicab distance, $d_T(P, Q) = |x_1 - x_2| + |y_1 - y_2|$, satisfies the four properties of a metric.

7. Using grid paper, plot each pair of points, A and B, and then graph the set of points, $\{P \mid d_T(A,P) + d_T(P,B) = d_T(A,B)\}$.

   (a) $A = (1,6), B = (6,1)$
   (b) $A = (-3,2), B = (4,1)$
   (c) $A = (-2,-4), B = (3,-1)$
   (d) $A = (-1,3), B = (-1,5)$

8. Determine the perimeter of the triangle given in each part.

   (a) Let $\triangle ABC$ be formed with points $A = (0,0)$, $B = (4,0)$ and $C = (2,4)$. Determine the taxicab perimeter of $\triangle ABC$.
   (b) Let $\triangle DEF$ be formed with points $D = (-2,-2)$, $E = (1,3)$ and $F = (-1,6)$. Determine the taxicab perimeter of $\triangle DEF$.

9. We can use trigonometry to calculate angles in Taxicab geometry. For example, if $G = (0,0)$, $H = (4,-2)$ and $I = (2,3)$, then $\angle IGH = \arctan \frac{3}{2} + \arctan \frac{1}{2} \approx 82.87°$. Determine the angles in each of the following triangles. [Note regarding trigonometric functions in Taxicab geometry: Since the tangent of an angle in a right-angled triangle is the "opposite side" over the "adjacent side," this function will return identical values in Taxicab and Euclidean geometries when the opposite and adjacent sides are positioned as horizontal or vertical displacements because these distances are the same in both geometries.]

   (a) $\triangle ABC$, where $A = (0,0)$, $B = (4,0)$, and $C = (2,4)$.
   (b) $\triangle DEF$, where $D = (-2,-2)$, $E = (1,3)$, and $F = (-1,6)$.
   (c) $\triangle GHI$, where $G = (0,0)$, $H = (4,-2)$, and $I = (2,3)$

10. For each given center C and radius r, use grid paper to draw the Taxicab circle.

    (a) $C = (0,0)$ and $r = 2$
    (b) $C = (-1,2)$ and $r = 3$

11. For each pair of points, A and B, use grid paper to graph the set of points $\{P \mid d_T(P,A) = d_T(P,B)\}$.

    (a) $A = (0,0)$ and $B = (4,2)$
    (b) $A = (0,0)$ and $B = (2,6)$
    (c) $A = (0,0)$ and $B = (0,4)$
    (d) $A = (0,0)$ and $B = (6,0)$
    (e) $A = (0,0)$ and $B = (4,4)$ (Be careful!)

12. For each of the following, use grid paper to sketch the point A and the line L passing through the given points. Locate the point (or points), P, on L that is (are) closest to A, and then use this to calculate $d_T(A,L)$.

    (a) $A = (-2,3)$ and L passes through the points $(0,-6)$ and $(4,0)$
    (b) $A = (-2,3)$ and L passes through the points $(0,6)$ and $(4,0)$
    (c) $A = (5,-2)$ and L passes through the points $(-1,-1)$ and $(8,2)$
    (d) $A = (5,-2)$ and L passes through the points $(-1,-1)$ and $(8,-4)$

(e) $A = (-1, 6)$ and L passes through the points $(-3, 0)$ and $(1, 4)$

(f) $A = (-1, 6)$ and L passes through the points $(-3, 0)$ and $(1, -4)$

13. Use Exercise 12 to determine a rule for finding the distance between a point P and a line L. Be sure to justify the rule. [*Hint: What happens as the radius of a circle with center P expands until it just touches line L? Consider cases based on the slope of L.*]

14. Let $A = (0, 0)$, $B = (3, 3)$ and $C = (4, 0)$.

(a) Using grid paper, graph the set of points in the interior of $\angle BAC$ equidistant from rays $\overrightarrow{AB}$ and $\overrightarrow{AC}$. [*Hint: Use the rule determined in Exercise 13.*]

(b) In Theorem 3.13, we proved that a point interior to an angle $\angle BAC$ lies on the angle's bisector if and only if it is equidistant from rays $\overrightarrow{AB}$ and $\overrightarrow{AC}$. Does this theorem still hold in Taxicab geometry? Explain.

15. An **ellipse** is the set of points such that the sum of the distances to two fixed points, A and B (the foci), is some positive constant, c. For each given A, B and c, use grid paper to draw a *Taxicab ellipse* with foci A and B. That is, find all points P such that $d_T(A, P) + d_T(B, P) = c$.

(a) $A = (0, 0)$, $B = (4, 0)$ and $c = 6$

(b) $A = (0, 0)$, $B = (2, 2)$ and $c = 8$

(c) $A = (0, 0)$, $B = (2, 1)$ and $c = 7$

16. A **hyperbola** is the set of points such that the shorter from the longer distance to two fixed points, A and B (the foci), is a positive constant, c. For each given A, B and c, use grid paper to draw a *Taxicab hyperbola* with foci A and B. That is, find all points P such that $|d_T(A, P) - d_T(B, P)| = c$.

(a) $A = (0, 0)$, $B = (4, 0)$ and $c = 2$

(b) $A = (0, 0)$, $B = (4, 2)$ and $c = 4$

(c) $A = (0, 0)$, $B = (2, 2)$ and $c = 4$ (Be careful)

(d) $A = (0, 0)$, $B = (2, 1)$ and $c = 1$ (Be careful)

5.2 Euclid's postulates in Taxicab geometry

Using our working understanding of lines and circles in Taxicab geometry, let's revisit Euclid's axioms in our new Taxicab world. As in the previous chapter, the Common Notions are valid here, too, as they are independent of the shape of our world. The five Postulates concern our main characters, lines and circles, which have changed in behavior in Taxicab geometry. We wish to determine if these geometric axioms are still valid here. Before reading further, you should carefully consider each postulate to determine whether or not it is valid in Taxicab geometry.

Postulate 1: To draw a straight line from any point to any point.

Given two points (x_1, y_1) and (x_2, y_2), it is clear that the line given by $y - y_1 = m(x - x_1)$ where $m = \dfrac{y_2 - y_1}{x_2 - x_1}$ passes through these two points (as it does in Euclidean geometry). Thus, this postulate holds. Moreover, the line is unique.

Postulate 2: To produce a finite straight line continuously in a straight line.

Since this holds in Euclidean geometry, and lines are the same, this postulate holds.

Postulate 3: To describe a circle with any center and distance.

If we are given a point, C, and a radius, r, we can certainly construct the set of points $\{P \mid d_T(P, C) = r\}$. Thus, this postulate holds.

Postulate 4: That all right angles are equal to one another.

Since angles are measured in the same manner as in Euclidean geometry, this postulate holds.

Postulate 5: That, if a straight line falling on two straight lines make the interior angles on the same side less than two right angles, the two straight lines, if produced indefinitely, meet on that side on which are the angles less than two right angles.

Again, since lines and angles are the same as their Euclidean counterparts, this postulate holds.

Therefore, we have established that all of Euclid's axioms are valid in Taxicab geometry. In the previous chapter, we saw that some of the Neutral geometry propositions of Book I failed in Spherical geometry even though Euclid's axioms were valid. This left us with the realization that Euclid's set of axioms must be incomplete. We'll keep our findings from the previous chapter in mind, and if propositions *different* from those in the previous chapter fail to hold in Taxicab geometry, then these will signal our discovery of other missing axioms that will help us amend Euclid's set of axioms to distinguish Euclidean geometry from Taxicab geometry.

5.3 Congruence schemes in Taxicab geometry

Triangles in Taxicab geometry are exactly the same as triangles in Euclidean geometry. Consider the following two triangles shown in the Figure 5.6. Triangle $\triangle ABC$ has vertices $A = (0, 1), B = (1, 0)$ and $C = (3, 2)$, while triangle $\triangle DEF$ has vertices $D =$

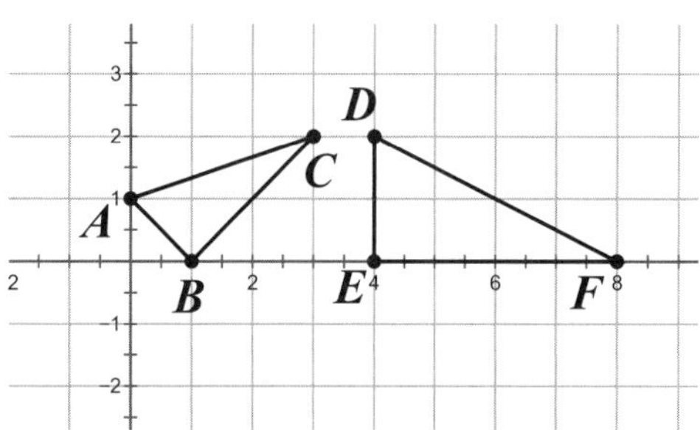

Figure 5.6. SAS does not hold in Taxicab geometry

$(4, 2), E = (4, 0)$ and $F = (8, 0)$. Now, $AB = DE = 2$, $BC = EF = 4$ and $\angle B = \angle E = 90°$, so the triangles have corresponding *side-angle-side*. However, $AC = 4$ while $DF = 6$. Thus, the triangles are not congruent, and Proposition I.4 [SAS] does not hold. This example demonstrates the distortion of side lengths and angles that can occur as a triangle is moved within the plane in Taxicab geometry, and thus, exposes the hidden flaw of the superposition technique employed by Euclid in the proof of SAS. We leave it to the reader to produce counterexamples which demonstrate the failure of Propositions I.8 [SSS] and I.26 [both ASA and AAS] in Taxicab geometry. As we shall see in the next chapter, SAS must be added to Euclid's set of axioms for Euclidean geometry.

Exercises 5.3

1. Use grid paper to find a counterexample for SSS in Taxicab geometry. That is, find two triangles $\triangle ABC$ and $\triangle DEF$ that have three pairs of congruent sides, but are not congruent triangles.

2. Use grid paper to find a counterexample for ASA in Taxicab geometry. That is, find two triangles $\triangle ABC$ and $\triangle DEF$ that have corresponding *angle-side-angle*, but are not congruent triangles.

3. Use grid paper to find a counterexample for AAS in Taxicab geometry. That is, find two triangles $\triangle ABC$ and $\triangle DEF$ that have corresponding *angle-angle-side*, but are not congruent triangles.

5.4 The rest of Neutral geometry

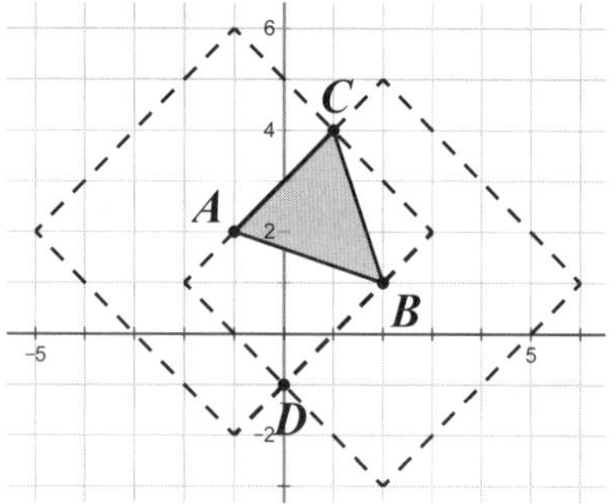

Figure 5.7. Proposition I.1 in Taxicab geometry

We will start with Proposition I.1: *On a given finite straight line to construct an equilateral triangle.* At first glance, this proposition appears to hold exactly as it does in the Euclidean plane. Consider Figure 5.7 where AB is the given line segment joining points $A = (-1, 2)$ and $B = (2, 1)$. The construction is exactly what we expect, with Taxicab circles of radius AB drawn at each endpoint of the line segment. Here, we see that these circles intersect at points C and D, and the shaded $\triangle ABC$ is an equilateral triangle as desired. The unexpected occurs when segment AB has slope ± 1. Try it for yourself, but as Figure 5.8 illustrates with $A = (0, 2)$ and $B = (2, 0)$, the intersection of the two Taxicab circles is not two points, but rather the two segments CD and EF. This produces infinitely many equilateral triangles from which to choose, two of which, $\triangle ABD$ and $\triangle ABG$, are shaded in the diagram.

Let's continue to Proposition I.2 (*To place at a given point [as an extremity] a straight line equal to a given straight line*). Since the proof of this proposition requires the construction of an equilateral triangle on a segment of arbitrary length, but does not require that the triangle be unique, this proposition and its proof hold. Here is a good place to discuss the difference between the existence of a geometric object and our

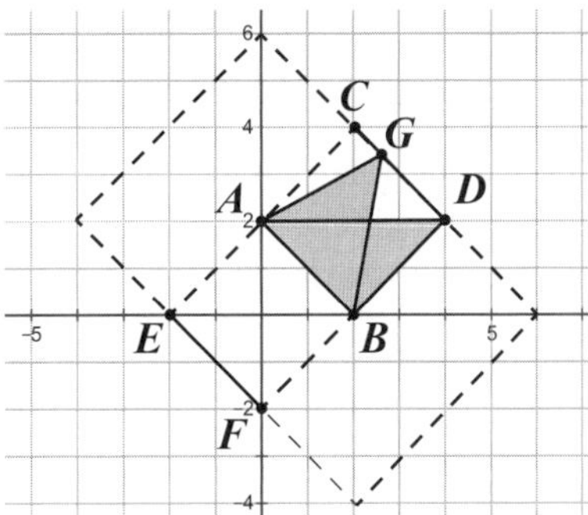

Figure 5.8. More than one equilateral triangle

ability to construct it using only an unmarked straightedge and compass. At the end of Chapter 3, we noted that just over a third of the 29 neutral propositions from Book I are constructions. For example, Proposition I.9 bisects an angle, Proposition 1.10 constructs the midpoint of a segment, Proposition I.11 constructs a perpendicular to a line through a point on the line, and Proposition I.12 constructs a perpendicular to a line through a point not on the line. Analytic geometry tells us that the angle bisector, midpoint and perpendicular lines from these propositions certainly exist in Taxicab geometry, but the question is whether or not we can construct any of them using only our tools of an unmarked straightedge and compass.

We leave it as an exercise for the reader to determine which of the remaining propositions hold in Taxicab geometry. What should strike the reader as truly amazing is how many of these propositions ultimately rely on SAS, and thus, fail to hold. Much of Euclidean geometry relies on this congruence scheme, which we will soon come to learn must be elevated (or demoted?) to axiom status.

Exercises 5.4

1. For Propositions I.3, I.5, I.6 and I.7 consider the following:

   - Does the proposition hold in Taxicab geometry?
   - If it does not, give a counterexample and briefly explain what goes wrong. Use grid paper to illustrate your counterexample.
   - If it does hold, does Euclid's proof work? If it does not, briefly explain what goes wrong. Note: You do not need to provide a valid proof.

2. In Proposition I.9, we bisect a given angle. Use grid paper to illustrate two examples in Taxicab geometry where the construction is followed: one where the angle is bisected, and one where it is not.

3. In Proposition I.10, we bisect a given segment. While Euclid's construction for I.10 uses Proposition 1.9, the following alternative construction also works in Euclidean geometry while avoiding the use of I.9:

   Alternative I.10 construction steps: Given segment AB, construct circles with centers A and B, each of radius AB. Let the two intersection points of these circles be C and D. Join CD. Let the intersection of AB and CD be E. Then E is the midpoint of AB.

Use grid paper to illustrate two examples in Taxicab geometry where this construction is followed: one where the segment is bisected, and one where it is not.

4. In Proposition I.11, we construct a perpendicular to a line through a point on the line. Use grid paper to illustrate two examples in Taxicab geometry where the construction is followed: one where the line is perpendicular, and one where it is not.

5. In Proposition I.12, we construct a perpendicular to a line through a point not on the line. While Euclid's construction for I.12 uses Proposition I.10, the following alternative construction also works in Euclidean geometry while avoiding the use of I.10:

   Alternative I.12 construction steps: Consider line AB with point C not on it. Let D be a random point on the side of AB opposite from C. Construct a circle with center C and radius CD. Let E and F be the intersections of this circle with AB. Construct circles with centers E and F, each of radius EF. Let G be the intersection of these two circles on the side of AB opposite from C. Then CG is perpendicular to AB.

   Use grid paper to illustrate two examples in Taxicab geometry where this construction is followed: one where the line is perpendicular, and one where it is not.

6. Follow the directions in Exercise 1 for Propositions I.13 through I.17.

7. Follow the directions in Exercise 1 for Propositions I.18 through I.21.

8. In Proposition I.22, we construct a triangle given three segments (where the length of any two is greater than the third). This construction still holds. Use grid paper to illustrate two examples in Taxicab geometry where this construction is followed, one where the construction determines a unique triangle, and one where it does not.

9. In Proposition I.23, we copy a given angle. Use grid paper to illustrate an example in Taxicab geometry where the construction is followed, but the angle is not copied.

10. Follow the directions in Exercise 1 for Propositions I.24, I.25, I.27, and I.28.

11. In Proposition I.31, given a line *BC* and a point *A*, not on the line, we construct a line through *A* that is parallel to *BC*. Speculate as to why, in Taxicab geometry, it is not possible to construct this parallel line using only an unmarked straightedge and compass.

6

Hilbert and Gödel

Figure 6.1. Kurt Gödel on the left and David Hilbert on the right.

German mathematician David Hilbert (1862–1943) received his Ph.D. in mathematics from the University of Königsberg in 1885 under the direction of Ferdinand von Lindemann (1852–1939), just three years after Lindemann proved that π is transcendental.[1] Within fifteen years, Hilbert was indisputably among the top mathematicians of the day, and was invited to give a major address at the International Congress of Mathematicians in 1900. Hilbert looked to the future of mathematics and famously presented his list of the most important open problems, a challenge to the mathematicians of his day. The list was purposely wide-ranging in its scope of mathematical fields, containing a rich variety of deep problems demonstrating the vitality and breadth of mathematical research at the turn of the century. Hilbert's list, which included such famous problems as the Continuum Hypothesis, the Riemann Hypothesis and Goldbach's Conjecture, either predicted or directed much of the mathematical research in the twentieth century, with most of the problems seeing their solution during

---

[1] The geometric significance of transcendental numbers is explained in Chapter 16.

105

that time. There have been other lists of unsolved problems that followed, but none as inspirational as Hilbert's. Most recently, in 2000, the Clay Institute named seven Millennium Prize Problems each worth one million dollars. One problem, the Poincaré Conjecture, has been solved by Russian mathematician Grigori Perelman.   Another problem, the Riemann Hypothesis, appeared on Hilbert's list and is conjectured to appear on lists of open problems for many centuries to come. Hilbert himself declared: "If I were to awaken after having slept for a thousand years, my first question would be: Has the Riemann hypothesis been proven?" [**26**]

Hilbert wanted to axiomatize all of mathematics, that is, to formalize the various branches of mathematics by developing the "best" axiomatic systems. Indeed, among the open problems on Hilbert's list, one concerned the formalization of mathematics, and another, physics. Published in 1899, his *Foundations of Geometry*, was his first attempt to do this for Euclidean geometry. Of course, determining the meaning of "best" requires a way to analyze and classify an axiomatic system. Hilbert coined the term **metamathematics** to describe the study of the properties of formal axiomatic systems, an apt name for mathematics about mathematics. Three fundamental properties of axiomatic systems described in this chapter are consistence, independence and completeness. These are the three properties that Hilbert wanted his axiomatic systems to possess. In particular, he claimed that the axiomatic system for Euclidean plane geometry that he gave at the turn of the century had all of the desirable properties. After discussing these properties and Hilbert's axioms, we will discover how metamathematics informs our understanding of the similarities and differences of Euclidean, Spherical and Taxicab geometries. Finally, we will learn how a young mathematician in Austria, Kurt Gödel (1906–1978), would exceed Hilbert's turn-of-the-century challenge.

6.1 Axiomatic systems

Before we can have axioms, it is clear from our initial findings in Chapter 2 that any axiomatic system must also have some *undefined* or **primitive** terms. In Euclid's case, the primitive terms are *point, line and plane*. Additionally, we need to have a term, say *lies on*, to describe how a point is related to a line. We will also say that a point is **incident** with a line to mean that the point *lies on* the line. The problem with undefined terms is that each reader may choose their own interpretation of these terms, though quite rightly, it seems unlikely that anyone will have an unusual interpretation of the terms point and line. Nevertheless, we create an interpretation of the axiomatic system when we imbue these primitive terms with meaning. Depending on the meaning, some axioms may make sense and some may not. In Spherical geometry, we had to make restrictions so that spherical triangles made sense. A **model** of an axiomatic system is an interpretation of the primitive terms in which all of the axioms "make sense," or are valid.

We have considered three geometries: Euclidean, Spherical and Taxicab. We have provided a set of axioms for Euclidean geometry but we have not specified a set of axioms for either Spherical or Taxicab. What we *have* given for both Spherical and Taxicab geometry is a model of each geometry, that is, an interpretation of the basic terms which provided a way for us to understand whether or not Euclid's axioms still held. The importance of a model cannot be understated. A model not only provides a way for us to envision an axiomatic system, but it can give us a much better understanding of the system. Since it is far beyond the scope of this book, we state the following metamathematical theorem relating models and theorems without a formal proof.

Theorem 6.1. *All theorems resulting from an axiomatic system are true in any model of the system.*

To help see this, we note that in an axiomatic system the proof of any theorem does not depend on any particular interpretation of the basic terms, but rather, its proof relies only on the relationships prescribed by the axioms (or previously proven theorems which, ultimately, rely on the axioms). For example, a proof of a proposition from Book I should rely only on the axioms or previously proven propositions. A diagram, which is a physical manifestation of our interpretation of the undefined terms, is not necessary for the proof. Thus, the proposition should hold in any model of the system since the axioms are valid in any model. So, the proofs and hence the theorems will also hold. On a humorous note, Hilbert famously said that we could replace "point, line and plane" with "chair, table and beer mug," and as long as they satisfy the axioms, then the theorems of geometry would sound silly but remain valid [**118**]. For readers interested in a more detailed discussion of these topics, we suggest Chapters 2 and 3 of *The Foundations of Mathematics*, by Raymond Wilder.

Let's describe the fundamental properties of axiomatic systems. Here, you may take the word "statement" to mean a declarative sentence in the language of our system. For example, Proposition I.16 is a statement in the system of Euclidean geometry.

The most important property of an axiomatic system is that the axioms do not contradict each other, and that they do not produce any contradictory statements. Thus, it should be impossible to prove both a statement and the negation of the statement within the system. This is called *consistency* and is defined as follows.

Definition 6.2. An axiomatic system is **consistent** if and only if no contradictions can be proven from the axioms.

While we have noted that Euclid made some omissions in his axioms, his axioms do not pose any contradictions to themselves. We can construct a simple example of an inconsistent system adapted from his axioms as follows.

Example 6.3. Inconsistent system
    undefined terms: point, line, through
    Axiom 1: Through any two points there is exactly one line.
    Axiom 2: Through any two points there are two distinct lines.

Like oxygen to human survival, consistency in any mathematical system is essential. Proving that a system is consistent is quite difficult, and we will reserve further discussion of this topic until Section 6.5. For now, we will assume that all systems under consideration are consistent.

The property of independence has to do with keeping the number of axioms to a minimum. When a set of axioms is independent, no axiom can be proven from the other axioms. This is a preferred property of an axiomatic system and is defined as follows.

Definition 6.4. An axiom is **independent** of the other axioms in the system if it cannot be proved from other axioms. In an axiomatic system, if each axiom is independent of the other axioms, the set of axioms is said to be *independent*.

A system that is not independent is not unsound or fallacious, it merely has some redundancy. Here's a small example to illustrate this point.

Example 6.5. Nonindependent system
    undefined terms: point, line, through
    Axiom 1: Through any two points there is at least one line.
    Axiom 2: Through any two points there is no more than one line.
    Axiom 3: Through any two points there is exactly one line.

Clearly, Axiom 3 is not independent of Axioms 1 and 2. If we prefer an independent set of axioms that contains the same information, then we have two choices: {Axiom 1, Axiom 2} or {Axiom 3}. Certainly, any statement that can be proved from the other axioms can be rightly placed with the theorems, and thus, does not need to be assumed. So, independence has to do with assuming no more than is necessary.

Where independence is concerned with streamlining the assumptions, completeness is concerned with the system being large enough to allow for the determination of the truth value of *any* statement that can be made in the language of the system.

Definition 6.6. An axiomatic system is **complete** if and only if every statement based on the undefined terms can either be proved or disproved from the axioms.

You might be wondering how we could possibly check *all* possible statements based on the undefined terms. It should come as no surprise that it's easier to demonstrate that a system is incomplete. Suppose A, B, C and D are all statements based on some primitive terms that are true in some model. So, they could all be called axioms. If Axiom D is independent of axiom set {A,B,C}, then Axiom D cannot be proved from the other axioms and, therefore, axiom set {A,B,C} is not complete. In fact, a set of axioms is complete if it is not possible to add an independent axiom to the set.

While it can be extremely difficult to determine whether or not a given system has these properties, models can be useful tools. As a direct result of Theorem 6.1, a model of an axiomatic system gives us a way to prove the independence of a statement. For example, all of Euclid's axioms (5 postulates and 5 common notions) hold in both our analytic model of Euclidean geometry and in Taxicab geometry. So, we have two models for the same set of ten axioms. By Theorem 6.1, all theorems resulting from these ten axioms must be true in both models. However, SAS holds in the first model, but not in the Taxicab model. This can only mean that the truth of SAS cannot be determined by these ten axioms. Therefore, **SAS is independent of Euclid's axioms.**

Models can also be used to demonstrate the incompleteness of an axiomatic system. If a statement based on the undefined terms is independent of the set of axioms, then the axiomatic system is incomplete. Since SAS is independent of Euclid's ten axioms, Euclid's axioms are incomplete and we will have to amend our set of axioms. As we observed in earlier chapters, this is one of several omissions in Euclid's assumptions that must be addressed. Let's take a look at a very small example to illuminate this discussion, or to make you hungry for pizza.

Example 6.7. Incomplete system
    undefined terms: pizza, topping, has
    Axiom 1: Every pizza has at least two toppings.
    Axiom 2: Every topping has at least two pizzas.
    Axiom 3: There is at least one pizza.

Each box below demonstrates a model for this system, where each large circle represents a pizza and each color, a type of topping.

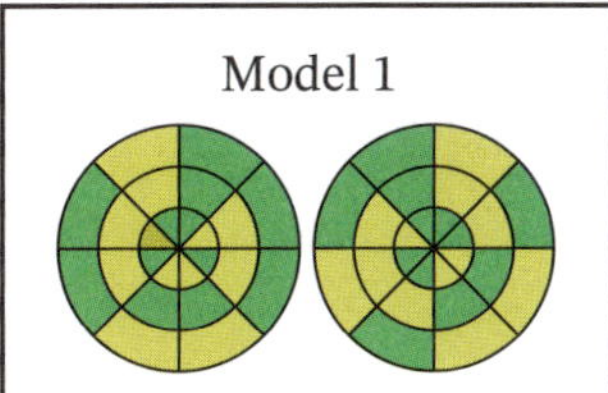

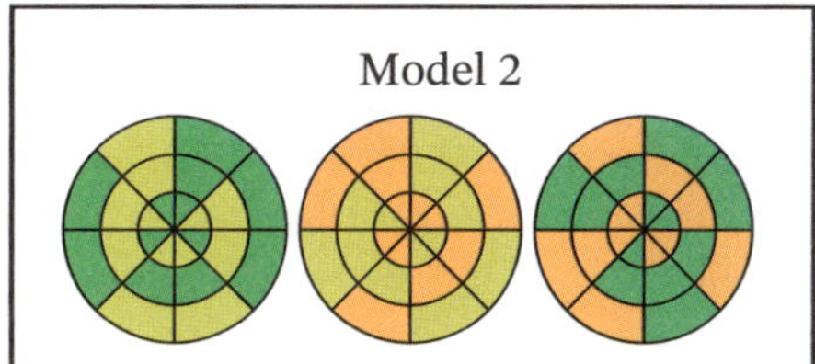

Check to make sure that both models satisfy the given set of axioms, then take a few minutes to produce another model for this system. Consider the following statement based on the undefined terms: *There are exactly three pizzas.* This statement is true for Model 2, and false for Model 1. Thus, the statement cannot be proven using only the axioms, and the system is incomplete.

We can also use Theorem 6.1 to discuss the resolution of one of the longest controversies in the history of mathematics. For nearly two thousand years, many mathematicians believed that Euclid's axiomatic system was not independent. Specifically, mathematicians were troubled with the complexity of the fifth postulate, the Parallel Postulate, and thought that it could be proven from the other four. Books were devoted to convincing a reader that it *had* to be provable, but in the 1800s three mathematicians discovered a new geometry that would finally put the problem to rest. The mathematicians were Carl Friedrich Gauss (1777–1855), János Bolyai (1802–1860) and Nikolai Lobachevsky (1792–1856), and their discovery was Hyperbolic geometry. Like Spherical and Taxicab geometries, Hyperbolic geometry is a non-Euclidean geometry. (We discuss this geometry in detail in Chapters 12 and 13.) A model was created for this new geometry in which Neutral geometry clearly held, but the fifth postulate did not. So, there were two models for Neutral geometry, one where the Parallel Postulate was valid and a new one where it was not. Thus, it was proven that the fifth postulate is independent of the others.

Now that we have an understanding of the essential and preferred properties of axiomatic systems, let's take a moment to gain a bit of historical perspective to see why these concerns became so important by the end of the nineteenth century. We have not followed the progression of these deep, abstract categorizations of important properties of axiomatic systems in chronological order. Terms such as primitive, model, complete and independent were only introduced to the mathematical vernacular within the past two centuries. There was no cause for concern about such things until the first non-Euclidean geometry was introduced in the nineteenth century. Gauss, Bolyai and Lobachevsky's geometry inspired a flurry of activity on the foundations of geometry in Germany and Italy. To set Euclidean geometry on its proper logical foundation, mathematicians set out to expose all hidden Euclidean assumptions and to develop a new complete and independent axiomatic system. Though we have focused on Hilbert's contribution to the field, there are many who made significant contributions and deserve credit. In 1882, German Moritz Pasch (1843–1930) presented his version of a complete and independent set of Euclidean axioms. At the University of Turin, Italians Giuseppe Peano (1858–1932) and Mario Pieri (1860–1913) also offered postulational developments of geometry in 1889 and 1899, respectively. David Hilbert's lectures at the University of Göttingen in 1898–1899 were published later in 1899 as *Grundlagen der Geometrie* (Foundations of Geometry). In this book, Hilbert presents a set of twenty axioms as the foundation for Euclidean geometry. Though it was Hilbert's axiomatic presentation that was to become widely accepted, all systems developed during this period shared common elements. The terms *point* and *line* were accepted as primitive or undefined terms and the distinction between *axioms* and *postulates* was abandoned, allowing these terms to be used interchangeably. Before we take a look at Hilbert's amendments to Euclid's axioms, let's explore the abstract properties of axiomatic systems by looking at a very small geometry.

6.2 A Four Point geometry

To better understand the concepts of independence and completeness, we investigate a very small finite geometry with a simple structure. In this set of axioms, our undefined terms are *point*, *line* and *on*. The term *on* will be used symmetrically to describe a relationship between points and lines in that we may say 'a point is on a line' or 'a line is on a point.'

Four Point geometry

Undefined terms: *point*, *line* and *on*.

Axiom 4P-1. There exist exactly four points.

Axiom 4P-2. Any two distinct points are on exactly one line.

Axiom 4P-3. Each line is on exactly two points.

To produce a visual representation of a line on points A and B in this geometry, we may take the usual interpretation of points and lines as $A \bullet\!\!-\!\!\!-\!\!\!-\!\!\bullet B$. Notice that there is no need to extend the line since it consists of exactly two points by Axiom 4P-3. Most importantly, though we connect points A and B with a drawn segment, we are merely indicating that, as a pair, they are on a line. This geometry allows for no points on the drawn segment between A and B. Alternatively, if we call this line ℓ and use set notation, then we would say that ℓ contains exactly two points and is represented by $\ell = \{A, B\}$. For this reason, we also say that a line *contains* a point. To add even more variety to our language, we may also write that a line *goes through* a point, a point *lies on* a line, or a point *is an element of* a line.

Distance has no meaning here, which is devastating for one of our main characters, the *circle*. Gone are our beloved angles and much of what we traditionally think of as geometry. There will be no compass and straightedge, no congruence or similarity, and no lengths or areas. This leaves us to wonder: what exactly remains? To help answer this question, we will begin by constructing a few models for this set of axioms. We will start with one that uses sets. By Axiom 4P-1, suppose A, B, C and D are the four points of our geometry. According to Axioms 4P-2 and 4P-3, we must consider all possible pairs of two distinct points as lines. Writing the lines as sets gives $\{A, B\}$, $\{A, C\}$, $\{A, D\}$, $\{B, C\}$, $\{B, D\}$ and $\{C, D\}$. We may also keep our standard geometric notation here and write them as AB, AC, AD, BC, BD and CD. Alternatively, since this geometry is so small, we could give a diagram as a model. This can take many forms, a few of which are shown in Figure 6.2.

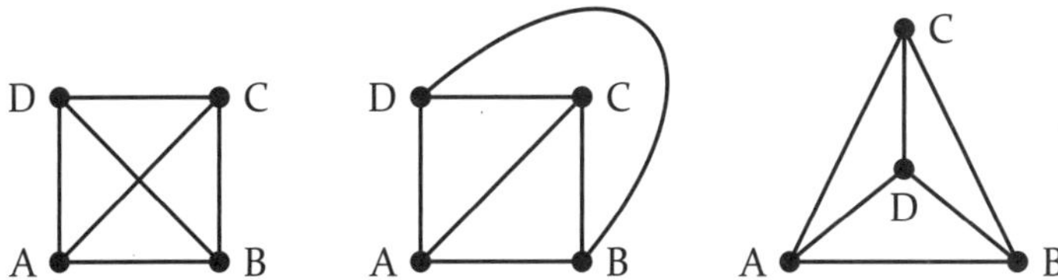

Figure 6.2. Diagrams to model Four Point geometry

It is important to notice that in the first diagram there is no point of intersection where lines $\{A, C\}$ and $\{B, D\}$ meet. Also, the second diagram illustrates our freedom to break the hegemony of the "straightness" we feel so compelled to impose on our lines. Although to be fair, the reader should have relaxed their inherent prejudices regarding such notions after tackling Spherical geometry. Lastly, notice that even though the diagrams (or models) look different, they all contain exactly the same information. The same cannot be said of the pizza models given for Example 6.7.

Now that we have a model for our axioms, let's show that the set of axioms for Four Point geometry is independent, meaning that our set of axioms is minimal in that no single axiom can be derived from the other two. If we give an example of a model where Axioms 4P-1 and 4P-2 hold, but Axiom 4P-3 does not, then this shows that Axiom 4P-3 cannot be proved using the other axioms, and is, thus, independent of Axioms 4P-1 and 4P-2. If we can show that each axiom is independent of the others, then the system is independent. We give three different models in Figure 6.3 where one of the axioms does not hold but the other two axioms are valid. In the first model there is one line, in the second there are two lines, and the third model has four lines, where one of the lines is $\{A, B, C\}$. Verifying the independence of the axioms using these models is left as an exercise for the reader.

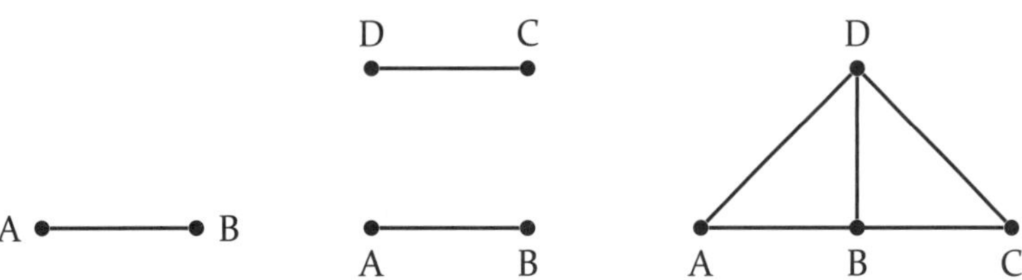

Figure 6.3. Models to verify independence of Four Point geometry axioms

Finally, we recall that an axiomatic system is complete if it is impossible to add a new independent axiom based on the undefined terms. The set of axioms for Four Point geometry is complete, but directly proving the completeness of any axiomatic system, even a tiny one, is often quite difficult. A far easier approach is to show that a set of axioms is *categorical*. An axiom system is **categorical** when it describes a unique mathematical object. This means that all models of the system are essentially the same, or, they all have the same structure. More technically, this means that all models are equivalent up to isomorphism.[2] As an example, all of the diagrams in Figure 6.2 are isomorphic since, as previously noted, they all contain exactly the same information. The pizza models given for Example 6.7 are not isomorphic. A deep result of metamathematics relates completeness and categoricalness as follows: A set of axioms that is categorical is necessarily complete. (For a more detailed discussion, see [44] or [65].) To show that the set of axioms for Four Point geometry is categorical, we need to show that that any bijection between two models of the Four Point geometry is an isomorphism. To do this we will need to understand our geometry a little bit better. In particular, we need to prove that the six lines we've drawn in each of our models are the only possible lines that can exist.

Theorem 6.8. *There are exactly six lines in Four Point geometry.*

Proof. By Axiom 4P-1 we have exactly four points, say A, B, C and D, and thus, $\binom{4}{2} = \frac{4!}{2!2!} = 6$ pairs of distinct points. Thus, by Axiom 4P-2 there must be at least 6 lines, namely AB, AC, AD, BC, BD and CD. How do we know that these lines are distinct? If any two of these were the same, then that line would be on at least three distinct points, which contradicts Axiom 4P-3. How do we know there cannot be another line? In order to reach a contradiction, let's assume that there is a seventh line, ℓ. By Axiom 4P-3, there are exactly two points on ℓ, and by Axiom 4P-2 they cannot both come from the set $\{A, B, C, D\}$. Thus, there must be a fifth point, contradicting Axiom 4P-1. Thus, there are exactly six lines.   □

This means that every model of Four Point geometry has four points and six lines, with two points on every line. What exactly would an isomorphism between two models of this geometry look like? In general, a bijection between two sets is a mapping from one set to the other that is both one-to-one and onto. Informally, we can think of this as matching each element in one set with an element in the other set in such a way that every element in each set has exactly one mate. In addition to being a bijection, our mapping must also preserve relations. While for many geometries this concept can be quite complicated, the only relation of Four Point geometry is given by the term "on." So, for this geometry, preserving the relation simply means that if a line joins two points in one model then the image of the line under the mapping must join the images of the same two points in the other model. Hence, to produce a bijection for this geometry, we will have to match up the four points in one model with the four points in the other model. There are 24 different ways to do this (Why?), and fortunately, all of them will preserve relations (Why?). Thus, our axiomatic system is categorical, and hence, complete.

Exercises 6.2

1. Use Figure 6.3 to demonstrate the independence of the axioms for Four Point geometry.

   (a) Use the first model to explain why Axiom 4P-1 is independent.
   (b) Use the second model to explain why Axiom 4P-2 is independent.

(c) Use the third model to explain why Axiom 4P-3 is independent.

2. Consider these axioms for Trois Point geometry.

Axiom Trois-1. There exist exactly three points.

Axiom Trois-2. Any two distinct points are on exactly one line.

Axiom Trois-3. Each line is on exactly two points.

(a) Construct a diagram to represent this geometry.

(b) Use a model to verify that Axiom Trois-1 is independent.

(c) Use a model to verify that Axiom Trois-2 is independent.

(d) Use a model to verify that Axiom Trois-3 is independent.

3. Consider these axioms for Three Point geometry.

Axiom 3P-1. There exist exactly three points.

Axiom 3P-2. Any two distinct points are on exactly one line.

Axiom 3P-3. Two distinct lines are on at least one point.

Axiom 3P-4. Not all points of the geometry are on the same line.

(a) Rewrite the axioms replacing line with *pizza*, and point with *topping*.

(b) Construct a diagram to represent this geometry. (We suggest using points and lines instead of toppings and pizzas.)

(c) Use a model to verify that Axiom 3P-1 is independent.

(d) Use a model to verify that Axiom 3P-2 is independent.

(e) Use a model to verify that Axiom 3P-3 is independent.

(f) Use a model to verify that Axiom 3P-4 is independent.

6.3 Hilbert's axioms for Euclidean plane geometry

Let's amend Euclid's axioms to account for the omissions that we have previously identified and some others that we have not. While there are many possible sets of amended axioms from which to choose, we will use the definitive set of axioms for Euclidean plane geometry, Hilbert's axioms. While Euclid uses both lines and circles in his set of postulates, Hilbert uses only lines. As Hilbert claims in the introduction to *Foundations of Geometry*, his axioms are both independent and complete.

> The following investigation is a new attempt to choose for geometry a simple and complete set of independent axioms and to deduce from these the most important geometrical theorems in such a manner as to bring out as clearly as possible the significance of the different groups of axioms and the scope of the conclusions to be derived from the individual axioms. [**74**]

The axioms presented here are adapted from Hilbert's axioms for Euclidean plane geometry. These are reprinted with permission of Open Court Publishing Company ©1971, a division of Carus Publishing Company, Chicago, IL, and can be found in Appendix E.

Undefined Terms. *point, line, plane, lie (or lie on), between, congruent*

If a point A lies on a line ℓ, then for variety of language we will also write "line ℓ goes through A," "on line ℓ is point A," "A is incident with line ℓ," "ℓ is incident with line A," "ℓ meets A," or "ℓ contains A." Here, "on," "through," "incident," "meets" and "contains" are synonymous ways to describe the relationship between a point and a line.

I. **Axioms of Incidence (Connection)**

    I.1 Through any two distinct points A and B there exists a line ℓ.

    I.2 Through any two distinct points A and B there exists no more than one line ℓ.

    I.3 On every line there exist at least two distinct points. There exist at least three points that do not lie on the same line.

    I.4 Through any three points A, B and C not on the same line there exists a plane.

We modify Hilbert's definition of *segment* to reflect the standard interpretation of this term.

Definition 6.9. Given two distinct points A and B on a line ℓ, the **segment** AB is the set of points consisting of A, B and all the points lying between them. The points A and B are the **endpoints** of the line segment.

II. **Axioms of Order (Betweenness)**

    II.1 If point B lies between points A and C, then A, B and C are distinct points of a line, and B lies between C and A.

    II.2 For two distinct points A and C, there is at least one point B on the line $\overleftrightarrow{AC}$ such that C lies between A and B.

    II.3 For three distinct points A, B and C on a line, there is one and only one point which lies between the other two.[3]

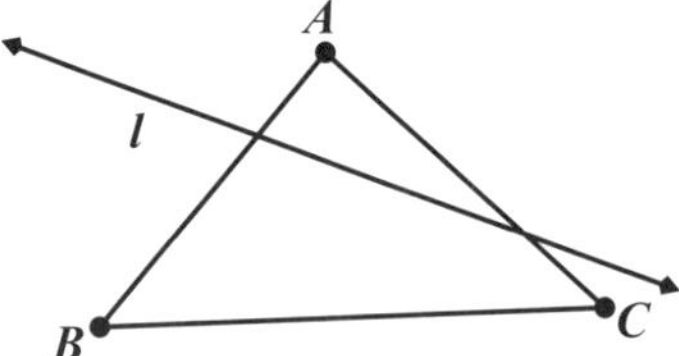

Figure 6.4. II.4 Pasch's Axiom

    II.4 (Pasch's Axiom) Let A, B and C be three distinct points that do not lie on a line, and let ℓ be a line in the plane that does not meet any of the points A, B or C. If line ℓ passes through a point of segment AB, it also passes through a point of segment AC or a point of segment BC.

---

[3]This is the version from Open Court's 1938 translation of Hilbert's axioms. The 1971 translation states: *Of any three points on a line there exists no more than one point that lies between the other two.* We note that the slightly stronger version we have chosen can be proven from this version.

We will use the standard notation for congruence, that is, we write $X \cong Y$ when X and Y are congruent geometric objects. We also give updated definitions of *ray* and *angle*.

Definition 6.10. Given two distinct points A and B on a line, the **ray** $\overrightarrow{AB}$ is the set of points consisting of the line segment AB and all points C such that B lies between A and C.

Hilbert defines an angle as a pair of rays on distinct lines emanating from the same point.

Definition 6.11. Let h and k be any two distinct rays emanating from point O and lying on distinct lines. The pair of rays h, k is called an **angle**, denoted by $\angle(h, k)$ or $\angle(k, h)$, and O is the vertex of the angle.

To match standard notation, the angle described by rays $\overrightarrow{AB}$ and $\overrightarrow{AC}$ lying on distinct lines will also be written as $\angle BAC$ or $\angle CAB$.

III. **Axioms of Congruence**

   III.1 If A and B are two distinct points on line ℓ, and if A' is a point on the same or on another line ℓ', then it is always possible to find a point B' on a given side of the line ℓ' such that $AB \cong A'B'$.

   III.2 For segments AB, $A'B'$ and $A''B''$, if $A'B' \cong AB$ and $A''B'' \cong AB$, then $A'B' \cong A''B''$, or briefly, if two segments are congruent to a third one they are congruent to each other.

   III.3 On a line ℓ, let AB and BC be two segments which except for B have no point in common. Furthermore, on the same or another line ℓ', let $A'B'$ and $B'C'$ be two segments which except for B' also have no point in common. If $AB \cong A'B'$ and $BC \cong B'C'$, then $AC \cong A'C'$.

   III.4 Given rays $\overrightarrow{AB}$ and $\overrightarrow{AC}$ which lie on distinct lines, and ray $\overrightarrow{A'B'}$, there is exactly one ray $\overrightarrow{A'C'}$ on each side of line $\overleftrightarrow{A'B'}$ such that $\angle B'A'C' \cong \angle BAC$. Every angle is congruent to itself.

   III.5 If for two triangles $\triangle ABC$ and $\triangle A'B'C'$ we have $AB \cong A'B'$, $AC \cong A'C'$, $\angle BAC \cong \angle B'A'C'$, then we also have $\angle ABC \cong \angle A'B'C'$.

IV. **Axiom of Parallels**

   IV.1 Let ℓ be a line and A a point not on line ℓ. Then there is at most one line in the plane, determined by ℓ and A, that passes through A and does not intersect ℓ.

V. **Axioms of Continuity**

   V.1 (Archimedes' Axiom) If AB and CD are any segments, then there exists a number n such that n segments CD constructed contiguously along ray $\overrightarrow{AB}$ starting at A, will pass beyond B.

   V.2 (Axiom of Line Completeness) An extension of a set of points on a line with its order and congruence relations that would preserve the relations existing among the original elements as well as the fundamental properties of line order and congruence that follows from Axioms I–III and from V.1 is impossible.

Note: We have omitted the last four Axioms of Incidence as they concern planes in solid geometry. Thus, while we have listed sixteen of Hilbert's axioms for Euclidean geometry, there are a total of twenty.

These axioms share some similarities with the Euclidean axioms, and they introduce new concepts of betweenness and continuity. The first two Axioms of Incidence include Euclid's first postulate and the uniqueness of the line in this postulate. The others concern the existence of noncollinear points. Based on the work of Moritz Pasch, the Axioms of Betweenness detail the arrangement of points on a line and include an axiom named after him. (Pasch codified his axioms while working on a different type of geometry, Projective geometry. We discuss Projective geometry in Chapter 14.)

The congruence axioms concern geometric equivalence of segments and angles. The first Axiom of Congruence is the equivalent of choosing a rigid compass. This is the axiom that will allow us to copy segments, and thus, replaces Euclid's Proposition I.3. Hilbert's second and third congruence axioms should remind the reader of Euclid's first two common notions. Hilbert's fourth Axiom of Congruence allows us to copy angles, and thus, replaces Euclid's Proposition I.23. Note that none of the familiar Euclidean triangle congruence schemes is included, even though we just concluded in the previous section that Euclid's axioms are incomplete without SAS. Hilbert's fifth Axiom of Congruence is nearly SAS, but while it gives congruence of the the remaining angles it does not give congruence of the final side. Ideally, a set of axioms will employ an economy of assumptions, and while SAS is not here, as we will see, it can be proven from this set of axioms.

The axiom concerning parallels is known as Playfair's Axiom, named for Scottish mathematician John Playfair (1748–1819) who worked on the Parallel Postulate problem and published his own edition of the first six books of Euclid's geometry. In the following chapter, we will show that, assuming Neutral geometry, Playfair's Axiom is logically equivalent to Euclid's Parallel Postulate.[4] The continuity axioms will allow us to conclude that our constructed lines and circles are unbroken, meaning they have no gaps or holes. This will allow us, for example, to justify the intersection of the two circles in the construction of an equilateral triangle in Proposition I.1.

We mentioned that, unlike Euclid's postulates, one of our main characters has been written out of this part of the script. Hilbert's axioms allow him to define a circle in terms of congruence rather than postulating one as Euclid did.

Hilbert's Definition of Circle. For an arbitrary point M, the totality of all points A, for which the segments MA are congruent to one another, is called a *circle*. M is called the *centre of the circle*.

Let's take a look at a few new consequences of Hilbert's axioms. Using Pasch's Axiom, Hilbert proves the following theorems of *plane separation* [**74**].

Hilbert's Theorem 8 [Plane Separation by Lines]. Every line ℓ that lies in a plane separates the remaining points of this plane into two regions having the following properties: Every point A of the one region determines with each point B of the other region a segment AB containing a point of the line ℓ. On the other hand, any two points A, A' of the same region determine a segment AA' containing no point of ℓ.

---

[4]With respect to Hilbert's axioms, Neutral geometry refers to the geometry that follows from all axioms except the Axiom of Parallels (IV.1).

We will call these two regions the two **sides** of our line ℓ. Given a point A on ℓ and another point B, not on ℓ, we can consider the ray $\overrightarrow{AB}$. Theorem 8 implies that all points, other than A, that lie on the ray $\overrightarrow{AB}$, lie on one side of ℓ. We leave the proof to the reader.

Hilbert's Theorem 9 [Plane Separation by Polygons]. Every polygon separates the points of the plane that are not on the polygon into two regions, the *interior* and the *exterior*, with the following property: If A is a point of the interior and B is a point of the exterior, then every polygonal segment that joins A with B has at least one point in common with the polygon. On the other hand if A, A' are two points of the interior and B, B' are two points of the exterior then there exist polygonal segments which join A and A' and others which join B and B', none of which have any point in common with the polygon. There exist lines that always lie entirely in the exterior of the polygon. However, there are no lines that lie entirely in the interior of the polygon.[5]

On a related note, Hilbert defines the set of **points interior to an acute angle** $\angle BAC$ as the collection of points lying at the intersection of the set of points on the angle-facing side of line AB with the set of points on the angle-facing side of line AC. We can use this notion in conjunction with Pasch's Axiom to justify an unstated assumption of Euclid that is found in the proof of Proposition I.10. Known as the Crossbar Theorem, it is the missing piece that explains why, in any triangle, an angle bisector must intersect the side lying opposite the angle it bisects.

Theorem 6.12 [Crossbar Theorem]. *If D is in the interior of angle $\angle BAC$, then ray $\overrightarrow{AD}$ intersects segment BC.*

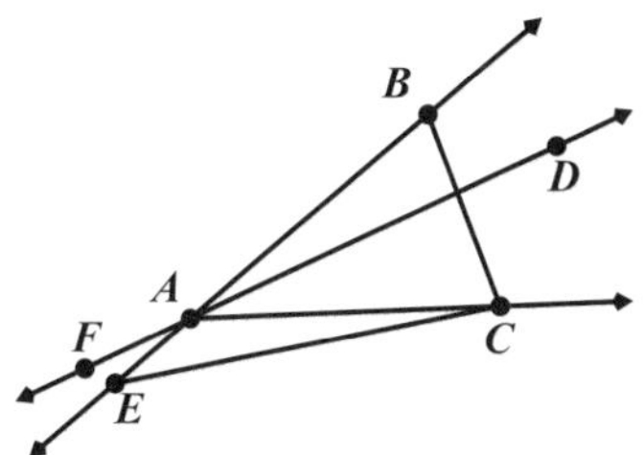

Figure 6.5. Proof of the Crossbar Theorem

Proof. Consider a point D that lies in the interior of angle $\angle BAC$. Construct E on the line $\overleftrightarrow{AB}$ such that A lies between E and B, as shown in Figure 6.5. Consider triangle $\triangle BEC$. Since $\overleftrightarrow{AD}$ intersects BE at A, then $\overleftrightarrow{AD}$ must cross either BC or CE by Pasch's Axiom. By construction, B and E are on opposite sides of $\overleftrightarrow{AC}$. Since B and D are on the same side of $\overleftrightarrow{AC}$, D and E are also on opposite sides of $\overleftrightarrow{AC}$. Therefore, by Hilbert's Theorem 8, all points on the ray $\overrightarrow{AD}$ are on the opposite side of $\overleftrightarrow{AC}$ as all points on segment CE. Thus, $\overrightarrow{AD}$ and CE do not intersect. Construct F on $\overleftrightarrow{AD}$ such that A is between F and D. We leave it to the reader to use a similar argument to show that

---

[5]By "polygonal segment that joins A with B," Hilbert means a path from A to B consisting of line segments, that is, a piecewise linear path.

$\overrightarrow{AF}$ cannot intersect segment CE. Therefore, $\overleftrightarrow{AD}$ must intersect BC. We leave it to the reader to show that ray $\overrightarrow{AF}$ cannot intersect BC, and therefore, ray $\overrightarrow{AD}$ must intersect BC, as desired. Note that $\overrightarrow{AD}$ cannot intersect BC at B, else two lines would intersect at two distinct points. $\qquad\square$

Recall that in Euclid's proof of Proposition I.1, two circles of radius $\overline{AB}$ were constructed at each endpoint of segment AB. Euclid assumed that these circles had a point of intersection but his axioms did not support this claim. In order to fix the gap in his reasoning, we need the *Circular Continuity Principle*. Though beyond the scope of this book, this, too, follows from Hilbert's axioms.

Circular Continuity Principle

(1) A line segment with one endpoint outside a given circle and the other endpoint inside the circle will intersect the circle exactly once.

(2) A circle passing through a point inside a given circle and a point outside that circle will intersect the given circle twice.

Given this principle, we can prove the following theorem:

Theorem 6.13. *Given a circle with center O and radius r, any ray $\overrightarrow{OA}$ emanating from O must intersect the circle.*

Proof.

Case 1. $OA = r$. In this case, A lies on the circle and is itself the point of intersection.

Case 2. $OA > r$. Then A lies outside of the circle and the Circular Continuity Principle implies that OA, and thus $\overrightarrow{OA}$, must intersect the circle.

Case 3. $OA < r$. By Archimedes' Axiom (Hilbert Axiom V.1), there is a number n and a point A' on the ray $\overrightarrow{OA}$ such that $OA' = n \cdot OA > r$. Thus, A' lies outside of the circle, so OA', and thus $\overrightarrow{OA}$, will intersect the circle. $\qquad\square$

We are now ready to fill in the gap in the proof of Proposition I.1, and show that the two circles do intersect.

Theorem 6.14. *The two circles in Proposition I.1 intersect.*

Proof. Note that A lies on the circle centered at B. By Axiom II.2, there exists a point E such that B is between A and E. By Theorem 6.13, ray $\overrightarrow{BE}$ must intersect the circle centered at B at a point F. Now, B lies between A and F, and by Hilbert's definition of a circle, $AB = BF$. Thus, $AF > AB$, so F lies outside the circle centered at A with radius AB, while A clearly lies inside this circle. Therefore, the Circular Continuity Principle guarantees that the two circles intersect at two points, C and D. $\qquad\square$

Earlier we noted that no triangle congruence scheme appears in Hilbert's axioms. While SAS is not taken as an axiom, Axiom III.5 is often referred to as SAS even though congruence of the triangles is never mentioned. This axiom only gives equivalence of

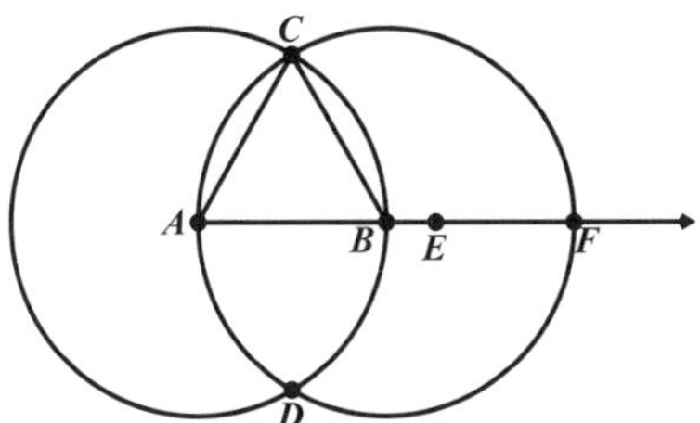

Figure 6.6. Proposition I.1

angles, stating that if for two triangles $\triangle ABC$ and $\triangle A'B'C'$ we have $AB = A'B'$, $AC = A'C'$, and $\angle BAC = \angle B'A'C'$, then the congruence $\angle ABC = \angle A'B'C'$ also holds. We can, however, prove Euclid's Proposition I.4 [SAS] using Hilbert's axioms, and thereby eliminate Euclid's method of superposition, i.e. "applying" one triangle to another.

Proposition I.4 [SAS]. *If for two triangles,* $\triangle ABC$ *and* $\triangle A'B'C'$, *we have* $AB = A'B'$, $AC = A'C'$, *and* $\angle BAC = \angle B'A'C'$, *then* $\angle ABC = \angle A'B'C'$, $\angle ACB = \angle A'C'B'$, *and* $BC = B'C'$ *also hold. Thus* $\triangle ABC \cong \triangle A'B'C'$.

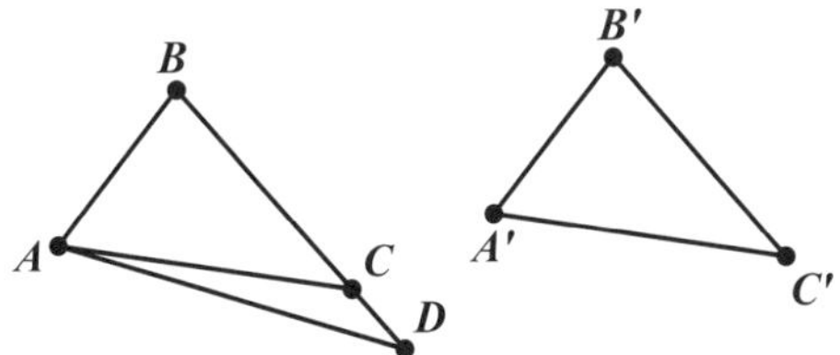

Figure 6.7. Proposition I.4 SAS

Proof. By Axiom III.5, we have $\angle ABC = \angle A'B'C'$. Consider triangles $\triangle ACB$ and $\triangle A'C'B'$. Then, again by Axiom III.5, we also have $\angle ACB = \angle A'C'B'$. We claim that $BC = B'C'$. Suppose for the sake of reaching a contradiction that this is not the case, that is, $BC \neq B'C'$. On ray $\overrightarrow{BC}$, construct a point D such that $BD = B'C'$ [Axiom III.1]. Join AD [Axiom I.1]. Consider triangles $\triangle BAD$ and $\triangle B'A'C'$. Since $BD = B'C'$, $BA = B'A'$ and $\angle ABD = \angle A'B'C'$, by Axiom III.5, we have $\angle BAD = \angle B'A'C'$. This implies $\angle BAD = \angle BAC$, which is impossible by Axiom III.4. Thus, we have $BC = B'C'$ as desired. $\square$

Since Euclid also uses superposition in his proof of Proposition I.8 [SSS], let's prove this congruence scheme using Hilbert's axioms. Note that Pappus' proof of Proposition I.5 is still valid, as it only requires Proposition I.4.

Proposition I.8 [SSS]. *If for two triangles,* $\triangle ABC$ *and* $\triangle DEF$, *we have* $AB = DE$, $AC = DF$, *and* $BC = EF$, *then* $\triangle ABC \cong \triangle DEF$.

Proof. By Axioms III.4 and III.1, there is a unique ray $\overrightarrow{AC'}$ such that $\angle C'AB = \angle FDE$ and $AC' = DF = AC$, where C' lies on the opposite side of $\overleftrightarrow{AB}$ as C. Since $DE = AB$, by I.4 (SAS), $\triangle DEF \cong \triangle ABC'$. Therefore, $BC' = EF = BC$. Because C' is on the opposite side of $\overleftrightarrow{AB}$ as C, the segment CC' must intersect $\overleftrightarrow{AB}$ at a point G. We consider three possible cases depending on where G lies.

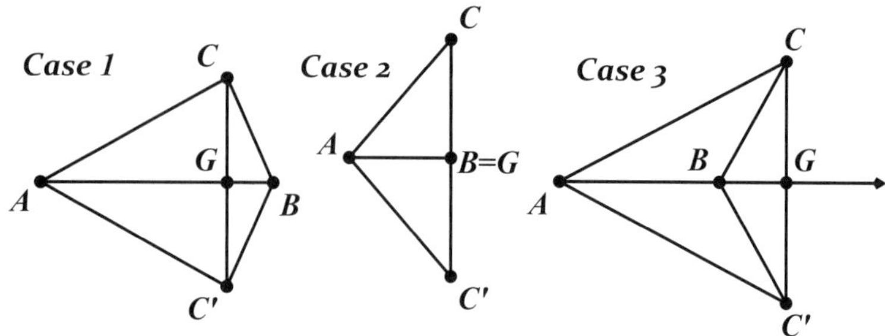

Figure 6.8. Proposition I.8 SSS

Case 1. G lies between A and B. In this case, consider isosceles triangles $\triangle CAC'$ and $\triangle CBC'$. By I.5, $\angle ACC' = \angle AC'C$ and $\angle BCC' = \angle BC'C$. But $\angle ACB = \angle ACC' + \angle BCC' = \angle AC'C + \angle BC'C = \angle AC'B$. Therefore, by I.4 (SAS), $\triangle ABC \cong \triangle ABC'$. Thus $\triangle ABC \cong \triangle DEF$.

Case 2. $G = A$ or $G = B$. WLOG, assume $G = B$. We leave this case as an exercise for the reader.

Case 3. G does not lie between A and B. Then either A lies between B and G, or B lies between A and G. WLOG, assume that B lies between A and G. We leave this case as an exercise for the reader. $\square$

One final consequence of Hilbert's axioms worth mentioning is a result that is beyond the scope of this book and would typically appear in a real analysis course. Axiom V.2 makes it possible to establish a one-to-one correspondence between the points of a segment and the real numbers, ensuring that lines have no gaps.

Exercises 6.3

1. Given a point A on a line ℓ, and another point B, not on ℓ, consider the ray $\overrightarrow{AB}$. Using Theorem 8, prove that all points on ray $\overrightarrow{AB}$ with the exception of A lie on one side of ℓ.

2. Explain why, in the proof of Theorem 6.12, ray $\overrightarrow{AF}$ cannot intersect segment CE.

3. Explain why, in the proof of Theorem 6.12, ray $\overrightarrow{AF}$ cannot intersect segment BC.

4. Prove Case 2 for Proposition I.8 [SSS].

5. Prove Case 3 for Proposition I.8 [SSS].

6.4 Spherical and Taxicab geometries

Let's examine Hilbert's axioms for Euclidean geometry within the context of Spherical and Taxicab geometries. In Chapters 4 and 5, we interpreted the terms "point, line, on" and "congruent" within these geometries. Recall, in all three geometries, segments AB and $A'B'$ are congruent if they have the same length. Similarly, angles $\angle ABC$ and $\angle A'B'C'$ are congruent if they have the same measure. Hilbert has given us the new undefined term "between." Thus, in order to work with the Axioms of Order, we need to give an interpretation for *betweenness* in the models for Spherical and Taxicab geometries. As we will use distance to interpret the meaning of *between two points*, let's review distance in these two geometries.

In Taxicab geometry, the distance between points $A = (x_1, y_1)$ and $B = (x_2, y_2)$ is defined as

$$d_T(A, B) = |x_1 - x_2| + |y_1 - y_2|.$$

In Spherical geometry, the distance between two nonantipodal points is the length of the minor arc of the unique great circle on which they lie, that is, AB. The distance between antipodal points is half the circumference of a great circle. This is summarized in the following equation:

$$d_S(A, B) = \begin{cases} AB, & \text{when } A \text{ and } B \text{ are nonantipodal} \\ \pi R, & \text{when } A \text{ and } B \text{ antipodal, where } R \text{ is the sphere's radius.} \end{cases}$$

In order to interpret betweenness in these geometries, perhaps we can start by taking our cue from ideas about what it means to travel from one point to another. When travelling from point A to C, we might say that B is between A and C if stopping at B adds no distance to the journey. In Taxicab and Spherical geometries, let's see what happens if we interpret betweenness in this way, summarized as follows: *If A, B and C are distinct points then B is between A and C if and only if $d(A, B) + d(B, C) = d(A, C)$.* A quick example in Taxicab geometry reveals a problem with this interpretation. The

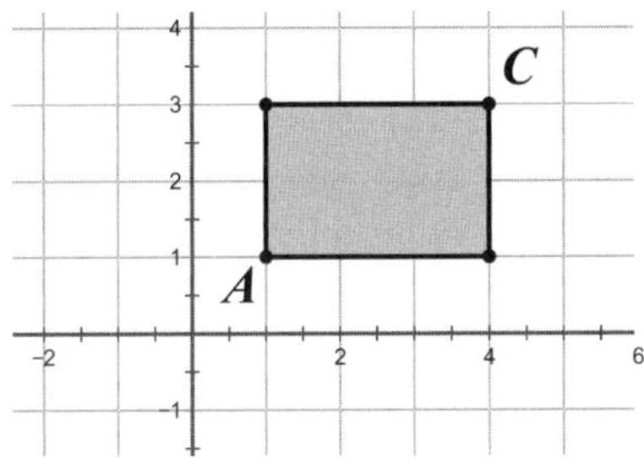

Figure 6.9. Set $\{B \mid d_T(A, B) + d_T(B, C) = d_T(A, C)\}$ is shaded

shaded region in Figure 6.9 shows all points B satisfying this interpretation of between A and C, but this certainly does not match our Euclidean interpretation of betweenness, We can fix this by requiring that the points all lie on the same line. Here's our new interpretation of betweenness:

*If A, B and C are distinct points on a line ℓ, then B is **between** A and C if and only if $d(A, B) + d(B, C) = d(A, C)$.*

With the added requirement of collinearity, the interpretation of betweenness is identical in Euclidean and Taxicab geometries. How about Spherical geometry? When A and B are nonantipodal, this matches our Euclidean interpretation of "between" and Hilbert's definition of "segment" since there is a unique minor arc joining them. Note that, in the nonantipodal case, we have $d_S(A, B) = AB < \pi R$. For antipodal points N and S on great circle g, $d_S(N, S) = \pi R$, where R is the radius of the sphere. Here's the problem: How do we interpret Hilbert's definition of segment NS when it could mean one of two different arcs with the same length? Since all points on g meet our interpretation of "between N and S," by Definition 6.9, NS must be the entire great circle g. This does not fit with our intuition about segments. For starters, we expect the length of a segment joining N and S to be the distance between N and S. Thus, for now we will say that both arcs of great circle g from N to S are line segments, each with a length of half of the circumference of the great circle. As we will see, these interpretations will be good enough for us to evaluate the validity of Hilbert's axioms in Spherical geometry.

Let's consider Hilbert's axioms in the context of Spherical geometry. We start with the Axioms of Incidence. Clearly Axiom I.1 holds, but for each pair of antipodal points N and S, there are infinitely many lines that contain them. Thus Axiom I.2 does not hold. We leave it to the reader to verify that the other three Axioms of Incidence do hold. Given our spherical context, it's important to note that anytime the word "plane" is used, we should replace it with "sphere." Axiom I.4 is the first time this occurs.

Next, we turn to the Axioms of Order. We start with Axiom II.2, noting that if N and S are antipodal points, there does not exist a B such that S is between N and B. Thus Axiom II.2 does not hold. If A, B and C are three points equally spaced on a great circle, then none of the three points lies between the other two. This contradicts our version of Axiom II.3. We leave it to the reader to verify that the two remaining axioms of Betweenness (II.1 and II.4) still hold.

We will finish with a discussion of Axiom III.4, leaving it to the reader to verify that the remaining Axioms of Congruence and the Axiom of Parallels hold for Spherical geometry. Regarding Axiom III.4, while we agree that each angle is congruent to itself on the sphere, a ray poses an interesting problem in this geometry since, at first glance, it appears to meet itself again at its vertex. However, when we review our interpretation of betweenness on a sphere, it's clear that when A and B are nonantipodal, ray $\overrightarrow{AB}$ must end at the point antipodal to A. So now, if we are given three rays, $\overrightarrow{AB}, \overrightarrow{AC}$ and $\overrightarrow{A'B'}$ as in Axiom III.4, we see that the axiom holds. (Why? $\overrightarrow{AB}$ and $\overrightarrow{AC}$ form a lune with angle $\angle BAC$. Given $\overrightarrow{A'B'}$, there is exactly one ray $\overrightarrow{A'C'}$ in each hemisphere formed by the great circle containing $\overrightarrow{A'B'}$ such that $\angle B'A'C' = \angle BAC$.) The fact that a ray is half the circumference of a great circle does pose a problem for one of the continuity axioms, Axiom V.1. This axiom says that if n copies of segment CD are placed contiguously on ray $\overrightarrow{AB}$ starting at A, there is a point E which lies on $\overrightarrow{AB}$ where $n \cdot CD = AE$, and B is between A and E. Since a ray extends only halfway around a great circle, we quickly run out of room for our copies of CD. For this reason, Axiom V.1 does not hold. It is with Archimedes' Axiom that Hilbert separates the concepts of infinite and boundless.

Having completed our discussion of Hilbert's axioms within Spherical geometry, let's recap what we have found. In Chapter 4, we gave an interpretation of the primitive terms and Euclid's axioms on the sphere which suggested that the propositions of Neutral geometry *should* hold. After showing that several of these propositions fail, it was clear that these geometries have demonstrable differences that could not be explained by Euclid's axioms, leading us to conclude that the set of Euclidean axioms must be deficient. As a complete and consistent set of independent axioms for Euclidean geometry, Hilbert's set provides all the necessary fixes. With only a subset of Hilbert's axioms valid on the sphere, we now have our explanation for the differences exhibited by these two geometries.

Let's take a look at Hilbert's axioms in our other non-Euclidean geometry, Taxicab geometry. We leave it to the reader to verify that all of the Axioms of Incidence and Order will hold, as well as the first four Axioms of Congruence, and the Axiom of Parallels. In the first fifteen axioms, the only axiom that fails to hold is Axiom III.5. We give the following counterexample, where triangle $\triangle ABC$ has $A = (0,2)$, $B = (2,0)$ and $C = (0,0)$, and triangle $\triangle A'B'C'$ has $A' = (3,2)$, $B' = (7,2)$ and $C' = (4,1)$. Then,

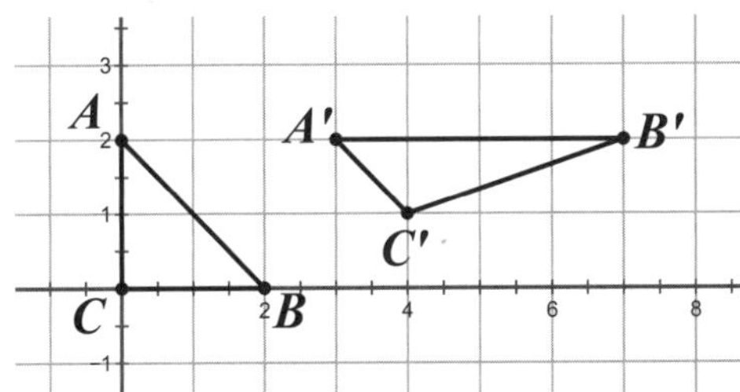

Figure 6.10. Counterexample for Axiom III.5 in Taxicab geometry

$AB = A'B' = 4$, $AC = A'C' = 2$ and $\angle BAC = \angle B'A'C' = 45°$, but $\angle ABC \neq \angle A'B'C'$. We cannot help but marvel at the difference this little axiom makes! It, alone, accounts for all of the differences between Taxicab geometry and Euclidean geometry that we saw in Chapter 5.

6.5 Gödel and consistency

We close this chapter by returning to the most important property of an axiomatic system, consistency. This property guarantees that there are no inherent contradictions in our system. In order to understand consistency, we need to say a bit more about models for axiomatic systems. A model, or interpretation of the primitive terms, can come in two flavors, **concrete** and **abstract**. In a concrete model the interpretations are based on the physical or real world, but in an abstract model the interpretations are based on some other axiomatic system. For example, our model for Taxicab geometry is abstract since our interpretation of the primitive terms is based on the analytic model of Euclidean geometry, that is, the Cartesian plane. Our model for Spherical geometry is also abstract since we envision the sphere existing in the analytic model of three-dimensional Euclidean geometry. Is the analytic model of Euclidean geometry concrete or abstract? It, too, is abstract, but this is by no means obvious, and we expect that this might take some convincing.

The Cartesian plane has infinitely many points. Our physical world is finite: at any given moment in time there are only finitely many atoms in our universe. So, there is nothing in existence in our physical world that can be used as an interpretation. Hence, the analytic model of Euclidean geometry is abstract and it is based on another axiomatic system, the real number system. Since the models for all of the (nonfinite) systems that we have considered, including Taxicab and Spherical geometry, rely on the analytic model of Euclidean geometry, these models are called *relatively consistent*, which is defined as follows.

Definition 6.15. An axiomatic system is **relatively consistent** if and only if its consistency can be proven assuming the consistency of another axiomatic system.

Since the analytic model of Euclidean geometry depends upon the real number system, this means that Euclidean geometry and the real number system are relatively consistent. In turn, the consistency of the real number system depends upon the consistency of the axioms of elementary arithmetic. They are all consistent relative to each other. A proof of the consistency of elementary arithmetic would cause a domino effect of consistency in geometry and analysis, and provide Hilbert with the missing piece in his quest to formalize all of the branches of mathematics. This unsolved problem held such importance to Hilbert that he listed it as the second problem on his list of open problems in 1900, just behind the Continuum Hypothesis.

These examples might lead the reader to assume that all systems of sufficient complexity to be of interest to mathematicians have abstract models. This was certainly the case until the end of the nineteenth century. During the time of renewed interest in the axiomatic development of geometry in the late 1800s, geometers became interested in minimizing the number of axioms and constructing interpretations of the undefined terms, or models, that would satisfy some, but not all, of the axioms. As geometers restricted their consideration to a mere handful of simply stated axioms, analysts grappled with the denseness of the real number line, and algebraists classified finite fields, it was only a matter of time before the unstated assumption of an infinite number of points was questioned. Throwing out this assumption gave rise to geometries of finitely many points, and Four Point geometry of Section 6.2 is one such example.

Finite geometries have concrete models since they can be interpreted in our physical world. For example, we could use four coins for points and six pencils for lines to produce a concrete model of Four Point geometry, where every geometric object corresponds to a physical object. For another example, consider the pizza and topping models in Example 6.7. Not only does the finiteness of these systems admit a concrete model, but additionally, the concrete model implies the consistency of the axioms. To understand why, suppose that a system with a concrete model is not consistent. This means that contradictory statements are implied by the axioms, and therefore, these contradictions exist in the model of physical objects. This is an impossibility as we do not allow for the existence of contradictions in the real world. For example, a coin cannot be on the pencil and not on the pencil at the same time, just as a topping cannot be both on and not on a pizza! We will state this as the following theorem.

Theorem 6.16. *Any system with a concrete model is consistent.*

This theorem still does not answer Hilbert's open problem, or more generally, the problem of consistency for a system with an abstract model. It was four decades after the first finite geometry appeared in print and three decades after Hilbert's challenge, that a young Austrian settled the issue. Hilbert's open question was answered and his hopes of formalizing the foundation of all branches of mathematics were dashed by Kurt Gödel, a mathematician at the University of Vienna who had published two revolutionary papers by the time he was 25 years old. Though greatly simplified here, Gödel's Theorem says that it is impossible to prove the consistency of the axioms of elementary arithmetic. Hence, by extension, it is impossible to prove the consistency of Euclidean geometry and the real number system. It's a house of cards, and Gödel pulled out the crucial supporting card on the bottom of the structure. Thus, the best we can hope for in these axiomatic systems is relative consistency. The theorem which states this result is known as Gödel's Incompleteness Theorem, a theorem which ranks at the top of many a mathematician's list of the most significant results of twentieth-century mathematics.

Theorem 6.17 [Gödel's Incompleteness Theorem, 1931]. *Any axiomatic system complicated enough to contain elementary arithmetic is at best relatively consistent.*

All, however, is not lost. As long as we are willing to believe that there are no inherent contradictions in the real number system then we can feel secure in using it as the basis for our other models. But do note that we have come full circle with the Calvin and Hobbes comic strip at the start of Chapter 2 since one of the greatest mathematicians of the twentieth century *proved* that mathematics begins with a leap of faith.

Exercises 6.5

1. Since the existence of a concrete model for a system guarantees the consistency of the system, we can demonstrate the consistency of the axioms for Four Point geometry by verifying that each axiom holds in a concrete model. Do this for each of the following models.

   (a) the set model where lines are given by subsets of points
   (b) one of the diagrams in Figure 6.2

7

Book I: Non-Neutral Geometry

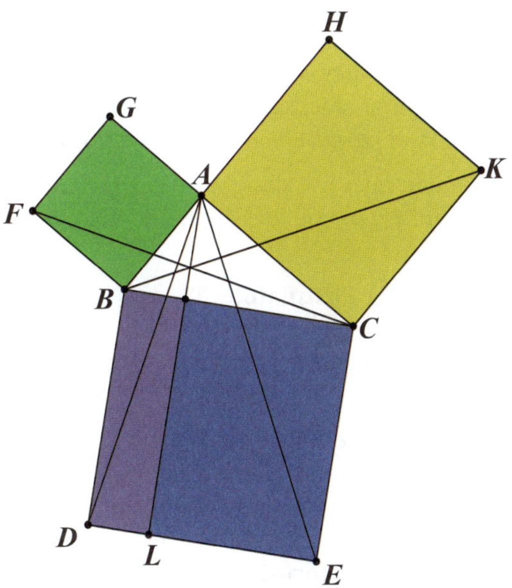

Figure 7.1. The windmill of the Pythagorean Theorem

A vacation is a break from our day-to-day routine. We might venture far from home, try some new adventure, or step outside the norm. The line and the circle have just had a vacation. Our main characters are world travelers now, and they had some mind-bending experiences along the way. Some circles were elevated in status to great circles. Circles got to see what it was like to live life "on the other side" as a line and as a square. Lines were circles but all circles were not lines. Vertical and horizontal lines felt shorter than their diagonal counterparts. Lines felt self-destructive, in a head-eating-the-tail kind of way. All in all, it looks like the circle had the better trip. Well, now, the party is over and they are ready to tackle life at home with the Parallel Postulate, where lines will be the star of the show again and circles will be toiling away in the background. With a new appreciation for the axiomatic systems that lay at the foundation of every branch of mathematics, we too leave behind the strange travel destinations of Spherical and Taxicab geometries to continue our story where we left off, in Book I of Euclid's *Elements*. The rest of the first book is devoted to theorems about *parallelism* and *area*, culminating in the Pythagorean Theorem and its converse. None of these results belongs to Neutral geometry.

As regards the set of axioms for our continued study of plane Euclidean geometry, with two sets of axioms at our disposal, we need to make a choice as to which set we will use when proving the propositions. At this point of the book, we could supplement Euclid's axioms with all of the other required axioms, including axioms of separation and SAS. Instead, we prefer to use *Euclid's* postulates and common notions so that we may continue to appreciate the elegance of his reasoning. So, if a proof relies on SAS we will still refer to it as Proposition I.4 or SAS. We will, however, revisit the updated axioms and related results from Chapter 6 whenever necessary.

7.1 Parallel lines

In Chapter 3, we proved all of the propositions from Book I that belong to Neutral geometry (I.1-I.28, I.31), that is, they do not rely on the Parallel Postulate. As we pick up where we left off, we cannot sidestep parallel lines any longer. So, we begin by recalling both the definition of *parallel lines* and Euclid's fifth postulate.

Definition. *Parallel straight lines* are straight lines which, being in the same plane and being produced indefinitely in both directions, do not meet one another in either direction.

Euclid's Postulate 5 [Parallel Postulate]. That, if a straight line falling on two straight lines make the interior angles on the same side less than two right angles, the two straight lines, if produced indefinitely, meet on that side on which are the angles less than two right angles.

In the nearly 2000 years of the futile search for a proof of the Parallel Postulate, mathematicians developed other propositions which are logically equivalent under the assumption of Neutral geometry. That is, any of these other statements could be substituted for Euclid's fifth postulate and, with this new statement taken as an axiom, the Parallel Postulate would become a proposition. The following theorem establishes the equivalence of the Parallel Postulate with four of its well-known reformulations.

Theorem 7.1. *Assuming Neutral geometry, the following five statements are all equivalent:*

- **Euclid's Postulate 5 [Parallel Postulate].** *That, if a straight line falling on two straight lines make the interior angles on the same side less than two right angles, the two straight lines, if produced indefinitely, meet on that side on which are the angles less than two right angles.*

- **Euclid's Proposition I.29.** *A straight line falling on parallel straight lines makes the alternate angles equal to one another, the exterior angle equal to the interior and opposite angle, and the interior angles on the same side equal to two right angles.*

- **Euclid's Proposition I.30.** *Straight lines parallel to the same straight line are also parallel to one another.*
- **Proclus' Axiom.** *If a line cuts one of two parallel lines, then it cuts the other.*
- **Playfair's Axiom.** *Through a point not on a given straight line, there exists at most one straight line that is parallel to the given line.*

Proof. To prove this theorem, we will show the following series of implications: Parallel Postulate $\Rightarrow$ I.29 $\Rightarrow$ I.30 $\Rightarrow$ Proclus' Axiom $\Rightarrow$ Playfair's Axiom $\Rightarrow$ Parallel Postulate. With this chain complete, any one of these statements follows from any other, and we have proven the theorem. We begin by following Euclid's proof of Proposition I.29.

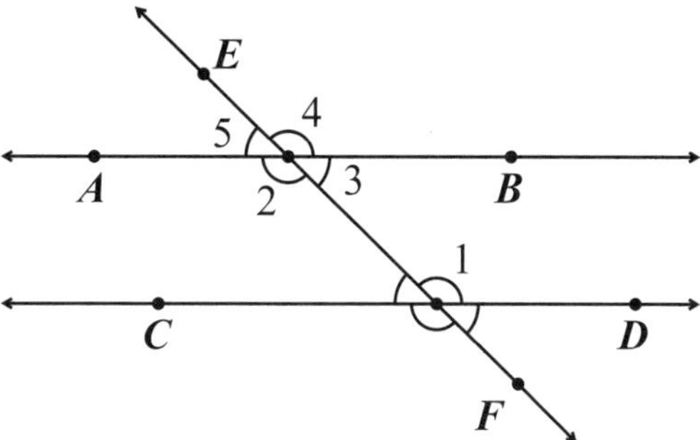

Figure 7.2. Euclid's Fifth Postulate $\Rightarrow$ I.29

(Euclid's Fifth Postulate $\Rightarrow$ I.29).

We assume that Euclid's Fifth Postulate holds. We are given parallel straight lines AB and CD with transversal EF. We wish to show that $\angle 1 = \angle 2$, $\angle 1 = \angle 4$ and $\angle 1 + \angle 3 =$ two right angles. For the sake of reaching an eventual contradiction, assume that $\angle 1 \neq \angle 2$. Since one of them must be larger, WLOG assume $\angle 2 > \angle 1$. Adding $\angle 3$ gives $\angle 2 + \angle 3 > \angle 1 + \angle 3$. By Proposition I.13, $\angle 2 + \angle 3 =$ two right angles. Therefore, $\angle 1 + \angle 3 <$ two right angles, and by Euclid's Fifth Postulate, AB and CD will meet. This contradicts our hypothesis that $AB \parallel CD$. Thus, $\angle 1 = \angle 2$.

By I.15, we have $\angle 2 = \angle 4$, hence $\angle 4 = \angle 1$. Adding $\angle 3$ gives $\angle 4 + \angle 3 = \angle 1 + \angle 3$. By I.13, $\angle 4 + \angle 3 =$ two right angles. Hence $\angle 1 + \angle 3$ equals two right angles, as desired.

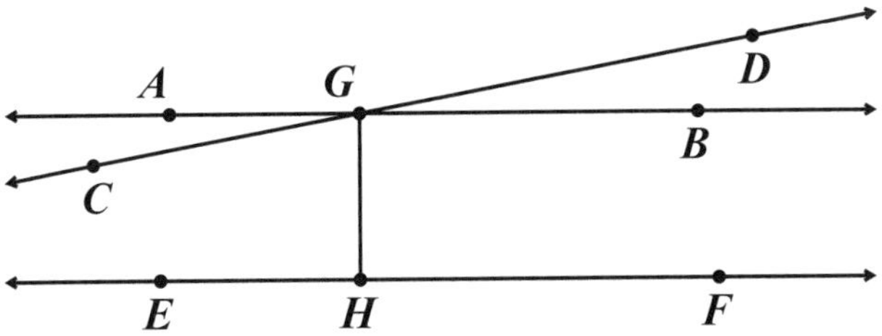

Figure 7.3. I.29 $\Rightarrow$ I.30

(I.29 ⇒ I.30).

We assume that Proposition I.29 holds. We are given lines AB, CD and EF where $AB \parallel EF$ and $CD \parallel EF$. We wish to show that $AB \parallel CD$. For the sake of reaching an eventual contradiction, assume $AB \nparallel CD$. Thus, AB and CD intersect at a point G. Using I.12, construct a perpendicular to EF from G, and label the intersection H. Since $AB \parallel EF$, by I.29, we have $\angle AGH = \angle FHG$. Similarly, since $CD \parallel EF$, then $\angle CGH = \angle FHG$ by I.29. By these two equations, we have $\angle AGH = \angle CGH$. This is a contradiction since the whole must be greater than the part. Thus, AB and CD are parallel, as desired.[1]

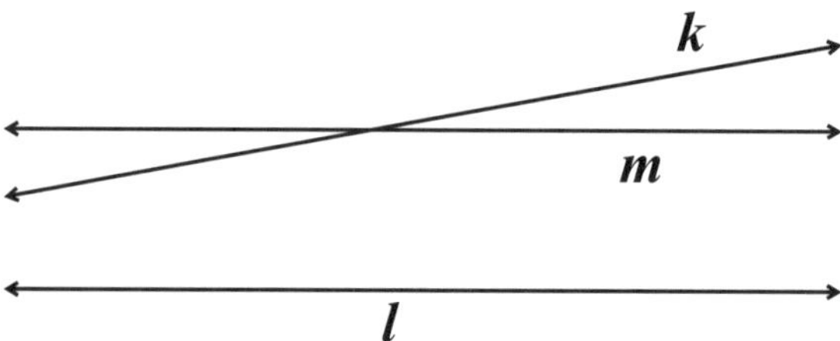

Figure 7.4. I.30 ⇒ Proclus' Axiom

(I.30 ⇒ Proclus' Axiom).

We assume that I.30 holds. We are given parallel lines l and m, and a line, k, that intersects m. To show that k will also intersect l, we assume that k does not. So, $k \parallel l$. But, then $k \parallel l$ and $m \parallel l$, yet k and m intersect. This contradicts I.30. Thus, k must intersect l.

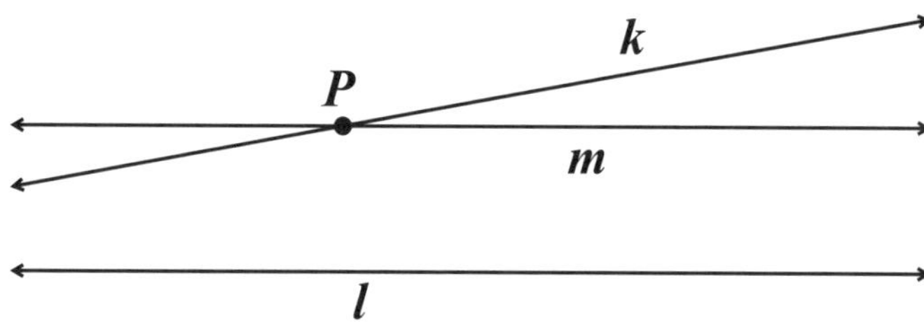

Figure 7.5. Proclus' Axiom ⇒ Playfair's Axiom

(Proclus' Axiom ⇒ Playfair's Axiom).

We assume that Proclus' Axiom holds. We are given a line, l, and a point, P, that is not on ℓ. We wish to show that there is at most one line through P that is parallel to l. To reach a contradiction, we will assume that more than one such parallel line exists: assume k and m are both lines through P that do not intersect l. Then, $m \parallel l$ and k intersects m at P, yet $k \parallel l$. This contradicts Proclus' Axiom. Thus, there exists at most one line through P that is parallel to l. (I.31 guarantees that one such line does, in fact, exist.)

---

[1]We do not follow Euclid's proof of Proposition I.30. After starting with lines AB and CD both parallel to EF, Euclid tacitly assumes Proclus' Axiom by allowing line GK to intersect all three lines.

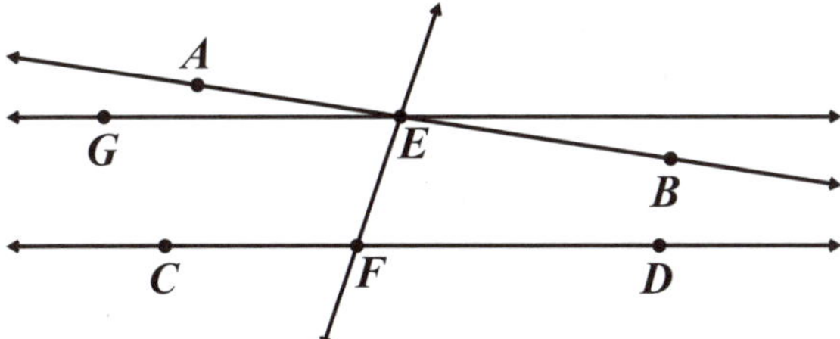

Figure 7.6. Playfair's Axiom $\Rightarrow$ Euclid's Fifth Postulate

(Playfair's Axiom $\Rightarrow$ Euclid's Fifth Postulate).

We assume that Playfair's Axiom holds. We are given straight lines AB and CD that are cut by a transversal EF, where E lies on AB, and F lies on CD. We assume that interior angles $\angle BEF$ and $\angle DFE$ sum to less than two right angles.

First, we will show that AB and CD must intersect. Since $\angle BEF + \angle AEF = $ two right angles, $\angle BEF + \angle AEF > \angle BEF + \angle DFE$. Subtracting $\angle BEF$ from both sides gives $\angle AEF > \angle DFE$. By I.23, construct $\angle GEF$ on EF so that $\angle GEF = \angle DFE$ and G lies on the opposite side of EF as B. By I.27, $GE \parallel CD$. Since AB and GE both pass through E, then AB must intersect CD by Playfair's Axiom.

Next, we must show that the intersection of AB and CD occurs on the same side of EF as B. We will assume that, instead, AB intersects CD on the same side of EF as A. Assume that rays $\overrightarrow{EA}$ and $\overrightarrow{FC}$ intersect at point H. Consider $\triangle HEF$. Note that $\angle HEF = \angle AEF > \angle DFE$. Also, $\angle DFE$ is the exterior angle to triangle $\triangle HEF$, and $\angle AEF$ is an opposite interior angle. This contradicts I.16 for $\triangle HEF$. Therefore, AB and CD must intersect on the same side of EF as B, as desired. $\square$

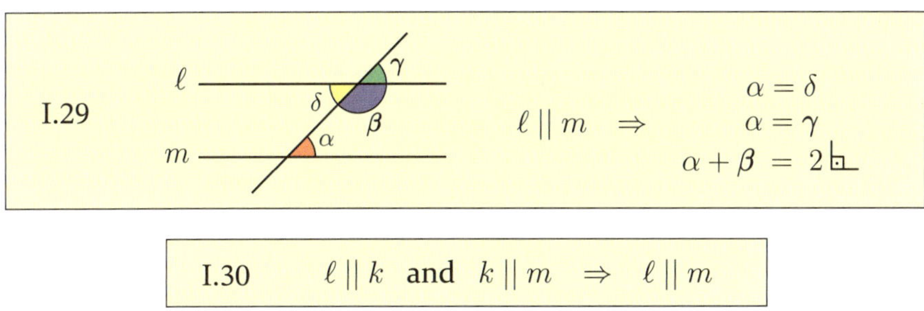

Euclid typically follows a proposition with its converse, when possible, as in Propositions I.5 and I.6, or I.13 and I.14. It is interesting to note that Proposition I.29 takes this one step further as it combines the converses of both I.27 and I.28, the last of the consecutively numbered Neutral geometry propositions.

Exercises 7.1

1. Directly prove each of the following. For each part, you may assume the propositions of Neutral geometry (Book I: Propositions 1 - 28, 31).

   (a) I.30 $\Rightarrow$ Playfair's Axiom.

   (b) Proclus's Axiom $\Rightarrow$ I.30.

   (c) Playfair's Axiom $\Rightarrow$ I.30.

   (d) Playfair's Axiom $\Rightarrow$ Proclus's Axiom.

2. Suppose points A and B are both on the same side of line ℓ. Prove: If line AB is parallel to ℓ, then A and B are equidistant from ℓ. [*Hint: Definition* 3.10 *will be helpful.*]

3. Suppose points A and B are both on the same side of line ℓ. Prove: If A and B are equidistant from ℓ, then line AB is parallel to ℓ.

4. Consider the rectangle $\square ABCD$.

   (a) Prove that opposite sides are equal, that is, $AD = BC$ and $AB = CD$. [*Hint: Exercise* 3.3.13 *may be useful here.*]

   (b) Prove that the diagonals are equal, that is, $AC = BD$.

7.2 Propositions I.32 and I.33

Ask your sibling, parent, grandparent, neighbor or friend this question: Can you tell me any fact about a triangle that you remember from high-school geometry? In our experience, the two most common answers to this question are the Pythagorean Theorem and the sum of the angles in a triangle is 180°. The former is Euclid's ultimate goal in Book I, and we will see it at the end of this chapter. The latter is Proposition I.32, though Euclid's statement refers to two right angles since he does not work in degrees and the right angle is his only angle of measure. This result lives squarely outside of Neutral geometry as its proof, given here, requires parallel lines and Proposition I.29. (For those readers keeping track of the proposition numbers, Proposition I.31 (*Through a given point to draw a straight line parallel to a given straight line*) does not require the use of Euclid's fifth postulate, and hence, appears with the Neutral geometry propositions in Chapter 3.)

Proposition I.32. *In any triangle, if one of the sides be produced, the exterior angle is equal to the two interior and opposite angles, and the three interior angles of the triangle are equal to two right angles.*

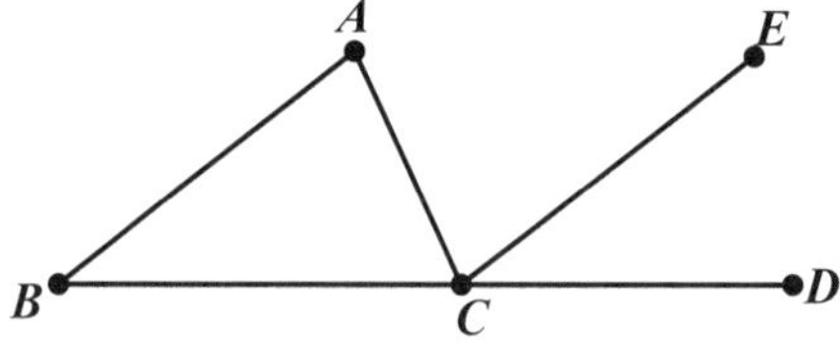

Figure 7.7. Proposition I.32

Proof. Let ABC be a triangle, and let one side of it BC be produced to D;

I say that the exterior angle ACD is equal to the two interior and opposite angles CAB, ABC, and the three interior angles of the triangle ABC, BCA, CAB are equal to two right angles.

For let CE be drawn through the point C parallel to the straight line AB. [I.31]

Then, since AB is parallel to CE, and AC has fallen upon them, the alternate angles BAC, ACE are equal to one another. [I.29]

Again, since AB is parallel to CE, and the straight line BD has fallen upon them, the exterior angle ECD is equal to the interior and opposite angle ABC. [I.29]

But the angle ACE was also proved equal to the angle BAC; therefore the whole angle ACD is equal to the two interior and opposite angles BAC, ABC.

Let the angle ACB be added to each; therefore the angles ACD, ACB are equal to the three angles ABC, BCA, CAB.

But the angles ACD, ACB are equal to two right angles; [I.13] therefore the angles ABC, BCA, CAB are also equal to two right angles.

Therefore etc. Q.E.D. □

By definition, a parallelogram is a quadrilateral with two sets of parallel sides. In the following proposition, Euclid proves that one set of parallel sides is sufficient as long as they are equal. Though this is our first proposition about parallelograms, the term never appears in the statement or its proof.

Proposition I.33. *The straight lines joining equal and parallel straight lines (at the extremities which are) in the same directions (respectively) are themselves also equal and parallel.*

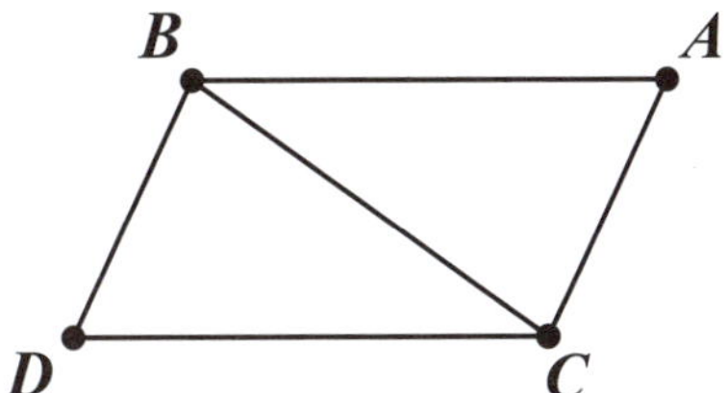

Figure 7.8. Proposition I.33

Proof. Let AB, CD be equal and parallel, and let the straight lines AC, BD join them (at the extremities which are) in the same directions (respectively); I say that AC, BD are also equal and parallel.

Let BC be joined.

Then, since AB is parallel to CD, and BC has fallen upon them, the alternate angles ABC, BCD are equal to one another. [I.29]

And, since AB is equal to CD, and BC is common, the two sides AB, BC are equal to the two sides DC, CB; and the angle ABC is equal to the angle BCD; therefore the base AC is equal to the base BD, and the triangle ABC is equal to the triangle DCB, and the remaining angles will be equal to the remaining angles respectively, namely those which the equal sides subtend; [I.4] therefore the angle ACB is equal to the angle CBD.

And, since the straight line BC falling on the two straight lines AC, BD has made the alternate angles equal to one another, AC is parallel to BD. [I.27]

And it was also proved equal to it.

Therefore etc. Q.E.D. □

I.33 $AB \parallel CD$ $AB = CD$ $\Rightarrow$ $AD \parallel BC$ $AD = BC$

Exercises 7.2

1. Give an updated version of Euclid's proof of each of the listed propositions. Be sure
   to justify each step, substitute mathematical symbols where appropriate, and include helpful diagrams as needed.

   (a) Proposition I.32   (b) Proposition I.33

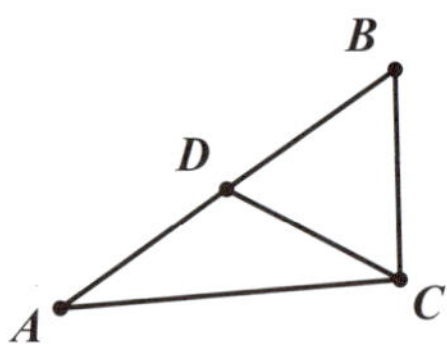

Figure 7.9. Exercise 7.2.2

2. In Figure 7.9, consider triangle $\triangle ABC$, where D is the midpoint of AB. If $AD = CD$,
   prove that angle $\angle ACB$ is a right angle.

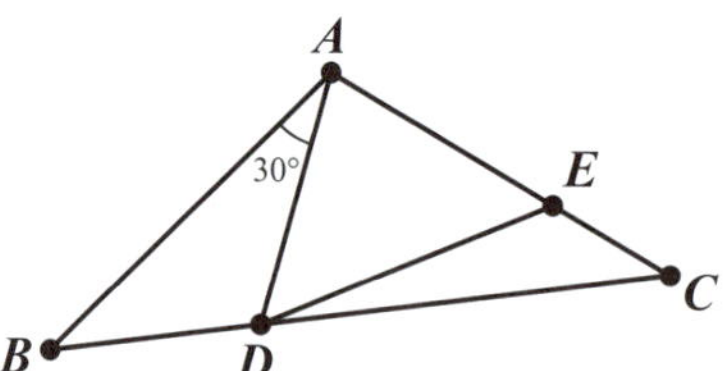

Figure 7.10. Exercise 7.2.3

3. In Figure 7.10, consider triangle $\triangle ABC$ where $AB = AC$, $AD = AE$ and $\angle BAD =$
   $30°$. Find $\angle CDE$.

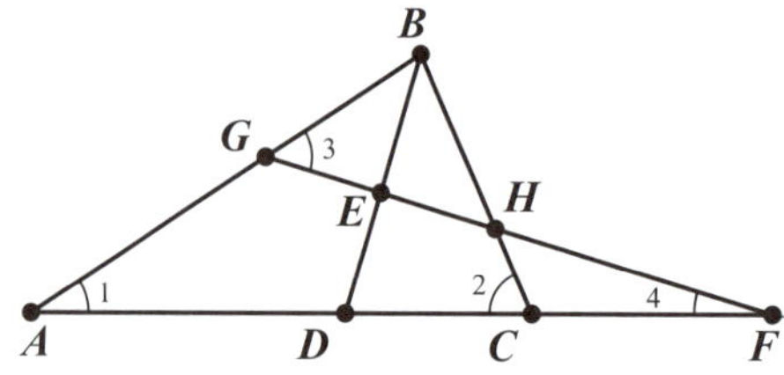

Figure 7.11. Exercise 7.2.4

4. Consider a scalene triangle $\triangle ABC$. Let D be the intersection of the angle bisector
   of $\angle ABC$ with side AC. Pick E on BD, and construct a perpedicular to BD through
   E. WLOG, assume that $\angle BDC$ is less than a right angle. Let F be the intersection

of this perpendicular with $\overrightarrow{AC}$, and let G be its intersection with AB. Prove that $\angle 3 = \frac{1}{2}(\angle 1 + \angle 2)$, as illustrated in Figure 7.11.[2]

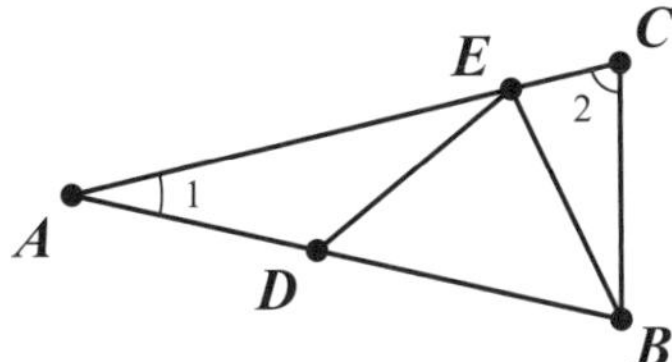

Figure 7.12. Exercise 7.2.5: A triangle in which the angles are in a 3:3:1 ratio

5. Consider an isosceles triangle $\triangle ABC$ with $AB = AC$. Assume that there are points D and E on AB and AC, respectively, such that $AD = DE = BE = BC$. Prove that $\angle 2 = 3 \cdot \angle 1$, and therefore, $\angle 1 = \frac{\pi}{7}$, as illustrated in Figure 7.12.[3]

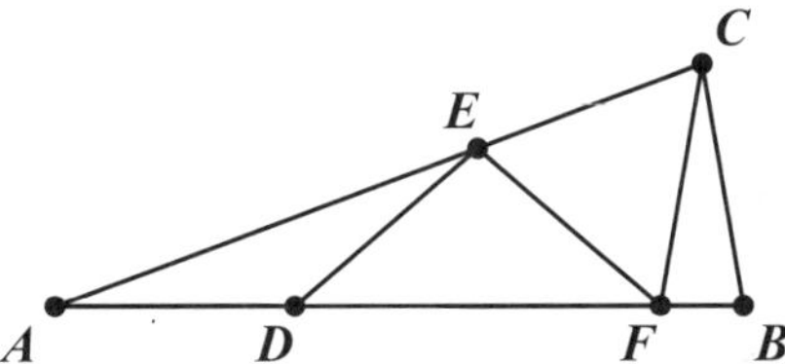

Figure 7.13. Exercise 7.2.6: A triangle in which the angles are in a 4:4:1 ratio

6. Consider an isosceles triangle $\triangle ABC$ with $AB = AC$. Assume that D and F lie on AB, and E lies on AC, such that $AD = DE = EF = FC = BC$, as illustrated in Figure 7.13.[4]

   (a) Prove that $\angle ABC = 4 \cdot \angle BAC$, and therefore, $\angle BAC = \frac{\pi}{9}$.

   (b) Prove that $\triangle CEF$ is equilateral.

7.3 Area

The word geometry is Greek in origin, stemming from the words for earth and measure. In his history text from the fifth century BCE, Herodotus theorized that the geometric arts of the Greeks developed out of practical concerns for land surveying in the agrarian civilization that developed along the fertile valley of the Nile River [71]. The annual

---

[2] This problem is from the 1954 Annual High School Mathematics Examination.

[3] This isosceles triangle appeared in "Zig-zag paths," a *Mathematical Gazette* article written by Archibald H. Finlay in 1959 [50]. Crockett Johnson rediscovered this triangle in his 1975 article on the construction of a regular heptagon [75].

[4] This isosceles triangle appeared in "Zig-zag paths," a *Mathematical Gazette* article written by Archibald H. Finlay in 1959 [50].

floods would necessitate the resurveying of land to determine the appropriate levying of taxes. We can be sure that these ancient practitioners were skilled in the measurement of physical quantities such as length, area and volume. Such practical concerns are not found in Euclid's *Elements*. Recall that in the first half of Book I, Euclid never measures the length of a line segment. At most, he compares the length of a segment to that of another segment. Likewise, Euclid does not measure angles; he compares all angles to a right angle. As such, the *Elements* exists at a distance from pragmatic concerns, and yet, we see its connection to the origin of the term geometry as it is a distillation of such practical problems into a purer form. The remaining propositions of Book I, including the Pythagorean Theorem, proceed in the same distilled manner, but here, Euclid turns his attention to **area**. Euclid does not give any of the familiar area formulas for a square, rectangle, parallelogram or triangle. Rather, he relates the area of one geometric figure to another. For example, he would describe the area of rectangle $ABCD$ as twice the area of $\triangle ABC$ rather than as the product of AB and BC. Before we consider Euclid's treatment of area, we take the distillation process one step further to the more formal concept of area.

Informally, we can think of area as a function that assigns a positive numerical value to a region, where larger regions are assigned larger values. As a formal concept, any area function must satisfy the following four postulates [**87**].

A-1   Area is considered a function from the set of polygonal regions to the real numbers. Using standard function notation, we have $Area : \mathcal{R} \to \mathbb{R}$, where $\mathcal{R}$ is the set of all polygonal regions and $\mathbb{R}$ is the set of all real numbers. We will write $Area(R)$ to denote the area of region R.

A-2   For every polygonal region R, $Area(R) > 0$.

A-3   *The Congruence Postulate.* If two triangular regions are congruent, then they have the same area.

A-4   *The Additivity Postulate.* If two polygonal regions intersect only in edges and vertices (or do not intersect at all), then the area of their union is the sum of their areas.

These four postulates define an area function up to a scalar multiple. This means that we would not violate these four axioms if we were to decide that the formula for the area of a triangle is simply the product of its base and its altitude rather than half that amount. Of course, this would have consequences for the areas of other objects. For example, it would imply that the area of a rectangle is twice the product of its base and its altitude. Likewise, it would mean that a square with side of length a would have area $2a^2$. While it may seem strange, this is a perfectly acceptable choice for an area function. Since we wish to obtain the area function that matches our elementary school formulas, we must add another area postulate.

A-5   If a square region has sides of length a, then its area is a^2.

With this last postulate, our area function for Euclidean geometry is completely determined. For example, let's consider a rectangle, which is defined as a right-angled quadrilateral in Section 2.2. Note that, in Exercise 3.3.13 we show that a rectangle is a parallelogram, and in Exercise 7.1.4 we show that opposite sides of a rectangle are equal. Thus, two lengths, a base and height, define a rectangle up to congruence. We can now use the area axioms to prove that a rectangle with base b and height h has area bh. Admittedly, this proof puts the horse ahead of the cart since Euclid does not construct a square until Proposition I.46, and does not justify its decomposition (shown in Figure 7.14) until Proposition II.4. Since we would like to develop the area formulas for these basic geometric figures as we encounter them in the upcoming propositions, we ask the reader to consider the following theorem even though its proof does not fit at this point of Euclid's Book I.

Theorem 7.2 [Area of a Rectangle]. *The area of a rectangular region is the product of its base and its altitude (height).*

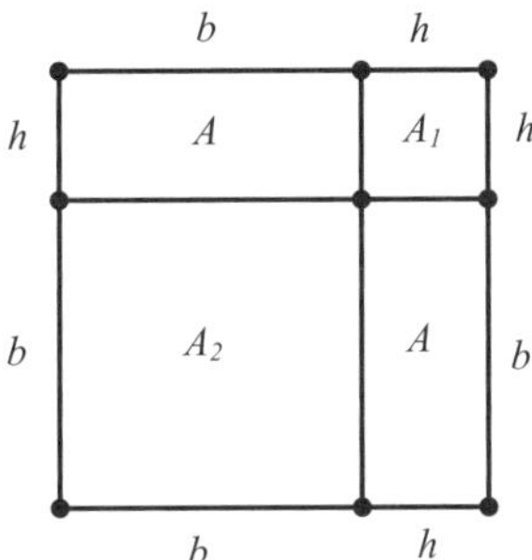

Figure 7.14. Area of a rectangular region

Proof.   Consider a rectangle with base b and height h. We will start by constructing a square whose side has length $b + h$. We leave it to the reader to give a careful construction of the decomposition of this large square into four regions consisting of two squares and two copies of our original rectangle as shown in Figure 7.14.

Thus we have

$$\begin{aligned}
(b + h)^2 &= 2A + A_1 + A_2 \\
b^2 + 2bh + h^2 &= 2A + h^2 + b^2 \\
2bh &= 2A
\end{aligned}$$

and $A = bh$ as desired. $\square$

Area in Taxicab geometry

In Section 5.1, we were warned that the Euclidean formula for the area of a square would not work for Taxicab geometry. Let's see why. Consider the square $\square ABCD$ with vertices $A = (0,2), B = (0,0), C = (2,0)$ and $D = (2,2)$. By area axiom A-5, the area of the square must be 4 square units since it has sides of length two. Consider placing a point, E, at $(1,1)$ and dividing the square into four congruent equilateral triangles $\triangle AEB, \triangle BEC, \triangle CED$ and $\triangle AED$. (All have sides of length two, and angles $45° - 45° - 90°$.) Each of these smaller triangles is half of a square with a side of length two in Taxicab geometry. Thus, each smaller triangle has an area of 2 square units. Adding the areas of the four smaller triangles gives a total area of 8 square units for the square.

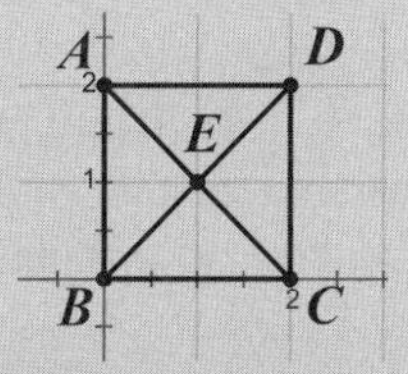

> But wait! This contradicts area axiom A-4 as the area of square $\square ABCD$ must be 4 square units.

Exercises 7.3

1. Explain how to construct the square in the proof of Theorem 7.2. Furthermore, show how to decompose this square into the two squares and two rectangles as illustrated in Figure 7.14. In particular, show that the two resulting nonsquare figures in this decomposition are, in fact, rectangles.

7.4　Propositions I.34 through I.41

Euclid uses the term *parallelogrammic area* for the first time in Proposition I.34, followed by the first use of **parallelogram** in Proposition I.35. The definition of a parallelogram as a quadrilateral whose opposite sides are parallel is clear from the proof of the second proposition. In the first proposition, he proves that opposite sides and opposite angles are equal to one another in any parallelogram. Additionally, he shows that the diagonal bisects the area, implying that a triangle has half the area of the parallelogram created by two of its adjacent sides. On a related note, Euclid's twenty-second definition specifies a rhomboid as a quadrilateral whose opposite sides and angles are equal but is neither right-angled nor equilateral. Therefore, Proposition I.34 implies that any parallelogram that is neither right-angled (rectangle) nor equilateral (rhombus) must be a rhomboid.

Proposition I.34. *In parallelogrammic areas the opposite sides and angles are equal to one another, and the diameter bisects the areas.*

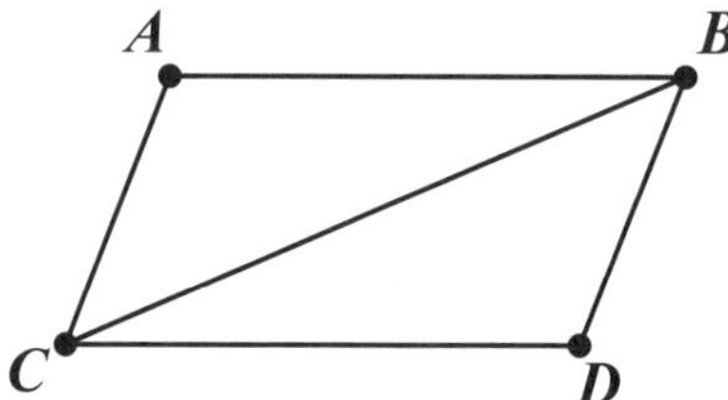

Figure 7.15.　Proposition I.34

Proof.　Let $ACDB$ be a parallelogrammic area, and BC its diameter; I say that the opposite sides and angles of the parallelogram $ACDB$ are equal to one another, and the diameter BC bisects it.

For, since AB is parallel to CD, and the straight line BC has fallen upon them, the alternate angles ABC, BCD are equal to one another. [I.29]

Again, since AC is parallel to BD, and BC has fallen upon them, the alternate angles ACB, CBD are equal to one another. [I.29]

Therefore ABC, DCB are two triangles having the two angles ABC, BCA equal to the two angles DCB, CBD respectively, and one side equal to one side, namely that adjoining the equal angles and common to both of them, BC; therefore they will also have the remaining sides equal to the remaining sides respectively, and the remaining angle to the remaining angle; [I.26] therefore the side AB is equal to CD, and AC to BD, and further the angle BAC is equal to the angle CDB.

And, since the angle ABC is equal to the angle BCD, and the angle CBD to the angle ACB, the whole angle ABD is equal to the whole angle ACD. [C.N. 2]

And the angle BAC was also proved equal to the angle CDB.

Therefore in parallelogrammic areas the opposite sides and angles are equal to one another.

I say, next, that the diameter also bisects the areas.

For, since AB is equal to CD, and BC is common, the two sides AB, BC are equal to the two sides DC, CB respectively; and the angle ABC is equal to the angle BCD; therefore the base AC is also equal to DB, and the triangle ABC is equal to the triangle DCB. [I.4]

Therefore the diameter BC bisects the parallelogram $ACDB$. Q.E.D. $\square$

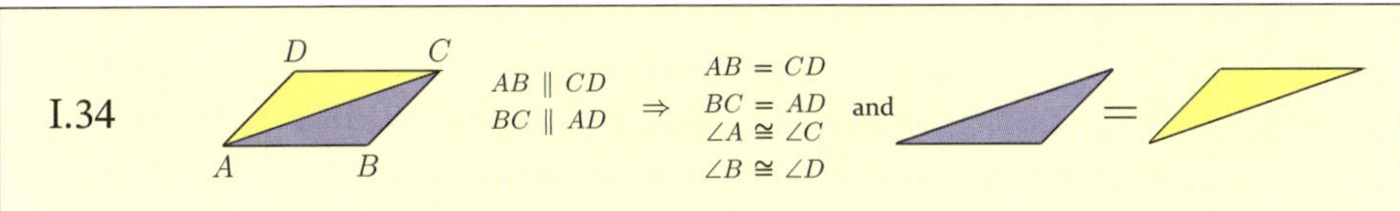

Most readers will find the last part of the proof a bit puzzling. After having used I.26 [ASA] to show that $\triangle ABC \cong \triangle DCB$, Euclid then goes back and uses I.4 [SAS] to show that "triangle $\triangle ABC$ is equal to the triangle $\triangle DCB$," meaning they have the same area. We find this last part of the proof to be unnecessary as we tend to apply the common-sense notion that congruent triangles have equal area. It is, however, the third area postulate of Section 7.3 which guarantees this.

Propositions I.35 through I.41 describe the relationships that exist between parallelograms and triangles that lie between the same parallel lines. There are several terms and phrases used here that warrant further clarification. When Euclid uses the term "equal," he means that the figures have the same area. Two triangles or parallelograms are "in the same parallels" when their vertices are on a pair of parallel lines. Two triangles or parallelograms are "on the same base" when they share a side. Two triangles or parallelograms are "on equal bases" when they have a congruent side. Suppose lines $\overleftrightarrow{AD}$ and $\overleftrightarrow{BC}$ are parallel in Figures 7.16 and 7.17. In Figure 7.16, triangles $\triangle ABC$ and $\triangle DBC$ are in the same parallels and on the same base. If $BC \cong EF$ then triangles $\triangle ABC$ and $\triangle AEF$ are in the same parallels and on equal bases. In Figure 7.17, parallelograms $\square ABCD$ and $\square EBCF$ are in the same parallels and on the same base. If $BC \cong HI$ then parallelograms $\square ABCD$ and $\square GHIJ$ are in the same parallels and on equal bases.

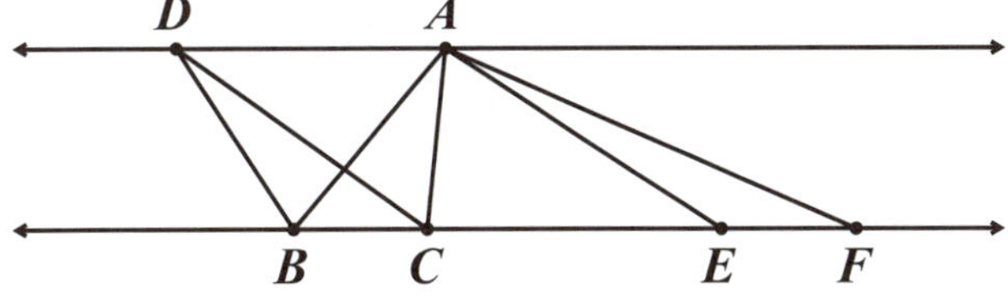

Figure 7.16. Triangles in the same parallels

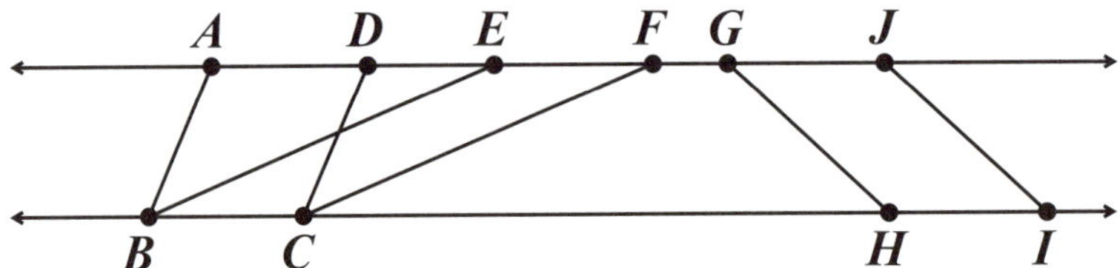

Figure 7.17. Parallelograms in the same parallels

In the propositions that follow in this section, we will prove that all triangles on equal bases in the same parallels have the same area. For example, triangles $\triangle ABC$, $\triangle DBC$ and $\triangle AEF$ of Figure 7.16 have the same area. We also show that a parallelogram in the same parallels and on equal bases with a triangle has twice the triangle's area. For example, parallelogram $\square EBCF$ has twice the area of triangle $\triangle GHI$ in Figure 7.17.

Proposition I.35. *Parallelograms which are on the same base and in the same parallels are equal to one another.*

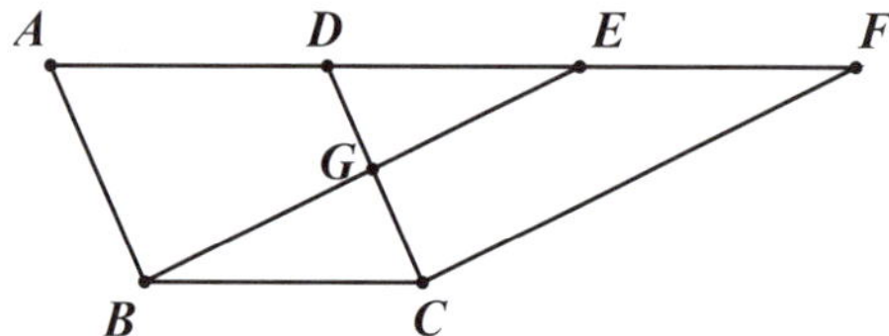

Figure 7.18. Proposition I.35

Proof.   Let $ABCD, EBCF$ be parallelograms on the same base BC and in the same parallels AF, BC; I say that $ABCD$ is equal to the parallelogram $EBCF$.

For, since $ABCD$ is a parallelogram, AD is equal to BC. [I.34]

For the same reason also EF is equal to BC, so that AD is also equal to EF; [C.N. 1] and DE is common; therefore the whole AE is equal to the whole DF. [C.N. 2]

But AB is also equal to DC; [I.34] therefore the two sides EA, AB are equal to the two sides FD, DC respectively, and the angle FDC is equal to the angle EAB, the exterior to the interior; [I.29] therefore the base EB is equal to the base FC, and the triangle EAB will be equal to the triangle FDC. [I.4]

Let DGE be subtracted from each; therefore the trapezium $ABGD$ which remains is equal to the trapezium $EGCF$ which remains. [C.N. 3]

Let the triangle GBC be added to each; therefore the whole parallelogram $ABCD$ is equal to the whole parallelogram $EBCF$. [C.N. 2]

Therefore etc. Q.E.D.   □

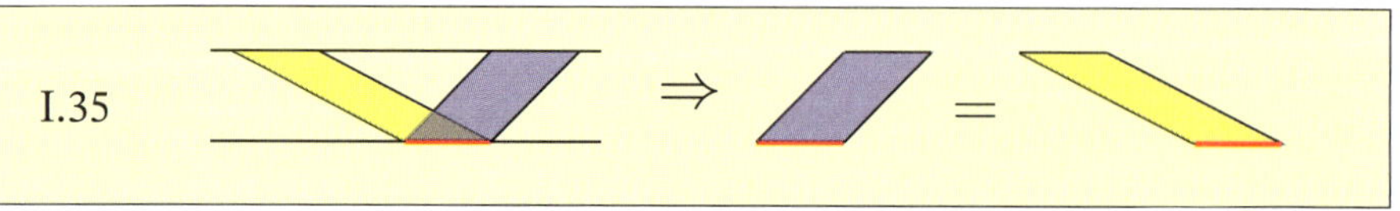

Euclid's proof depends on the order of the points A, D, E and F. Specifically, he assumes that D lies between A and E, but there are other cases to consider. For instance, we should consider the case where AD and EF overlap each other as shown in Figure 7.19. We leave it to the reader to prove this case.

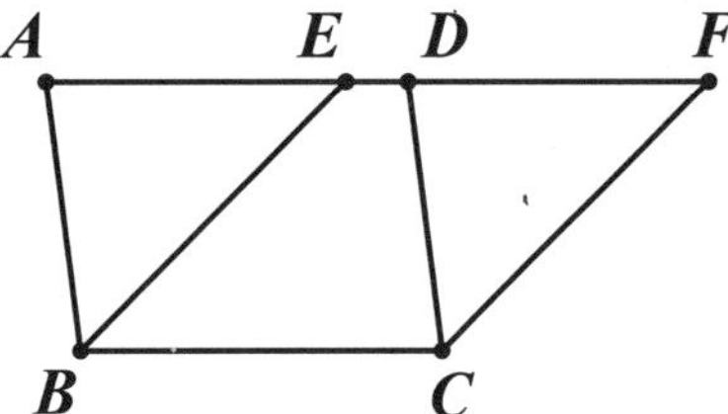

Figure 7.19. Alternative figure for Proposition I.35

In combination with the two previous propositions, the area formula for a rectangle that we developed from the area postulates in Section 7.3 pays some dividends. First, combining Proposition I.35 with Theorem 7.2 gives the area formula for a parallelogram.

Theorem 7.3 [Area of a Parallelogram]. *The area of a parallelogram with base b and altitude h is the product of its base and its altitude, or bh.*

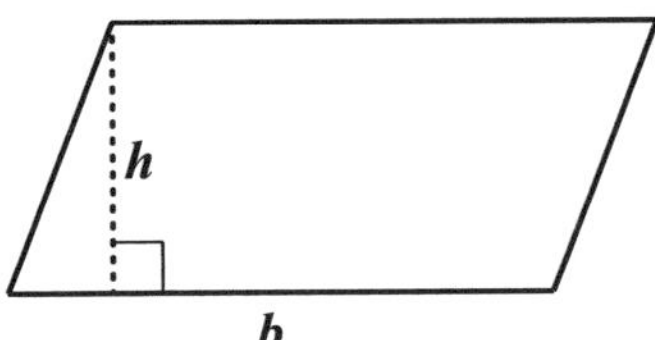

Figure 7.20. Area of a parallelogram is *bh*

Next, combining Theorem 7.3 with Proposition I.34 gives the area formula for a triangle.

Theorem 7.4 [Area of a Triangle]. *The area of a triangle with base b and altitude h is half of the product of its base and its altitude, or $\frac{1}{2}bh$.*

We leave the area formula for the trapezoid as an exercise.

Theorem 7.5 [Area of a Trapezoid]. *The area of a trapezoid with parallel bases b_1 and b_2, and altitude h, is half the product of its altitude and the sum of its bases, or $\frac{1}{2}(b_1 + b_2)h$.*

Euclid's proof of Proposition I.36 does not rely on the figure and holds regardless of whether or not the two parallelograms overlap.

Proposition I.36. *Parallelograms which are on equal bases and in the same parallels are equal to one another.*

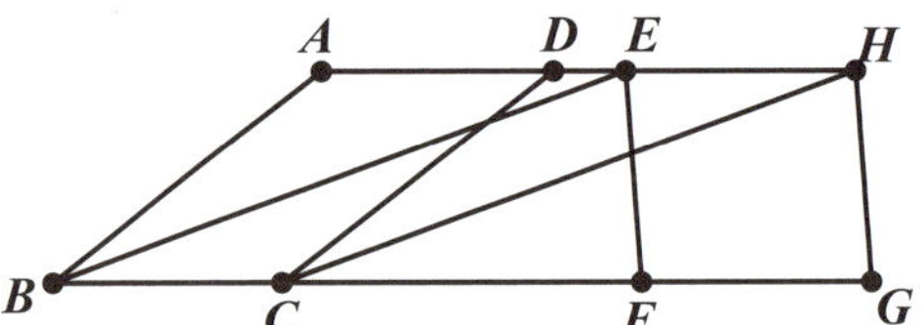

Figure 7.21. Proposition I.36

Proof. Let $ABCD, EFGH$ be parallelograms which are on equal bases BC, FG and in the same parallels AH, BG; I say that the parallelogram $ABCD$ is equal to $EFGH$.

For let BE, CH be joined.

Then, since BC is equal to FG while FG is equal to EH, BC is also equal to EH. [C.N. 1]

But they are also parallel.

And EB, HC join them; but straight lines joining equal and parallel straight lines (at the extremities which are) in the same directions (respectively) are equal and parallel. [I.33]

Therefore $EBCH$ is a parallelogram. [I.34]

And it is equal to $ABCD$; for it has the same base BC with it, and is in the same parallels BC, AH with it. [I.35]

For the same reason also $EFGH$ is equal to the same $EBCH$; [I.35] so that the parallelogram $ABCD$ is also equal to $EFGH$. [C.N. 1]

Therefore etc. Q.E.D. □

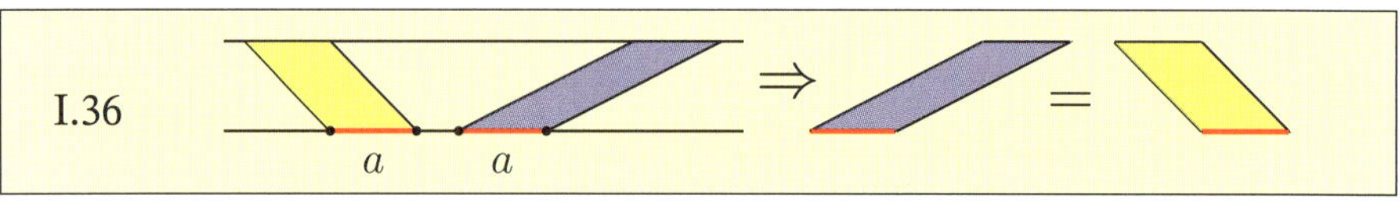

Proposition I.37. *Triangles which are on the same base and in the same parallels are equal to one another.*

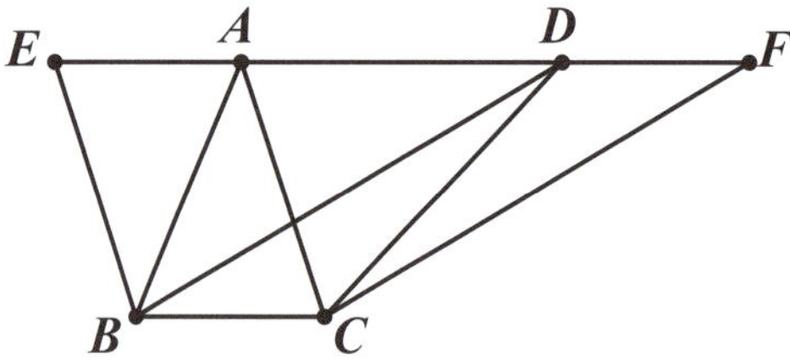

Figure 7.22. Proposition I.37

Proof. Let ABC, DBC be triangles on the same base BC and in the same parallels AD, BC; I say that the triangle ABC is equal to the triangle DBC.

Let AD be produced in both directions to E, F; through B let BE be drawn parallel to CA, [I.31] and through C let CF be drawn parallel to BD. [I.31]

Then each of the figures $EBCA, DBCF$ is a parallelogram; and they are equal,

for they are on the same base BC and in the same parallels BC, EF. [I.35]

Moreover the triangle ABC is half of the parallelogram $EBCA$; for the diameter AB bisects it. [I.34]

And the triangle DBC is half of the parallelogram $DBCF$; for the diameter DC bisects it. [I.34]

[But the halves of equal things are equal to one another.]

Therefore the triangle ABC is equal to the triangle DBC.

Therefore etc. Q.E.D.   □

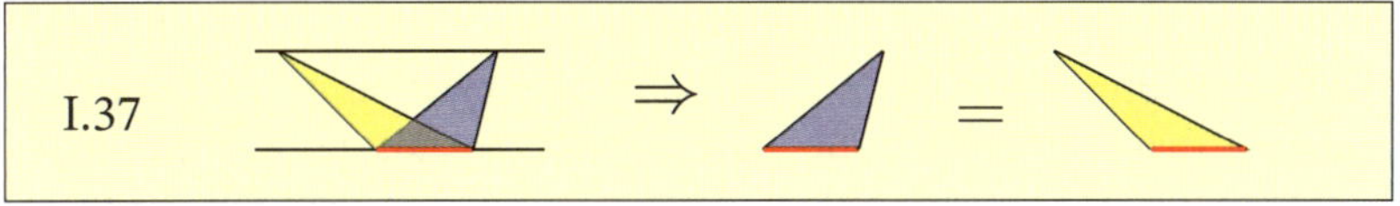

For the next proposition, Euclid loosens the previous hypotheses to show that the areas of two triangles in the same parallels are equal when the bases are equal. The proof can be found in Appendix D.

Proposition I.38. *Triangles which are on equal bases and in the same parallels are equal to one another.*

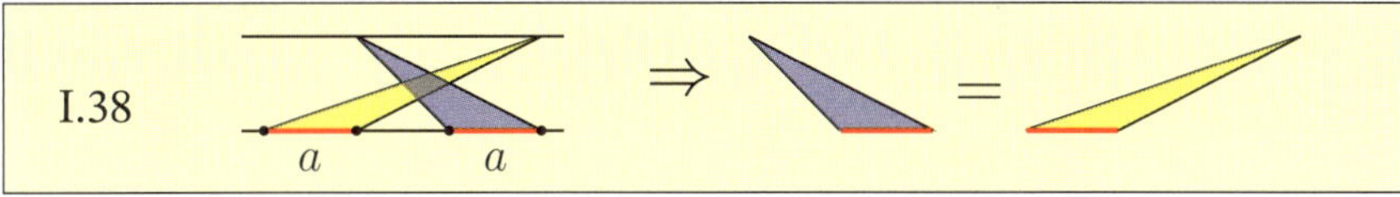

Proposition I.39 is the converse of I.37, though we must specify that the triangles lie on the same side of line BC.

Proposition I.39. *Equal triangles which are on the same base and on the same side are also in the same parallels.*

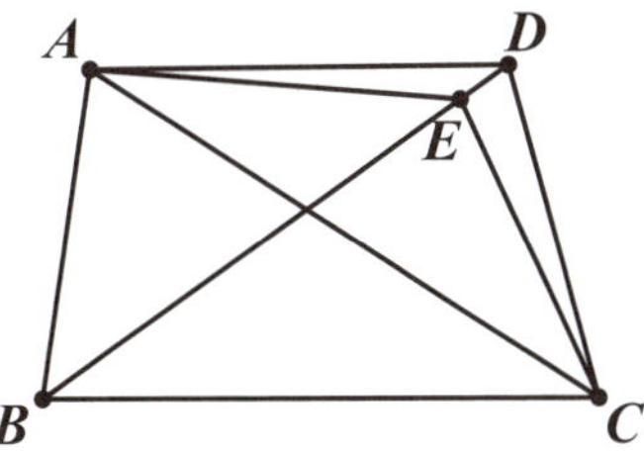

Figure 7.23. Proposition I.39

Proof. Let ABC, DBC be equal triangles which are on the same base BC and on the same side of it; [I say that they are also in the same parallels.]

And [For] let AD be joined; I say that AD is parallel to BC.

For, if not, let AE be drawn through the point A parallel to the straight line BC, [I.31] and let EC be joined.

Therefore the triangle ABC is equal to the triangle EBC; for it is on the same base BC with it and in the same parallels. [I.37]

But ABC is equal to DBC; therefore DBC is also equal to EBC, [C.N. 1] the greater to the less: which is impossible.

Therefore AE is not parallel to BC.

Similarly we can prove that neither is any other straight line except AD; therefore AD is parallel to BC.

Therefore etc. Q.E.D.   □

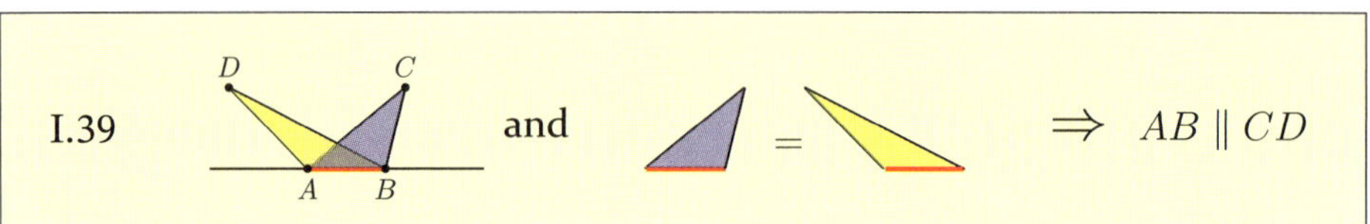

Note that for this proof to work, E must be the intersection of the ray $\overrightarrow{BD}$ with the line through A that is parallel to BC. The proof does not, however, rely on Figure 7.23. (Which of the logically equivalent statements listed in Theorem 7.1 most clearly guarantees the existence of this point of intersection, E?)

In Proposition I.40, Euclid loosens the hypotheses of the previous proposition to show that equal triangles are in the same parallels when the bases are equal. The proof can be found in Appendix D.

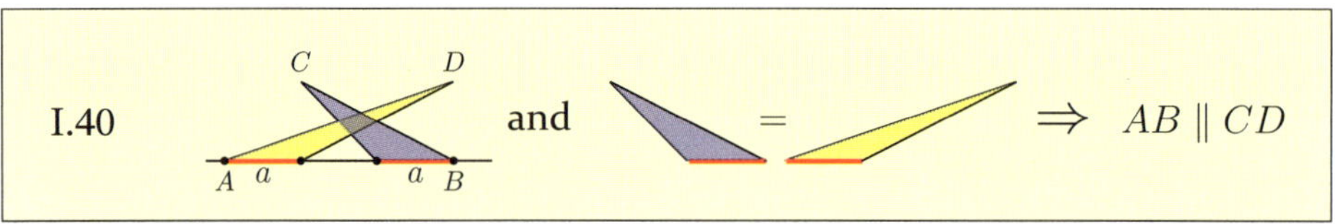

While Propositions I.39 and I.40 follow the same progression as I.37 and I.38, from a hypothesis of same bases to that of equal bases, the statement, diagram and proof of I.40 do not exhibit the same careful attention to detail as other Euclidean propositions. Furthermore, Euclid never uses I.40 in the *Elements*. In the course of 2000 years of a book which was copied and recopied time and again, there is evidence of marginal notes of earlier editions appearing in the main text of later editions. Scholars of Greek mathematics believe this proposition to be an addition after Euclid's time.

For the final proposition of this section, we have a result with the same flavor as the first proposition in this section, I.34. Here, we see that a parallelogram has twice the area of any triangle with the same base when it lies in the same parallel lines. The statement, diagram and miniature pictorial version are given here, while the proof can be found in Appendix D.

Proposition I.41. *If a parallelogram have the same base with a triangle and be in the same parallels, the parallelogram is double of the triangle.*

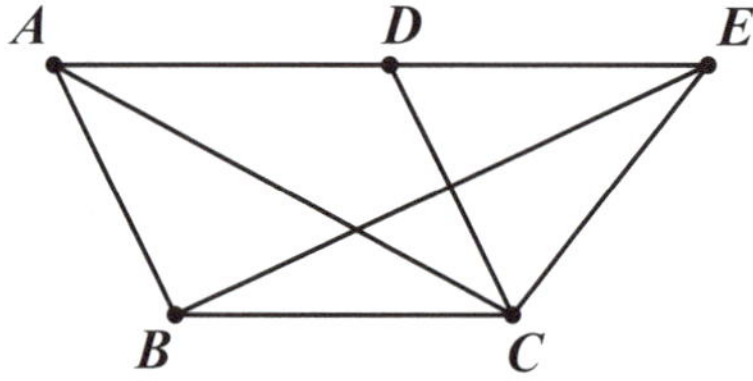

Figure 7.24. Proposition I.41

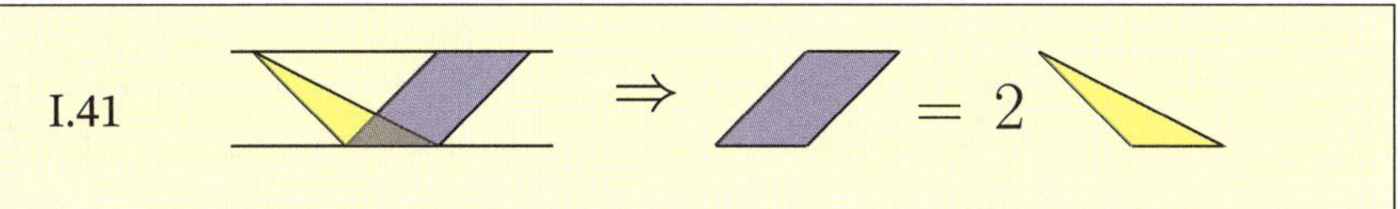

Exercises 7.4

1. Give an updated version of Euclid's proof of each of the listed propositions. Be sure to justify each step, substitute mathematical symbols where appropriate, and include helpful diagrams as needed.

   (a) Proposition I.34

   (b) Proposition I.35 [Be sure to include both cases.]

   (c) Proposition I.36

   (d) Proposition I.37

   (e) Proposition I.38

   (f) Proposition I.39

   (g) Proposition I.40 [Do not assume that the two triangles share a vertex.]

   (h) Proposition I.41

2. When Euclid claims "But the halves of equal things are equal to one another." in the proof of Proposition I.37, he utilizes a nonexistent lemma. State the missing lemma and give a geometric proof to justify this step.

3. Prove Theorem 7.4.

4. Prove Theorem 7.5, the formula for the area of a trapezoid. [*Hint: For a clever approach that avoids cases which depend upon the shape of the trapezoid, construct a larger, related parallelogram between the parallels formed by the bases of the trapezoid.*]

5. Let $\square ABCD$ be a parallelogram. Prove: Diagonals AC and BD bisect each other.

6. Let $\square ABCD$ be a parallelogram. Prove: $\square ABCD$ is a rectangle iff $AC = BD$.

7.5 Propositions I.42 through I.46

In Propositions I.33 through I.41, Euclid's focus is the comparison of areas of geometric figures. Starting with Proposition I.42, the focus shifts to the construction of a figure equal in area to some given figure. The move towards the construction of these figures in the next four propositions may seem completely unmotivated, but Euclid is making steady progress towards proving the Pythagorean Theorem. Starting with a given triangle and a given angle, Proposition I.42 gives the construction of a parallelogram equal in area to the triangle, and containing the angle. Proposition I.43 establishes the equality of two particular parallelograms. Proposition I.44 further restricts the construction of the parallelogram in I.42 by specifying a required side length in addition to an angle. Finally, in Proposition I.45 we construct a parallelogram equal in area to a given polygon. To help understand the motivation for this progression, it is necessary to introduce the concept of *quadrature*, or *squaring*, of a figure.

Definition 7.6. The **quadrature**, or **squaring**, of a figure in the plane is the construction, using only an unmarked straightedge and compass, of a square that has the same area as the original figure. If the construction is possible, we say that the figure is **quadrable**, or **squarable**.

In essence, quadrature is the fulfillment of the desire to impose symmetry on irregularly shaped figures. We make our first step towards this goal in Proposition I.45 when we construct a rectangle equal in area to any given polygon. So, any polygon is "rectangularable," but this is not as appealing as being squarable. In Book II, Euclid gives the construction of a square equal in area to a given rectangle. When this quadrature result is combined with I.45, we see that every polygon in the plane is quadrable. Having tackled all possible rectilineal figures, the Greeks also tried to square other geometric shapes. One of the three famous construction problems of antiquity was to determine whether or not a circle is quadrable. There were many attempts made on this problem over the two millennia that it went unsolved. As we will see in Chapter 16, Hippocrates of Chios (ca. 470–410 BCE) managed to square a very specific crescent-shaped figure created from two circles, but it wasn't until 1882 that Lindemann (Hilbert's Ph.D. advisor) proved that it is, in fact, impossible to square the circle. We will have to wait until then to discuss this story as well as the other famous "impossible constructions."

Proposition I.42. *To construct, in a given rectilineal angle, a parallelogram equal to a given triangle.*

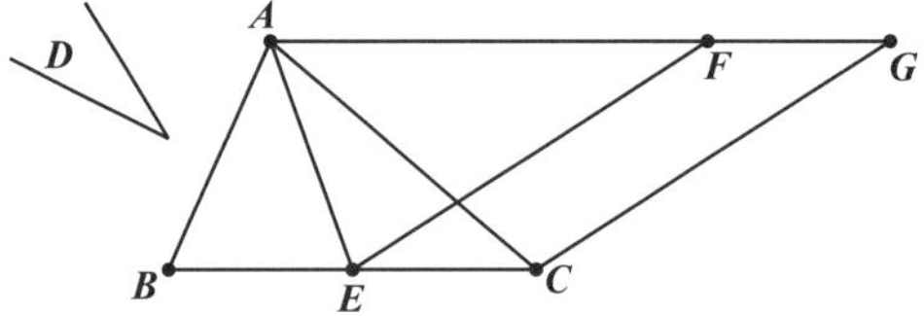

Figure 7.25. Proposition I.42

Proof.   Let *ABC* be the given triangle, and *D* the given rectilineal angle; thus it is required to construct in the rectilineal angle *D* a parallelogram equal to the triangle *ABC*.

Let *BC* be bisected at *E*, and let *AE* be joined; on the straight line *EC*, and at the point *E* on it, let the angle *CEF* be constructed equal to the angle *D*; [I.23] through *A* let *AG* be drawn parallel to *EC*, and [I.31] through *C* let *CG* be drawn parallel to *EF*.

Then *FECG* is a parallelogram.

And, since *BE* is equal to *EC*, the triangle *ABE* is also equal to the triangle *AEC*, for they are on equal bases *BE*, *EC* and in the same parallels *BC*, *AG*; [I.38] therefore the triangle *ABC* is double of the triangle *AEC*.

But the parallelogram *FECG* is also double of the triangle *AEC*, for it has the same base with it and is in the same parallels with it; [I.41] therefore the parallelogram *FECG* is equal to the triangle *ABC*.

And it has the angle *CEF* equal to the given angle *D*.

Therefore the parallelogram *FECG* has been constructed equal to the given triangle *ABC*, in the angle *CEF* which is equal to *D*. Q.E.F.   □

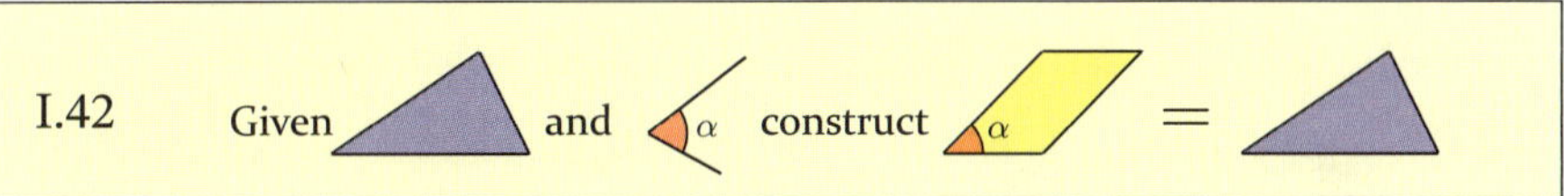

The following proposition introduces new language referring to smaller parallelograms found within a parallelogram, specifically, those about the diameter. In Figure 7.26, for example, $\square AEKH$ and $\square KGCF$ are parallelograms about diameter AC of the large parallelogram. The unshaded small parallelograms, $\square EBGK$ and $\square HKFD$, are the complements of the parallelograms about the diameter.

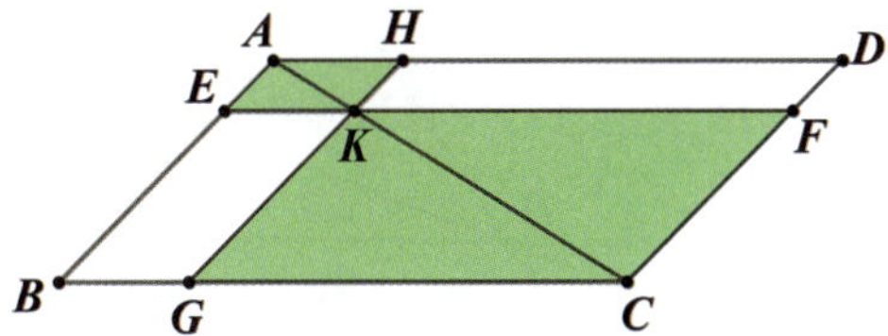

Figure 7.26. Parallelograms about the diameter

Proposition I.43. *In any parallelogram the complements of the parallelograms about the diameter are equal to one another.*

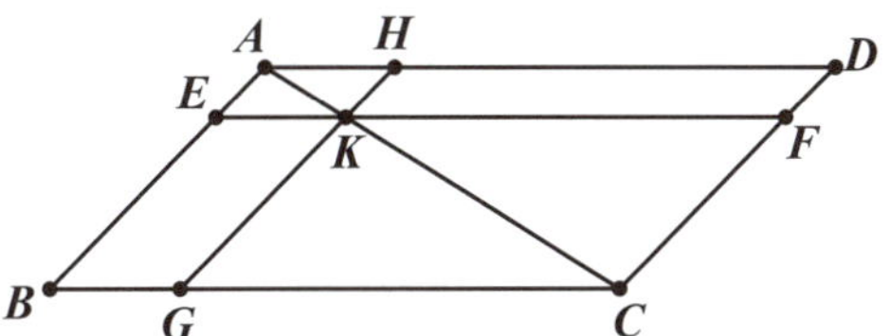

Figure 7.27. Proposition I.43

Proof. Let $ABCD$ be a parallelogram, and AC its diameter; and about AC let EH, FG be parallelograms, and BK, KD the so-called complements;

I say that the complement BK is equal to the complement KD.

For, since $ABCD$ is a parallelogram, and AC its diameter, the triangle ABC is equal to the triangle ACD. [I.34]

Again, since EH is a parallelogram, and AK is its diameter, the triangle AEK is equal to the triangle AHK.

For the same reason the triangle KFC is also equal to KGC.

Now, since the triangle AEK is equal to the triangle AHK, and KFC to KGC,

the triangle AEK together with KGC is equal to the triangle AHK together with KFC. [C.N. 2]

And the whole triangle ABC is also equal to the whole ADC; therefore the complement BK which remains is equal to the complement KD which remains. [C.N. 3]

Therefore etc. Q.E.D.   □

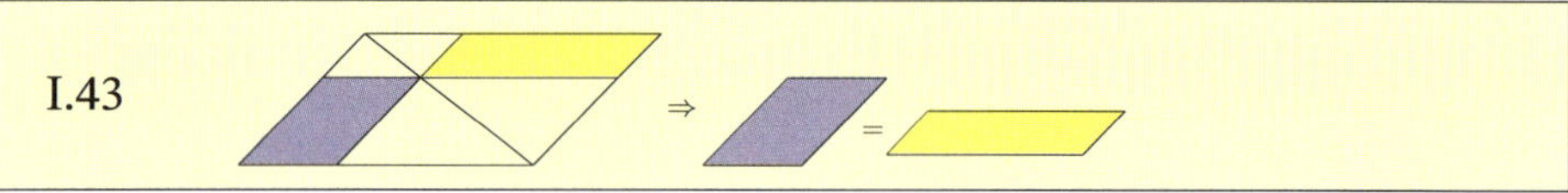

Proposition I.44. *To a given straight line to apply, in a given rectilineal angle, a parallelogram equal to a given triangle.*

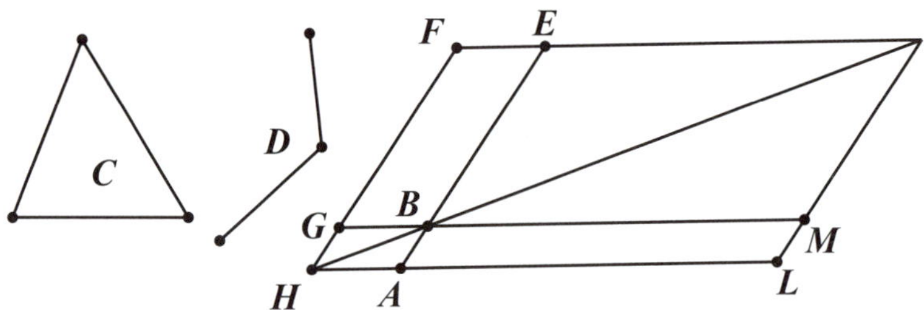

Figure 7.28. Proposition I.44

Proof.   Let AB be the given straight line, C the given triangle and D the given rectilineal angle; thus it is required to apply to the given straight line AB, in an angle equal to the angle D, a parallelogram equal to the given triangle C.

Let the parallelogram $BEFG$ be constructed equal to the triangle C, in the angle EBG which is equal to D [I.42]; let it be placed so that BE is in a straight line with AB; let FG be drawn through to H, and let AH be drawn through A parallel to either BG or EF. [I.31]

Let HB be joined.

Then, since the straight line HF falls upon the parallels AH, EF, the angles AHF, HFE are equal to two right angles. [I.29] Therefore the angles BHG, GFE are less than two right angles; and straight lines produced indefinitely from angles less than two right angles meet; [Post. 5] therefore HB, FE, when produced, will meet.

Let them be produced and meet at K; through the point K let KL be drawn parallel to either EA or FH, [I.31] and let HA, GB be produced to the points L, M.

Then $HLKF$ is a parallelogram, HK is its diameter, and AG, ME are parallelograms, and LB, BF the so-called complements, about HK; therefore LB is equal to BF. [I.43]

But BF is equal to the triangle C; therefore LB is also equal to C. [C.N. 1]

And, since the angle GBE is equal to the angle ABM, [I.15] while the angle GBE is equal to D, the angle ABM is also equal to the angle D.

Therefore the parallelogram LB equal to the given triangle C has been applied to the given straight line AB, in the angle ABM which is equal to D.

Q.E.F.   □

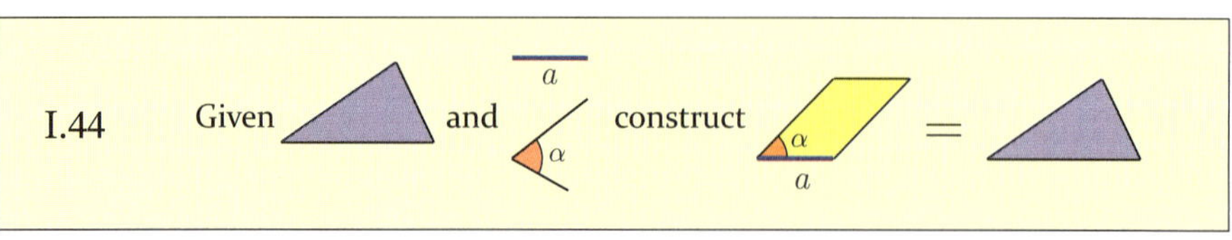

The statement of Proposition I.45 asserts that we can construct a parallelogram equal in area to any rectilineal figure. However, for the proof, Euclid chooses an arbitrary quadrilateral instead of a general polygon. Though it is clear that Euclid's proof can be extended to incorporate a polygon with more sides, an updated proof for an arbitrary n-sided figure requires mathematical induction.

Proposition I.45. *To construct, in a given rectilineal angle, a parallelogram equal to a given rectilineal figure.*

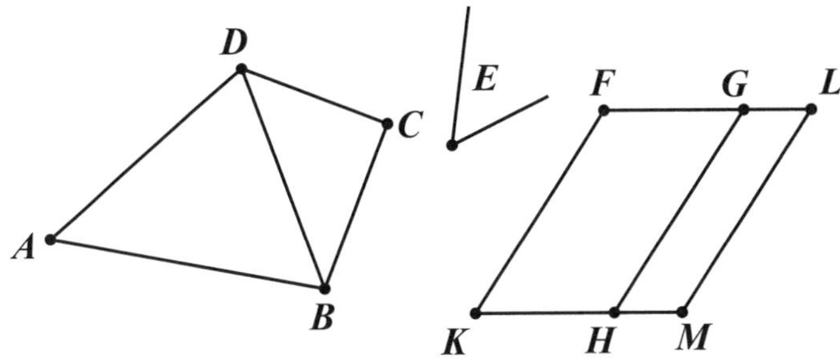

Figure 7.29. Proposition I.45

Proof.   Let $ABCD$ be the given rectilineal figure and E the given rectilineal angle; thus it is required to construct, in the given angle E, a parallelogram equal to the rectilineal figure $ABCD$.

Let DB be joined, and let the parallelogram FH be constructed equal to the triangle ABD, in the angle HKF which is equal to E; [I.42] let the parallelogram GM equal to the triangle DBC be applied to the straight line GH, in the angle GHM which is equal to E. [I.44]

Then, since the angle E is equal to each of the angles HKF, GHM, the angle HKF is also equal to the angle GHM. [C.N. 1]

Let the angle KHG be added to each; therefore the angles FKH, KHG are equal to the angles KHG, GHM.

But the angles FKH, KHG are equal to two right angles; [I. 29] therefore the angles KHG, GHM are also equal to two right angles.

Thus, with a straight line GH, and at the point H on it, two straight lines KH, HM not lying on the same side make the adjacent angles equal to two right angles; therefore KH is in a straight line with HM. [I.14]

And, since the straight line HG falls upon the parallels KM, FG, the alternate angles MHG, HGF are equal to one another. [I.29]

Let the angle HGL be added to each; therefore the angles MHG, HGL are equal to the angles HGF, HGL. [C.N. 2]

But the angles MHG, HGL are equal to two right angles; [I. 29] therefore the angles HGF, HGL are also equal to two right angles. [C.N. 1] Therefore FG is in a straight line with GL. [I.14]

And, since FK is equal and parallel to HG, [I.34] and HG to ML also, KF is also equal and parallel to ML; [C.N. 1; I.30] and the straight lines KM, FL join them (at their extremities); therefore KM, FL are also equal and parallel. [I.33] Therefore $KFLM$ is a parallelogram.

And, since the triangle ABD is equal to the parallelogram FH, and DBC to GM, the whole rectilineal figure $ABCD$ is equal to the whole parallelogram $KFLM$.

Therefore the parallelogram $KFLM$ has been constructed equal to the given rectilineal figure $ABCD$, in the angle FKM which is equal to the given angle E.

Q.E.F.

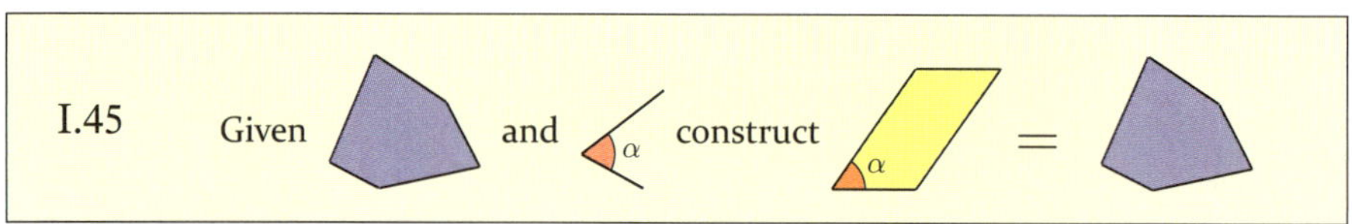

You may recall that Book I begins with the construction of an equilateral triangle. Here, we have our second, and only other, construction of a *regular polygon* in Book I: the square. While this construction could have appeared earlier in the book, Euclid places it exactly where he needs it, directly before the main result of the first book, the Pythagorean Theorem.

Definition 7.7. A **regular polygon** is a convex polygon with equal sides and equal interior angles.

Recall from Chapter 4 that a figure is **convex** if the line segment joining any two points of the figure lies entirely in the figure. Euclid waits until Book IV to construct another regular polygon, the pentagon.

Proposition I.46. *On a given straight line to describe a square.*

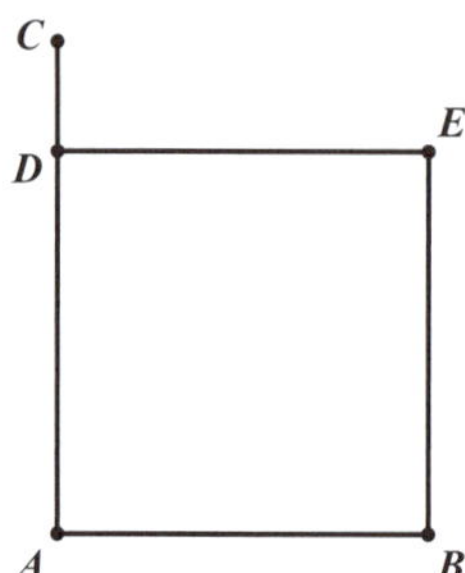

Figure 7.30. Proposition I.46

Proof.   Let AB be the given straight line; thus it is required to describe a square on the straight line AB.

Let AC be drawn at right angles to the straight line AB from the point A on it [I.11], and let AD be made equal to AB; through the point D let DE be drawn parallel to AB, and through the point B let BE be drawn parallel to AD. [I.31]

Therefore $ADEB$ is a parallelogram; therefore AB is equal to DE, and AD to BE. [I.34]

But AB is equal to AD; therefore the four straight lines BA, AD, DE, EB are equal to one another; therefore the parallelogram $ADEB$ is equilateral.

I say next that it is also right-angled.

For, since the straight line AD falls upon the parallels AB, DE, the angles BAD, ADE are equal to two right angles. [I.29]

But the angle BAD is right; therefore the angle ADE is also right.

And in parallelogrammic areas the opposite sides and angles are equal to one another; [I.34] therefore each of the opposite angles ABE, BED is also right. Therefore $ADEB$ is right-angled.

And it was also proved equilateral.

Therefore it is a square; and it is described on the straight line AB.

Q.E.F.   □

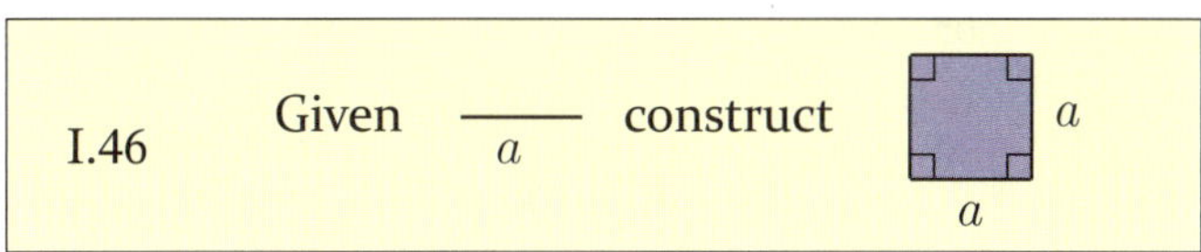

Exercises 7.5

1. Give an updated version of Euclid's proof of each of the listed propositions. Be sure to justify each step, substitute mathematical symbols where appropriate, and include helpful diagrams as needed.

   (a) Proposition I.42   (d) Proposition I.45

   (b) Proposition I.43   (e) Proposition I.46

   (c) Proposition I.44

2. Give a construction of a square on a segment that is different from Euclid's. Make sure that your proof uses only propositions prior to I.46. Be sure to prove that your construction has equal sides and four right angles.

7.6 The Pythagorean Theorem

Book I ends with two of the most well-known results in geometry, the Pythagorean Theorem and its converse. In algebraic terms, the Pythagorean Theorem states: *If $\triangle ABC$ is a right triangle with right angle $\angle BAC$ and sides of lengths a, b, c (as shown in Figure 7.31), then $a^2 = b^2 + c^2$. Its converse states: If triangle $\triangle ABC$ with sides of lengths a, b, c has the property that $a^2 = b^2 + c^2$, then $\angle BAC$ is a right angle.* There is evidence that the ancient civilizations of Babylonia and China used these results for measurement and construction. Legend also has it that the builders of the great pyramids employed ropes to construct right angles. As shown in Figure 7.32, when thirteen knots are placed at equal intervals in a length of rope, if one person secures the first and last knot together, two others can pull the rope taut at distances of three and four knots away to form a right angle. While other civilizations understood how to apply the theorem or its converse, the Pythagoreans may have been the first to provide a general proof. Though there are now at least 367 proofs of the Pythagorean Theorem, including one given by United States President James A. Garfield, Euclid's proof is certainly one of the most famous [**85**].

Proposition I.47 [The Pythagorean Theorem]. *In right-angled triangles the square on the side subtending the right angle is equal to the squares on the sides containing the right angle.*

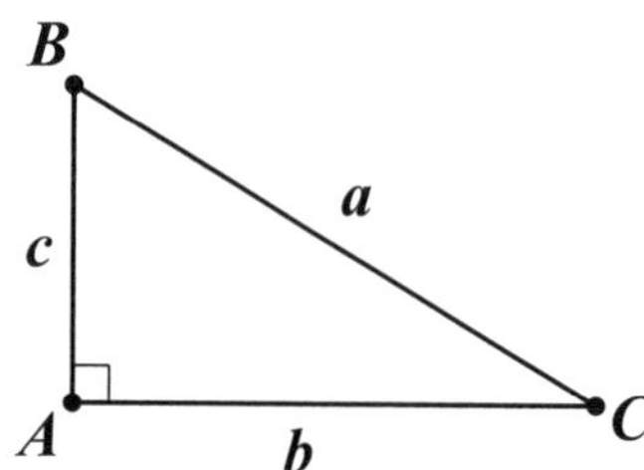

Figure 7.31. Pythagorean
Theorem: $a^2 = b^2 + c^2$

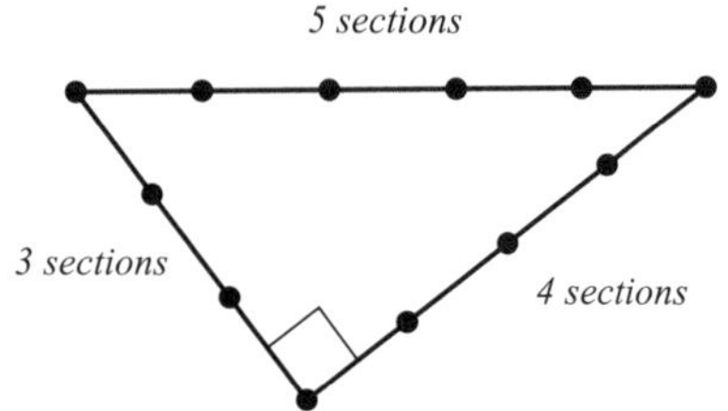

Figure   7.32. Egyptian
method for right angle
construction

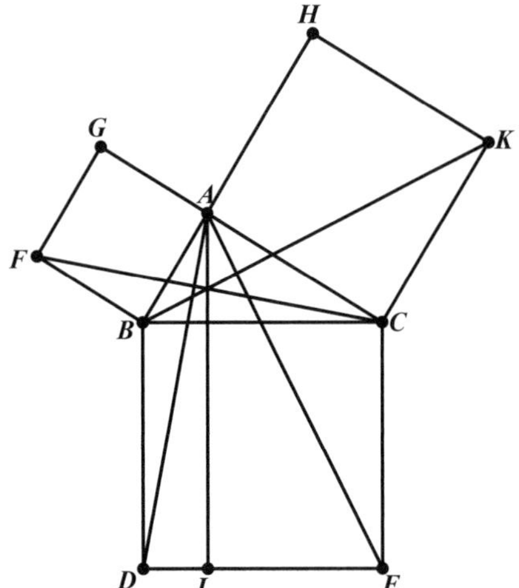

Figure 7.33. Prop I.47: The Pythagorean Theorem

Proof.   Let ABC be a right-angled triangle having the angle BAC right;

I say that the square on BC is equal to the squares on BA, AC.

For let there be described on BC the square $BDEC$, and on BA, AC the squares GB, HC; [I.46] through A let AL be drawn parallel to either BD or CE, and let AD, FC be joined.

Then, since each of the angles BAC, BAG is right, it follows that with a straight line BA, and at the point A on it, the two straight lines AC, AG not lying on the same side make the adjacent angles equal to two right angles; therefore CA is in a straight line with AG. [I.14]

For the same reason BA is also in a straight line with AH.

And, since the angle DBC is equal to the angle FBA: for each is right: let the angle ABC be added to each; therefore the whole angle DBA is equal to the whole angle FBC. [C.N. 2]

And, since DB is equal to BC, and FB to BA, the two sides AB, BD are equal to the two sides FB, BC respectively, and the angle ABD is equal to the angle FBC; therefore the base AD is equal to the base FC, and the triangle ABD is equal to the triangle FBC. [I.4]

Now the parallelogram BL is double of the triangle ABD, for they have the same base BD and are in the same parallels BD, AL. [I.41]

And the square GB is double of the triangle FBC, for they again have the same base FB and are in the same parallels FB, GC. [I.41]

[But the doubles of equals are equal to one another.] Therefore the parallelogram BL is also equal to the square GB.

Similarly, if AE, BK be joined, the parallelogram CL can also be proved equal to the square HC; therefore the whole square $BDEC$ is equal to the two squares GB, HC. [C.N. 2]

And the square $BDEC$ is described on BC, and the squares GB, HC on BA, AC.

Therefore the square on the side BC is equal to the squares on the sides BA, AC.

Therefore etc. Q.E.D. □

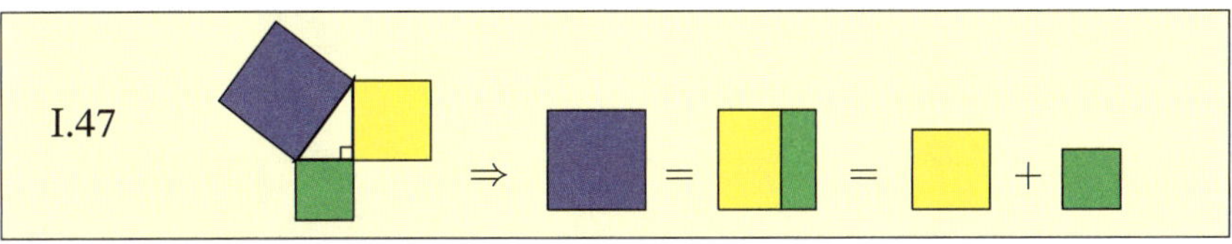

The diagram employed by Euclid is often described as a windmill. By coloring the areas of the squares, the shaded version shown in Figure 7.1 highlights this famous windmill diagram and accentuates the difference between the Greek view and the current view of this theorem. We tend to remember the theorem as an algebraic equation, but for Euclid, it is a relationship between the squares on the sides of a right triangle. As such, this theorem is perfectly in line with the others in the last third of Book I as it, too, relates the area of one figure to another, well, two others. Euclid's statement of the theorem concerns areas and his proof is completely geometric, but many of those other 366 proofs of the Pythagorean Theorem employ algebra. We will have more to say about the relationship between geometry and algebra in Chapter 8. In Section 9.3, we take a look at Proposition IV.31 which generalizes the Pythagorean Theorem by allowing for figures other than squares to be constructed on the sides of a right triangle.

Proposition I.48 [The Pythagorean Theorem's Converse]. *If in a triangle the square on one of the sides be equal to the squares on the remaining two sides of the triangle, the angle contained by the remaining two sides of the triangle is right.*

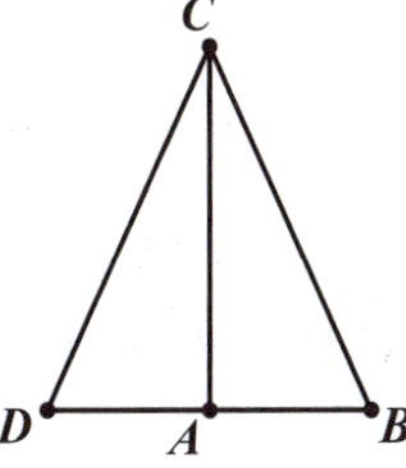

Figure 7.34. Prop I.48: Converse of the Pythagorean Theorem

Proof. For in the triangle ABC let the square on one side BC be equal to the squares on the sides BA, AC;

I say that the angle BAC is right.

For let AD be drawn from the point A at right angles to the straight line AC, let AD be made equal to BA, and let DC be joined.

Since DA is equal to AB, the square on DA is also equal to the square on AB.

Let the square on AC be added to each; therefore the squares on DA, AC are equal to the squares on BA, AC.

But the square on DC is equal to the squares on DA, AC, for the angle DAC is right; [I.47] and the square on BC is equal to the squares on BA, AC, for this is the hypothesis; therefore the square on DC is equal to the square on BC, so that the side DC is also equal to BC.

And, since DA is equal to AB, and AC is common, the two sides DA, AC are equal to the two sides BA, AC; and the base DC is equal to the base BC; therefore the angle DAC is equal to the angle BAC. [I.8] But the angle DAC is right; therefore the angle BAC is also right.

Therefore etc. Q.E.D.   □

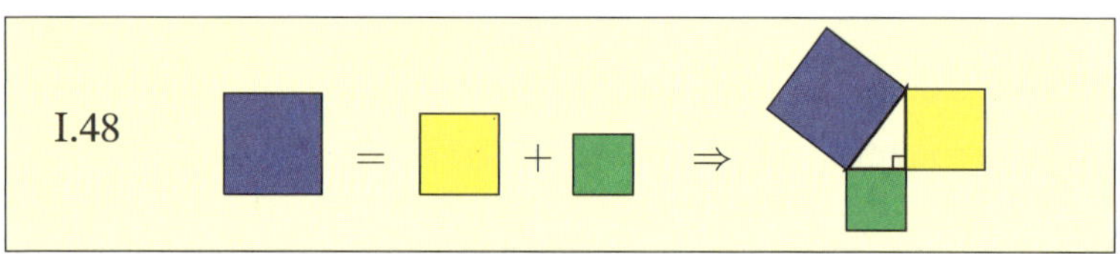

Euclid calls upon the Pythagorean Theorem to prove its converse. This is also true of converses Propositions I.13 and I.14, but it is not always the case in mathematics. For example, while Proposition I.29 contains the converse of I.27, the proof of I.29 does not rely on I.27. Interestingly, we do not know of the existence of a proof of I.48 using only the pre-Pythagorean propositions. Put another way, could the order of these propositions be reversed? If so, would the proof of the Pythagorean Theorem rely on its converse?

Figure 7.35. Babylonian tablet Plimpton 322

Thousands of clay tablets dating from the Babylonian Period (ca. 1900-1700 BCE) were discovered in the eighteenth and nineteenth centuries in present-day Iraq, many preserving mathematics. Figure 7.35 captures the most famous mathematical tablet,

Plimpton 322, as it appeared on display at the Institute for the Study of the Ancient World in New York City in 2011. Dating from approximately 1700 BCE, this 5 inch × 3.5 inch Babylonian tablet consists of fifteen lines of numbers arranged in four columns. The column on the far right is a numbering of the lines from 1 to 15. The first three columns of numbers are of interest to historians and mathematicians alike as the numbers exhibit a Pythagorean relationship. Triples of integers $\{a, b, c\}$ such that

$$c^2 = a^2 + b^2$$

are known as **Pythagorean triples**. The fifteen such triples preserved on the tablet include $\{119, 120, 169\}$ as well as the whopper $\{12709, 13500, 18541\}$. While we do not know the method used to construct these triples, it is clear that the Babylonians were not simply guessing.

In Lemma 1 to Proposition 29 of Book X, Euclid gives his own method for generating Pythagorean triples. Translating his geometric result to our modern algebraic language, he starts with perfect squares p^2 and q^2, where $p > q$ and p and q have the same parity (either both even or both odd), then proves that

$$p^2 q^2 + \left(\frac{p^2 - q^2}{2}\right)^2 = \left(\frac{p^2 + q^2}{2}\right)^2.$$

An equivalent method is to pick positive integers m and n with $m > n$ and note that

$$(2mn)^2 + (m^2 - n^2)^2 = (m^2 + n^2)^2.$$

We leave it to the reader to verify this equality, but once established, it follows that $\{m^2 - n^2, \, 2mn, \, m^2 + n^2\}$ forms a triple. For example, $m = 2$ and $n = 1$ produces the well-known triple $\{3, 4, 5\}$, while $m = 12$ and $n = 5$ gives triple $\{119, 120, 169\}$. We leave it to the reader to determine m and n for triple $\{12709, 13500, 18541\}$.

A final question remains. Will this method generate all possible Pythagorean triples, or are there some which do not conform to the algebraic formula? The answer wasn't known until Leonhard Euler (1707–1783) tackled the problem [**41**]. Fortunately, any triple that cannot be generated by this method is simply a multiple of a triple formed by the method. For example, the triple $\{15, 20, 25\}$ cannot be generated by an m and n, but it is a multiple of the triple $\{3, 4, 5\}$ which, as shown above, is generated by this method. Stated more formally, if $\{a, b, c\}$ is a Pythagorean triple with no common factors (other than 1), then there exists a pair of integers m and n with no common factors (other than 1) such that $\{a, b, c\} = \{m^2 - n^2, 2mn, m^2 + n^2\}$. Thus, all other Pythagorean triples are simply multiples of those generated in this way.

Exercises 7.6

1. Give an updated version of Euclid's proof of Proposition I.47. Be sure to justify each step, substitute mathematical symbols where appropriate, and include helpful diagrams as needed. Fill in the details that Euclid skips at the end, namely that the parallelogram CL has the same area as the square $\square ACKH$.

2. Give an updated version of Euclid's proof of Proposition I.48. Be sure to justify each step, substitute mathematical symbols where appropriate, and include helpful diagrams as needed.

3. Using algebra, show that $(2mn)^2 + (m^2 - n^2)^2 = (m^2 + n^2)^2$.

4. Find m and n for the Pythagorean triple $\{12709, 13500, 18541\}$.

5. Given a square of side length b, construct a square of double its area. Use dynamic geometry software such as Geometer's Sketchpad® or GeoGebra to determine the simplest construction. Give all construction steps as well as the justification.

6. Construct a square whose area equals the sum of three given squares. Use dynamic geometry software such as Geometer's Sketchpad® or GeoGebra to determine the simplest construction. Give all construction steps as well as the justification.

7. The following proof of the Pythagorean Theorem is attributed to the Thābit ibn Qurra (836–901) and is recounted by Heath.

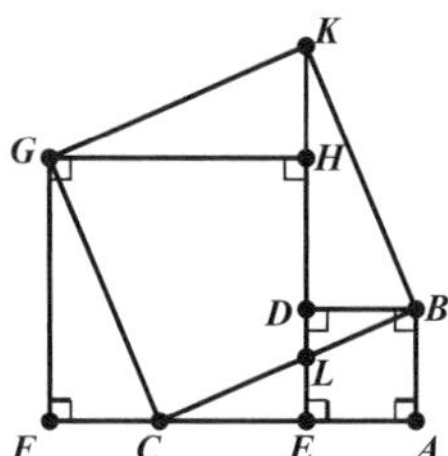

Figure 7.36. Exercise 7.6.7: Thābit ibn Qurra's proof of the Pythagorean Theorem

Let $\triangle ABC$ be given with right angle at A. Construct square $\square ABDE$ on side AB; produce AC to F such that $EF = AC$. Construct square $\square EHGF$ on side EF, and produce DH to K such that $DK = AC$, as illustrated in Figure 7.36.

   (a) Prove that triangles $\triangle BAC$, $\triangle CFG$, $\triangle KHG$ and $\triangle BDK$ are all congruent.

   (b) Prove that quadrilateral $BCGK$ is a square.

   (c) Decompose $\square ABDE$ and $\square EFGH$ into nonoverlapping pieces that, when rearranged, form square $\square BCGK$. Explain why this proves the Pythagorean Theorem.

8. The following proof of the Pythagorean Theorem is attributed to Leonardo da Vinci (1452–1519).

Let $\triangle ABC$ be given with right angle at B. Construct squares $\square ABED$, $\square BCGF$ and $\square CAKH$ on the sides of $\triangle ABC$. On segment HK, construct triangle $\triangle HKL$ such that $KL = BC$ and $LH = AB$, as illustrated in Figure 7.37. Construct segments BD, BG and BL.

   (a) Prove that triangles $\triangle ABC$, $\triangle EBF$ and $\triangle HLK$ are all congruent.

   (b) Prove that BD and BG lie in a straight line.

   (c) Prove that quadrilaterals $ADGC$, $EDGF$, $ABLK$ and $HLBC$ are all congruent. [*Hint: Use congruence scheme SASAS for quadrilaterals from Exercise* 3.1.8.]

   (d) Explain why hexagons $ADEFGC$ and $ABCHLK$ have equal area. From this, prove the Pythagorean Theorem.

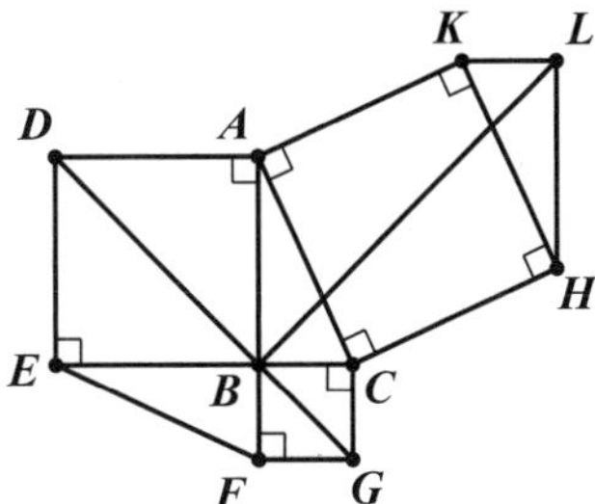

Figure 7.37. Exercise 7.6.8:   Leonardo da Vinci's proof of the Pythagorean Theorem

9. The following proof of the Pythagorean Theorem is attributed to Congressman (later President) James A. Garfield, who published it in the *New England Journal of Education* in 1876.

   Let $\triangle ABC$ be given with right angle at A. On hypotenuse BC, construct a "half-square," $\triangle BCE$, that is, construct a right angle $\angle BCE$ such that $BC = CE$. Construct a perpendicular from E to $\overrightarrow{AC}$, and let the intersection be D, as illustrated in Figure 7.38.

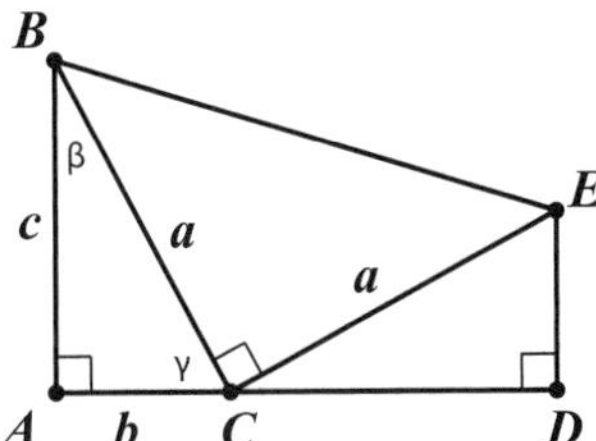

Figure 7.38. Exercise 7.6.9:   James A. Garfield's proof of the Pythagorean Theorem

   (a) Prove that $\triangle ABC \cong \triangle DCE$.
   (b) Explain why quadrilateral $BADE$ is a trapezoid.
   (c) Find the area of trapezoid $BADE$ in two different ways: first, by using the area formula for a trapezoid, and second, by decomposing the trapezoidal area into the sum of the areas of its three component triangles. From this, prove the Pythagorean Theorem.

10. Find the error in the following "proof" of this clearly false proposition.

    Proposition: Every triangle is isosceles.

    Proof.   Let $\triangle ABC$ be given. We will show that $AB = AC$. Let D be the midpoint of BC and let E be the intersection of the angle bisector for $\angle BAC$ and the perpendicular bisector of BC as shown in Figure 7.39.

    Construct EF perpendicular to AB and EG perpendicular to AC, where F and G lie on AB and AC, respectively.

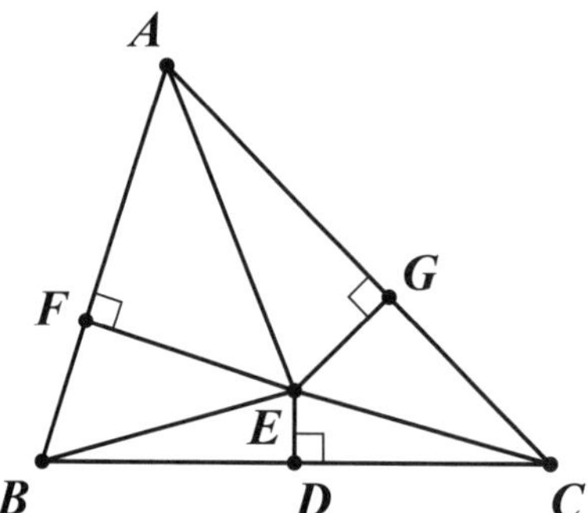

Figure 7.39. Exercise 7.6.10: Are all triangles isosceles?

Then, $EF = EG$, by Theorem 3.13.

Thus, $AF = AG$, by I.47.

Futhermore, $EB = EC$, by Theorem 3.5.

Hence, $BF = CG$, again by I.47.

Therefore, $AB = AC$ by CN 2. $\square$

8

Book II: Geometric Algebra

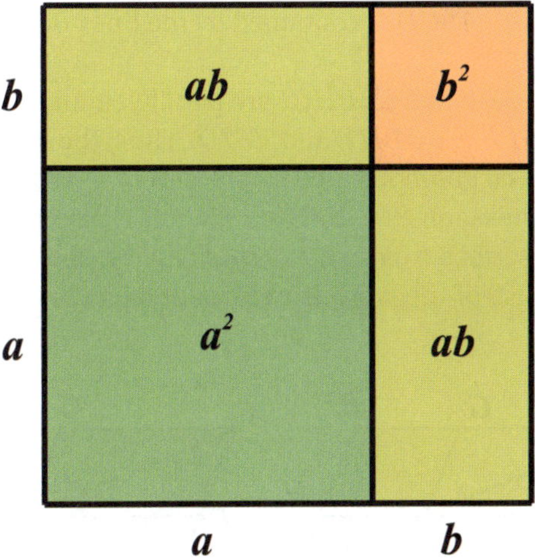

Figure 8.1. Proposition II.4: $(a + b)^2 = a^2 + 2ab + b^2$

The propositions of Book II are often referred to as **geometric algebra**, that is, the representation of algebraic concepts through geometric figures. An example of this can be found in Figure 8.1 which gives a geometric representation of an algebraic equation that can be verified by a high-school student. It's important to note that the variables, arithmetic operations and equation demonstrated here were not part of the lexicon of the Greek mathematicians. The now familiar symbols of the algebraic arts were developed slowly over the centuries, and were not universally applied until after the invention of the printing press. For example, the convention of using x as an unknown variable is thought to have been used first by Descartes in the seventeenth century. The Greeks solved algebraic problems through geometry. Thus, for Euclid an unknown term takes its representation as the length of a line segment which, as in the figure, may be part of a side of some geometric object. Likewise, the Pythagorean Theorem

159

relates the areas of the squares on the sides of a right triangle. Naturally, this presents a roadblock to working with negative quantities. Though we have seen subtraction take the form of shorter segments removed from longer segments, the reverse is impossible.

All but two of the propositions in Book II are more easily understood by the modern reader as algebraic equations. Propositions II.11 and II.14 are the lone exceptions, as they are constructions. In particular, Propositions II.12 and II.13 are equivalent to the *Law of Cosines*, a natural extension of the Pythagorean Theorem. Since our readers are well-versed in the algebraic arts, we will state and discuss the first ten propositions of this book, but not prove them. We will, however, state and prove the last four propositions as they hold the most significant results. In addition to the new constructions, they provide an adequate sampling of geometric algebra proofs.

As Euclid begins Book II, he gives a few new definitions.

(1) Any rectangular parallelogram is said to be *contained* by the two straight lines containing the right angle.

(2) And in any parallelogrammic area let any one whatever of the parallelograms about its diameter with the two complements be called a *gnomon*.

In Euclid's language, rectangle $\square ABCD$ is contained by the lines AB and BC. Likewise, we could say that $\square ABCD$ is contained by the lines BC and CD, CD and DA, or DA and AB.

In Figure 8.2, $\square AGFE$ and $\square FHCI$ are parallelograms about the diameter. The complements of $\square AGFE$ are $GBHF$ and $EFID$. Thus, the shaded L-shaped hexagon $ABHFID$ on the right is a gnomon since it consists of $\square AGFE$ and its complements. The shaded L-shaped hexagon $BCDEFG$ on the left is also a gnomon. In Greek, the word gnomon once described a carpenter's square, an L-shaped tool used for drawing right angles. Euclid's use of the word is more general as his parallelograms need not be right-angled.

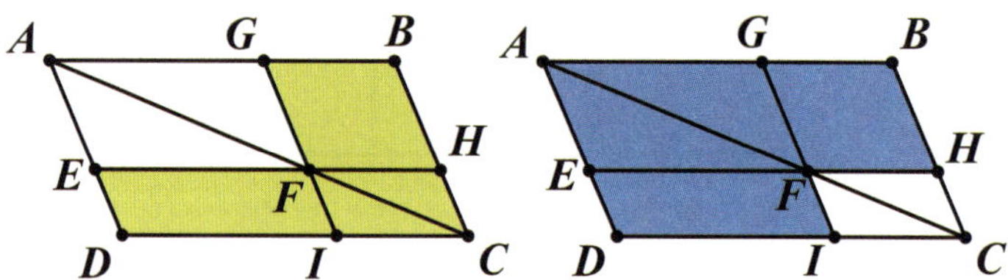

Figure 8.2. Gnomons

8.1 Proposition II.1 through II.10

For the first ten propositions, in addition to stating the proposition, we include the first line of each proof as it greatly helps to clarify the meaning of the proposition. We will rewrite Propositions II.1 and II.4 as algebraic equations and leave the other propositions as exercises for the reader.

Proposition II.1. *If there be two straight lines, and one of them be cut into any number of segments whatever, the rectangle contained by the two straight lines is equal to the rectangles contained by the uncut straight line and each of the segments.*

Start of the Proof. Let A, BC be two straight lines, and let BC be cut at random at the points D, E; I say that the rectangle contained by A, BC is equal to the rectangle contained by A, BD, that contained by A, DE and that contained by A, EC. ...

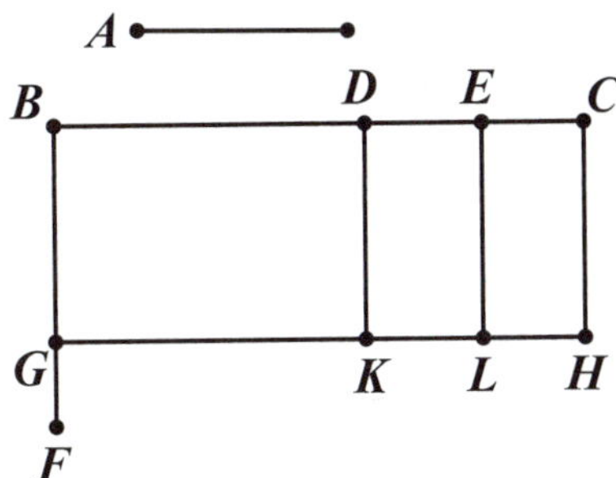

Figure 8.3. Euclid's diagram for Proposition II.1

As in Euclid's first definition, the phrase "the rectangle contained by the two straight lines" means any rectangle that can be constructed from the given lengths A and BC. Thus, the rectangle shown in Figure 8.3 is contained by BC and BG, where BG is equal to given length A. We give a revised version of Euclid's diagram in Figure 8.4 where lengths a, b, c and d take the place of segments BG, BD, DE and EC, respectively. With these substitutions, we see that Euclid's first proposition proves the equivalent algebraic identity $a(b + c + d) = ab + ac + ad$. This is a specific case of the general distributive property:

$$a(b_1 + b_2 + b_3 + \cdots + b_n) = ab_1 + ab_2 + ab_3 + \cdots + ab_n.$$

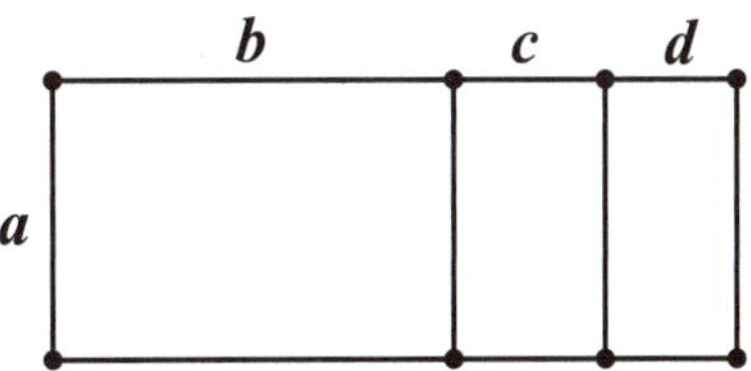

Figure 8.4. Revised diagram for Proposition II.1

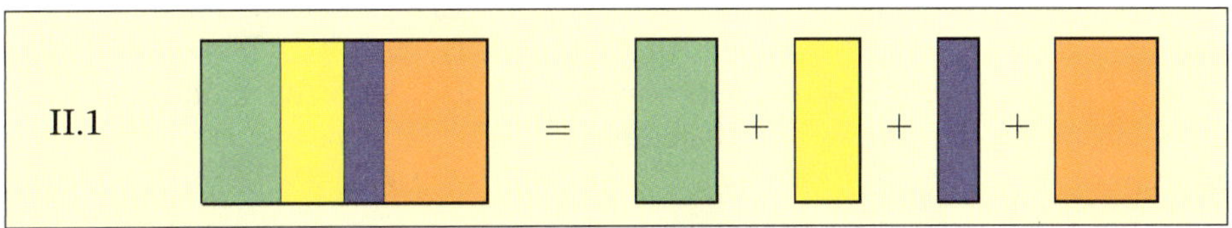

Proposition II.2. *If a straight line be cut at random, the rectangles contained by the whole and both of the segments are equal to the square on the whole.*

Start of the Proof. For let the straight line AB be cut at random at the point C; I say that the rectangle contained by AB, BC together with the rectangle contained by BA, AC is equal to the square on AB. ...

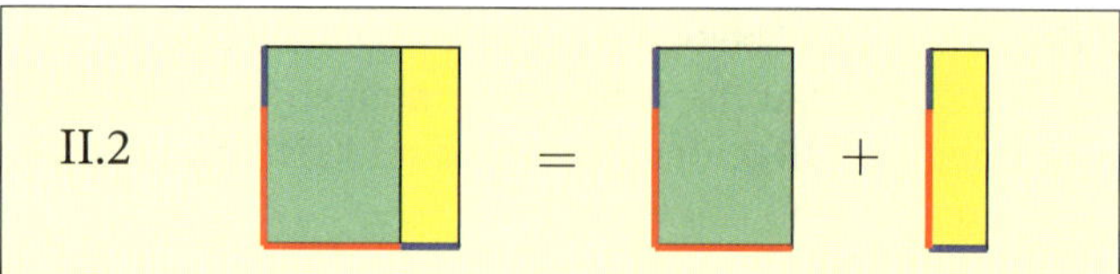

Proposition II.3. *If a straight line be cut at random, the rectangle contained by the whole and one of the segments is equal to the rectangle contained by the segments and the square on the aforesaid segment.*

    Start of the Proof. For let the straight line AB be cut at random at C; I say that the rectangle contained by AB, BC is equal to the rectangle contained by AC, CB together with the square on BC. ...

Proposition II.4. *If a straight line be cut at random, the square on the whole is equal to the squares on the segments and twice the rectangle contained by the segments.*

    Start of the Proof. For let the straight line AB be cut at random at C; I say that the square on AB is equal to the squares on AC, CB and twice the rectangle contained by AC, CB. ...

Proposition II.4 merits further discussion. With $a = AC$ and $b = CB$, it can be seen as a statement of the familiar algebraic result

$$(a + b)^2 = a^2 + 2ab + b^2.$$

The large square found in the miniature pictorial for II.4 and in Figure 8.1, both modified versions of Euclid's diagram, are often used to justify the result to algebra students. It can also be used to give a visual representation of the process of completing the square. To make it look like the formula seen by many an algebra student, redraw the diagram replacing a with the unknown x. Now, erase the square of area b^2 in the upper right corner, making the large square incomplete. Algebraically, the portion that remains is $x^2 + xb + xb = x^2 + 2bx$. In algebra, we learn to "take half of the middle coefficient and square it" in order to complete the square. Since the middle coefficient is $2b$, this means that adding b^2 will complete the square, which is obvious from the diagram.

Proposition II.5. *If a straight line be cut into equal and unequal segments, the rectangle contained by the unequal segments of the whole together with the square on the straight line between the points of section is equal to the square on the half.*

    Start of the Proof. For let a straight line AB be cut into equal segments at C and into unequal segments at D; I say that the rectangle contained by AD, DB together with the square on CD is equal to the square on CB. ...

Proposition II.6. *If a straight line be bisected and a straight line be added to it in a straight line, the rectangle contained by the whole with the added straight line and the added straight line together with the square on the half is equal to the square on the straight line made up of the half and the added straight line.*

> *Start of the Proof.* For let a straight line AB be bisected at the point C, and let a straight line BD be added to it in a straight line; I say that the rectangle contained by AD, DB together with the square on CB is equal to the square on CD. ...

Proposition II.7. *If a straight line be cut at random, the square on the whole and that on one of the segments both together are equal to twice the rectangle contained by the whole and the said segment and the square on the remaining segment.*

> *Start of the Proof.* For let a straight line AB be cut at random at the point C; I say that the squares on AB, BC are equal to twice the rectangle contained by AB, BC and the square on CA. ...

Proposition II.8. *If a straight line be cut at random, four times the rectangle contained by the whole and one of the segments together with the square on the remaining segment is equal to the square described on the whole and the aforesaid segment as on one straight line.*

> *Start of the Proof.* For let a straight line AB be cut at random at the point C; I say that four times the rectangle contained by AB, BC together with the square on AC is equal to the square described on AB, BC as on one straight line. ...

Proposition II.9. *If a straight line be cut into equal and unequal segments, the squares on the unequal segments of the whole are double of the square on the half and of the square on the straight line between the points of section.*

> *Start of the Proof.* For let a straight line AB be cut into equal segments at C, and into unequal segments at D; I say that the squares on AD, DB are double of the squares on AC, CD. ...

Proposition II.10. *If a straight line be bisected, and a straight line be added to it in a straight line, the square on the whole with the added straight line and the square on the added straight line both together are double of the square on the half and of the square described on the straight line made up of the half and the added straight line as on one straight line.*

> *Start of the Proof.* For let a straight line AB be bisected at C, and let a straight line BD be added to it in a straight line; I say that the squares on AD, DB are double of the squares on AC, CD. ...

Exercises 8.1

1. In Proposition II.2, letting $a = AC$ and $b = CB$, rewrite the proposition as an algebraic equation using only a and b.

2. In Proposition II.3, letting $a = AC$ and $b = CB$, rewrite the proposition as an algebraic equation using only a and b.

3. In Proposition II.5, letting $a = AD$ and $b = DB$, rewrite the proposition as an algebraic equation using only a and b. WLOG, assume that $a > b$.

4. In Proposition II.6, letting $a = AC$ and $b = BD$, rewrite the proposition as an algebraic equation using only a and b.

5. In Proposition II.7, letting $a = AC$ and $b = CB$, rewrite the proposition as an algebraic equation using only a and b.

6. In Proposition II.8, letting $a = AC$ and $b = CB$, rewrite the proposition as an algebraic equation using only a and b.

7. In Proposition II.9, letting $a = AD$ and $b = DB$, rewrite the proposition as an algebraic equation using only a and b. WLOG, assume that $a > b$.

8. In Proposition II.10, letting $a = AC$ and $b = BD$, rewrite the proposition as an algebraic equation using only a and b.

9. The following proof of the Pythagorean Theorem is an example of a **dissection** proof. While its origins are unclear, it is very old, possibly dating back to Pythagoras. Here we consider two ways of dividing up a square with sides of length $a + b$, as illustrated in Figure 8.5. For the square on the left, we join the dividing points between the segments of length a and those of length b on each side of the square to form a tilted quadrilateral with sides of length c. For the square on the right, we decompose it into a square with side a, a square with side b, and two rectangles with sides a and b, exactly as we did in Proposition II.4. We then bisect the two rectangles, creating four congruent right triangles.

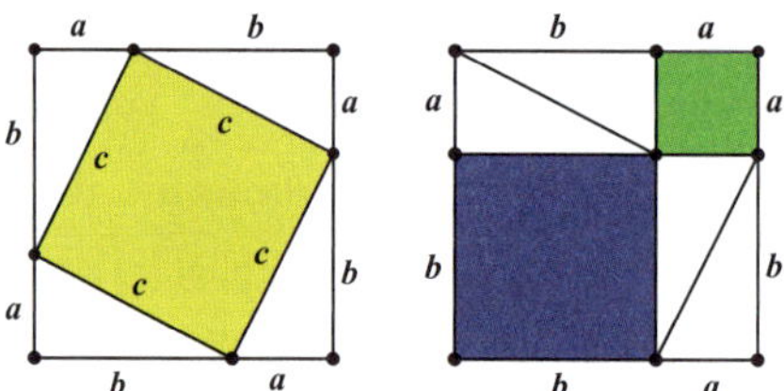

Figure 8.5. Exercise 8.1.9

(a) Explain why all eight triangles are congruent.

(b) Prove that in the left figure, the quadrilateral with sides c is a square.

(c) Using these two figures, prove the Pythagorean Theorem.

10. To "prove" the Pythagorean Theorem, the Hindu mathematician Bhāskara (1114–ca. 1185) simply drew four congruent right triangles inwards, one on each side of the square on the hypotenuse, as illustrated in Figure 8.6, and said, "See!"

(a) Label the hypotenuse c and the legs a and b, with $a > b$. Prove that the large quadrilateral with sides c is a square.

(b) Prove that the small quadrilateral inside is also a square. Find the length of its sides in terms of a and b.

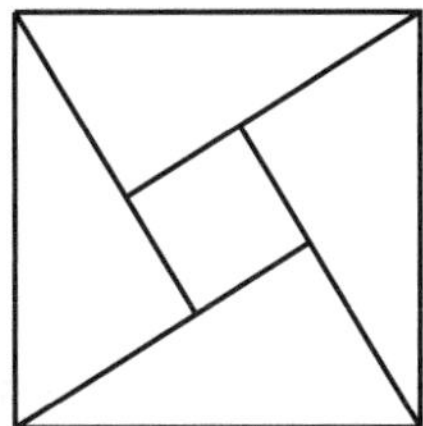

Figure 8.6. Exercise 8.1.10

(c) Calculate the area of the large square two different ways, first using the formula for area and second as the sum of the smaller square and four triangles. From this, prove the Pythagorean Theorem.

11. The following is an outline of the proof of the quadratic formula using the process of completing the square. Fill in any gaps and justify each step.

$$
\begin{aligned}
ax^2 + bx + c &= 0, \\
x^2 + \frac{b}{a}x &= -\frac{c}{a}, \\
\left(x + \frac{b}{2a}\right)^2 &= \frac{b^2 - 4ac}{4a^2}.
\end{aligned}
$$

8.2 Propositions II.11 through II.14

Proposition II.11 turns our focus away from geometric algebra to a new construction. Given line segment AB, Euclid uses his geometric tools to mark the point H on AB such that $AB \cdot HB = AH \cdot AH = (AH)^2$. If we let $a = AB$ and $x = AH$, then algebraically we wish to solve for x such that $a(a - x) = x^2$, or equivalently, $x^2 + ax = a^2$. This is equivalent to the construction of what has come to be known as the **golden ratio**. Usually denoted by ϕ, the golden ratio (or golden mean) is a number, specifically $\phi = 1.6180339887...$, such that two segments which are in the same proportion as ϕ are *divinely proportioned*. More specifically, two quantities p and q, where $p > q$, are **divinely proportioned** when

$$
\frac{p}{q} = \frac{p + q}{p}. \tag{8.1}
$$

(Note that we have $\phi = \frac{p}{q}$.) There has been much written on the aesthetic properties of the golden mean. Perhaps explaining its divine nature, Italian mathematician Luca Pacioli (1445–1517) saw the hand of god in this ratio. Rectangles with adjacent sides in proportion with the golden ratio (golden rectangles) are said to be most pleasing to the eye, and it is claimed that Greek architecture was influenced by this proportion as evidenced by its surviving structures.

To see how Euclid's proposition is related to the golden mean, we can simply rearrange the equation given above, $x^2 + ax = a^2$, into $\frac{a}{x} = \frac{x+a}{a}$. Euclid's construction identifies the point H on the given segment AB where the entire segment, $AB = a$, is proportional to the constructed segment, $AH = x$, in the golden ratio. Before we take a look at Euclid's proof we can put our algebraic tools to use, applying the quadratic formula to solve for the positive root, giving

$$x = \frac{a}{2}(\sqrt{5} - 1).$$

(8.2)

When we solve this equation for the ratio of segment AB to the subsegment AH, we find it to be divine since

$$\frac{AB}{AH} = \frac{a}{x} = \frac{1 + \sqrt{5}}{2} = \phi = 1.6180339887\ldots.$$

We give Euclid's construction here, supplemented with a modern algebraic proof to show that the construction is valid. For Euclid, this proposition is essential to his construction of a regular pentagon in Book IV.

Proposition II.11. *To cut a given straight line so that the rectangle contained by the whole and one of the segments is equal to the square on the remaining segment.*

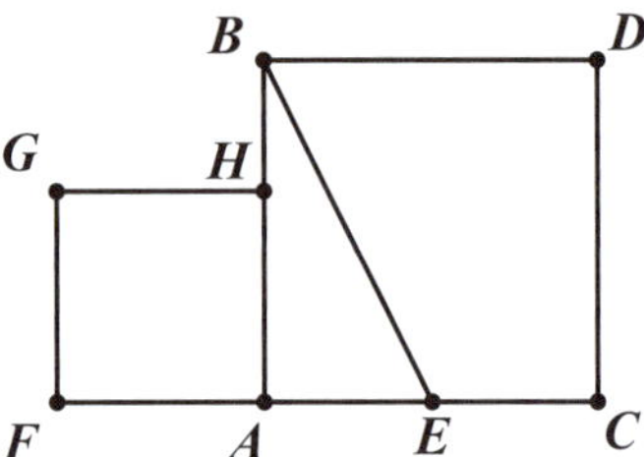

Figure 8.7. Proposition II.11

Proof. *Euclid's Construction:* Let AB be the given segment.
  Construct the square $ABDC$ on segment AB. [I.46]
  Bisect AC at the point E. [I.10]
  Join BE [Post. 1]
  Extend EA to a point F such that $EF = EB$. [Post. 2 & I.3]
  Construct the square $AFGH$ so that H lies on AB. [1.46]
  Claim: H is the point on AB such that $AB \cdot HB = (AH)^2$.

Algebraic justification: Let $a = AB$ and $x = AH$. It suffices to show that $x = \frac{a}{2}(\sqrt{5} - 1)$.
Since E bisects AC, we have $EA = \frac{a}{2}$. By the Pythagorean Theorem,

$$EF = EB = \sqrt{(AB)^2 + (EA)^2} = \sqrt{a^2 + \left(\frac{a}{2}\right)^2} = \frac{a\sqrt{5}}{2}.$$

But then $x = AH = AF = (EF - EA) = \frac{a\sqrt{5}}{2} - \frac{a}{2} = \frac{a}{2}(\sqrt{5} - 1)$, as desired. $\square$

II.11	Given		construct		such that		

The next two propositions are equivalent to the **Law of Cosines**, as shown in Figure 8.8, which states that in any triangle with sides of length a, b and c, we have

$$a^2 = b^2 + c^2 - 2bc \cos A,$$

where A is the angle opposite side a. As Euclid did not employ trigonometric functions, he breaks the proposition into two cases. In Proposition II.12 (see Figure 8.9), given triangle $\triangle ABC$ with obtuse angle $\angle BAC$, he proves that

$$BC^2 = AB^2 + AC^2 + 2AC \cdot AD,$$

where D is a point constructed on $\overleftrightarrow{AC}$ such that $BD \perp \overleftrightarrow{AC}$. To see that this is the Law of Cosines, notice that angles $\angle BAD$ and $\angle BAC$ are supplementary and

$$\cos(\angle BAD) = \frac{AD}{AB}.$$

Therefore,

$$\cos(\angle BAC) = \cos(\pi - \angle BAD) = -\cos(\angle BAD) = -\frac{AD}{AB},$$

or equivalently, $AD = -AB \cos(\angle BAC)$ as required. In Proposition II.13 (see Figure 8.11), given triangle $\triangle ABC$ with acute angle $\angle ABC$, he proves that

$$AC^2 = AB^2 + BC^2 - 2BC \cdot BD.$$

Here, we note that

$$\cos(\angle ABD) = \frac{BD}{AB}.$$

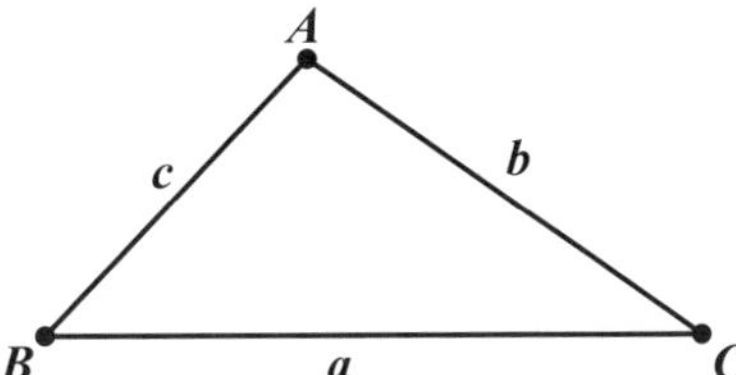

Figure 8.8. Law of Cosines: $a^2 = b^2 + c^2 - 2bc \cos A$

Proposition II.12. *In obtuse-angled triangles the square on the side subtending the obtuse angle is greater than the squares on the sides containing the obtuse angle by twice the rectangle contained by one of the sides about the obtuse angle, namely that on which the perpendicular falls, and the straight line cut off outside by the perpendicular towards the obtuse angle.*

Euclid's Proof. Let ABC be an obtuse-angled triangle having the angle BAC obtuse, and let BD be drawn from the point B perpendicular to CA produced;

I say that the square on BC is greater than the squares on BA, AC by twice the rectangle contained by CA, AD.

For, since the straight line CD has been cut at random at the point A, the square on DC is equal to the squares on CA, AD and twice the rectangle contained by CA, AD. [II.4]

Let the square on DB be added to each; therefore the squares on CD, DB are equal to the squares on CA, AD, DB and twice the rectangle CA, AD.

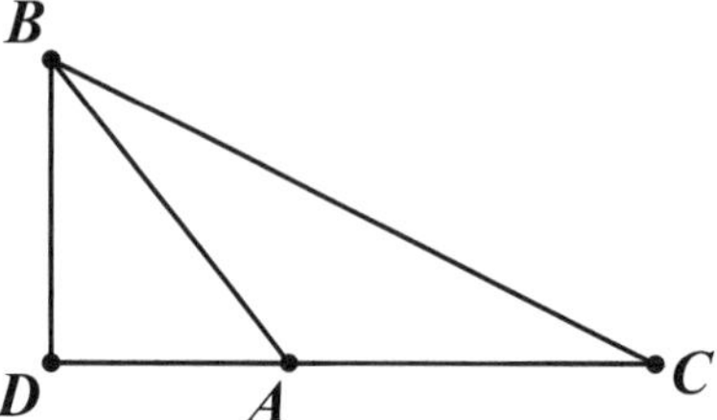

Figure 8.9. Proposition II.12

But the square on CB is equal to the squares on CD, DB, for the angle at D is right; [I.47]

and the square on AB is equal to the squares on AD, DB; [I.47]

therefore the square on CB is equal to the squares on CA, AB and twice the rectangle contained by CA, AD;

so that the square on CB is greater than the squares on CA, AB by twice the rectangle contained by CA, AD.

Therefore etc. Q.E.D.   □

An updated proof is as follows:

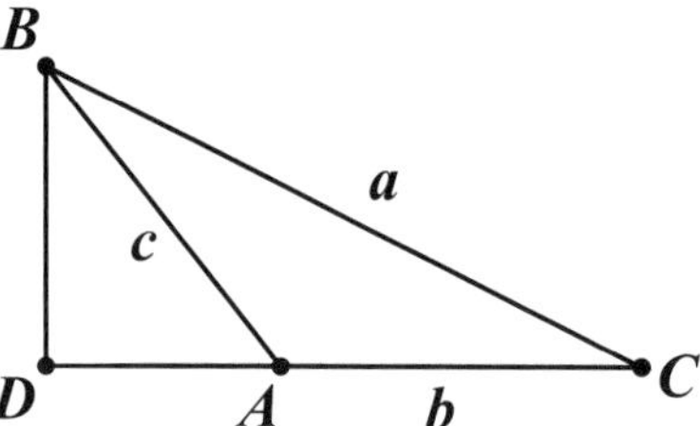

Figure 8.10. Revised diagram for Proposition II.12

Revised Proof.   Consider the triangle $\triangle ABC$ with $\angle BAC$ obtuse. Let $a = BC, b = AC$ and $c = AB$.

By I.12, construct a perpendicular from B to $\overleftrightarrow{AC}$ and let the intersection be called D.

Clearly, $DC^2 = (b + AD)^2 = b^2 + 2b \cdot AD + AD^2$.

Adding BD^2 to each side, we have $DC^2 + BD^2 = b^2 + 2b \cdot AD + AD^2 + BD^2$.

Since $\angle BDC$ is a right angle, by the Pythagorean Theorem [I.47], $a^2 = DC^2 + BD^2$ and $c^2 = AD^2 + BD^2$.

Substituting, we have $a^2 = b^2 + c^2 + 2b \cdot AD$, as desired.   □

Proposition II.13. *In acute-angled triangles the square on the side subtending the acute angle is less than the squares on the sides containing the acute angle by twice the rectangle contained by one of the sides about the acute angle, namely that on which the perpendicular falls, and the straight line cut off within by the perpendicular towards the acute angle.*

Euclid's Proof.   Let ABC be an acute-angled triangle having the angle at B acute, and let AD be drawn from the point A perpendicular to BC;

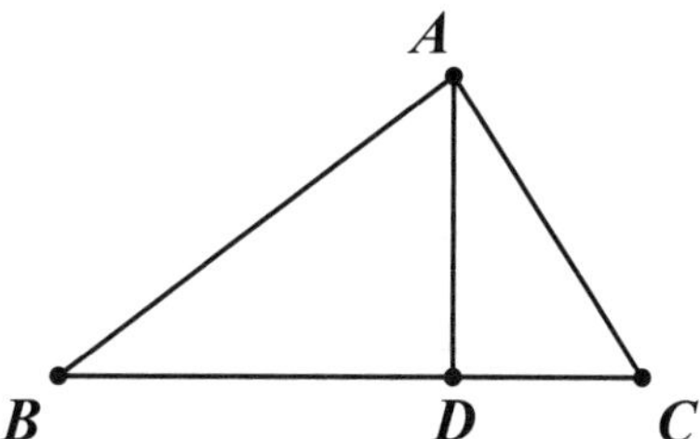

Figure 8.11. Proposition II.13

I say that the square on AC is less than the squares on CB, BA by twice the rectangle contained by CB, BD.

For, since the straight line CB has been cut at random at D, the squares on CB, BD are equal to twice the rectangle contained by CB, BD and the square on DC. [II.7]

Let the square on DA be added to each; therefore the squares on CB, BD, DA are equal to twice the rectangle contained by CB, BD and the squares on AD, DC.

But the square on AB is equal to the squares on BD, DA, for the angle at D is right; [I.47]

and the square on AC is equal to the squares on AD, DC;

therefore the squares on CB, BA are equal to the square on AC and twice the rectangle CB, BD,

so that the square on AC alone is less than the squares on CB, BA by twice the rectangle contained by CB, BD.

Therefore etc. Q.E.D. $\square$

We leave it as an exercise for the reader to provide an updated proof.

As discussed in Section 7.5, the final proposition in Book II establishes the quadrability of a general polygon. Starting with a polygon, Euclid first uses Proposition I.45 to construct a rectangle of area equal to that of the polygon. He then gives the construction for a square with area equal to the rectangle. Once again, we give Euclid's construction, but supplement with an updated algebraic proof to show that the construction is valid.

Proposition II.14. *To construct a square equal to a given rectilineal figure.*

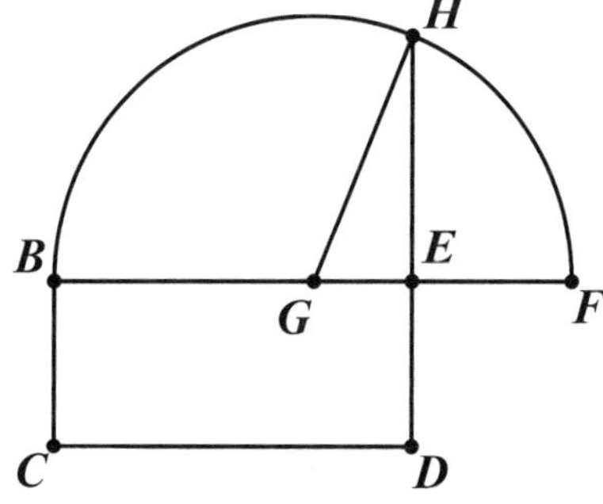

Figure 8.12. Proposition II.14

Proof. *Euclid's Construction:* Starting with the polygon A, construct a rectangle $\square BCDE$ that has the same area as A. [I.45]

    If $BE = ED$, then we are done. If not, then WLOG, assume $BE > ED$.

    Extend $\overrightarrow{BE}$ to a point F such that $EF = ED$. [Post. 2 & I.3]

    Bisect BF at G. [I.10]

    Draw a circle with center G and radius $GB = GF$. [Post. 3]

    Extend $\overrightarrow{DE}$ and let the intersection with the circle be H. Join GH.

    Claim: The area of the square on HE equals the area of the rectangle $\square BCDE$

Algebraic Justification: Since $\angle GEH$ is a right angle, $GH^2 = HE^2 + GE^2$. [I.47] But then:

$$
\begin{aligned}
HE^2 &= GH^2 - GE^2 \\
&= (GH + GE)(GH - GE) \\
&= (BG + GE)(GF - GE) \\
&= BE \cdot EF \\
&= BE \cdot DE. \qquad\qquad \square
\end{aligned}
$$

Exercises 8.2

1. Consider triangle $\triangle ABC$ with $\alpha = \dfrac{2\pi}{3}$, $b = 2$ and $c = 1$. Use the Law of Cosines to solve for a.

2. Consider triangle $\triangle ABC$ with $\alpha = \dfrac{\pi}{3}$, $b = 2$ and $c = 2$. Solve the triangle.

3. Consider triangle $\triangle ABC$ with $\alpha = \dfrac{\pi}{4}$, $b = 3$ and $c = 3$. Solve the triangle.

4. Follow parts (a) through (f) to produce an updated proof of Proposition II.13.

  (a) The proof begins:

> Let ABC be an acute-angled triangle having the angle at B acute, and let AD be drawn from point A perpendicular to BC;
> I say that the square on AC is less than the squares on CB, BA by twice the rectangle contained by CB, BD.

Here Euclid has set up the statement of what he wishes to prove. If we let the segment AB have length c, the segment AC have length b, and the segment BC have length a, rewrite this statement as an equation using a, b, and c, as well as BD.

  (b) Euclid then continues by starting the proof:

> For, since the straight line CB has been cut at random at D, the squares on CB, BD are equal to twice the rectangle contained by CB, BD and the square on DC. [II.7]

Write this as an equation using a, BD, and DC. Then using the fact that $a = BD + DC$, verify that it is true. [*Hint: Start on one side of the equation and produce the other.*]

(c) Next he states:

> Let the square on DA be added to each; therefore the squares on CB, BD, DA are equal to twice the rectangle contained by CB, BD and the squares on AD, DC.

Write this as an equation using a, BD, DC, and AD.

(d) He continues:

> But the square on AB is equal to the squares on BD, DA, for the angle at D is right; [I.47] and the square on AC is equal to the squares on AD, DC;

Write these two equations using b, c, AD, BD and DC. What theorem is Euclid using at this point?

(e) He finishes:

> therefore the squares on CB, BA are equal to the square on AC and twice the rectangle CB, BD, so that the square on AC alone is less than the squares on CB, BA by twice the rectangle contained by CB, BD. Therefore etc. Q.E.D.

Write these two equations in order, and explain why each of them holds.

(f) Recall that, in a right triangle, the cosine of an acute angle is given by the ratio of adjacent over hypotenuse. Using this, write the cosine of B in terms of lengths in the diagram. Substitute for BD in the equation that Euclid proved above and simplify to get the Law of Cosines.

8.3 Quadrature on the sphere

Let's briefly revisit Spherical geometry in order to prove the spherical version of Proposition II.14. Here, we assume that we are on a unit sphere. First, recall from Theorem 4.10 that the area of quadrilateral $ABCD$ with vertex angles α, β, γ and δ, where $0 < \alpha, \beta, \gamma, \delta < \pi$, is given by $(\alpha + \beta + \gamma + \delta) - 2\pi$. Next, we generalize the definitions of a rectangle and a square.

Definition 8.1. A **spherical rectangle** is an equiangular quadrilateral with angles less than π.

Definition 8.2. A **spherical square** is an equilateral spherical rectangle.

Since the area of a spherical rectangle with angle α is $4\alpha - 2\pi$, the vertex angle must be greater than $\pi/2$. Much like spherical triangles, a spherical square of any appropriately sized angle can be constructed. As it is the Spherical analogue to Proposition I.45, we rightly name it I.45S.

Theorem 8.3 [I.45S]. *To construct a spherical square given an angle between $\dfrac{\pi}{2}$ and π.*

Proof. Given angle α between $\dfrac{\pi}{2}$ and π, by Theorem 4.21 we may construct isosceles triangle $\triangle ABC$ with angles $\alpha/2$, $\alpha/2$ and $\alpha = \angle ABC$. Construct two circles of radius AB, one centered at A and the other at C. Let D be the intersection point of the circles other than B. By SSS, we have $\triangle ABC \cong \triangle ADC$. Thus, quadrilateral $ABCD$ is a spherical square with angle α. $\qquad\square$

We now give the spherical version of Proposition II.14.

Theorem 8.4 [II.14S]. *To construct a spherical square equal to a given spherical rectangle.*

Proof.   Suppose the given rectangle has angle α. The area of this rectangle is $4\alpha - 2\pi$. Construct a square with angle α. It will have the same area. $\square$

Proposition II.14 is the culmination of Euclid's effort to square not just a rectangle, but any polygon. We can nearly match Euclid's result in Spherical geometry. First, we show that triangles are quadrable in Theorem 8.5, then generalize to convex polygons in Theorem 8.6 to produce the analogous result.

Theorem 8.5. *To construct a spherical square equal to a given spherical triangle.*

Proof.   Let spherical triangle $\triangle ABC$ be given with angles α_1, α_2 and α_3. By Theorem 4.8, the area of $\triangle ABC$ is equal to $\alpha_1 + \alpha_2 + \alpha_3 - \pi$. Let

$$\beta = \left(\sum_{i=1}^{3} \frac{\alpha_i}{4} \right) + \frac{\pi}{4}.$$

Note that we can quarter any angle by bisecting twice. Thus, we can construct β. Moreover, since $\pi < \sum_{i=1}^{3} \alpha_i < 3\pi$ by Corollary 4.9 (the Spherical analogue of I.32), we have $\frac{\pi}{2} < \beta < \pi$. Construct a square with angle β by Theorem 8.3. The area of this square equals

$$4\beta - 2\pi = \left(\sum_{i=1}^{3} \alpha_i \right) + \pi - 2\pi = area(\triangle ABC).   \qquad \square$$

We finish this chapter by establishing the quadrability of any convex spherical polygon.

Theorem 8.6. *To construct a spherical square equal to a convex spherical polygon.*

Proof.   In a convex spherical polygon, each vertex angle is less than π. Consider the spherical polygon with n sides and angles, α_i ($1 \leq i \leq n$), where each α_i is less than π. We will assume that $n \geq 4$. By Theorem 4.11, the area of this polygon is given by $\left(\sum_{i=1}^{n} \alpha_i \right) - (n-2)\pi$. Notice that $\sum_{i=1}^{n} \alpha_i > (n-2)\pi$ since area is positive. Let

$$\beta = \left(\sum_{i=1}^{n} \frac{\alpha_i}{4} \right) - \frac{(n-4)\pi}{4}.$$

Since we can quarter any angle by bisecting twice, we can construct angle β. Moreover, since $(n-2)\pi < \sum_{i=1}^{n} \alpha_i < n\pi$, we have $\frac{\pi}{2} < \beta < \pi$. Construct a square with angle β by Theorem 8.3. The area of this square equals

$$4\beta - 2\pi = \left(\sum_{i=1}^{n} \alpha_i \right) - (n-2)\pi,$$

as desired. $\square$

This result is analogous to Euclid's quadrature result, but certainly not as comprehensive due to the assumption of convexity. Unlike Euclidean geometry, we cannot square all spherical polygons since the area of a nonconvex polygon may be greater than 2π, the maximum area of any spherical square.

9

Book VI: Similarity

Figure 9.1. Golden rectangle

While the focus of the first two books of the *Elements* is clearly the *line*, Euclid shifts his gaze to the *circle* in Book III. "It's about time!," our neglected supporting character whines. The poor circle has played second fiddle to the line since Chapter 5 and is looking forward to this reversal of fortune. Turnabout is fair play, and while we are happy to oblige our curvy friend, we kindly request a skosh of patience as we take a detour to introduce *similarity*.

Two polygons are called similar when their angles and sides meet a specific set of conditions. Two triangles, for example, are similar when they have three congruent angles. Though Euclid does not introduce similarity until Book VI, some proofs in the first four books would be significantly easier were we able to utilize similarity results. Several of the propositions in Book III, in particular, are so much simpler that we find it best to skip ahead to Book VI. Consequently, the reign of the line continues for one more chapter as we lay the foundation for the circle's uprising in the next.

Of the 33 propositions of Book VI, we consider a selection of those necessary for understanding the most important similarity results. In doing so, we prove the familiar triangle similarity scheme, *AAA similarity*, as well as the two lesser-known triangle similarity schemes, *SSS* and *SAS similarity*. We end the chapter with a generalization of the Pythagorean Theorem.

9.1 Book V: Ratio and proportion

As a precursor to his results on similarity, Euclid needs twenty-five propositions in Book V consisting of a geometric treatment of ratio and proportion. These results are independent of the first four books, and not unlike Book II, the algebra underlying these geometric results boils down to basic manipulations. Listed below are two examples with updated notation.

Proposition V.16. *If* $\dfrac{a}{b} = \dfrac{c}{d}$, *then* $\dfrac{a}{c} = \dfrac{b}{d}$.

Proposition V.24. *If* $\dfrac{a}{b} = \dfrac{c}{d}$ *and* $\dfrac{p}{b} = \dfrac{q}{d}$, *then* $\dfrac{a+p}{b} = \dfrac{c+q}{d}$.

Clearly, these propositions reduce to easy algebraic techniques, but when presented as geometry problems, both the language and the geometric treatment of ratios present unnecessary obstacles, and we discover a newfound appreciation for the brevity and specificity of our algebraic language. As an example, Proposition V.24 is as follows: *If a first magnitude have to a second the same ratio as a third has to a fourth, and also a fifth have to the second the same ratio as a sixth to the fourth, the first and fifth added together will have to the second the same ratio as the third and sixth have to the fourth.* Since any high-school student is familiar with such well-defined notions of ratios and proportions, we skip all of Book V in favor of relying on our considerable algebraic skills.

9.2 Similarity

Let's begin with Euclid's definitions at the start of Book VI.[1] Note that we will use the term **altitude** interchangeably with Euclid's term *height* given in Definition 4.

(1) *Similar rectilineal figures* are such as have their angles severally equal and the sides about the equal angles proportional.

(3) A straight line is said to have been *cut in extreme and mean ratio* when, as the whole line is to the greater segment, so is the greater to the less.

(4) The *height* of any figure is the perpendicular drawn from the vertex to the base.

From Definition 1, we have triangles $\triangle ABC$ and $\triangle DEF$ are **similar** if $\angle A = \angle D$, $\angle B = \angle E$, $\angle C = \angle F$ and

$$\frac{AB}{DE} = \frac{AC}{DF} = \frac{BC}{EF}.$$

We write $\triangle ABC \sim \triangle DEF$ to denote this similarity. The third definition revisits the construction given in Proposition II.11 where AB is cut at H so that $\dfrac{AB}{AH} = \dfrac{AH}{HB}$. With AH designated as the greater segment, this equation precisely matches the statement of Definition 3 where the whole is to the greater segment as the greater is to the smaller.

---

[1] Heath does not translate Definition 2 as he says it conveys "no intelligible meaning."

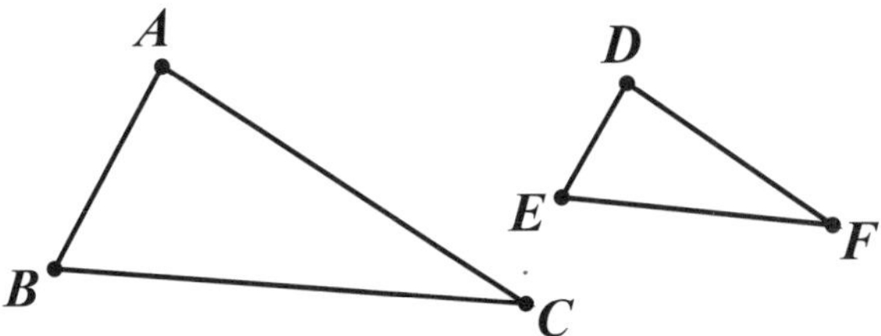

Figure 9.2. Similar triangles

From Section 8.2, we recognize the ratio of the whole segment to the greater as the golden ratio. The phrase "cut in extreme and mean ratio" was the common way for a Greek mathematician to refer to this useful proportion. It wasn't until the 1400s that it became known as the divine proportion, and the 1800s when it was first referred to as the golden mean.

Golden Rectangles

A **golden rectangle** is a rectangle with longer side, ℓ, and shorter side, w, divinely proportioned. By Equation 8.1, we have $\dfrac{\ell}{w} = \dfrac{\ell+w}{\ell}$. This means that, in addition to the $\ell \times w$ rectangle, the $(\ell + w) \times \ell$ rectangle is also golden. With a little algebra, we can identify another golden rectangle. Since $\dfrac{\ell}{w} = \dfrac{\ell+w}{\ell}$, we have $\ell^2 = w(\ell + w)$ which can be rearranged into $\ell^2 - \ell w = w^2$. Factoring and dividing gives the following relationship:

$$\frac{\ell}{w} = \frac{w}{\ell - w}.$$

Therefore, the $w \times (\ell - w)$ rectangle is also golden.

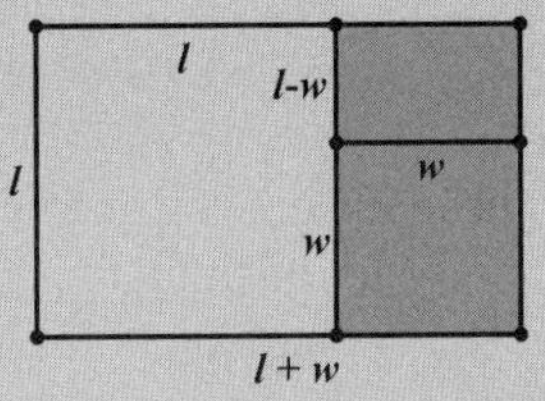

In the figure, we see that the two smaller golden rectangles are nested like matryoshka dolls inside the largest $(\ell + w) \times \ell$ golden rectangle. Notice the pattern that develops. The $(\ell + w) \times \ell$ rectangle can be decomposed into a square (of size $\ell \times \ell$) and the smaller shaded golden rectangle. The shaded golden rectangle can be decomposed into a square (of size $w \times w$) and another smaller golden rectangle. Our algebra shows that a smaller golden rectangle exists inside all golden rectangles. Thus, we can continue this process indefinitely, creating a nested sequence of smaller and smaller golden rectangles. This is demonstrated in Figure 9.1. These rectangles are all similar to each other since the sides of every rectangle are in the same ratio, the golden ratio!

The three triangle similarity schemes of Book VI appear as its fourth, fifth and sixth propositions. There is one preliminary result required, Proposition VI.2, and we give an updated version of Euclid's proof here.

Proposition VI.2. *If a straight line be drawn parallel to one of the sides of a triangle, it will cut the sides of the triangle proportionally; and, if the sides of the triangle be cut proportionally, the line joining the points of section will be parallel to the remaining side of the triangle.*

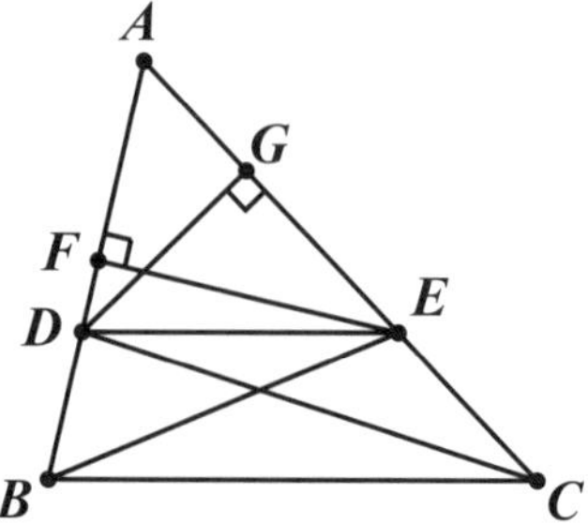

Figure 9.3. Proposition VI.2

Proof.   Consider triangle $\triangle ABC$ with D on AB and E on AC. We wish to show that $DE \parallel BC$ if and only if

$$\frac{AD}{DB} = \frac{AE}{EC}.$$

($\implies$) : Assume that $DE \parallel BC$. By I.37, the area of $\triangle BDE$ is equal to the area of $\triangle CDE$. Adding the area of $\triangle ADE$ to each of these, we have $Area(\triangle ABE) = Area(\triangle ACD)$. Dividing both sides by $Area(\triangle ADE)$ gives

$$\frac{Area(\triangle ABE)}{Area(\triangle ADE)} = \frac{Area(\triangle ACD)}{Area(\triangle ADE)}.$$

Let EF and DG be altitudes of $\triangle ADE$. Then,

$$Area(\triangle ADE) = \frac{1}{2}AD \cdot EF = \frac{1}{2}AE \cdot DG,$$

$$Area(\triangle ABE) = \frac{1}{2}AB \cdot EF,$$

and

$$Area(\triangle ACD) = \frac{1}{2}AC \cdot DG.$$

Substituting appropriately and cancelling the coefficients gives

$$\frac{AB \cdot EF}{AD \cdot EF} = \frac{AC \cdot DG}{AE \cdot DG}.$$

Simplifying, we have

$$\frac{AB}{AD} = \frac{AC}{AE}.$$

Since $\dfrac{AB}{AD} = \dfrac{AD + DB}{AD} = 1 + \dfrac{DB}{AD}$ and $\dfrac{AC}{AE} = \dfrac{AE + EC}{AE} = 1 + \dfrac{EC}{AE}$, substituting and cancelling the ones produces

$$\frac{DB}{AD} = \frac{EC}{AE},$$

or equivalently,

$$\frac{AD}{DB} = \frac{AE}{EC},$$

as desired.

($\impliedby$) : We leave this direction as an exercise for the reader. $\square$

For greater clarity, we give straightforward, revised versions of both Euclid's statement and his proof of Proposition VI.4, the AAA triangle similarity scheme, as Euclid's proof involves a complicated construction that is not entirely justified. We distinguish the similarity scheme abbreviations by adding the similarity symbol above the letters, for example, $\widetilde{AAA}$ and $\widetilde{SAS}$.

Proposition VI.4 [$\widetilde{AAA}$]. *Equiangular triangles are similar.*

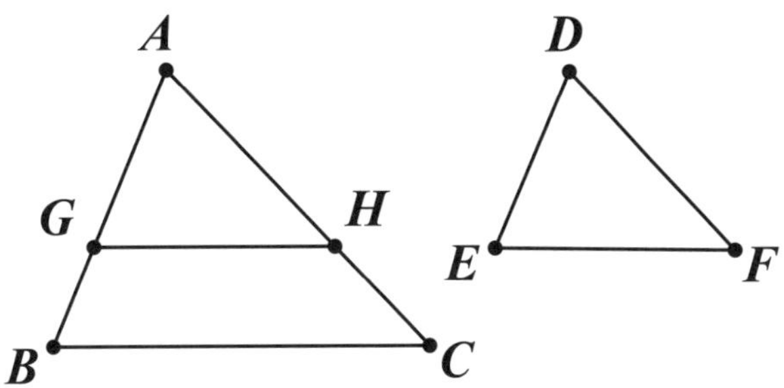

Figure 9.4. Proposition VI.4

Proof. Consider triangles $\triangle ABC$ and $\triangle DEF$. We will assume that $\angle A = \angle D$, $\angle B = \angle E$ and $\angle C = \angle F$. We wish to show that

$$\frac{AB}{DE} = \frac{AC}{DF} = \frac{BC}{EF}.$$

If $AB = DE$, then by ASA [I.26] the triangles are congruent, and hence, similar (every ratio is 1).

If $AB \neq DE$, then WLOG, assume $AB > DE$. Construct G on AB such that $AG = DE$ [I.3]. Draw a line through G that is parallel to BC [I.31]. Let the intersection of this line with AC be H [Pasch's Axiom]. By I.29, $\angle AGH = \angle ABC$. Since $\angle ABC = \angle DEF$ by hypothesis, then $\angle AGH = \angle DEF$. Hence, by ASA [I.26] we have $\triangle AGH \cong \triangle DEF$, and thus, $AH = DF$. Since $GH \parallel BC$, by VI.2 we have

$$\frac{GB}{AG} = \frac{HC}{AH}.$$

Adding 1 to both sides of the equation gives

$$\frac{GB}{AG} + \frac{AG}{AG} = \frac{HC}{AH} + \frac{AH}{AH}.$$

Simplifying produces

$$\frac{AB}{AG} = \frac{AC}{AH}.$$

By substitution, we have

$$\frac{AB}{DE} = \frac{AC}{DF}.$$

A similar argument produces

$$\frac{AC}{DF} = \frac{BC}{EF}. \qquad \qquad \square$$

The fact that two pairs of equal angles suffice for similar triangles is an immediate corollary to $\widetilde{AAA}$.

Corollary 9.1. [$\widetilde{AA}$]. *If two triangles have two pairs of equal angles, then they are similar.*

Proof. By I.32, the third pair of angles are also equal. Thus, we may apply $\widetilde{AAA}$. $\square$

We end this section with the remaining similarity schemes, $\widetilde{SSS}$ and $\widetilde{SAS}$, and leave their updated proofs as an exercise for the reader.

Proposition VI.5 [$\widetilde{SSS}$]. *If two triangles have their sides proportional, the triangles will be equiangular and will have those angles equal which the corresponding sides subtend.*

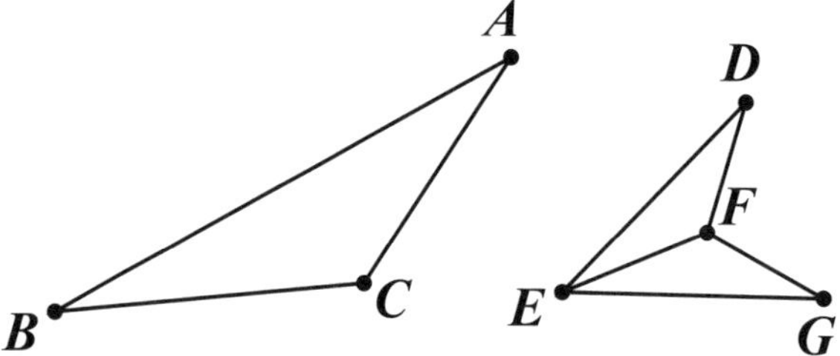

Figure 9.5. Proposition VI.5

Proof. Let ABC, DEF be two triangles having their sides proportional, so that, as AB is to BC, so is DE to EF, as BC is to CA, so is EF to FD, and further, as BA is to AC, so is ED to DF; I say that the triangle ABC is equiangular with the triangle DEF, and they will have those angles equal which the corresponding sides subtend, namely the angle ABC to the angle DEF, the angle BCA to the angle EFD, and further the angle BAC to the angle EDF.

For on the straight line EF, and at the points E, F on it, let there be constructed the angle FEG equal to the angle ABC, and the angle EFG equal to the angle ACB; [I.23] therefore the remaining angle at A is equal to the remaining angle at G. [I.32]

Therefore the triangle ABC is equiangular with the triangle GEF.

Therefore in the triangles ABC, GEF the sides about the equal angles are proportional, and those are corresponding sides which subtend the equal angles; [VI.4] therefore, as AB is to BC, so is GE to EF.

But, as AB is to BC, so by hypothesis is DE to EF; therefore, as DE is to EF, so is GE to EF. [V.11]

Therefore each of the straight lines DE, GE has the same ratio to EF; therefore DE is equal to GE. [V.9]

For the same reason DF is also equal to GF.

Since then DE is equal to EG, and EF is common, the two sides DE, EF are equal to the two sides GE, EF; and the base DF is equal to the base FG; therefore the angle DEF is equal to the angle GEF, [I.8] and the triangle DEF is equal to the triangle GEF, and the remaining angles are equal to the remaining angles, namely those which the equal sides subtend. [I.4]

Therefore the angle DFE is also equal to the angle GFE, and the angle EDF to the angle EGF.

And, since the angle FED is equal to the angle GEF, while the angle GEF is equal to the angle ABC, therefore the angle ABC is also equal to the angle DEF.

For the same reason the angle ACB is also equal to the angle DFE, and further, the angle at A to the angle at D; therefore the triangle ABC is equiangular with the triangle DEF.

Therefore etc. Q.E.D.   □

Proposition VI.6 [$\widetilde{SAS}$].   *If two triangles have one angle equal to one angle and the sides about the equal angles proportional, the triangles will be equiangular and will have those angles equal which the corresponding sides subtend.*

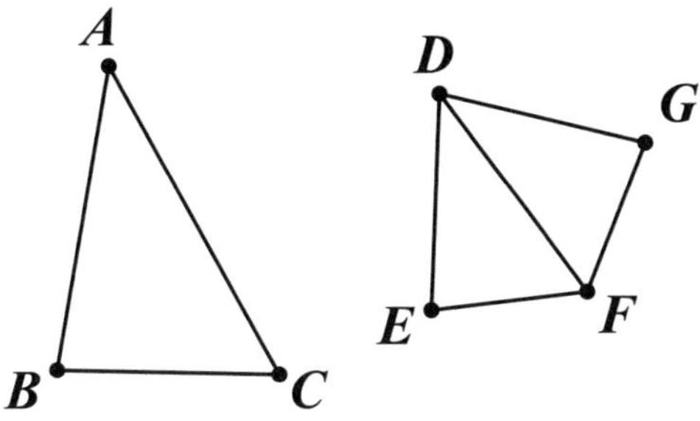

Figure 9.6. Proposition VI.6

Proof.   Let ABC, DEF be two triangles having one angle BAC equal to one angle EDF and the sides about the equal angles proportional, so that, as BA is to AC, so is ED to DF; I say that the triangle ABC is equiangular with the triangle DEF, and will have the angle ABC equal to the angle DEF, and the angle ACB to the angle DFE.

For on the straight line DF, and at the points D, F on it, let there be constructed the angle FDG equal to either of the angles BAC, EDF, and the angle DFG equal to the angle ACB; [I.23] therefore the remaining angle at B is equal to the remaining angle at G. [I.32]

Therefore the triangle ABC is equiangular with the triangle DGF.

Therefore, proportionally, as BA is to AC, so is GD to DF. [VI.4]

But, by hypothesis, as BA is to AC, so also is ED to DF; therefore also, as ED is to DF, so is GD to DF. [V.11]

Therefore ED is equal to DG; [V.9] and DF is common; therefore the two sides ED, DF are equal to the two sides GD, DF; and the angle EDF is equal to the angle GDF; therefore the base EF is equal to the base GF, and the triangle DEF is equal to the triangle DGF, and the remaining angles will be equal to the remaining angles, namely those which the equal sides subtend. [I.4]

Therefore the angle DFG is equal to the angle DFE, and the angle DGF to the angle DEF.

But the angle DFG is equal to the angle ACB; therefore the angle ACB is also equal to the angle DFE.

And, by hypothesis, the angle BAC is also equal to the angle EDF; therefore the remaining angle at B is also equal to the remaining angle at E; [I.32] therefore the triangle ABC is equiangular with the triangle DEF.

Therefore etc. Q.E.D.   □

You may recall from Chapter 4 that AAA_S is a spherical triangle *congruence* scheme. In Spherical geometry, similar triangles *are* congruent.

Exercises 9.2

1. Prove that similarity is transitive for triangles. That is, if $\triangle ABC \sim \triangle DEF$ and $\triangle DEF \sim \triangle GHI$, then $\triangle ABC \sim \triangle GHI$.

2. For the second part of the proof of Proposition VI.2, assume that

$$\frac{AD}{DB} = \frac{AE}{EC}$$

   and show that $DE \parallel BC$.

3. Let triangle $\triangle ABC$ be cut by a line parallel to one of its sides. Prove that the triangle formed is similar to the original triangle.

4. Let $ABCD$ be an arbitrary quadrilateral and let $E, F, G,$ and H be the midpoints of AB, BC, CD and AD, respectively. Prove that $EFGH$ is a parallelogram.

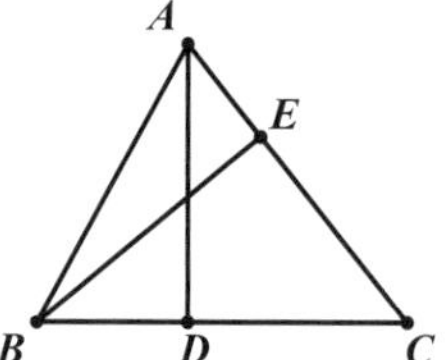

Figure 9.7. Exercise 9.2.5

5. In Chapter 7, we showed that the area of a triangle is given by the formula $\frac{1}{2}bh$. We did not, however, prove that this is well-defined. That is, if we have base b_1 with corresponding height h_1 for the triangle, and a second base b_2 with corresponding height h_2 for the same triangle, then $b_1 h_1 = b_2 h_2$. Consider triangle $\triangle ABC$. Construct $AD \perp BC$ and $BE \perp AC$ as illustrated in Figure 9.7. Let $b_1 = BC, h_1 = AD, b_2 = AC$ and $h_2 = BE$. Using similar triangles, prove that $b_1 h_1 = b_2 h_2$.

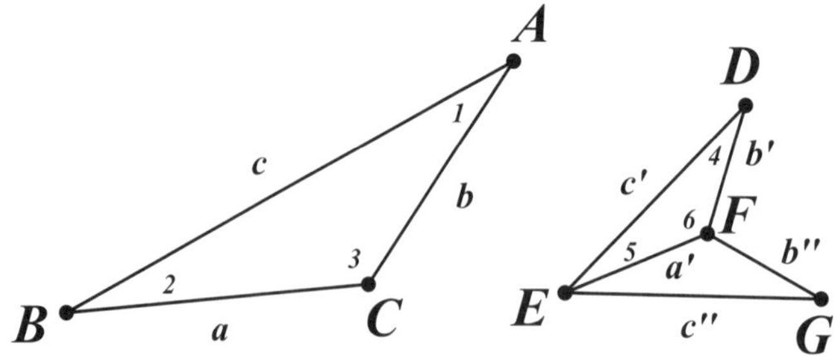

Figure 9.8. Exercise 9.2.6: Labels for updated proof of Proposition VI.5

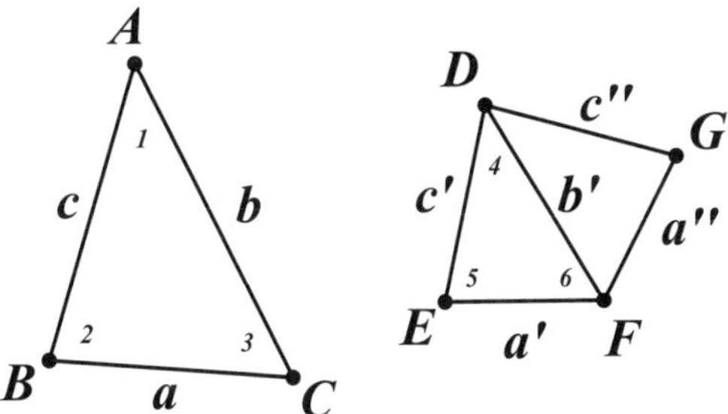

Figure 9.9. Exercise 9.2.7: Labels for updated proof of Proposition VI.6

6. Give an updated version of Euclid's proof of Proposition VI.5 [$\widetilde{SSS}$]. Be sure to justify each step, substitute mathematical symbols where appropriate, and include helpful diagrams as needed. Label the diagram as shown in Figure 9.8.

7. Give an updated version of Euclid's proof of Proposition VI.6 [$\widetilde{SAS}$]. Be sure to justify each step, substitute mathematical symbols where appropriate, and include helpful diagrams as needed. Label the diagram as shown in Figure 9.9.

8. Consider parallelogram $\square ABCD$ as shown in Figure 9.10, where E and F are the midpoints of AB and CD, respectively. Prove that AF and CE trisect the diagonal BD.

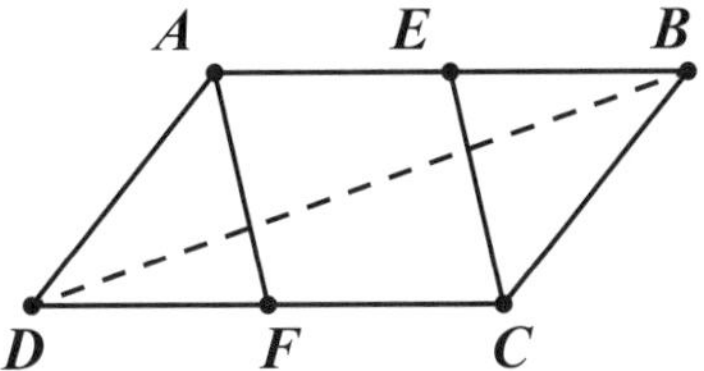

Figure 9.10. Exercise 9.2.8: AF and CE trisect the diagonal BD

9. Consider triangle $\triangle ABC$ as shown in Figure 9.11, where D is the midpoint of BC, E is the midpoint of BD, and $BC = 2AB$. Prove that AD bisects angle $\angle CAE$. [*Hint: Construct F on AB such that DF $\parallel$ AC.*]

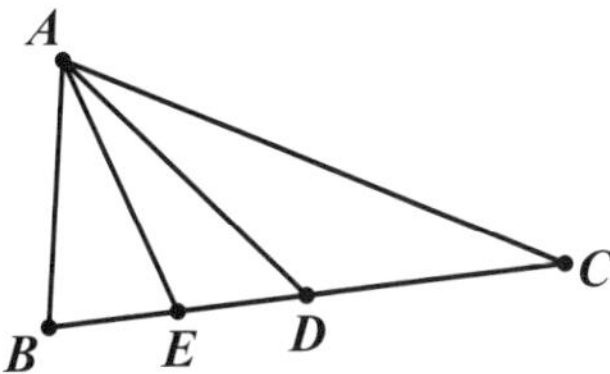

Figure 9.11. Exercise 9.2.9: AD bisects angle $\angle CAE$

10. Consider parallelogram $\square ABCD$ as shown in Figure 9.12, where F lies on AD, and E is the intersection of $\overrightarrow{BF}$ and $\overrightarrow{CD}$. Prove that $\triangle ABF \sim \triangle DEF$.

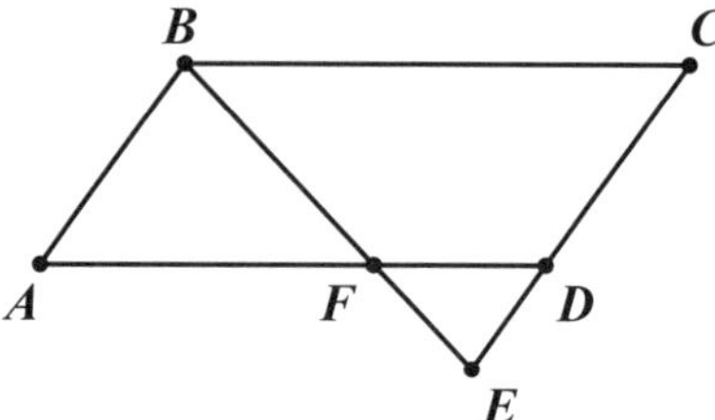

Figure 9.12. Exercise 9.2.10: $\triangle ABF \sim \triangle DEF$

9.3 A generalized Pythagorean Theorem

The Pythagorean Theorem is the penultimate proposition in the first book of the *Elements*. It is a purely geometric result relating the square on the hypotenuse to the two smaller squares on the legs. The proof is accomplished by decomposing the square on the hypotenuse into two subrectangles, one subrectangle equal in area to the square on one of the legs, and the other subrectangle equal to the square on the other leg. A quintessential question in mathematics is whether a proof applies to a larger set of objects. In the case of the Pythagorean Theorem, a natural question to ask is whether this statement is true if the squares are replaced with other regular n-gons. More generally, since any two regular n-gons are similar for a given n (equal angles and proportional sides), is the statement true for similar polygons? As we will see, these generalizations of the Pythagorean Theorem are true, but the proof technique given in Book I is not easily generalized to establish these extensions. Similarity is the proper tool for the job, and Proposition VI.8 provides the first preliminary result. This proposition identifies the three similar triangles found in any right-angled triangle, and the proof is left as an exercise for the reader.

Proposition VI.8. *If in a right-angled triangle a perpendicular be drawn from the right angle to the base, the triangles adjoining the perpendicular are similar both to the whole and to one another.*

In Exercise 9.3.5 we ask the reader to use this result in order to give an independent proof of the Pythagorean Theorem with a decidedly algebraic flavor, that is, $BC^2 = AB^2 + AC^2$ for $\triangle BAC$ where $\angle A$ is right. This algebraic proof of the Pythagorean Theorem is easily generalized to the cases of similar triangles or regular n-gons. Before we look at these generalizations, we must first establish a relationship between the areas of two similar triangles or two regular n-gons. First, referring to the similar triangles in Figure 9.2, we show that the triangular areas are in the same proportion to one another as are the squares of the corresponding sides, that is, $\dfrac{Area(\triangle ABC)}{Area(\triangle DEF)} = \dfrac{(AB)^2}{(DE)^2}$. Regarding the terminology in the proposition statement, the phrase "triangles are to one another" refers to the ratio of the triangular areas, and "duplicate ratio of their sides" refers to squaring, not doubling. We give an updated version of Euclid's proof.

Proposition VI.19. *Similar triangles are to one another in the duplicate ratio of the corresponding sides.*

Proof. Let triangles $\triangle ABC$ and $\triangle DEF$ be similar, where $\dfrac{AB}{DE} = \dfrac{AC}{DF} = \dfrac{BC}{EF}$ and $\angle B = \angle E = \beta$. We leave it to the reader to show that the area of $\triangle ABC$ is given by

$Area(\triangle ABC) = \frac{1}{2}(AB)(BC)\sin\beta$. Similarly, the area of $\triangle DEF$ is $Area(\triangle DEF) = \frac{1}{2}(DE)(EF)\sin\beta$.

Therefore,

$$\frac{Area(\triangle ABC)}{Area(\triangle DEF)} = \frac{\frac{1}{2}(AB)(BC)\sin\beta}{\frac{1}{2}(DE)(EF)\sin\beta} = \frac{AB}{DE}\cdot\frac{BC}{EF} = \frac{AB}{DE}\cdot\frac{AB}{DE} = \frac{(AB)^2}{(DE)^2},$$

as desired. $\square$

Next, we show that the same result holds in the more general case of regular n-gons.

Theorem 9.2. *If P_1 and P_2 are regular n-gons with sides, s_1 and s_2, respectively, then*

$$\frac{Area(P_1)}{Area(P_2)} = \frac{s_1^2}{s_2^2}.$$

Proof. We leave it to the reader to prove that the area of a regular n-gon, with a side of length s, is given by

$$Area(\text{regular } n\text{-gon}) = \frac{ns^2}{4\tan\left(\frac{\pi}{n}\right)}.$$

For a fixed n, the quantity $\dfrac{n}{4\tan\left(\frac{\pi}{n}\right)}$ is a constant. Thus, if P_1 and P_2 are regular n-gons with sides, s_1 and s_2, respectively, we have

$$\frac{Area(P_1)}{Area(P_2)} = \frac{s_1^2}{s_2^2},$$

as desired. $\square$

Euclid generalizes this result to the case of two similar polygons in Proposition VI.20. His proof resembles that of Proposition I.45 where the statement refers to any polygon but the proof falls short as he only tackles the case of a quadrilateral. Here, Euclid only provides a proof for similar pentagons. Even without the induction required for general polygons, the proof for the pentagon is complicated as it involves decomposing similar pentagons into their component similar triangles. Furthermore, we cannot easily provide an updated algebraic proof as we did for the two previous theorems since there is no simple area formula for a general polygon. Consequently, we state this proposition without proof.

Proposition VI.20. *If P_1 and P_2 are similar polygons with corresponding sides, s_1 and s_2, respectively, then $\dfrac{Area(P_1)}{Area(P_2)} = \dfrac{s_1^2}{s_2^2}$.*

Euclid uses this result near the end of Book VI to revisit the Pythagorean Theorem in Proposition VI.31, where he shows that the same relationship established for squares in I.47 exists for any polygons constructed on the sides of a right triangle, as long as those polygons are similar. Interestingly, there is nothing unique about the squares constructed on the sides of the right triangle in I.47. These figures could be regular octagons or similar dodecagons. Figure 9.13 shows four different sets of similar polygons constructed on the sides of right triangle $\triangle ABC$. The three equilateral

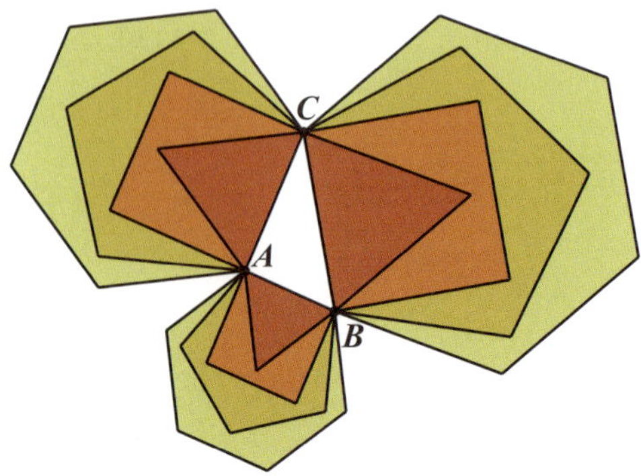

Figure 9.13. A generalized Pythagorean Theorem

triangles are similar to each other, as are the squares, the regular pentagons and the regular hexagons. By the Generalized Pythagorean Theorem (VI.31), the largest hexagon is equivalent to the two smaller hexagons, for example. The same relationship holds for the triangles or the pentagons because they are similar. We end our discussion of similarity with an updated proof of this beautiful result.

Proposition VI.31 [Generalized Pythagorean Theorem]. *In right-angled triangles the figure on the side subtending the right angle is equal to the similar and similarly described figures on the sides containing the right angle.*

Proof.   Consider triangle $\triangle ABC$ with right angle $\angle BAC$. Let P_1, P_2 and P_3 be similar polygons with corresponding sides $s_1 = BC$, $s_2 = AC$ and $s_3 = AB$, respectively. By Proposition VI.20, we have

$$\frac{Area(P_2)}{Area(P_1)} = \frac{s_2^2}{s_1^2} \text{ and } \frac{Area(P_3)}{Area(P_1)} = \frac{s_3^2}{s_1^2}.$$

Rewriting, we have $Area(P_2) \cdot s_1^2 = Area(P_1) \cdot s_2^2$, and $Area(P_3) \cdot s_1^2 = Area(P_1) \cdot s_3^2$. Using Proposition VI.8, we leave it to the reader to prove that $BC^2 = AC^2 + AB^2$, that is, $s_1^2 = s_2^2 + s_3^2$ (Exercise 9.3.5). Multiplying this equation by $Area(P_1)$ gives

$$Area(P_1) \cdot s_1^2 = Area(P_1) \cdot s_2^2 + Area(P_1) \cdot s_3^2.$$

Substituting, we have

$$Area(P_1) \cdot s_1^2 = Area(P_2) \cdot s_1^2 + Area(P_3) \cdot s_1^2.$$

Dividing by s_1^2 gives

$$Area(P_1) = Area(P_2) + Area(P_3),$$

as desired. $\square$

Furthermore, and as a precursor to our imminent focus on the circle, it is interesting to note that the Generalized Pythagorean Theorem can also be applied to circles, where the "figure on a side" of the right triangle is the circle of radius (or diameter) equal to the side.

Exercises 9.3

1. Let $\triangle ABC$ be a right triangle with right angle $\angle BAC$. Let $AD \perp BC$ as shown in Figure 9.14. Prove Euclid's Proposition VI.8. That is, prove that $\triangle ABC \sim \triangle DBA \sim \triangle DAC$.

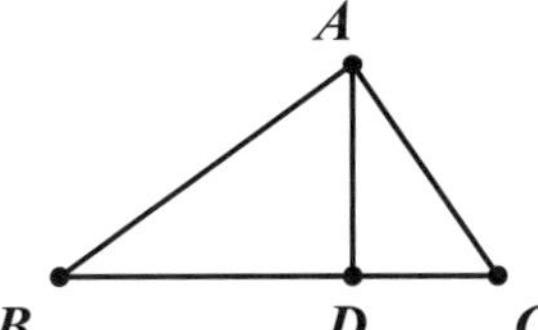

Figure 9.14. Exercise 9.3.1: Proposition VI.8

2. Prove that in a right triangle, the square of the altitude to the hypotenuse equals the product of the segments it determines on the hypotenuse.

3. Consider $\triangle ABC$, where $\angle ABC = \beta$. Prove that the area of triangle $\triangle ABC$ is given by $A = \frac{1}{2}(AB)(BC)\sin\beta$.

4. Prove that the area of a regular n-gon, with a side of length s, is given by the formula:

$$Area = \frac{ns^2}{4\tan\left(\frac{\pi}{n}\right)}.$$

(Note: when $n = 3$, we get the familiar formula for the area of an equilateral triangle which is $\frac{s^2\sqrt{3}}{4}$.)

5. Consider triangle $\triangle ABC$ with right angle $\angle BAC$. Using Proposition VI.8, prove that $BC^2 = AC^2 + AB^2$.

10

Book III: Circles

Figure 10.1. Used with permission from the United States Olympic Committee.

Sonnets of a Geometer: The Circle

Few things are perfect: we bear Eden's scar;
Yet faulty man was godlike in design
That day when first, with stick and length of twine,
He drew me on the sand. Then what could mar
His joy in that obedient mystic line;
And then, computing with a zeal divine,
He called π 3-point-14159
And knew my lovely circuit $2\,\pi\,r$!

A circle is a happy thing to be—
Think how the joyful perpendicular
Erected at the kiss of tangency
Must meet my central point, my avatar!
They talk of 14 points: yet only 3
Determine every circle: **Q.E.D.**

– Christopher Morley, 1920 [88]

What could be a more appropriate start to a chapter dedicated to the circle than a poem celebrating the perfection of our long neglected main character? Morley details a good number of the circle's finer points in a mere fourteen lines. His greater purpose, however, is political and topical, as indicated by the "14 points" which serve as a counterpoint to the three noncollinear points that determine a unique circle. Morley, of Quaker heritage, refers to President Woodrow Wilson's Fourteen Points speech to Congress in 1918 in which he outlined a program to broker world peace at the end of World War I. The fourteenth point of his plan was a proposal for the League of Nations which was established soon thereafter in 1920, and was replaced twenty-six years later by the United Nations after the end of World War II. The Platonic ideal of the circle, perfect in form, "lovely," "happy," and "joyful," sits in contrast to the imperfection of humanity at war.

Greek geometers and philosophers were similarly enthralled with the figure of a circle, and it is perfectly aligned with their demonstrated prejudice for symmetry: from art and architecture to music and mathematics, a symmetric object is a thing of beauty. If a figure lacks symmetry, geometers have a method for imposing it! The process of quadrature provides a technique for replacing any irregularly shaped rectilinear figure with a perfectly symmetric square of the same size. With its symmetry about infinitely many lines through its center, the circle could be the finest of all Euclidean figures. From an aesthetic viewpoint, we should not be surprised if there were a process called "circulature," but there is not. We will, however, discuss what happens when we try to impose symmetry through quadrature on this most symmetric of objects in Chapter 16. Until then, "nothing will mar [our] joy" as the spotlight rests squarely on "that happy thing to be" and we take a tour of Euclid's third book.

How does an author of more than 50 books come to rhapsodize the circle? Christopher Morley's father was mathematician Frank Morley (1860–1937), a professor at Haverford College who went on to chair the mathematics department at Johns Hopkins University, where he directed fifty doctoral dissertations and served as president of the American Mathematical Society. His mother, Lilian, was a violinist and a poet. His youngest brother, Frank V. Morley, received his Ph.D. at Oxford under the direction of G. H. Hardy. His father is best known for Morley's Trisector Theorem: *The three points of intersection of adjacent trisectors of the angles in a triangle form an equilateral triangle.* In the given figure, equilateral $\triangle A'B'C'$ is the Morley triangle of $\triangle ABC$.

10.1 Definitions

We start with Euclid's definitions of Book III, although this time we give updated terms as needed.

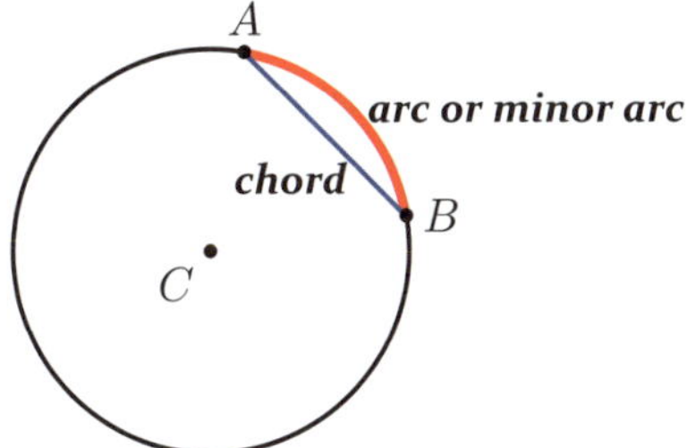

Figure 10.2. Arc and chord of a circle

- A **radius** of a circle is a line segment joining the center of the circle to a point on it.

- **Equal circles** are circles with equal radii. Equal circles may also be called **congruent circles**.

- A **chord** of a circle is a line segment joining two of its points. See Figure 10.2 for an example.

- A line that coincides with a chord, intersecting a circle at two points, is called a **secant** line. In Figure 10.2, $\overleftrightarrow{AB}$ is a secant line.

- A **diameter** is a chord containing the center of the circle.

- Distinct points A and B on a circle divide the circle into two pieces. The shorter of these is called the **minor arc**, and the longer, the **major arc**. If these arcs are equal, they are called **semicircles**. It takes three points to denote an arc unambiguously. Thus, $\overarc{ADB}$ denotes the arc of the circle between A and B through point D. It has become common to label the minor arc simply as $\overarc{AB}$. [Note: Here, Euclid uses the terms **less circumference** and **greater circumference** instead of minor arc and major arc.] See Figures 10.2 or 10.3 for examples.

Figure 10.3. Major arc $\overarc{AEB}$ with shaded greater segment, and minor arc $\overarc{ADB}$ with shaded less segment

- A **segment** of a circle is the region trapped between a chord and one of the arcs determined by its endpoints. Euclid calls the segment determined by a chord and its minor arc the **less segment**, and the segment determined by a chord and its major arc the **greater segment**. See Figure 10.3 for an example.

- A **central angle** of a circle is an angle $\angle ACB$ where A and B lie on the circle and C is its center. Every nondiameter chord and every minor arc has a corresponding central angle that is less than $180°$, and we say that the chord or minor arc **subtends** the central angle. See Figure 10.4 for an example. Notation $\angle ACB$ has referred to an angle whose measure is less than $180°$ since Chapter 2, and the definition of central angle respects this condition. However, in a circle we would like to consider angles with measure greater than 180°. For example, in Figure 10.3, if minor arc $\overset{\frown}{ADB}$ subtends a central angle $\angle ACB$ of measure 140°, then we would like to say that major arc $\overset{\frown}{AEB}$ subtends an angle at the center of measure $360° - \angle ACB = 220°$. In general, major arcs subtend angles that are greater than 180°. Euclid refers to such angles, which correspond to both major and minor arcs, as **angles at the center**. Angles that are between 180° and 360° are sometimes referred to as **reflex angles**.

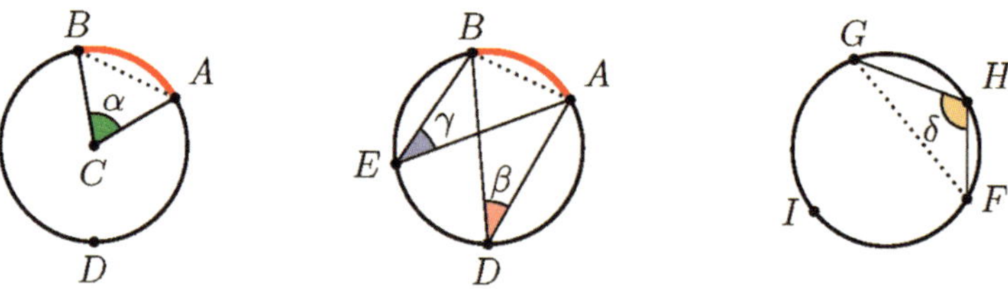

Figure 10.4. Central angle α, and inscribed angles β, γ and δ

- Given three distinct points A, B and D that lie on a circle, $\angle ADB$ is an **inscribed angle**, or an **angle at the circumference**, that intercepts chord AB. We also say that angle $\angle ADB$ **is subtended by** the chord AB. When D is not a point on arc $\overset{\frown}{AB}$, we say that $\angle ADB$ intercepts the arc $\overset{\frown}{AB}$, or that $\angle ADB$ is an angle subtended by the arc $\overset{\frown}{AB}$. In Figure 10.4, $\angle AEB$ is an inscribed angle that intercepts chord AB and arc $\overset{\frown}{AB}$. Angles $\angle ADB$ and $\angle AEB$ are examples of angles that are subtended by chord AB. Also, $\angle FHG$ is an inscribed angle that intercepts chord FG and arc $\overset{\frown}{FIG}$.

- An **angle in a segment** determined by $\overset{\frown}{ADB}$ is any inscribed angle $\angle AXB$ where X lies on arc $\overset{\frown}{ADB}$. In Figure 10.4, $\angle AEB$ is an angle in the segment determined by $\overset{\frown}{ADB}$, but $\angle FHG$ is not an angle in the segment determined by $\overset{\frown}{FIG}$.

- A **sector** of a circle is the region bounded between two radii and an arc joining them. See Figure 10.5 for an example.

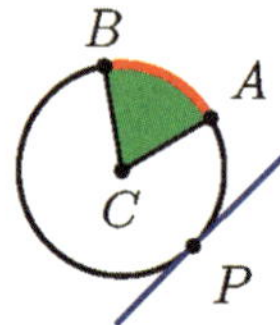

Figure 10.5. Sector and tangent line

- An infinitely extended *line is* **tangent** *to a circle* if and only if it touches the circle at exactly one point. See Figure 10.5 for an example.

- *Two circles are* **tangent** if they intersect in exactly one point. If one circle lies inside the other, the circles are **internally tangent**, otherwise they are **externally tangent**. See Figure 10.6 for an example.

- A polygon is **cyclic** if all of its vertices lie on a circle. See Figure 10.6 for an example.

Figure 10.6. Cyclic quadrilateral, and internally and externally tangent circles

Exercises 10.1

1. Prove that the center of a circle is unique. To do this, suppose circle C has two centers O and O' that lie inside C. Let A and B be the intersections of $\overleftrightarrow{OO'}$ with circle C, and reach a contradiction. Include a diagram with your proof.

2. Prove that a line can only intersect a circle in either one or two points. To do this, assume line ℓ intersects a circle with center O at three distinct points $A, B,$ and C, then find a contradiction. Include a diagram with your proof.

3. In a circle with center O, consider diameters AC and BD. Prove that the quadrilateral $ABCD$ is a rectangle. Include a diagram with your proof.

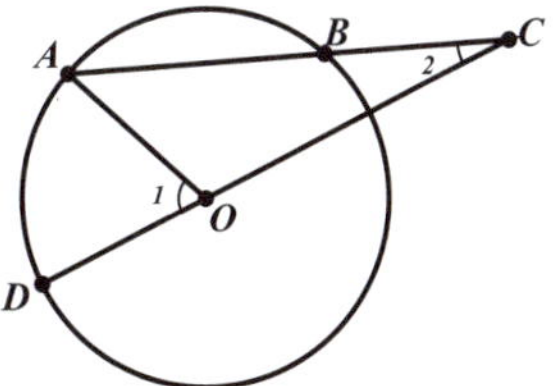

Figure 10.7. Exercise 10.1.4

4. In a circle with center O, a chord AB is extended to a point C such that $BC = OA$, the radius of the circle. Let D be the intersection of $\overrightarrow{CO}$ with the circle, as illustrated in Figure 10.7. Prove that $\angle 1 = 3 \cdot \angle 2$.

10.2 Tangency

We have selected over half of the thirty-seven propositions of Book III for our consideration. As the circle takes the limelight, we begin with the first, a construction which turns the tables on the third postulate. Here, we find the center of a given circle.

Proposition III.1. *To find the centre of a given circle.*

Proof. Let ABC be the given circle; thus it is required to find the centre of the circle ABC.

Let a straight line AB be drawn through it at random, and let it be bisected at the point D; from D let DC be drawn at right angles to AB and let it be drawn through to E; let CE be bisected at F; I say that F is the centre of the circle ABC.

For suppose it is not, but, if possible, let G be the centre, and let GA, GD, GB be joined.

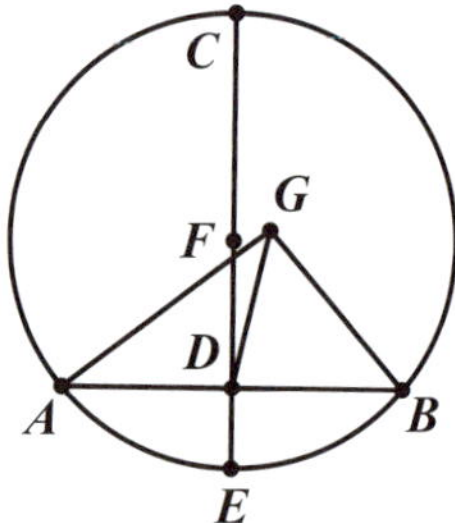

Figure 10.8. Proposition III.1

Then, since *AD* is equal to *DB*, and *DG* is common, the two sides *AD*, *DG* are equal to the two sides *BD*, *DG* respectively; and the base *GA* is equal to the base *GB*, for they are radii;

therefore the angle *ADG* is equal to the angle *GDB*. [I.8]

But, when a straight line set up on a straight line makes the adjacent angles equal to one another, each of the equal angles is right; [Def. 10] therefore the angle *GDB* is right.

But the angle *FDB* is also right; therefore the angle *FDB* is equal to the angle *GDB*, the greater to the less: which is impossible.

Therefore *G* is not the centre of the circle *ABC*.

Similarly we can prove that neither is any other point except *F*. Therefore the point *F* is the centre of the circle *ABC*. Q.E.F.   □

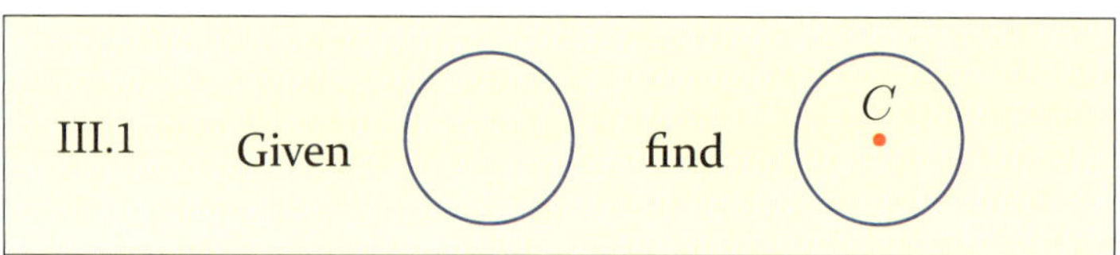

The third proposition states that a diameter (or radius) will bisect a chord that is not a diameter if and only if it is perpendicular to that chord.

Proposition III.3. *If in a circle a straight line through the centre bisect a straight line not through the centre, it also cuts it at right angles; and if it cut it at right angles, it also bisects it.*

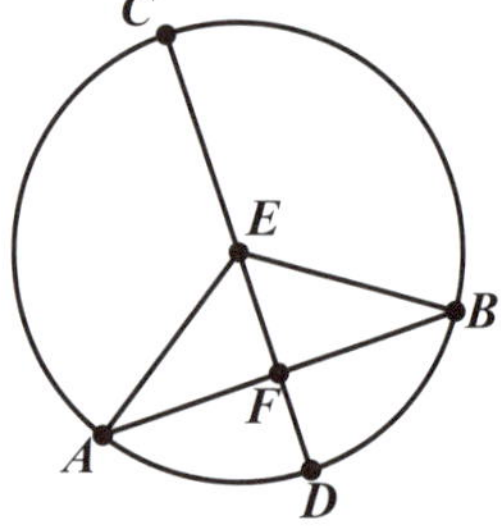

Figure 10.9. Proposition III.3

Proof.   Let *ABC* be a circle, and in it let a straight line *CD* through the centre bisect a straight line *AB* not through the centre at the point *F*; I say that it also cuts it at right angles.

For let the centre of the circle *ABC* be taken, and let it be *E*; let *EA*, *EB* be joined.

Then, since *AF* is equal to *FB*, and *FE* is common, two sides are equal to two sides;

and the base *EA* is equal to the base *EB*;

therefore the angle *AFE* is equal to the angle *BFE*. [I.8]

But, when a straight line set up on a straight line makes the adjacent angles equal to one another, each of the equal angles is right; [Def. 10]

therefore each of the angles *AFE*, *BFE* is right.

Therefore *CD*, which is through the centre, and bisects *AB* which is not through the centre, also cuts it at right angles.

Again, let *CD* cut *AB* at right angles; I say that it also bisects it, that is, that *AF* is equal to *FB*.

For, with the same construction, since *EA* is equal to *EB*, the angle *EAF* is also equal to the angle *EBF*. [I.5]

But the right angle *AFE* is equal to the right angle *BFE*, therefore *EAF*, *EBF* are two triangles having two angles equal to two angles and one side equal to one side, namely *EF*, which is common to them, and subtends one of the equal angles;

therefore they will also have the remaining sides equal to the remaining sides; [I.26] therefore *AF* is equal to *FB*.

Therefore etc. Q.E.D.   □

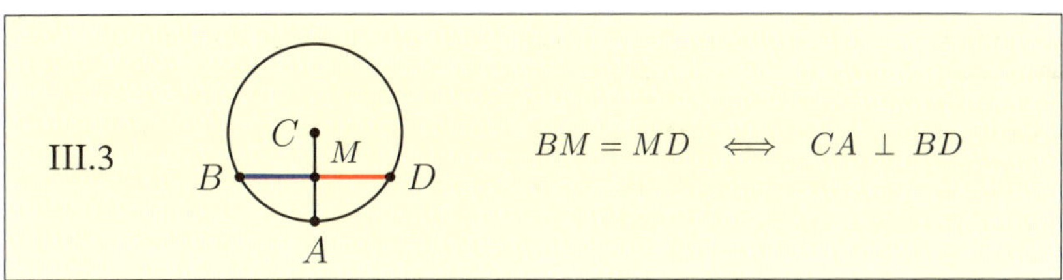

We skip ahead to the sixteenth through the nineteenth propositions which concern tangent lines to circles. The first of these contains the only appearance of nonrectilinear angles in the *Elements*, the subject of Definition 8 given in Chapter 2. This type of angle occurs at the meeting point of two curves which are not both straight lines, and here, appropriately, one of those curves is the circle. Since we have no need to consider this angle formed by a circle and an intersecting line, we omit part of the statement of III.16 to give an abbreviated version which states that a line perpendicular to a given radius (at its endpoint) must be tangent to the circle. [For more on this type of nonrectilineal angle, read about horn angles.] Euclid proves this by first showing that the perpendicular cannot intersect the circle twice, then showing that no other line can fit between the perpendicular and the circle. The second half of the proof serves to reinforce our intuitive understanding of the concept of a tangent line to a curve. Proposition III.18 asserts the converse; that is, a tangent line to a circle must be perpendicular to the radius whose endpoint is the point of tangency. It is unusual for Euclid to separate a statement and its converse with an intermediary proposition, especially when it is not needed for the proof of the converse. He does, however, need III.16 for the construction given in III.17 where he produces a tangent to a circle through a given point outside the circle. The last of these tangent line results is the rather obvious corollary to III.18

that a perpendicular to a tangent line constructed through the point of tangency will contain the center of the circle.

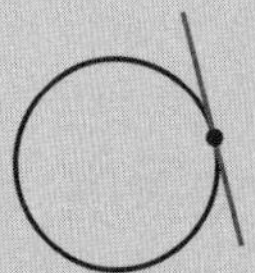

Horn angles and nonstandard analysis

In Proposition III.16 Euclid considers the angle formed at the intersection of a circle and its tangent line. Known as a **horn angle**, it was the subject of much controversy for Greek geometers. Euclid essentially states that the horn angle is smaller than any rectilineal angle. However, he does not go so far as to say that its measure is zero, which prompts a fundamental question: How do we measure an infinitesimally small angle? The difficulty encountered in providing an answer to this question is reflected in the fact that the controversy raged on for two thousand years. A satisfying answer requires a great deal of exposition as the historical path to a full-blown mathematical theory of infinitesimals travels through the analytic geometers working before the discovery of calculus, through Newton and Leibniz's calculus at the end of the 1600s, and ends with Abraham Robinson's nonstandard analysis in the 1960s. Those who remember their calculus definitions may be satisfied by the following synopsis of the eventual resolution of the problem presented by these angles: an angle formed at an intersection of two curves in a plane is defined as the rectilineal angle formed by the tangent lines to the curves at the point of intersection. How is this connected to infinitesimals? The slope of a tangent line to a curve is a derivative. A derivative is defined as a limit, and infinitesimals form the foundation for limits.

Proposition III.16. *The straight line drawn at right angles to the diameter of a circle from its extremity will fall outside the circle, and into the space between the straight line and the circumference another straight line cannot be interposed.*

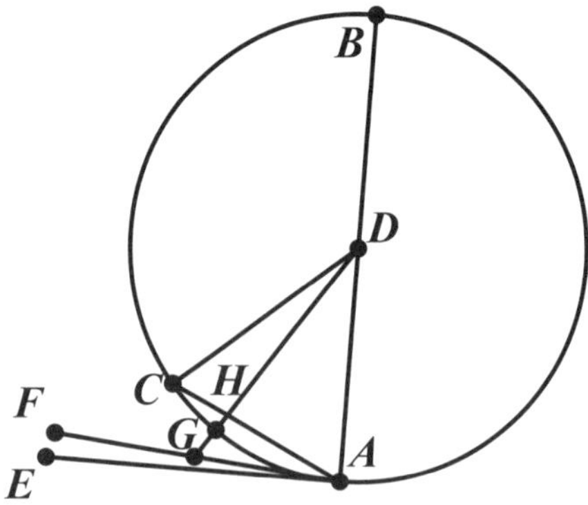

Figure 10.10. Proposition III.16

Proof. Let *ABC* be a circle about *D* as centre and *AB* as diameter; I say that the straight line drawn from *A* at right angles to *AB* from its extremity will fall outside the circle.

For suppose it does not, but, if possible, let it fall within as *CA*, and let *DC* be joined.

Since DA is equal to DC, the angle DAC is also equal to the angle ACD. [I.5]

But the angle DAC is right; therefore the angle ACD is also right: thus, in the triangle ACD, the two angles DAC, ACD are equal to two right angles: which is impossible. [I.17]

Therefore the straight line drawn from the point A at right angles to BA will not fall within the circle.

Similarly we can prove that neither will it fall on the circumference; therefore it will fall outside.

Let it fall as AE; I say next that into the space between the straight line AE and the circumference CHA another straight line cannot be interposed.

For, if possible, let another straight line be so interposed, as FA, and let DG be drawn from the point D perpendicular to FA.

Then, since the angle AGD is right, and the angle DAG is less than a right angle, AD is greater than DG. [I.19]

But DA is equal to DH; therefore DH is greater than DG, the less than the greater: which is impossible.

Therefore another straight line cannot be interposed into the space between the straight line and the circumference. Q.E.D.   □

Proposition III.17. *From a given point to draw a straight line touching a given circle.*

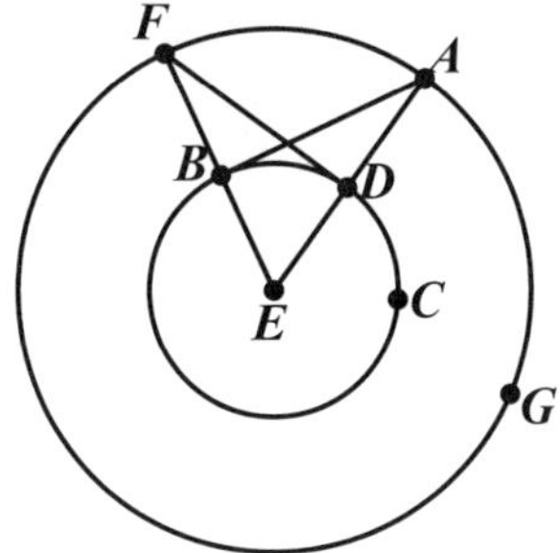

Figure 10.11. Proposition III.17

Proof.   Let A be the given point, and BCD the given circle; thus it is required to draw from the point A a straight line touching the circle BCD.

For let the centre E of the circle be taken; [III.1]

let AE be joined, and with centre E and distance EA let the circle AFG be described;

from D let DF be drawn at right angles to EA, and let EF, AB be joined;

I say that AB has been drawn from the point A touching the circle BCD.

For, since E is the centre of the circles BCD, AFG, EA is equal to EF, and ED to EB;

therefore the two sides AE, EB are equal to the two sides FE, ED: and they contain a common angle, the angle at E;

therefore the base DF is equal to the base AB, and the triangle DEF is equal to the triangle BEA, and the remaining angles to the remaining angles; [I.4] therefore the angle EDF is equal to the angle EBA.

But the angle *EDF* is right; therefore the angle *EBA* is also right.

Now *EB* is a radius; and the straight line drawn at right angles to the diameter of a circle, from its extremity, touches the circle; [III.16, Por.] therefore *AB* touches the circle *BCD*.

Therefore from the given point *A* the straight line *AB* has been drawn touching the circle *BCD*.

Q.E.F. □

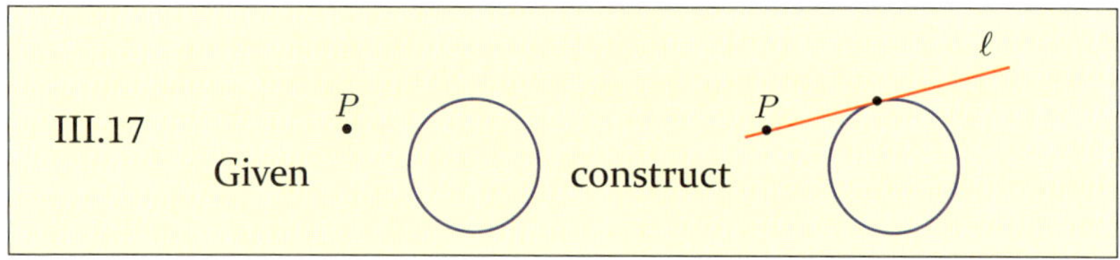

Proposition III.18. *If a straight line touch a circle, and a straight line be joined from the centre to the point of contact, the straight line so joined will be perpendicular to the tangent.*

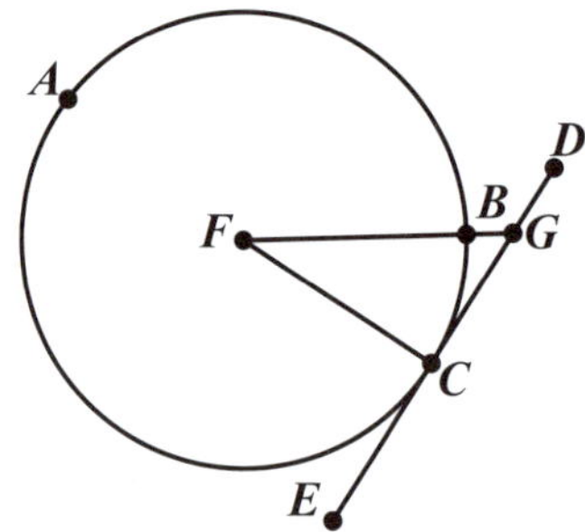

Figure 10.12. Proposition III.18

Proof. For let a straight line *DE* touch the circle *ABC* at the point *C*, let the centre *F* of the circle *ABC* be taken, and let *FC* be joined from *F* to *C*;

I say that *FC* is perpendicular to *DE*.

For, if not, let *FG* be drawn from *F* perpendicular to *DE*.

Then, since the angle *FGC* is right, the angle *FCG* is acute; [I.17]

and the greater angle is subtended by the greater side; [I.19]

therefore *FC* is greater than *FG*.

But *FC* is equal to *FB*;

therefore *FB* is also greater than *FG*, the less than the greater: which is impossible.

Therefore *FG* is not perpendicular to *DE*.

Similarly we can prove that neither is any other straight line except *FC*; therefore *FC* is perpendicular to *DE*.

Therefore etc. Q.E.D. □

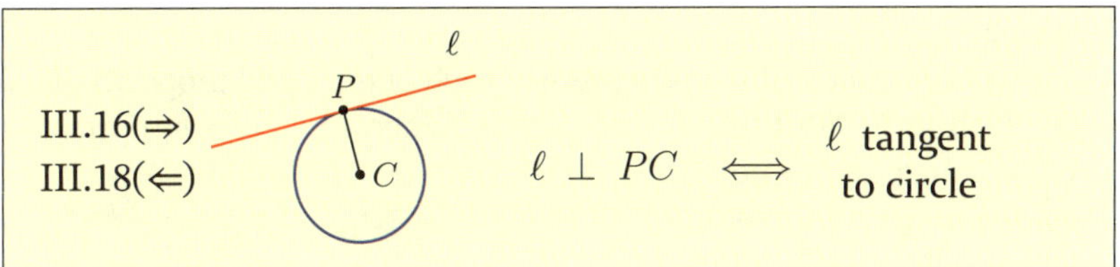

It is with this proposition that we can appreciate one of the lines of Christopher Morley's sonnet celebrating the properties of the circle at the start of the chapter. He alludes to Proposition III.18 when he writes "the joyful perpendicular Erected at the kiss of tangency Must meet my central point, my avatar!"

Proposition III.19. *If a straight line touch a circle, and from the point of contact a straight line be drawn at right angles to the tangent, the centre of the circle will be on the straight line so drawn.*

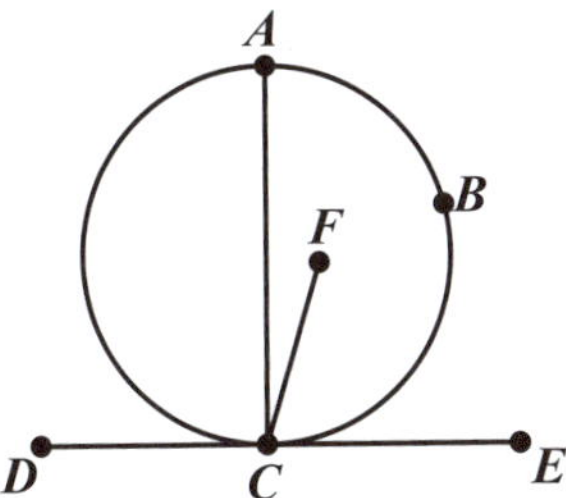

Figure 10.13. Proposition III.19

Proof. For let a straight line DE touch the circle ABC at the point C, and from C let CA be drawn at right angles to DE; I say that the centre of the circle is on AC.

For suppose it is not, but, if possible, let F be the centre, and let CF be joined.

Since a straight line DE touches the circle ABC, and FC has been joined from the centre to the point of contact, FC is perpendicular to DE; [III.18] therefore the angle FCE is right.

But the angle ACE is also right; therefore the angle FCE is equal to the angle ACE, the less to the greater: which is impossible.

Therefore F is not the centre of the circle ABC.

Similarly we can prove that neither is any other point except a point on AC.

Therefore etc. Q.E.D.   □

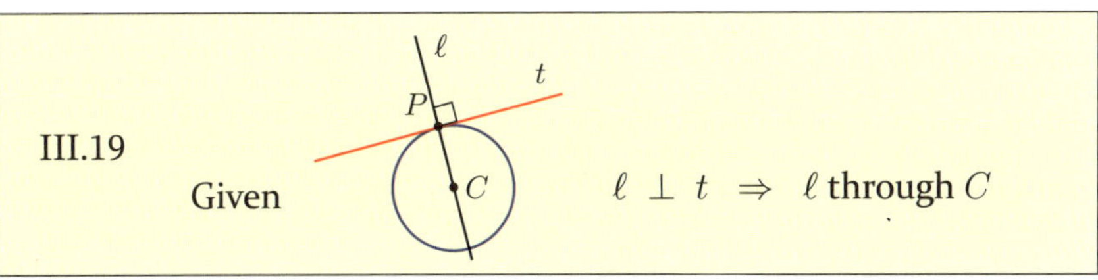

Exercises 10.2

1. Give an updated version of Euclid's proof of each of the listed propositions. Be sure to justify each step, substitute mathematical symbols where appropriate, and include helpful diagrams as needed.

   (a) Proposition III.1   (b) Proposition III.3

2. Prove that in a circle, the radius perpendicular to a chord bisects that chord's central angle.

3. Suppose A and B are the contact points of the tangents to a circle from a point C outside it. Prove that $AC = BC$.

4. Prove that the straight line that joins the center of a circle to the intersection of two of its tangents bisects the angle between these tangents.

5. Prove that the tangent lines at the endpoints of a diameter of a circle must be parallel.

6. Give an updated version of Euclid's proof of each of the listed propositions. Be sure to justify each step, substitute mathematical symbols where appropriate, and include helpful diagrams as needed.

   (a) Proposition III.16   (c) Proposition III.18
   (b) Proposition III.17   (d) Proposition III.19

7. Prove that if two circles are internally tangent, then the line joining their centers contains the point of tangency.

8. Prove that if two circles are externally tangent, then the line joining their centers contains the point of tangency.

9. Prove that if two circles are tangent to each other, then they have a common tangent line at their point of tangency.

10. Given two lines that intersect at a point A, construct a circle that is tangent to both lines. Is this circle unique?

11. An equilateral triangle is inscribed in a circle of radius 1 (each vertex lies on the circle). One of the sides of the triangle is then removed. Find the radius of the largest circle that can be constructed inside the region created by the remaining two sides of the triangle and the minor arc that they determine, as shown in Figure 10.14.

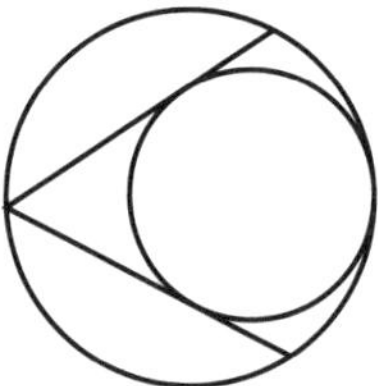

Figure 10.14. Exercise 10.2.11

10.3 Arcs, chords and angles

We leave tangency for some results relating central angles, inscribed angles and their corresponding arcs. The first states that the angle at the center is twice its corresponding inscribed angle. As shown in Figure 10.15, there are two cases to consider. In the first case, the center of the circle lies in the interior of inscribed angle $\angle BAC$, and in the second, it lies outside $\angle BAC$. In recent years, this proposition has come to be known as the Star Trek Theorem due to the resemblance of the diagram in the first case to the fictional Starfleet insignia. We include the proof of the first case and leave the second as an exercise. While we restrict the term *central angle* for those less than 180°, this theorem and its proof still hold when the inscribed angle is obtuse, and hence, the angle at the center is greater than 180°.

Proposition III.20 [Star Trek Theorem]. *In a circle the angle at the centre is double of the angle at the circumference, when the angles have the same circumference as base.*

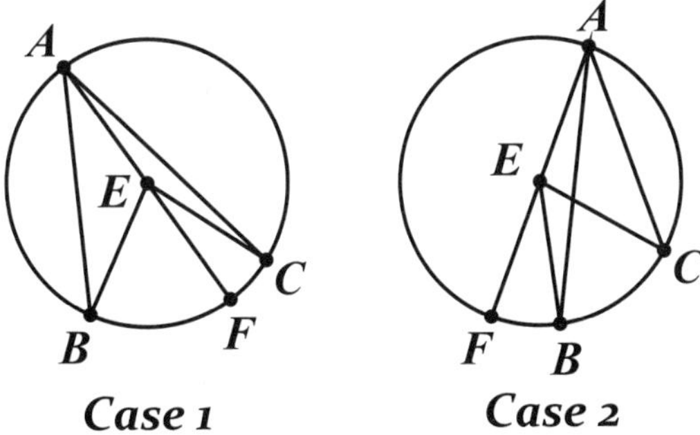

Figure 10.15. Two cases for Proposition III.20

Proof. Let ABC be a circle, let the angle BEC be an angle at its centre, and the angle BAC an angle at the circumference, and let them have the same circumference BC as base; I say that the angle BEC is double of the angle BAC.

For let AE be joined and drawn through to F.

Then, since EA is equal to EB,

the angle EAB is also equal to the angle EBA; [I.5]

therefore the angles EAB, EBA are double of the angle EAB.

But the angle BEF is equal to the angles EAB, EBA; [I.32]

therefore the angle BEF is also double of the angle EAB.

For the same reason the angle FEC is also double of the angle EAC.

Therefore the whole angle BEC is double of the whole angle BAC.

Q.E.D.

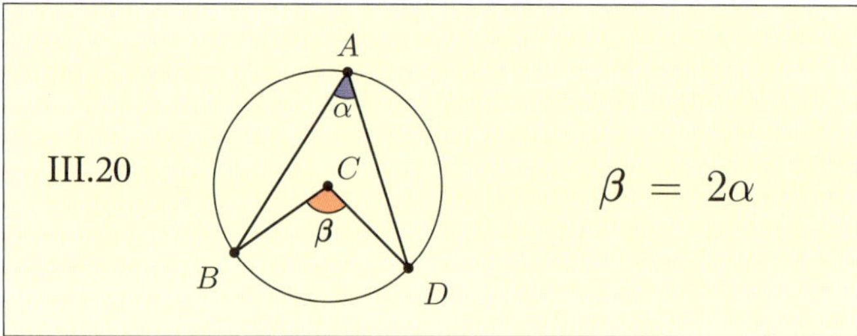

Proposition III.21 is a corollary to III.20, stating that inscribed angles subtended by the same arc (be it major or minor) must be congruent.

Proposition III.21. *In a circle the angles in the same segment are equal to one another.*

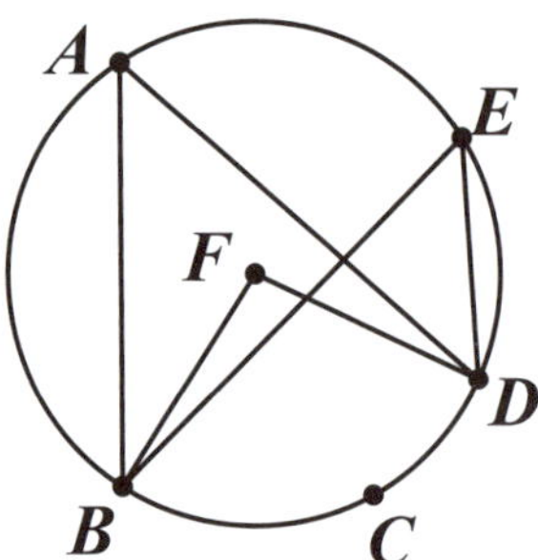

Figure 10.16. Proposition III.21

Proof. Let $ABCD$ be a circle, and let the angles BAD, BED be angles in the same segment $BAED$; I say that the angles BAD, BED are equal to one another.

For let the centre of the circle $ABCD$ be taken, and let it be F; let BF, FD be joined.

Now, since the angle BFD is at the centre, and the angle BAD at the circumference, and they have the same circumference BCD as base,

therefore the angle BFD is double of the angle BAD. [III.20]

For the same reason the angle BFD is also double of the angle BED;

therefore the angle BAD is equal to the angle BED.

Therefore etc. Q.E.D.   □

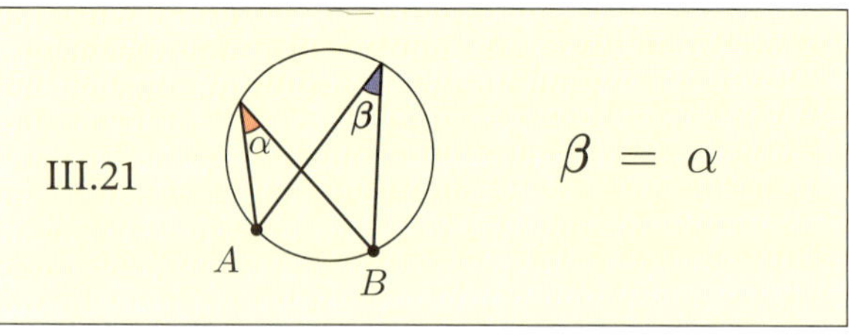

In Proposition III.22 we have our first result about cyclic polygons: in a cyclic quadrilateral, the opposite angles sum to two right angles.

Ptolemy and the Table of Chords

Another well-known result about cyclic polygons is due to a Greek astronomer and fellow Alexandrian who lived four hundred years after Euclid's time, Claudius Ptolemy (ca. 100–178). Ptolemy played the role of Euclid for the field of astronomy by authoring the *Mathematical Collection*, a thirteen book work detailing the Greek model of the universe. Not unlike the *Elements*, it was the bible of astronomy for 1500 years until Copernicus replaced its geocentric model with a heliocentric model. Owing to its masterpiece status, over time it became known as the *Almagest*, which translates as, "The Greatest." In it, he described the motion of the sun, moon and the planets, which the Greeks thought to be circular. To do this, Ptolemy needed Euclid's geometry, but he also needed to make many calculations of angles, chords and arcs. Nowadays, calculators, phones and computers provide an embarrassment of riches for such trigonometric calculations. Ptolemy, however, needed to consult the table of chords in his first book, essentially a table of values for the sine function. To complete his table of chords, he needed the half-angle formula, the sine value for a few angles, and an equation relating the sides and diagonals of a cyclic quadrilateral which has come to be known as Ptolemy's Theorem. It states, *"Given any cyclic quadrilateral, the product of the diagonals equals the sum of the products of the opposite sides."* For *ABCD* as shown in Figure 10.17, this statement is the equation

$$AC \cdot BD = AD \cdot BC + AB \cdot CD.$$

The proof of Ptolemy's Theorem is much simpler with our similarity results. As an exercise, the reader is asked to provide a proof for this theorem. While we will see another cyclic quadrilateral result as Euclid's thirty-fifth proposition, Ptolemy's Theorem is more famous.

Proposition III.22. *The opposite angles of quadrilaterals in circles are equal to two right angles.*

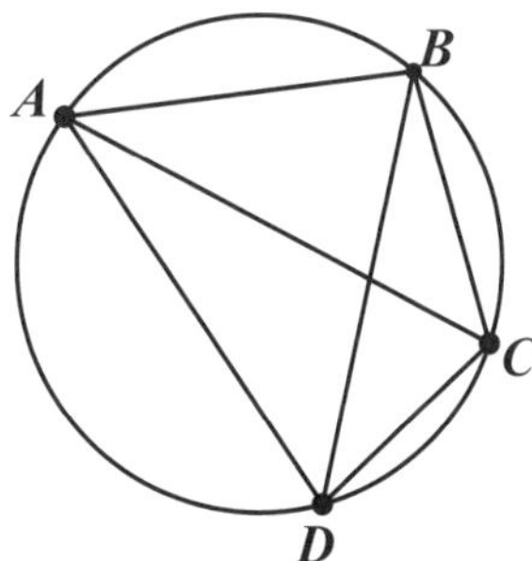

Figure 10.17. Proposition III.22

Proof. Let *ABCD* be a circle, and let *ABCD* be a quadrilateral in it;
   I say that the opposite angles are equal to two right angles.
   Let *AC*, *BD* be joined.
   Then, since in any triangle the three angles are equal to two right angles, [I.32] the three angles *CAB*, *ABC*, *BCA* of the triangle *ABC* are equal to two right angles.
   But the angle *CAB* is equal to the angle *BDC*, for they are in the same segment *BADC*; [III.21]

and the angle ACB is equal to the angle ADB, for they are in the same segment $ADCB$;

therefore the whole angle ADC is equal to the angles BAC, ACB.

Let the angle ABC be added to each; therefore the angles ABC, BAC, ACB are equal to the angles ABC, ADC.

But the angles ABC, BAC, ACB are equal to two right angles;

therefore the angles ABC, ADC are also equal to two right angles.

Similarly we can prove that the angles BAD, DCB are also equal to two right angles.

Therefore etc. Q.E.D. □

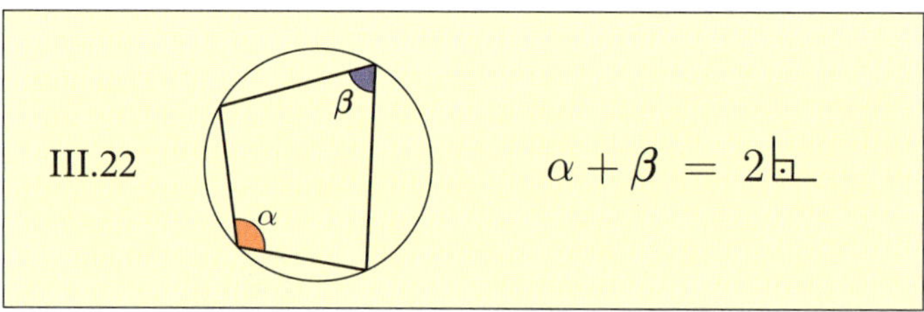

We incorporate Propositions III.26 through III.29 into the Circle Equivalence Theorem. This theorem details the obvious relationship between corresponding chords, central angles, minor and major arcs, and the inscribed angles they subtend.

Circle Equivalence Theorem [III.26, 27, 28, 29]. Consider equal circles centered at C and C', respectively, with points A, B, D and E on the circle centered at C, and points A', B', D' and E' on the other. Assume $\widehat{AB}$ is a minor arc through E but not D, and $\widehat{A'B'}$ is a minor arc through E' but not D'. The following are equivalent:

(1) $\widehat{AB} \cong \widehat{A'B'}$ minor arcs are congruent

(2) $\widehat{ADB} \cong \widehat{A'D'B'}$ major arcs are congruent

(3) $AB \cong A'B'$ chords are congruent

(4) $\angle ACB \cong \angle A'C'B'$ central angles are congruent

(5) $\angle ADB \cong \angle A'D'B'$ inscribed angles subtended by minor arcs are congruent

(6) $\angle AEB \cong \angle A'E'B'$ inscribed angles subtended by major arcs are congruent

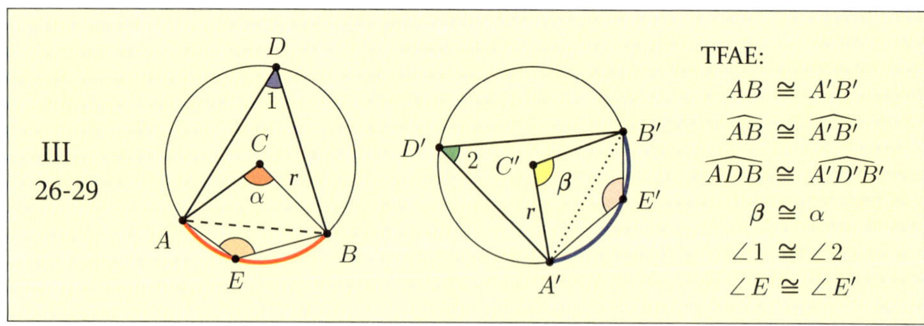

Proving that part (3) holds if and only if (4) holds, as well as (4) iff (5), and (5) iff (6), are straightforward exercises left for the reader. We cannot show that (1) is equivalent to either (3) or (4) because we do not have any tools to establish the equivalence of two arcs. Euclid's justification relies on Proposition III.24, the proof of which relies on superposition. Instead, we adopt the following definition.

Definition 10.1. Minor arcs are **congruent** if their corresponding central angles are congruent. Specifically, given equal circles centered at C and C', respectively, with points A, B on the circle centered at C, and points A', B' on the other, the minor arcs $\widehat{AB}$ and $\widehat{A'B'}$ are congruent if and only if their corresponding central angles are congruent, that is, $\angle ACB \cong \angle A'B'C'$. Two major arcs are **congruent** if their respective minor arcs are congruent.

By adopting this definition, we have (1) iff (2) iff (4). A reader may object, as the more obvious definition would be to say that two arcs are congruent if they have the same length. While this will also be the case, we postpone our discussion of the length of an arc until the end of this chapter.

Proposition III.30 provides the steps necessary to bisect an arc.

Proposition III.30. *To bisect a given circumference.*

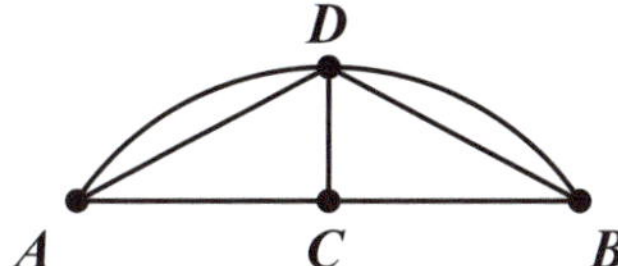

Figure 10.18. Proposition III.30

Proof. Let ADB be the given circumference; thus it is required to bisect the circumference ADB.

Let AB be joined and bisected at C; from the point C let CD be drawn at right angles to the straight line AB, and let AD, DB be joined.

Then, since AC is equal to CB, and CD is common, the two sides AC, CD are equal to the two sides BC, CD; and the angle ACD is equal to the angle BCD, for each is right; therefore the base AD is equal to the base DB. [I.4]

But equal straight lines cut off equal circumferences, the greater equal to the greater, and the less to the less; [III.28] and each of the circumferences AD, DB is less than a semicircle; therefore the circumference AD is equal to the circumference DB.

Therefore the given circumference has been bisected at the point D. Q.E.F. $\square$

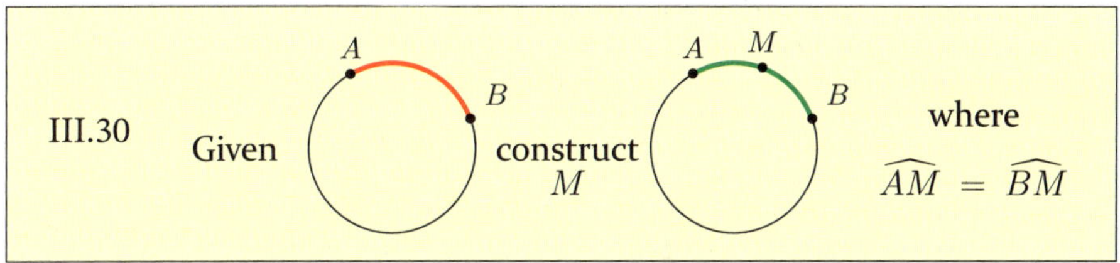

Proposition III.31, which states that an angle inscribed in a semicircle is a right angle, is attributed to Thales of Miletus (ca. 624–ca. 547 BCE). Often referred to as the "Father of Geometry," he is credited with proving that the base angles of an isosceles triangle are equal (I.5) and that vertical angles are equal (I.15). We will see the converse of Thales' Theorem in Exercise 11.2.4. Euclid's statement of III.31 is longer as he shows more than we do here.

Proposition III.31 [Thales' Theorem]. *In a circle the angle in the semicircle is right, that in a greater segment less than a right angle, and that in a less segment greater than a right angle.*

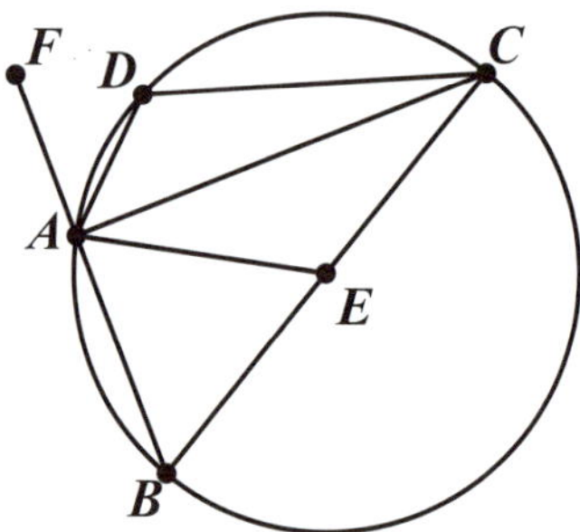

Figure 10.19. Proposition III.31

Proof. Let $ABCD$ be a circle, let BC be its diameter, and E its centre, and let BA, AC, AD, DC be joined;

I say that the angle BAC in the semicircle BAC is right, the angle ABC in the segment ABC greater than the semicircle is less than a right angle, and the angle ADC in the segment ADC less than the semicircle is greater than a right angle.

Let AE be joined, and let BA be carried through to F.

Then, since BE is equal to EA, the angle ABE is also equal to the angle BAE. [I.5]

Again, since CE is equal to EA, the angle ACE is also equal to the angle CAE. [I.5]

Therefore the whole angle BAC is equal to the two angles ABC, ACB.

But the angle FAC exterior to the triangle ABC is also equal to the two angles ABC, ACB; [I.32]

therefore the angle BAC is also equal to the angle FAC; therefore each is right; [Def. 10]

therefore the angle BAC in the semicircle BAC is right.

Next, since in the triangle ABC the two angles ABC, BAC are less than two right angles, [I.17] and the angle BAC is a right angle, the angle ABC is less than a right angle; and it is the angle in the segment ABC greater than the semicircle.

Next, since $ABCD$ is a quadrilateral in a circle, and the opposite angles of quadrilaterals in circles are equal to two right angles, [III.22] while the angle ABC is less than a right angle, therefore the angle ADC which remains is greater than a right angle; and it is the angle in the segment ADC less than the semicircle.

Q.E.D.   □

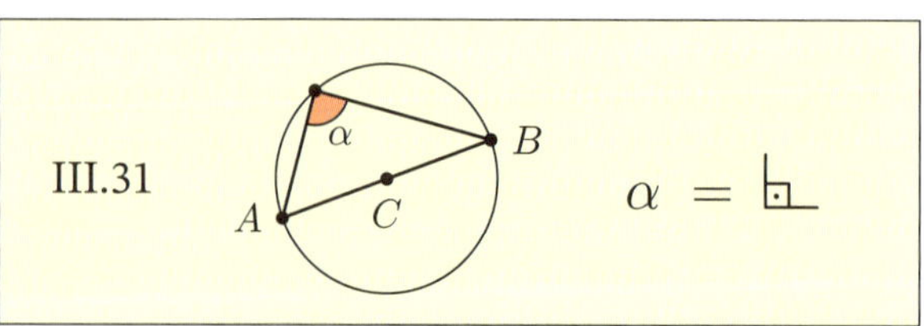

For greater clarity and better notation, we state Proposition III.32 with updated terminology and give a revised proof.

Proposition III.32. *Given a circle with chord BD and a tangent line EF which intersects the circle at B, if A is any point on the circle lying on the opposite side of $\overleftrightarrow{BD}$ as F, then the inscribed angle $\angle BAD$ is equal to the angle $\angle DBF$. Similarly, if C is any point on the circle lying on the opposite side of $\overleftrightarrow{BD}$ as E, then the inscribed angle $\angle BCD$ is equal to the angle $\angle DBE$.*

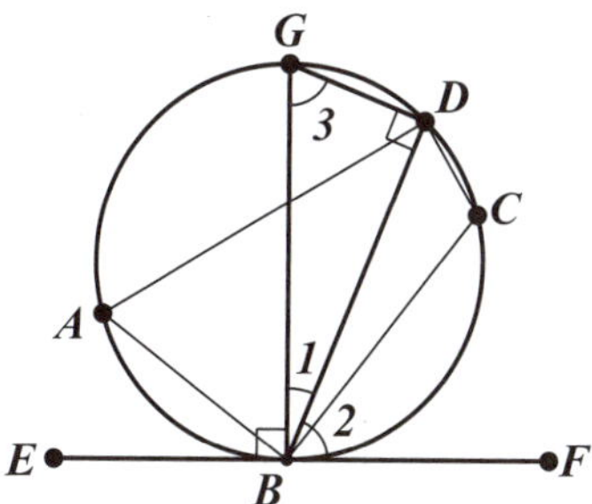

Figure 10.20. Proposition III.32

Proof.   Let BD be a chord of a circle and EF be tangent to the circle at B. WLOG, assume that $\angle FBD$ is less than a right angle. Let A be a point on the circle lying on the opposite side of $\overleftrightarrow{BD}$ as F.

Using I.11, construct $GB \perp EF$. Then by III.19, GB is a diameter of the circle. Since all inscribed angles of the arc $\overparen{BD}$ are congruent [III.21], we have $\angle BAD = \angle BGD$. Therefore, it suffices to show that $\angle BGD = \angle DBF$, or equivalently $\angle 2 = \angle 3$.

By III.31, $\angle GDB$ is a right angle. Thus $\angle 1 + \angle 3$ equals a right angle [I.32]. By construction $\angle 1 + \angle 2$ also equals a right angle. Subtracting equals from equals, we have $\angle 2 = \angle 3$, as desired.

Moreover, since $ABCD$ is a cyclic quadrilateral, by III.22, opposite angles $\angle BCD$ and $\angle BAD$ sum to two right angles. By I.13, $\angle EBD + \angle DBF$ also equals two right angles, hence $\angle BCD = \angle EBD$. $\square$

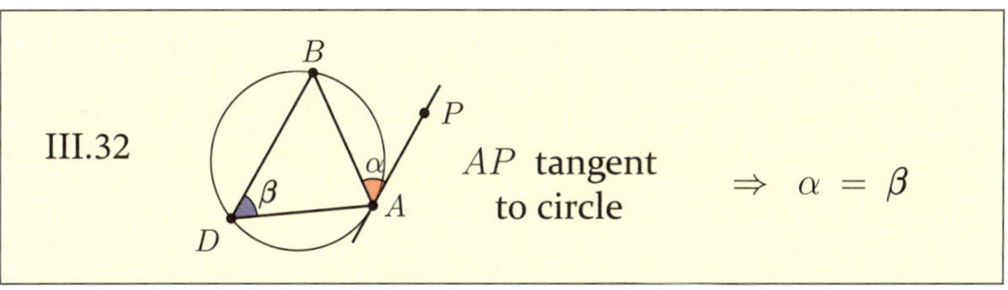

Though Euclid did not include it, we note that the converse of this theorem also holds, and we leave the proof as an exercise.

Theorem 10.2 [Converse of III.32]. *Given a circle with chord BD and distinct line $\overleftrightarrow{BF}$, if A is any point on the circle lying on the opposite side of $\overleftrightarrow{BD}$ as F, then if $\angle BAD = \angle DBF$, BF is tangent to the circle.*

Exercises 10.3

1. Give an updated version of Euclid's proof of each of the listed propositions. Be sure to justify each step, substitute mathematical symbols where appropriate, and include helpful diagrams as needed.

   (a) Proposition III.20   (c) Proposition III.22

   (b) Proposition III.21

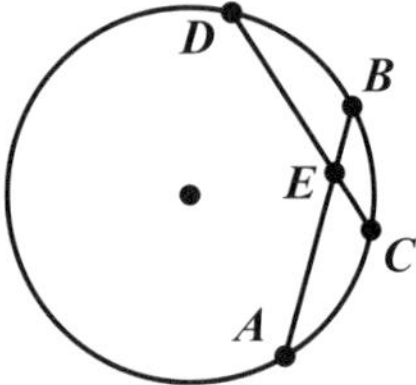

Figure 10.21. Exercise 10.3.2: Intersecting equal chords

2. Assume that equal chords AB and CD intersect at a point E, as shown in Figure 10.21. Prove that $BE = CE$ and $AE = DE$.

3. Given a circle with chords AB and CD, if the lines $\overleftrightarrow{AB}$ and $\overleftrightarrow{CD}$ intersect at a point, E, that lies outside of the circle, prove that $\angle ACE = \angle DBE$ and $\angle BCE = \angle DAE$. Include a figure with your proof.

4. Prove the second case for Proposition III.20. That is, let BC be an arc, A a point not on it and E the center of the circle where E lies outside of angle $\angle BAC$. Prove that angle $\angle BEC$ is twice the inscribed angle $\angle BAC$.

5. Consider equal circles centered at C and C', respectively, with points A, B, D and E on the circle centered at C, and points A', B', D' and E' on the other. Assume $\overarc{AB}$ is a minor arc through E but not D, and $\overarc{A'B'}$ is a minor arc through E' but not D'. Prove each of the following parts of the Circle Equivalence Theorem.

   (a) Show that chords are congruent if and only if central angles are congruent, that is, $AB \cong A'B'$ iff $\angle ACB \cong \angle A'C'B'$.

   (b) Show that central angles are congruent if and only if inscribed angles subtended by the minor arcs are congruent, that is, $\angle ACB \cong \angle A'C'B'$ iff $\angle ADB \cong \angle A'D'B'$.

   (c) Show that inscribed angles of the minor arcs are congruent if and only if inscribed angles subtended by the major arcs are congruent, that is, $\angle ADB \cong \angle A'D'B'$ iff $\angle AEB \cong \angle A'E'B'$.

6. Give an updated version of Euclid's proof of each of the listed propositions. Be sure to justify each step, substitute mathematical symbols where appropriate, and include helpful diagrams as needed.

   (a) Proposition III.30   (b) Proposition III.31

7. Consider a circle with points A, B and C, where $AB = AC = BC$. Let D be a point on arc BC. Construct segments AD, BD and CD, as illustrated in Figure 10.22. Prove that $AD = BD + CD$. [*Hint: Construct E on AD such that $DE = DC$. Prove that $\triangle CDE$ is an equilateral triangle.*]

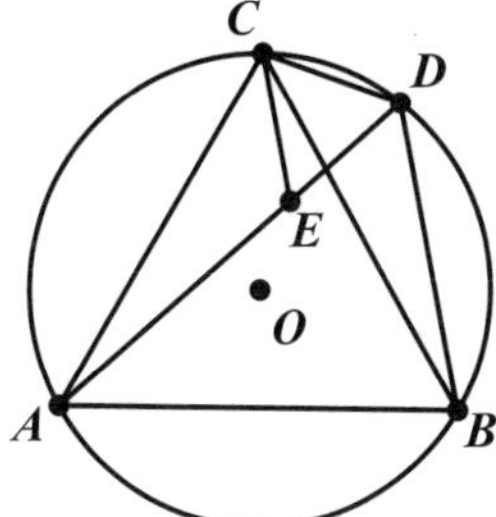

Figure 10.22. Exercise 10.3.7

8. Consider a circle, c, with center, O, and diameter AB. Let CD be any chord that is perpendicular to AB. Let the angle bisector of $\angle OCD$ intersect c at a point E. Prove that $EO \perp AB$. [*Hint: Extend $\overrightarrow{CO}$ to a chord CF, then construct segment DF, as illustrated in Figure 10.23.*]

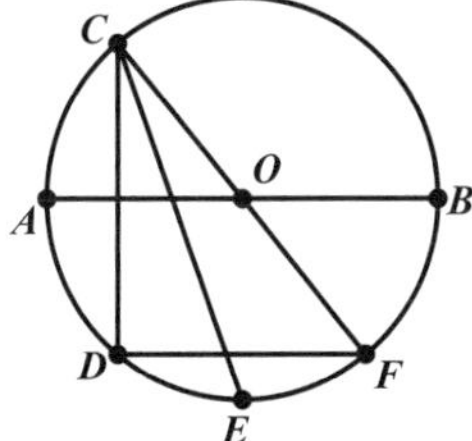

Figure 10.23. Exercise 10.3.8

9. Consider a circle with center O and diameter AB. Let C be the midpoint of AO and construct a chord DE through C such that $DC \perp AB$. Construct a circle with center C and radius CD and let its intersection with CB be F. Join EF and let $\overrightarrow{DF}$ intersect the circle at G, as illustrated in Figure 10.24. Prove that EG has length $r\sqrt{2}$, where r is the radius of the circle.

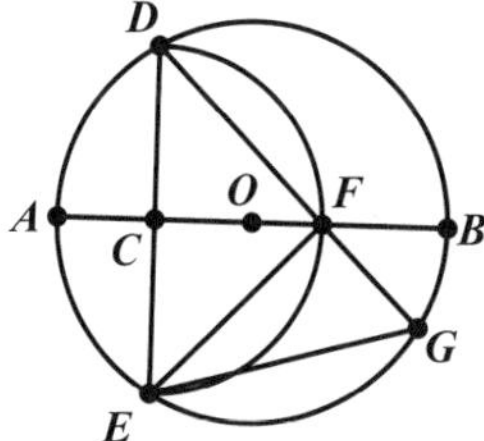

Figure 10.24. Exercise 10.3.9

10. Prove the converse of Proposition III.32. Be sure to justify each step.

11. Given cyclic quadrilateral $ABCD$ with $AD = BC$, prove that $\angle A = \angle B$ and $\angle C = \angle D$.

12. Given cyclic quadrilateral $ABCD$ with $AD = BC$, prove that the diagonals are equal, that is $AC = BD$.

13. Prove that every cyclic rhombus is a square.

14. Prove Ptolemy's Theorem: *Given any cyclic quadrilateral, the product of the diagonals equals the sum of the products of the opposite sides.* For cyclic quadrilateral $ABCD$ as shown in Figure 10.25, the theorem says

$$AC \cdot BD = AD \cdot BC + AB \cdot CD.$$

To begin, construct E on AC such that $\angle ABE = \angle DBC$.

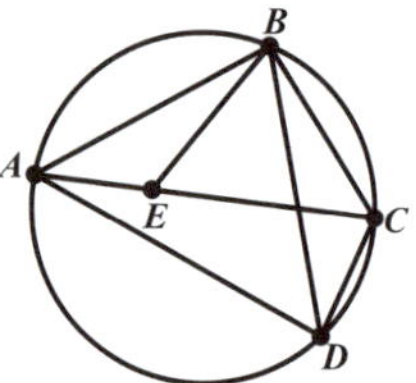

Figure 10.25. Exercise 10.3.14

(a) Justify why it is possible to construct such an E.

(b) Show that $\triangle ABE \sim \triangle DBC$.

(c) Show that $\triangle ABD \sim \triangle EBC$.

(d) Using algebra combined with the appropriate ratios of sides, finish the proof.

15. Use Ptolemy's Theorem (Exercise 14) to prove the Pythagorean Theorem.

10.4 Area Propositions: III.35 through III.37

The Intersecting Chords Theorem, Proposition III.35, is an area theorem establishing the equivalence of two rectangles produced by the segments of intersecting chords. As shown in Figure 10.26, given a circle with chords AC and BD whose intersection E lies within the circle, then $AE \cdot EC = BE \cdot ED$. If we connect adjacent vertices to form $ABCD$, then like Ptolemy's Theorem, this can be interpreted as a statement about the diagonals of a cyclic quadrilateral. We leave the proof as an exercise. The Intersecting Chords Theorem is also true when the secant lines meet outside the circle. This related theorem is called the *Intersecting Secants Theorem*, and appears as Exercise 10.4.4.

Proposition III.35 [Intersecting Chords Theorem]. *If in a circle two straight lines cut one another, the rectangle contained by the segments of the one is equal to the rectangle contained by the segments of the other.*

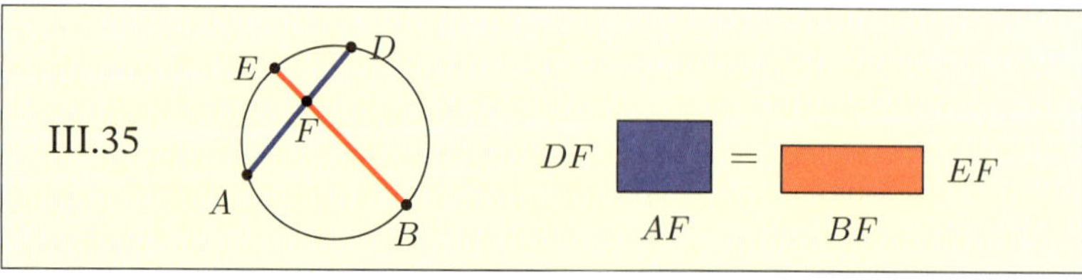

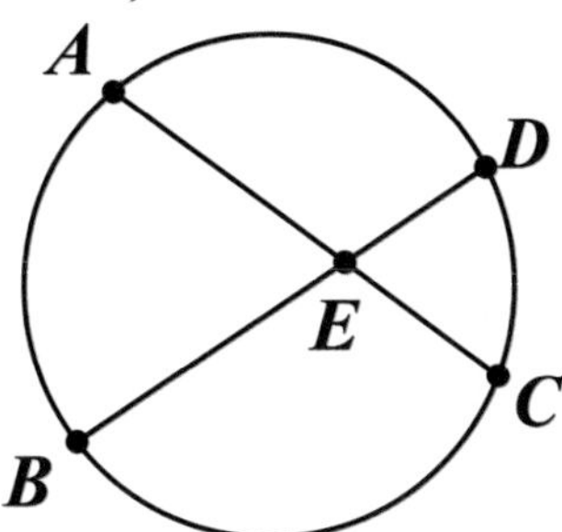

Figure 10.26. Proposition III.35

Book III concludes with Propositions III.36 and its converse, III.37, both stated here with updated notation for better clarity. Like III.35, the first is an area result, with this proposition equating a square on a segment of a tangent line with a rectangle on segments of an extended chord. First, we give a slightly revised version of Euclid's proof of III.36 using updated notation and requiring two cases. As a follow-up, we demonstrate the power of similarity by giving a completely revised proof that is significantly shorter. The similarity proof clearly would not have been possible for Euclid since it was the subject of a later book.

Proposition III.36. *Given a point D that lies outside of a given circle, if BD is tangent to the circle at B and ACD is a line that intersects the circle at two points, producing the chord AC, then $BD^2 = AD \cdot CD$.*

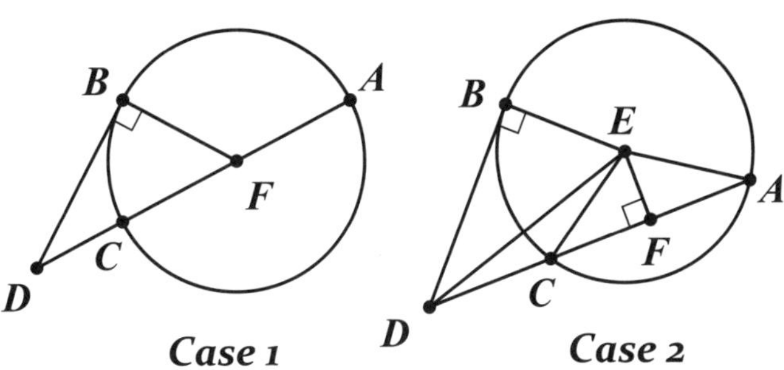

Figure 10.27. Proposition III.36

Revised Proof.   Let D be a point outside of a given circle with BD tangent to the circle at B and let ACD be a line that intersects the circle at two points, producing the chord AC.

Case 1. The center of the given circle, call it F, lies on AC. By III.18, angle $\angle DBF$ is a right angle. By the Pythagorean Theorem, we have

$$BD^2 + BF^2 = DF^2 = (AD - AF)^2.$$

Since $AF = BF$, this gives

$$BD^2 = (AD - BF)^2 - BF^2 = AD^2 - 2AD \cdot BF.$$

Since $AC = 2BF$, we have

$$BD^2 = AD(AD - 2BF) = AD(AD - AC) = AD \cdot CD,$$

as desired.

Case 2. The center of the given circle does not lie on AC. Let E be the center of our circle, and construct F on AC such that $EF \perp AC$ [1.12]. Since $\angle DBE$ is right [III.18], we may apply the Pythagorean Theorem to $\triangle DBE$, producing

$$BD^2 = DE^2 - BE^2.$$

Since $\angle EFD$ is also a right angle, applying the Pythagorean Theorem to $\triangle EFD$ gives

$$DE^2 = DF^2 + EF^2.$$

Combining these equations, we have

$$BD^2 = DF^2 + EF^2 - BE^2. \tag{10.1}$$

Since $DF = CD + CF$, we have

$$DF^2 = (CD + CF)^2 = CD^2 + 2CD \cdot CF + CF^2 = CF^2 + CD(2CF + CD). \tag{10.2}$$

By III.3, $AF = CF$, and thus, $2CF + CD = AD$. Substituting this in equation (10.2) gives $DF^2 = CF^2 + CD \cdot AD$. Substituting this in equation (10.1) produces

$$BD^2 = CD \cdot AD + CF^2 + EF^2 - BE^2.$$

Applying I.47 one last time, we consider the right triangle $\triangle EFC$ and note that

$$CE^2 = EF^2 + CF^2.$$

As $BE = CE$ (both radii), $CF^2 + EF^2 = BE^2$ and

$$BD^2 = CD \cdot AD,$$

as desired. $\square$

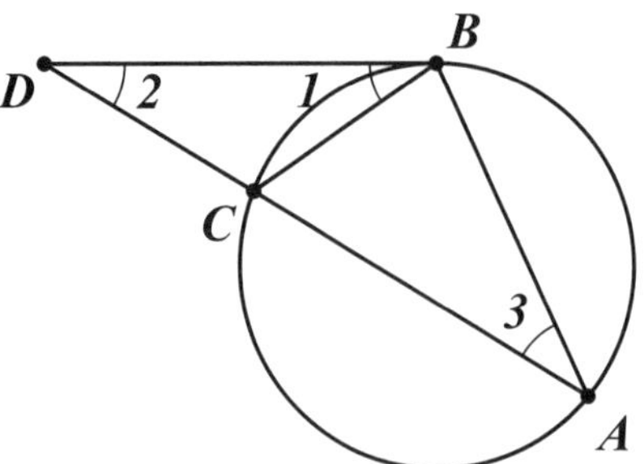

Figure 10.28. Updated proof of Proposition III.36 using similarity

Proof using similarity.   Let D be a point outside a given circle with BD tangent to the circle at B and let ACD be a line that intersects the circle at two points, producing the chord AC. Consider the triangles $\triangle DCB$ and $\triangle DBA$. Since BD is tangent to the circle, $\angle 1 = \angle 3$ by III.32. Also, both triangles share $\angle 2$. Since these triangles have two equal angles, $\triangle DCB \sim \triangle DBA$ by $\widetilde{AA}$. Thus,

$$\frac{BD}{CD} = \frac{AD}{BD},$$

and $BD^2 = AD \cdot CD$, as desired. $\square$

We give Euclid's proof of the converse, Proposition III.37, and leave it to the reader to provide an updated version of his proof, as well as a proof using similarity.

Proposition III.37. *Given a point D that lies outside of a given circle, if BD intersects the circle at B and ACD is a line that intersects the circle at two points, producing the chord AC, then if $BD^2 = AD \cdot CD$, BD is tangent to the circle.*

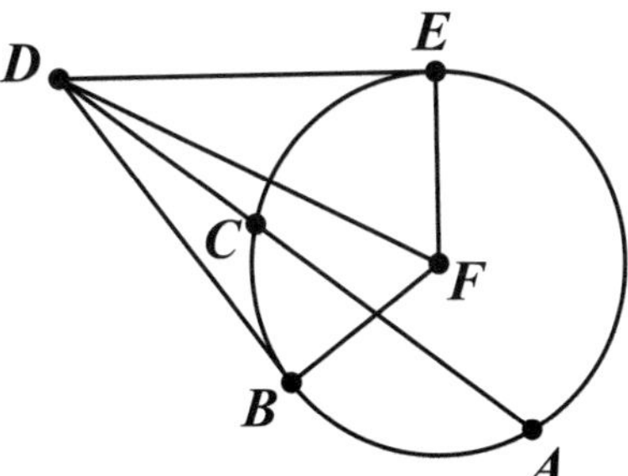

Figure 10.29. Proposition III.37

Euclid's Proof.   For let a point D be taken outside the circle ABC; from D let the two straight lines DCA, DB fall on the circle ACB; let DCA cut the circle and DB fall on it; and let the rectangle AD, DC be equal to the square on DB.

I say that DB touches the circle ABC.

For let DE be drawn touching ABC; let the centre of the circle ABC be taken, and let it be F; let FE, FB, FD be joined.

Thus the angle FED is right. [III.18]

Now, since DE touches the circle ABC, and DCA cuts it, the rectangle AD, DC is equal to the square on DE. [III.36]

But the rectangle AD, DC was also equal to the square on DB; therefore the square on DE is equal to the square on DB;

therefore DE is equal to DB.

And FE is equal to FB; therefore the two sides DE, EF are equal to the two sides DB, BF; and FD is the common base of the triangles;

therefore the angle DEF is equal to the angle DBF. [I.8]

But the angle DEF is right; therefore the angle DBF is also right.

And FB produced is a diameter; and the straight line drawn at right angles to the diameter of a circle, from its extremity, touches the circle; [III.16] therefore DB touches the circle.

Similarly this can be proved to be the case even if the centre be on AC.
Therefore etc. Q.E.D.

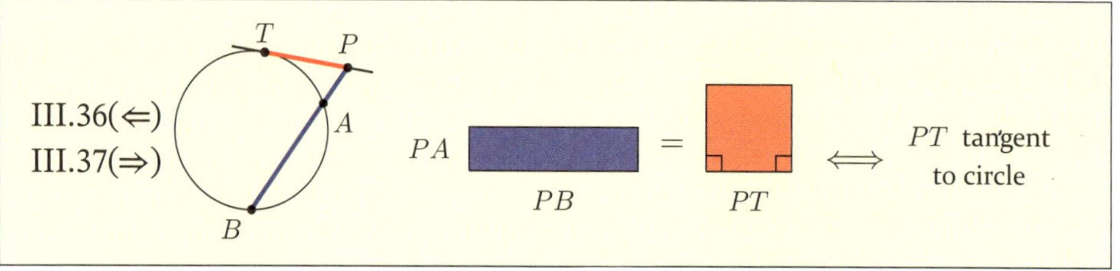

> ### Neutral and non-Neutral Geometry of the Circle
>
> Recall that Neutral geometry consists of theorems that do not require the Parallel Postulate. Some of the propositions of Book III fall under this heading, while others are non-Neutral. Since there is no easy line of demarcation for the circle propositions, we presented the selected results in numerical order. Since a classification of these results will be useful to us in Chapter 13, we do so now. Of note, Propositions III.26 and III.27 both consist of two statements, one relating central angles and their corresponding arcs, and the other relating inscribed angles and their corresponding arcs. Consequently, only one-half of each of these propositions is Neutral, as specified below.
>
> Neutral Book III Propositions: 1, 3, 16, 17, 18, 19, 26 (central angles), 27 (central angles), 28, 29
>
> Non-Neutral Book III Propositions: 20, 21, 22, 26 (inscribed angles), 27 (inscribed angles), 31, 32, 35, 36, 37

Exercises 10.4

1. Give a proof of Proposition III.35 using similarity.

2. Give an updated version of Euclid's proof of Proposition III.37. Be sure to justify each step, substitute mathematical symbols where appropriate, and include helpful diagrams as needed.

3. Give a proof of Proposition III.37 using similarity.

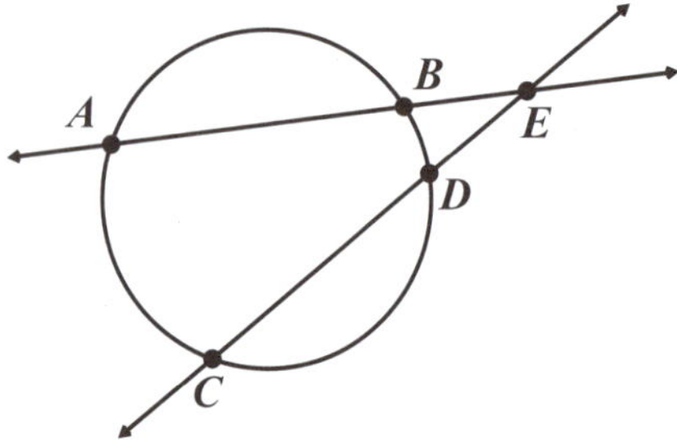

Figure 10.30. Exercise 10.4.4: Intersecting Secants Theorem

4. Prove the **Intersecting Secants Theorem**: Given a circle with chords AB and CD, if the lines $\overleftrightarrow{AB}$ and $\overleftrightarrow{CD}$ intersect at a point, E, that lies outside of the circle, as illustrated in Figure 10.30, prove that $EA \cdot EB = EC \cdot ED$. [Notice that, unlike the Intersecting Chords Theorem, here EA and EB are overlapping segments, as are EC and ED.]

5. Use the results of Section 10.4 to give a different solution to Exercise 10.2.3, restated here: Suppose A and B are the contact points of the tangents to a circle from a point C outside it. Prove that $AC = BC$.

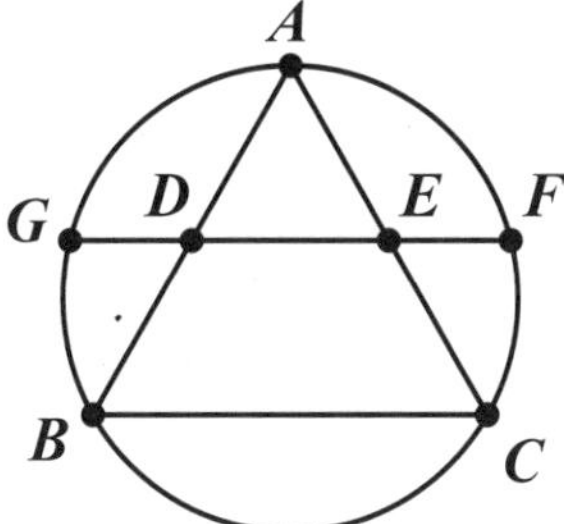

Figure 10.31. Exercise 10.4.6

6. Consider cyclic equilateral triangle $\triangle ABC$ in circle c. Bisect AB and AC at D and E, respectively. Extend DE to intersect c at F and G as shown in Figure 10.31. Prove that DE/EF is the golden ratio.

7. Consider a circle c with center A and radius $AB = r$. Construct a perpendicular CB to AB such that $CB = 2r$. Construct the line $\overleftrightarrow{AC}$ and let the intersections with c be D and E, as illustrated in Figure 10.32. Construct F on CB so that $CF = CD$. Prove that

$$CF^2 = BC \cdot BF,$$

and thus, we have constructed the *golden ratio* on BC.

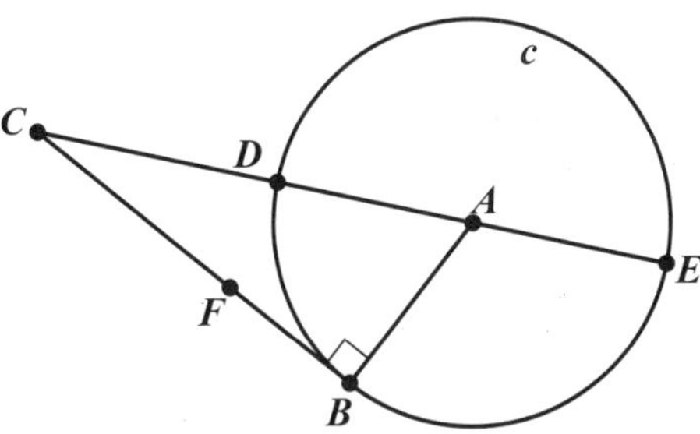

Figure 10.32. Exercise 10.4.7

8. Consider a circle c with chord AB. Let C be the midpoint of AB and let DE and FG be two other chords passing through C, where D and F lie on the same side of AB, as illustrated in Figure 10.33. Let H and J be the intersections of EF and DG with

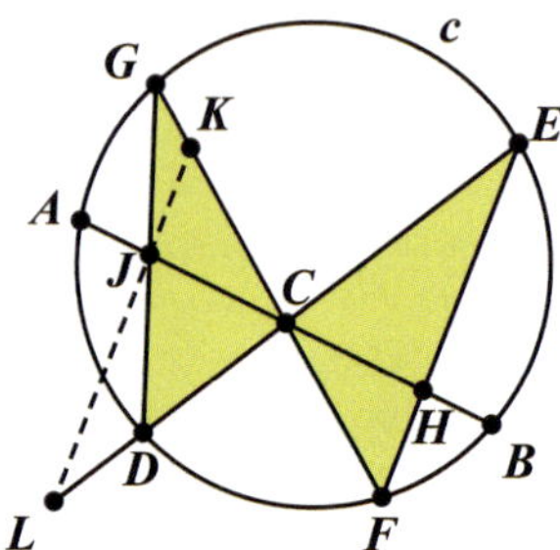

Figure 10.33. Exercise 10.4.8: The Butterfly Theorem

AB, respectively. Then $HC = JC$. The shaded triangles in the illustration give this result its name, the **Butterfly Theorem**.

The following proof was given in the 1827 book *Geometrical Problems* by Mile Bland. Note that the proof assumes that EF and DG are not parallel. In the case where EF and DG are parallel, C must be the center of the circle, and thus, we have $\triangle GJC \cong \triangle FHC$.

Proof. Construct the parallel line to $\overleftrightarrow{EF}$ through J, and let the intersections with $\overleftrightarrow{GF}$ and $\overleftrightarrow{DE}$ be K and L, respectively.

(a) Explain why $\triangle GKJ \sim \triangle LDJ$.

(b) Show that $KJ \cdot LJ = (AC - JC)(AC + JC)$.

(c) Explain why $\triangle CJK \sim \triangle CFH$ and $\triangle CJL \sim \triangle CHE$.

(d) Show that
$$\frac{KJ \cdot LJ}{JC^2} = \frac{EH \cdot FH}{HC^2}.$$

(e) Using the previous equations, show that
$$\frac{AC^2 - JC^2}{JC^2} = \frac{AC^2 - HC^2}{HC^2},$$
and hence, $JC = HC$, as desired. $\square$

10.5 The circumference of a circle & π

As a final note for this chapter dedicated to the circle, it may surprise you that, while the result was well-known at the time, nowhere in the *Elements* does Euclid *prove* the fundamental result that the ratio of the circumference of a circle to its diameter is constant. That is, given any two circles with circumferences *Circumference*$_1$ and *Circumference*$_2$, and diameters *diameter*$_1$ and *diameter*$_2$, respectively, we have
$$\frac{Circumference_1}{diameter_1} = \frac{Circumference_2}{diameter_2}.$$

It is this constant ratio, true for all circles, that has been denoted with the Greek letter π since the eighteenth century. We will have more to say about π in Chapter 11. For now, we will show that this equation holds, and therefore, the constant represented by π exists.

This equation is an algebraic description of the similarity that exists between any two circles, and by similarity, we only mean that they share the same shape. It also explains the circumference formula, $2\pi r$, that appears in the sonnet at the start of this chapter. Our familiarity with the formula belies the fact that this is a very deep result that requires us to venture beyond classical geometry. We can, however, give an informal modern proof to establish this result without getting too far afield. To do this, we will inscribe regular polygons in our circle, meaning every vertex of the polygon lies on the circle, and then we will take a limit. Though the task of constructing a regular polygon inside a given circle is left to Chapter 11, for now, we simply assume that such polygons exist, and that they have the same center as the circle in which they are inscribed. While we only appeal to the intuitive idea of a limit, please refer to any elementary calculus textbook for an explanation of the theory of limits.

Theorem 10.3. *If circles centered at O and O' have radii r and r', respectively, and circumferences, c and c', respectively, then*

$$\frac{c}{2r} = \frac{c'}{2r'}.$$

The main idea for the proof will be familiar to any calculus student: to use an object we *know* how to measure, a line segment, in order to determine the length of the curve that we do not know how to measure. To properly estimate the curve of the circle, we need many tiny line segments produced in a very systematic way. This is an example of a **piecewise linear approximation**. The more tiny pieces we have, the closer our piecewise linear curve resembles the circle, and the closer our approximation gets to the actual length of the circle. Figure 10.34 shows a systematic progression of regular n-gons inscribed in a circle. At this size, it is difficult to distinguish the circle from the regular 24-gon. Therefore, the perimeter of the regular n-gon is a good approximation for the circumference of the circle when n is large.

Figure 10.34. Inscribed regular 3-gon, 6-gon, 12-gon, 24-gon and 48-gon

For any $n \geq 3$, let P_n be a regular n-gon inscribed in the circle centered at O with AB as one side of the n-gon. Note that A and B lie on the circle, and the perimeter of the polygon, p_n, is $n \cdot AB$ since it is regular. Likewise, let P_n' be a regular n-gon inscribed in the circle centered at O', with $A'B'$ as one side of the n-gon. The perimeter of the regular polygon centered at O' is $p_n' = n \cdot A'B'$. An example illustrating P_8 and P_8' is shown in Figure 10.35.

When n is large, the perimeter of the n-gon is a good approximation for the circumference of the circle, that is $c \approx p_n$ and $c' \approx p_n'$. Thus, for large n we have

$$\frac{c}{c'} \approx \frac{p_n}{p_n'} = \frac{n \cdot AB}{n \cdot A'B'} = \frac{AB}{A'B'}. \tag{10.3}$$

Figure 10.35. Inscribed regular 8-gons

Because P_n is a regular n-gon, we have that $\angle O \cong \angle O'$. (We will see in the next chapter why both equal $\frac{360}{n}$.) Also, since $AO = BO = r$ and $A'O' = B'O' = r'$, we have $\frac{AO}{BO} = \frac{A'O}{B'O} = 1$. Thus, $\triangle AOB \sim \triangle A'O'B'$ by $\widetilde{SAS}$, which gives $\frac{AB}{A'B'} = \frac{AO}{A'O'} = \frac{r}{r'}$. Substituting this into Equation (10.3) yields $\frac{c}{c'} \approx \frac{r}{r'}$ for large n. Letting $n \to \infty$, in the limit we have the desired result of Theorem 10.3,

$$\frac{c}{c'} = \frac{r}{r'}, \quad \text{or} \quad \frac{c}{2r} = \frac{c'}{2r'}.$$

We can use the same technique to determine a ratio relating the measure of a *piece* of the circumference to the radius. Here, instead of the entire circumference, we would consider the length of an arc, $length(\overset{\frown}{AB})$. In the sector corresponding to this arc, we would bisect the arc and then bisect those half-arcs, then bisect the quarter-arcs, continually bisecting in order to piecewise approximate the length of the arc with tiny line segments. This would produce the following similar result.

Theorem 10.4. *Let circles centered at O and O' have radii r and r', respectively, and minor arcs $\overset{\frown}{AB}$ and $\overset{\frown}{A'B'}$, respectively. If $\angle AOB \cong \angle A'O'B'$, then*

$$\frac{length(\overset{\frown}{AB})}{r} = \frac{length(\overset{\frown}{A'B'})}{r'}.$$

A corollary to this theorem is that since, by definition, congruent minor arcs $\overset{\frown}{AB}$ and $\overset{\frown}{A'B'}$ have $r = r'$ and $\angle AOB \cong \angle A'O'B'$, these arcs will also have the same length, namely $length(\overset{\frown}{AB}) = length(\overset{\frown}{A'B'})$. The same definition gives us the obvious way to extend these results to major arcs since each major arc is the complement of a minor arc with respect to the full circumference of a circle. Additionally, since congruent arcs have equal length, we can use $\overset{\frown}{AB}$ to denote both the arc and the length of the arc whenever there is no cause for confusion.

Though Euclid does not include ratios relating arcs and radii in the *Elements*, in his final proposition in Book VI he does establish a proportion between arcs and the angles they subtend. It is oddly situated since it does not rely on any of the previous propositions in the sixth book. The proof relies on III.27 and III.20, but it appears in Book VI since it is a statement about a ratio for which Euclid requires the fifth book. For us, this proposition is perfectly placed with the other circle results.

Proposition VI.33. *In equal circles angles have the same ratio as the circumferences on which they stand, whether they stand at the centres or at the circumferences.*

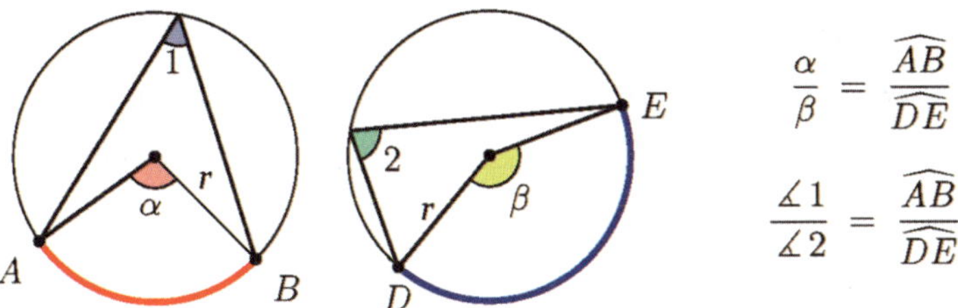

Figure 10.36. Proposition VI.33

We omit Euclid's proof as it relies on a complicated definition of magnitudes in the same ratio, and ultimately, falls short of its intended mark. Instead, as per Theorem 10.3, if we agree that the ratio of the circumference of any circle to its diameter is the number we call π, then we know that a full angle at the center of 360° corresponds to an arc of length $2\pi r$. Likewise, a central angle of 60° corresponds to an arc of length $\frac{1}{6}(2\pi r)$. We may consider the measure of any arc of a circle as a percentage of its circumference, and the corresponding angle at the center is the same percentage of 360°.

Another Alexandrian mathematician just after Euclid's time, Eratosthenes of Cyrene (276–194 BCE), famously made use of this result to calculate the circumference of the earth. Since any two cities live on a great circle, the distance of the arc between them is to the circumference of the earth, as the angle at the center subtended by the arc is to the angle of a full circle, or 360 degrees. With the Egyptian cities of Alexandria and Syene, this equation becomes

$$\frac{\alpha}{360°} = \frac{\text{distance between Alexandria and Syene}}{\text{circumference of the earth}},$$

as illustrated in the diagram. The distance between the cities was known to be approximately 5000 stades, most likely determined by surveyors trained as human pedometers to make such measurements by counting as they walked with equal steps. One stade, or 300 Royal Egyptian cubits, is approximately 516.7 feet. The only missing piece of the equation is a measure for the central angle, α. Determining this angle seems as impossible as travelling to 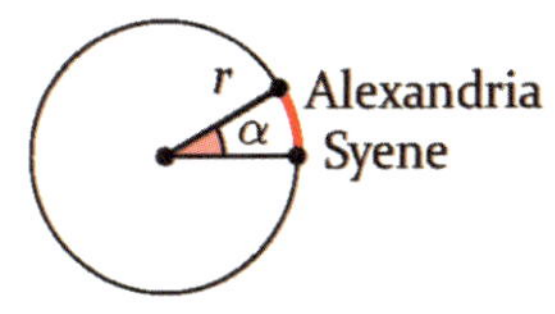 the center of the earth, but a bit of ingenuity and a Book I proposition shows that a little geometry goes a long way.

Syene sits on the Tropic of Cancer, the northernmost latitude at which the sun appears directly overhead at the summer and winter solstice. This was confirmed by looking into a deep well at noon on the summer solstice and seeing a perfect circle of sunlight reflected in the water below, the wall casting no shadow. Alexandria is very nearly due north of Syene, and at noon on the same day, Eratosthenes measured the angle of a shadow created by a stick. Assuming that the rays of sunlight arrive in parallel, as illustrated in the diagram, the ray passing through S travels to the center of the circle, and the ray hitting the top of the dotted line (Eratosthenes' shadow-casting stick at Alexandria) casts a shadow on the circle (the earth).

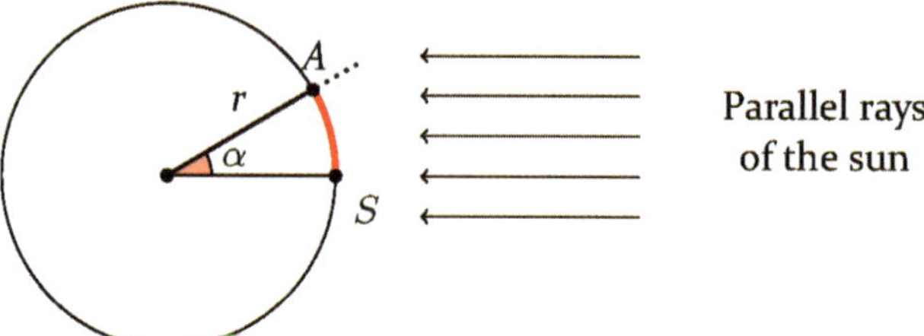

Figure 10.37. Eratosthenes' calculation of the circumference of the earth

Paring the diagram to its essential information, we have the following diagram of a transversal crossing parallel lines, where β is the angle created by the shadow of the stick at Alexandria.

By Proposition I.29, we know that $\alpha = \beta$, and Eratosthenes measured the angle of the shadow, β, as $7\frac{1}{5}^{\circ}$. Solving the equation, the circumference of the earth is approximately 250,000 stades. This is equivalent to 24,465 miles, about 1.75% short of the current measure of 24,901 miles.

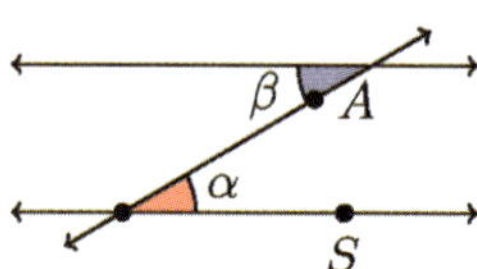

11

Book IV: Circles & Polygons

Figure 11.1. A ring of tangent circles in a hexagon

The connections and interplay between geometric constructions and art have long been explored by mathematicians and artists alike. Both Islamic and Gothic decorations in the Middles Ages used regular polygons. Often these designs incorporate the constructions of circles in regular polygons or regular polygons constructed in circles. The Baghdad mathematician and astronomer Abū al-Wafū'al Būzjānī (940–997) wrote a textbook entitled *A Book on the Geometrical Constructions Necessary to the Artisan*. In it he includes methods of constructing regular polygons in and about given circles as well as constructing various polygons inside given polygons. The German artist Albrecht Dürer (1471–1528) wrote a treatise on constructions using a straightedge and compass, in which he includes Ptolemy's construction of a regular pentagon. In her 1912 manuscript *A Source Book of Problems for Geometry, Based upon Industrial Design and Architectural Ornament*, Chicago author and high-school instructor Mabel Sykes (1868–1938) illustrates tiles, floor designs, and gothic arches and windows based on geometric constructions. The design in Figure 11.1 is one such example. It is a hexagon constructed in a circle followed by six circles constructed in the six equilateral triangles making up the hexagon and ending with a final circle constructed in the middle and has been used in Gothic cathedrals. The American architect Frank Lloyd Wright (1867–1959) chose a cross within a circle inscribed in a red square as his logo. More recently, the Phillips Exeter Academy Library, designed by Philadelphian architect Louis Kahn (1901–1974), and the Calakmul Corporate Building, designed by Mexican architect Agustín Hernández Navarro (1924–), are two spectacular examples of buildings employing the motif of circles inscribed in squares.

All sixteen propositions in Book IV are constructions. In all but two, we either construct a circle around or inside a given polygon, or construct a polygon around or inside a given circle. In the process, Euclid augments his ability to construct regular polygons, adding regular pentagons, hexagons and 15-gons to his toolbox.

> As part of the larger push for mathematics education reform at the start of the twentieth century, the American Federation of Teachers of Mathematics and Natural Sciences and the National Education Association jointly commissioned a group of educators to review the high-school geometry curriculum in 1908. The National Committee of Fifteen on the Geometry Syllabus consisted of seven university professors and eight high-school instructors hailing from schools across the United States. After four years of work, the committee proposed a new geometry syllabus that was adopted by the NEA in 1912 [**49**]. Mabel Sykes, a mathematics instructor at Bowen High School in Chicago, was the only woman to serve on this committee. Though biographical information on Sykes is scarce, her passion for teaching mathematics is evident in her books on geometry and algebra, and her articles on pedagogy. In her *Source Book of Problems for Geometry*, she uses complex and beautiful architectural designs, in her own words, "the best in historic ornament," as the starting point for over 1800 exercises on proof, construction and computation techniques. (See Exercise 11.3.13 for a few examples.) As she writes in the preface to this volume, "Geometry gives, as no other subject can give, an appreciation of form as it exists in the material world, and of the dependence of one form upon another" [**114**].

11.1 Definitions

We begin with four of the seven definitions given by Euclid at the beginning of Book IV. These all concern either a figure **inscribed in** another figure, or a figure **circumscribed about** another figure, where one of the figures is a circle and the other is a polygon.

(3) A rectilineal figure is said to be inscribed in a circle when each angle of the inscribed figure lies on the circumference of the circle.

(4) A rectilineal figure is said to be circumscribed about a circle, when each side of the circumscribed figure touches the circumference of the circle.

(5) Similarly, a circle is said to be inscribed in a figure when the circumference of the circle touches each side of the figure in which it is inscribed.

(6) A circle is said to be circumscribed about a figure when the circumference of the circle passes through each angle of the figure about which it is circumscribed.

As demonstrated in the figures, a polygon is inscribed in a circle when each of its vertices lies on the circle, and circumscribed about a circle when each of its sides is tangent to the circle. Similarly, a circle is inscribed in a polygon when it is tangent to each side of the polygon, and is circumscribed about a polygon when it passes through each vertex of the polygon. Using updated terminology, we will refer to an inscribed circle as an **incircle**, and a circumscribed circle as a **circumcircle**. We need one more definition.

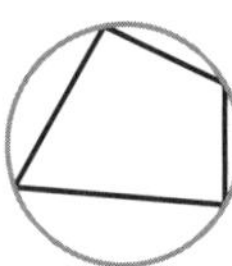

Quadrilateral inscribed in a circle, and circle circumscribed about a quadrilateral

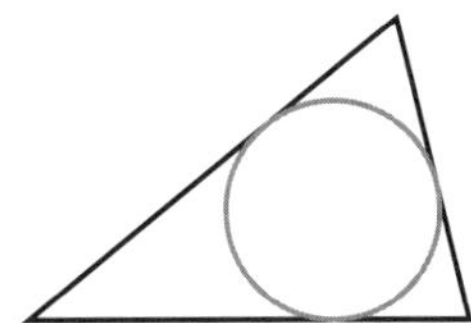

Circle inscribed in a triangle, and triangle circumscribed about a circle

Definition 11.1. Three (or more) lines are **concurrent** if they all meet at a single point, and the intersection is called the **point of concurrency.**

As we progress through Book IV we will see polygons of an increasing number of sides. Propositions IV.2 through IV.5 concern triangles, and IV.6 though IV.9 involve squares. Proposition IV.10 takes a step back to triangles to give the construction of a $36°-72°-72°$ isosceles triangle, also known as a **golden triangle**, which we need in order to construct a pentagon. Propositions IV.11 through IV.14 regard pentagons, and IV.15 and IV.16 give the constructions of a regular hexagon and a regular 15-gon, respectively. We will be able to venture a bit further than Euclid to show that his later proofs apply more generally to regular n-gons. We begin with triangles.

11.2 Circles & triangles

Proposition IV.2. *In a given circle to inscribe a triangle equiangular with a given triangle.*

Proof. Let ABC be the given circle, and DEF the given triangle; thus it is required to inscribe in the circle ABC a triangle equiangular with the triangle DEF.

Let GH be drawn touching the circle ABC at A [III.16, Por.];

on the straight line AH, and at the point A on it, let the angle HAC be constructed equal to the angle DEF, and on the straight line AG, and at the point A on it, let the angle GAB be constructed equal to the angle DFE; [I.23] let BC be joined.

Then, since a straight line AH touches the circle ABC, and from the point of contact at A the straight line AC is drawn across in the circle, therefore the angle HAC is equal to the angle ABC in the alternate segment of the circle. [III.32]

But the angle HAC is equal to the angle DEF; therefore the angle ABC is also equal to the angle DEF.

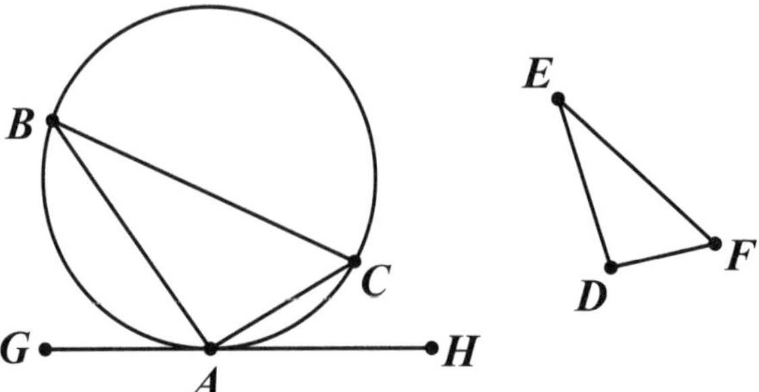

Figure 11.2. Proposition IV.2

For the same reason the angle ACB is also equal to the angle DFE; therefore the remaining angle BAC is also equal to the remaining angle EDF. [I.32]

Therefore in the given circle there has been inscribed a triangle equiangular with the given triangle. Q.E.F. □

In this chapter, we will not follow a Book IV proof with the miniature version of the proposition. Instead, since the resulting diagrams are identical, we will group the miniatures of the propositions in which a polygon is inscribed in a circle with those in which a circle is circumscribed about a polygon at the end of the chapter in Figure 11.45. Likewise, in Figure 11.46, propositions in which a polygon is circumscribed about a circle are grouped with those in which a circle is inscribed in a polygon.

Before continuing to Proposition IV.4, we pause to give the following theorem and the corresponding definition it inspires. We also recall from Theorem 3.13 that a point interior to an angle lies on the angle's bisector if and only if it is equidistant from the rays of the angle.

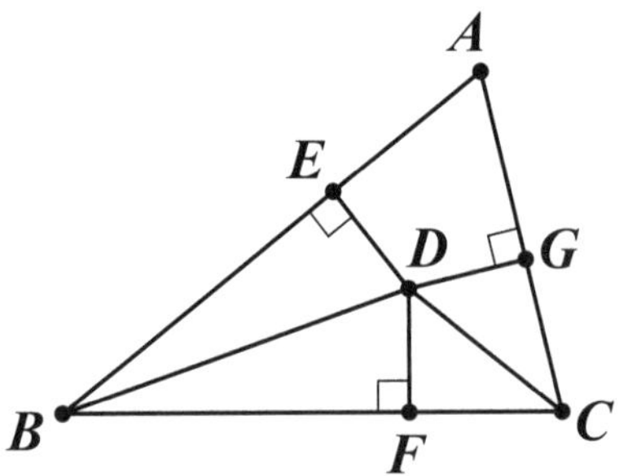

Figure 11.3. Theorem 11.2

Theorem 11.2. *Given any triangle, the three angle bisectors of the triangle are concurrent. Their point of intersection is called the* **incenter**.

Proof. Consider a triangle $\triangle ABC$. Let the angle bisectors of $\angle ABC$ and $\angle ACB$ intersect at a point D. Construct E, F, G on AB, BC, AC, respectively such that $DE \perp AB$, $DF \perp BC$ and $DG \perp AC$. As D lies on the angle bisector of $\angle ABC$, we have $DE = DF$. Similarly, since D lies on the angle bisector of $\angle ACB$, $DF = DG$. Since

$DE = DG$, D is equidistant from the sides of $\angle BAC$. Thus D must also lie on its angle bisector. Hence the three angle bisectors are concurrent. $\square$

There are some mathematics results that are surprising, and the fact that these three lines meet at a point is as shocking to us as it must have been to the Greeks. A point of concurrency is like a confluence of events, curious, unexpected and improbable. When they discovered that the three angle bisectors always met at a point, they must have thought that they had found *the* center of a triangle. Any such thought would soon be abandoned as they found three other points of concurrency related to a triangle! As we shall see in Proposition IV.4, the *incenter* is the first such point we encounter in the *Elements*. It is equidistant from all three sides of the triangle, and hence, is so named since it is the center of the inscribed circle for any triangle.

Proposition IV.4. *In a given triangle to inscribe a circle.*

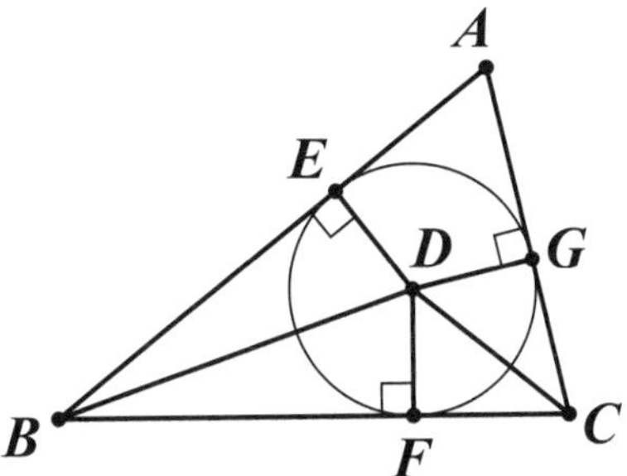

Figure 11.4. Proposition IV.4

Proof. Consider triangle $\triangle ABC$. Let the angle bisectors of $\angle ABC$ and $\angle ACB$ intersect at a point D. Construct E, F, G on AB, BC, AC, respectively such that $DE \perp AB, DF \perp BC$ and $DG \perp AC$. Construct a circle with center D and radius DE. By the previous theorem, we have $DE = DF = DG$. Thus the circle will be tangent to all three sides as desired. $\square$

Heron's formula

A fellow Alexandrian who is thought to have lived three centuries after Euclid, Heron (ca. 10–ca. 75) is known for a formula for the area of a triangle that bears his name. The formula appeared as the eighth proposition in his three-book treatise, *Metrica*. This work contained formulas for measuring area of plane figures, surface area and volumes of solids, and methods for approximating square roots. In all likelihood, Heron, like Euclid before him, was gathering the known results of his day into one work. This area formula for the triangle is distinctive in that it requires no angle or altitude, relying only on the lengths of the three sides. Heron's formula states that the area of $\triangle ABC$ with sides of length a, b, and c is given by

$$Area(\triangle ABC) = \sqrt{s(s-a)(s-b)(s-c)},$$

where s is the **semiperimeter** of the triangle, namely $s = \frac{1}{2}(a + b + c)$. We mention this result here since the first step of his proof is to construct the inscribed circle, producing a diagram like Figure 11.4, and then showing that three pairs of triangles are congruent, namely $\triangle BDE \cong \triangle BDF$, $\triangle CDF \cong \triangle CDG$, and $\triangle ADE \cong \triangle ADG$. Far from obvious, his proof requires cleverly extending BC by a length equal to AG, constructing a larger quadrilateral (containing only a part of $\triangle ABC$) that he shows to be cyclic, exploiting the similarity of two pairs of triangles, and for us, using a bit of algebra. We guide the reader through this proof as an exercise.

After he inscribes a circle in a given triangle, Euclid circumscribes a circle about a triangle. This construction reveals our second point of concurrency where the perpendicular bisectors of the sides of the triangle meet. Before continuing to Euclid's proposition, we give the relevant theorem and definition. Recall from Theorem 3.5 that a point lies on the perpendicular bisector of a segment if and only if it is equidistant from its endpoints.

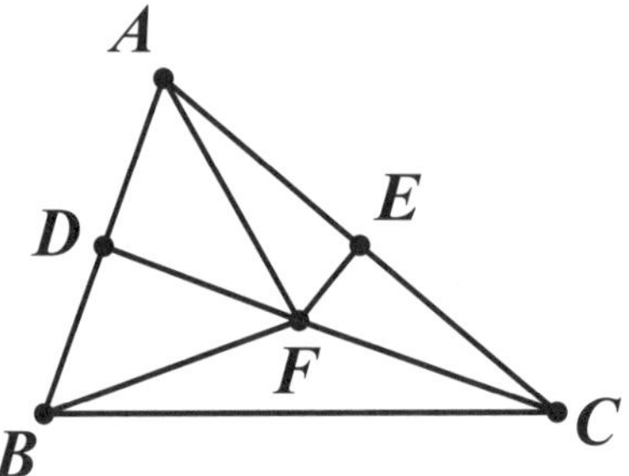

Figure 11.5. Theorem 11.3

Theorem 11.3. *Given any triangle, the three perpendicular bisectors of the sides of the triangle are concurrent. The point of concurrency is called the* **circumcenter**.

Proof.   Consider the triangle $\triangle ABC$. Let D, E be the midpoints of AB, AC, respectively. Let the perpendicular bisectors of AB and AC intersect at a point F. Since F lies on the perpendicular bisector of AB, $AF = BF$. Similarly, F lies on the perpendicular bisector of AC, so $AF = CF$. Thus $BF = CF$ and F must lie on the perpendicular bisector of BC. Hence the three perpendicular bisectors are concurrent. $\square$

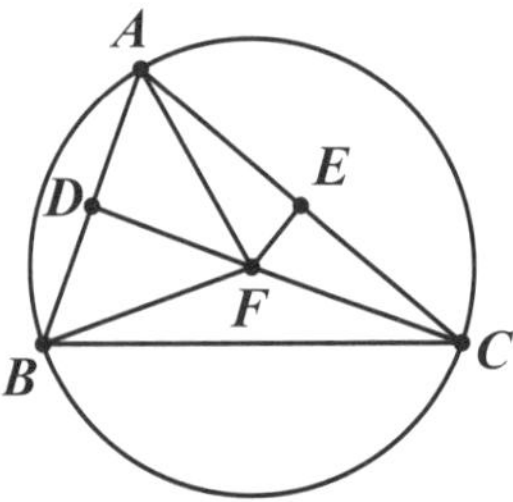

Figure 11.6. Proposition IV.5

Note that while Figure 11.5 shows F in the interior of $\triangle ABC$, F can also lie outside of the triangle, or on one of its sides. Our proof, however, does not depend on the position of F. The *circumcenter* gets its name as it is equidistant from the three vertices, and hence is the center of the circumscribed circle for the triangle. Thus, we have given the construction required for Proposition IV.5.

Proposition IV.5. *About a given triangle to circumscribe a circle.*

> ### Morley's sonnet, revisited
>
> It is with the construction of the circle circumscribing a triangle that we can now fully appreciate the remaining geometric result incorporated into Christopher Morley's poem celebrating the circle at the start of Chapter 10. When he ends with "only 3 [points] Determine every circle," he refers to this construction. Any three noncollinear points determine a triangle; the circle which circumscribes it, as constructed in Proposition IV.5, is unique. Furthermore, the rhyme scheme of Morley's sonnet, where each line ending on the same rhyme is given the same letter, only requires three letters. As traditional sonnet rhyming schemes require more than three letters, we can only assume that this subtle integration of "three" was an intentional choice.

Given three noncollinear points, it is clear from the construction in IV.5 that the circumcircle for the triangle defined by these points is the unique circle passing through the points. We state this as a corollary for future reference.

Corollary 11.4. *There is a unique circle passing through any three noncollinear points.*

We leave the proof of the following related lemma as an exercise for the reader.

Lemma 11.5. *The circumcenter and incenter of a triangle coincide if and only if the triangle is equilateral.*

After constructing the inscribed and circumscribed circles for a triangle, and in the process, proving two concurrency properties, Euclid shifts his focus to the square. We will pick up where he left off in the next section, but for now, we choose to examine the two other points of concurrency that are just as surprising as the incenter and circumcenter. First, recall that the *altitude* of a triangle is the perpendicular segment joining a vertex to its opposite side, and the *median* of a triangle is the segment which joins a vertex to the midpoint of its opposite side. As if two points of concurrency were not enough, both the altitudes and the medians are concurrent! Here, we give the related theorems and their corresponding definitions. We also encourage the reader to explore these claims with dynamic geometric software such as Geometer's Sketchpad® or GeoGebra.

Theorem 11.6. *Given any triangle, the three altitudes of the triangle are concurrent at a point called the* **orthocenter**.

Proof. Consider triangle $\triangle ABC$. The idea behind the proof is to create a larger triangle $\triangle DEF$ in such a way that the altitudes of our original $\triangle ABC$ are the perpendicular bisectors of our new $\triangle DEF$. We can then use Theorem 11.3 to justify that the three

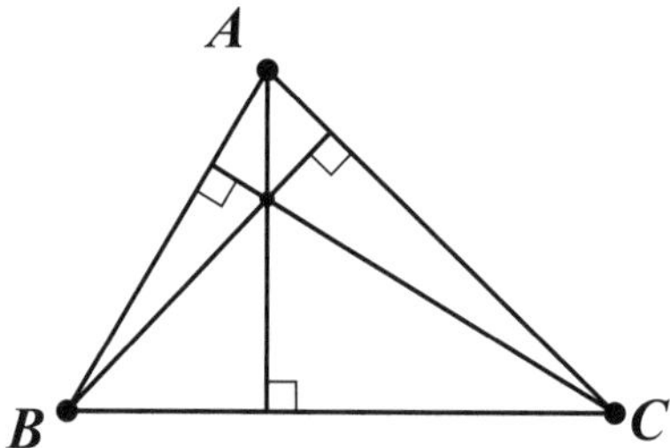

Figure 11.7. Orthocenter

lines are concurrent. We began the proof, but leave it to the reader to finish. Using I.31, construct a line ℓ through A that is parallel to BC. In a similar fashion, construct m through B parallel to AC, and n through C that is parallel to AB. Let D be the intersection of ℓ and m, E be the intersection of m and n, and F be the intersection of ℓ and n. (Why must these lines intersect?) Claim: $\triangle DEF$ is the desired triangle. $\square$

When the orthocenter is distinct from the vertices of the triangle, these four points form an **orthocentric set**, so named because each of the four points is the orthocenter of the triangle formed by the other three points. For example, if $\triangle ABC$ has orthocenter H which is distinct from its vertices, then A is the orthocenter of $\triangle HBC$, B is the orthocenter of $\triangle AHC$, and C is the orthocenter of $\triangle ABH$. We leave it as an exercise for the reader to show this.

The orthocenter is *not* at the center of a circle related to a given triangle, and thus we do not encounter it in Euclid's fourth book. The orthocenter first appears in Greek mathematics in Archimedes' book of lemmas, *Liber Assumptorum*, though it is not highlighted as such. It is used in his proof that the two shaded circles shown in the diagram, each defined by its three points of tangency, are equal for any point C on AB. Because of this connection, prior to the mid-nineteenth century, the orthocenter was called the Archimedean point.

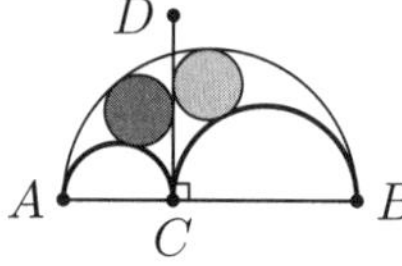

The Greeks knew of a fourth point of concurrency, the *centroid*, which occurs at the intersection of the medians. In physics or calculus, this point may be called the center of gravity, or center of mass, as it is the balancing point for a triangle of uniform density. Centroid calculations can be found in the works of Archimedes, though the first proof of the concurrency is attributed to Heron.

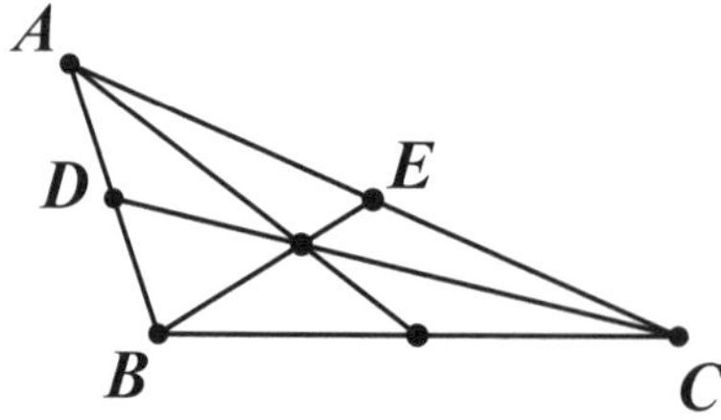

Figure 11.8. Centroid

Theorem 11.7. *Given any triangle, the three medians of the triangle are concurrent at a point called the* **centroid***.*

Proof. Consider triangle $\triangle ABC$. The idea behind the proof is to show that any two medians must intersect in a point that is two-thirds of the way from a vertex to its opposite side. Since this holds for any pair of medians, all three medians must coincide. We begin the proof, but leave it to the reader to finish. Using I.10, bisect AB and AC at D and E, respectively. Join DE. By Proposition VI.2, we have $DE \parallel BC$. Construct BE and CD, and let their intersection be F. Claim: $\triangle ABC \sim \triangle ADE$, and thus, $DE = \frac{1}{2}BC$. Additionally, $\triangle DEF \sim \triangle CBF$, and thus, $BF = \frac{2}{3}BE$ and $CF = \frac{2}{3}CD$. $\qquad\square$

From the proof, we have an immediate corollary.

Corollary 11.8. *The centroid of any triangle is located two-thirds of the distance from any vertex to the midpoint on the opposite side.*

It is interesting to note that a circle has only one center, but for the Greeks, in a way, the triangle has four. Even if we acknowledge that these "centers" lack the infinite lines of symmetry that a circle's center possesses, we must admit that the mere fact that these four points of concurrency exist for every triangle really is astounding. We should mention that, unlike a circle's center, some of them do not have to live in the interior of the triangle. Once again, we encourage the reader to develop their intuition regarding the possible arrangement of these points through the use of dynamic geometric software. Change the shape of a triangle to see which of these four points can live outside the triangle, and form a conjecture as to which must live inside. Consider the arrangement of these points when the triangle is equilateral. *Spoiler alert: In an equilateral triangle, the circumcenter, incenter, orthocenter and centroid are all the same point. Furthermore, if any two of these points coincide, then the triangle must be equilateral.* We leave both of these proofs as exercises.

We pick up the thread of this story of the triangle about 1700 years after Heron. The number of years separating the last result from the next is indicative of two things, the development of analytic geometry and the sheer genius of the mathematician at work, Leonhard Euler. Euler takes the improbable one step further, showing that three of these unexpected points of concurrency are collinear! If the concurrence of lines related to triangles is astounding, then the collinearity of three of these four points is wondrous. In 1763, Euler shows that the circumcenter, orthocenter and centroid are collinear, and that the distance from the centroid to the orthocenter is twice the distance from the centroid to the circumcenter. Like Archimedes glossing over the importance of the orthocenter, this discovery was merely a stepping stone in Euler's solution of the larger problem of reconstructing a triangle given its four points of concurrency. Though Euler does not appear to appreciate the importance of his discovery, generations of mathematicians to follow certainly did, and this line has come to be known as the **Euler line**. As the reader will show in Exercise 11.2.13, when $\triangle ABC$ is equilateral, all four points of concurrency coincide and there is no Euler line. Though Euler's proof, published in 1767, utilized similar triangle results, the Law of Cosines, and the Pythagorean Theorem, it also relied on coordinate geometry with quartic equations and roots of cubics. We give a classical proof that does not require analytic geometry.

Theorem 11.9 [Euler Line Theorem]. *Given any nonequilateral triangle, its circumcenter, orthocenter and centroid are collinear, lying on a line called the* **Euler line.**

Proof. Consider nonequilateral triangle $\triangle ABC$, with circumcenter F and centroid G. As noted above, and as the reader will show in Exercise 11.2.14, F and G must be distinct. There are two cases to consider.

Case 1. $\triangle ABC$ is isosceles. The reader will show in Exercise 11.2.12 that all four points of concurrency are collinear.

Case 2. $\triangle ABC$ is scalene. Let D be the midpoint of BC. Then $FD \perp BC$ and the median, AD, contains G. Construct E on $\overleftrightarrow{BC}$ such that $AE \perp \overleftrightarrow{BC}$. Thus AE is an altitude of $\triangle ABC$. By I.28, we have $AE \parallel FD$. Note that median AD, and hence, centroid G, which is on all three medians, both lie between these two distinct parallel lines. By Proclus' Axiom, the ray $\overrightarrow{FG}$ must intersect $\overleftrightarrow{AE}$ at a point, H.

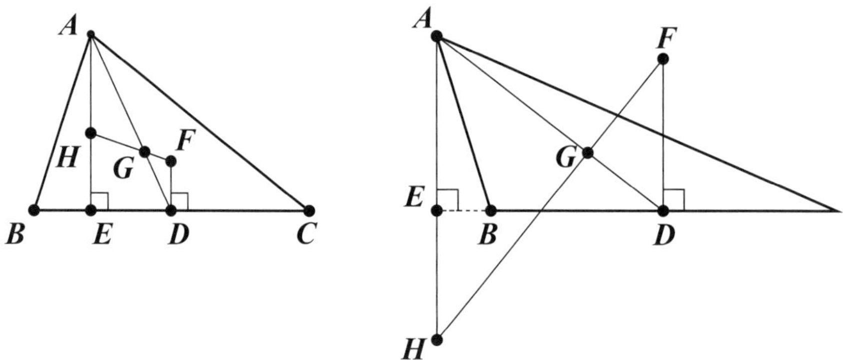

Figure 11.9. Two possible configurations for Euler line FGH

Claim: H is the orthocenter of $\triangle ABC$. To show this, we consider triangles $\triangle AHG$ and $\triangle DFG$. By I.29, $\angle AHG \cong \angle DFG$ and $\angle HAG \cong \angle FDG$. By I.15, $\angle AGH \cong \angle DGF$. Therefore, by $\widetilde{AAA}$, $\triangle AHG \sim \triangle DFG$. By Corollary 11.8, we have $AG = 2GD$. Thus, $HG = 2FG$.

Finishing this proof depends on changing our perspective. The argument that ray $\overrightarrow{FG}$ intersects altitude $\overleftrightarrow{AE}$ can easily be modified to show that $\overrightarrow{FG}$ must also intersect the altitude from B to AC. Furthermore, it will intersect this altitude at a point, H', where $H'G = 2FG$. Thus, H coincides with H', and since it lies on both altitudes, is the orthocenter. □

From concurrence of lines to collinearity of the points of concurrency, the triangle and its related centers are certainly taking the starring role at the start of this chapter. To be fair, we could devote several chapters to the triangle, but instead, we have one more result relating triangles and circles before returning to Euclid's fourth book. Nearly sixty years after Euler discovered the collinearity of the circumcenter, orthocenter and centroid, three mathematicians were simultaneously investigating another property of a triangle related to these points and the Euler line, and in a moment of life imitating art, their independent lines of investigation led to the same discovery, producing a point of concurrence in the physical world. Two French mathematicians, Charles-Julien Brianchon (1783–1864) and Jean-Victor Poncelet (1788–1867), and a young German, Karl Wilhelm Feuerbach (1800–1834), all discovered a circle containing nine significant points related to any given triangle. Instead of a collinearity theorem, they produced an equally, if not more, impressive "cocircularity" theorem, if there were such a word. (Points that lie on the same circle are said to be **concyclic**.) Suppose we have $\triangle ABC$ with orthocenter H, centroid G and circumcenter F as shown in Figure 11.10. Consider the following collection of points related to $\triangle ABC$, points that were all well-known to two millennia's worth of geometers: the midpoint of each side of the triangle (A_G, B_G, C_G), the foot of each altitude (A_T, B_T, C_T), and the midpoint of each segment joining a vertex to the orthocenter (A_H, B_H, C_H).

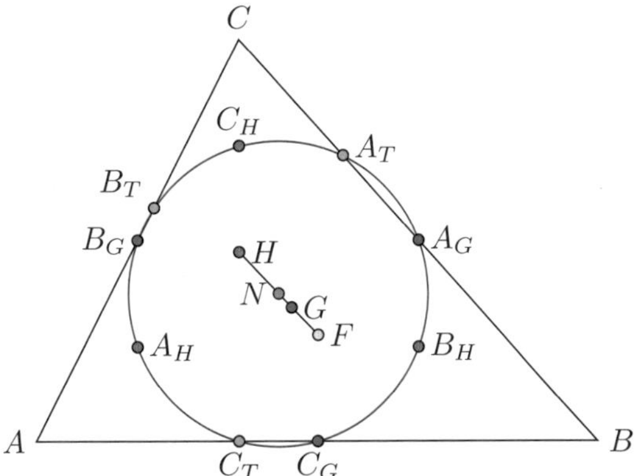

Figure 11.10. Thirteen points related to $\triangle ABC$

There is one more point shown in the diagram that has not been identified yet, N. Brianchon, Poncelet and Feuerbach all noticed that these nine points all lie on a circle centered at N. As noted earlier for the Euler line, if the triangle is equilateral or isosceles, then some of these points may coincide and the circle may not have nine distinct points. Regardless, this instance of concyclic points has come to be known as the Nine-point circle. Brianchon and Poncelet published their results in 1821 and Feuerbach in 1822. As if this were not astonishing enough, not only does the center of the circle lie on the Euler line, it is the midpoint of the segment joining the orthocenter and the circumcenter! The Nine-point circle is also related to the inscribed and circumscribed circles of $\triangle ABC$. The radius of the Nine-point circle is half the radius of the circumscribed circle, and the inscribed circle is internally tangent to the Nine-point circle. Even though Feuerbach was the last to publish, it is his name that is most often associated with this result, most likely because he proved some additional results regarding the tangency of the Nine-point circle. The point of tangency where the Nine-point circle meets the inscribed circle is called the Feuerbach point.

We need one lemma in order to prove that these points are concyclic. It is the converse of the Intersecting Secants Theorem as given in Exercise 10.4.4.

Lemma 11.10 [Converse of Intersecting Secants Theorem]. *Consider distinct segments AC and AE with common endpoint A. Let B be a point on AC, and D a point on*

AE. If $AB \cdot AC = AD \cdot AE$, then quadrilateral BCED is cyclic, that is, points B, C, D and E lie on a unique circle.

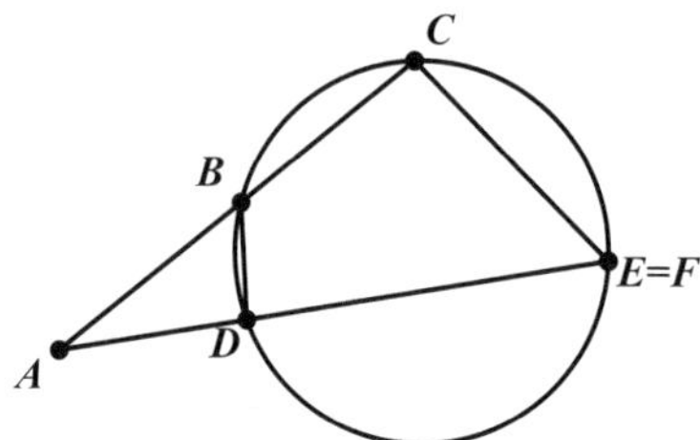

Figure 11.11. Converse of Intersecting Secants Theorem: Four points on a circle

Proof. By Corollary 11.4, let c be the unique circle joining the three noncollinear points B, C and D. Since AC intersects c at two points, A must lie outside of c.

Claim: The ray $\overrightarrow{AD}$ will also intersect c at two points. For the sake of reaching an eventual contradiction, suppose AD is tangent to c at D. By III.36, $AD^2 = AB \cdot AC$. Since $AB \cdot AC = AD \cdot AE$ by hypothesis, then $AD^2 = AD \cdot AE$ and we have $AD = AE$. Thus, $D = E$, which is a contradiction.

Therefore, ray $\overrightarrow{AD}$ intersects circle c at another point, say F. By III.17, construct tangent to circle c at A, intersecting the circle at T. Applying III.36 to AC gives $AT^2 = AB \cdot AC$. Applying III.36 to AF gives $AT^2 = AD \cdot AF$. Thus, $AD \cdot AF = AB \cdot AC$. Since $AB \cdot AC = AD \cdot AE$ by hypothesis, then $AF = AE$. Thus, $F = E$. Therefore, E also lies on c, which means $BCED$ is a cyclic quadrilateral. $\qquad\square$

The following proof of the Nine-point Circle Theorem was given by R. D. Bohannan in 1884.

Theorem 11.11 [Nine-point Circle Theorem]. *Given any triangle, the midpoints of the three sides, the feet of the three altitudes, and the three midpoints of the line segments connecting the orthocenter to the vertices lie on a circle, called the **Nine-point circle.***

Proof. Consider triangle $\triangle ABC$. Let A_T be the foot of the altitude from A to BC, and let A_G be the midpoint of the side opposite A. Define B_T, B_G, C_T and C_G similarly. Let H be the orthocenter, and let A_H be the midpoint of the segment AH. Define B_H and C_H similarly.

Consider triangles $\triangle BAA_T$ and $\triangle BCC_T$. Each contains a right angle and they share angle $\angle B$. Thus by $\widetilde{AA}$, $\triangle BAA_T \sim \triangle BCC_T$. Therefore,

$$\frac{BA_T}{BC_T} = \frac{BA}{BC} = \frac{2 \cdot BC_G}{2 \cdot BA_G} = \frac{BC_G}{BA_G}.$$

Cross-multiplying, we have $BA_T \cdot BA_G = BC_T \cdot BC_G$. By Lemma 11.10, the points A_T, A_G, C_T and C_G all lie on a unique circle c.

Next consider triangles $\triangle CHA_T$ and $\triangle CBC_T$. Once again, each contains a right angle and they share $\angle BCC_T = \angle HCA_T$. By $\widetilde{AA}$, $\triangle CHA_T \sim \triangle CBC_T$. Therefore,

$$\frac{CC_T}{CA_T} = \frac{CB}{CH} = \frac{2 \cdot CA_G}{2 \cdot CC_H} = \frac{CA_G}{CC_H}.$$

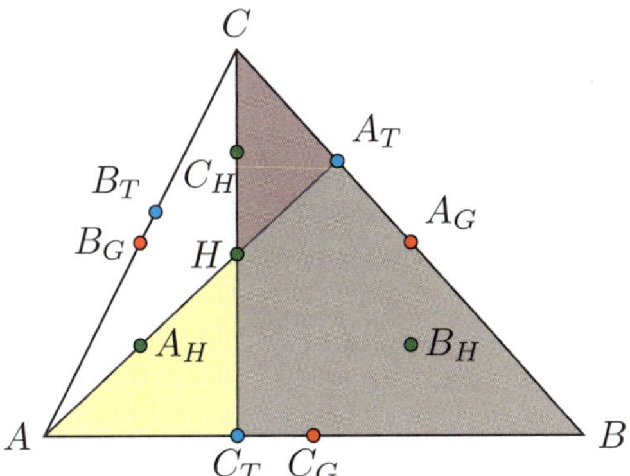

Figure 11.12. Proof of the Nine-point Circle Theorem

Cross-multiplying, we have $CC_H \cdot CC_T = CA_T \cdot CA_G$. Again, by Lemma 11.10, the points A_T, A_G, C_T and C_H lie on a unique circle. Note, however, that A_T, A_G, and C_T lie on c. Therefore, C_H must also lie on c.

We have shown that the five points A_T, A_G, C_T, C_G and C_H all lie on a circle. The remainder of the proof relies on changing our perspective. Reversing the roles of A and C, we will have A_T, A_G, C_T, C_G and A_H also lie on a circle. Since these two circles share four points, they must be the same. Thus, A_T, A_G, A_H, C_T, C_G and C_H are all on the same circle c. Changing perspective one last time where we would start over by replacing Cs with Bs, a similar argument will show that A_T, A_G, A_H, B_T, B_G and B_H all lie on the same circle. Finally, since both circles share the three points A_T, A_G and A_H, all nine points must lie on c. $\qquad\qquad\square$

Exercises 11.2

1. Let r_1 be the radius of the circumscribed circle for an equilateral triangle and let r_2 be the radius of the inscribed circle for the same triangle. Find the ratio r_1/r_2. See Figure 11.13.

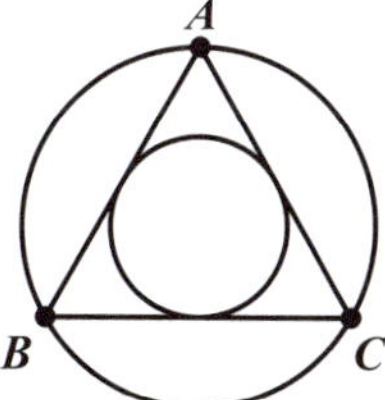

Figure 11.13. Exercise 11.2.1

2. Suppose we are given one leg of a right triangle, along with the radius of its inscribed circle. Carefully explain how to reconstruct the right triangle (up to congruence).

3. Give a construction for the circumscribed circle about a rectangle $\square ABCD$. Prove that your construction works.

4. Prove the converse of Thales' Theorem (III.31): The circumcenter of a right triangle is the midpoint of its hypotenuse. [*Hint: For right triangle $\triangle ABC$ with midpoint of its hypotenuse, M, show that $AM = CM = BM$.*]

5. Use the converse of Thales' Theorem (III.31) as given in Exercise 4 to show the following: Given cyclic quadrilateral $ABCD$ with $AB = AD$ and $BC = DC$, prove that the diagonal AC is a diameter of the circumscribed circle around $ABCD$.

6. Determine a condition for when the circumcenter lies outside of the triangle. Justify your answer.

7. Determine a condition for when the orthocenter lies outside of the triangle. Justify your answer.

8. Finish the proof of Theorem 11.6 to show that the altitudes of a triangle are concurrent. Be sure to include a diagram.

9. Given triangle $\triangle ABC$ with orthocenter H, prove that A is the orthocenter of triangle $\triangle HBC$.

10. Finish the proof of Theorem 11.7 to show that the medians of a triangle are concurrent. Be sure to include a diagram.

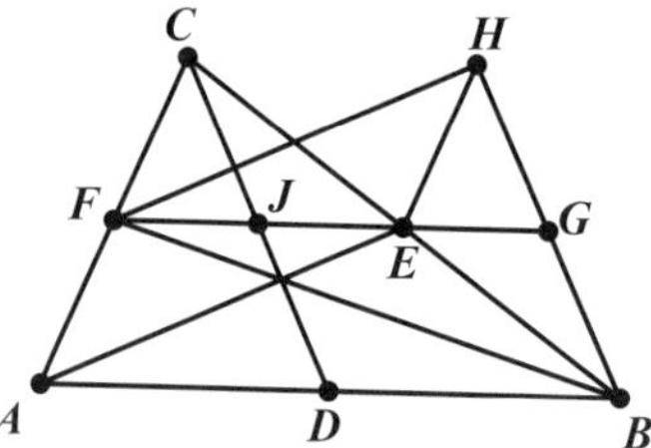

Figure 11.14. Exercise 11.2.11

11. Consider triangle $\triangle ABC$ with medians AE, BF and CD. Construct FH parallel to AE such that $FH = AE$. Construct segments BH and HE. Let the intersection of CD and FE be J, and let FE be extended to meet BH in G, as illustrated in Figure 11.14. Prove each of the following statements.

   (a) Quadrilateral $AFED$ is a parallelogram.
   (b) Ray $\overrightarrow{HE}$ intersects AB at D.
   (c) Quadrilateral $ACHD$ is a parallelogram.
   (d) $CD = HB$.
   (e) Quadrilateral $BGJD$ is a parallelogram.
   (f) $FG = \frac{3}{4}AB$.
   (g) FG is the median of $\triangle BFH$.

12. Prove that for an isosceles triangle, the incenter, circumcenter, orthocenter and centroid are collinear. (We note that while it is more difficult to show, the converse is also true.)

13. Prove that for an equilateral triangle, the incenter, circumcenter, orthocenter and centroid all coincide. (Note that this is half of the proof of Lemma 11.5.)

14. Given a triangle, if any two of the incenter, circumcenter, orthocenter and centroid coincide, then the triangle must be equilateral. We break this into the six possible pairs of points.

   (a) Prove that if the circumcenter and the incenter of a triangle coincide, then the triangle is equilateral. (Note that this is half of the proof of Lemma 11.5.)

   (b) Prove that if the circumcenter and orthocenter of a triangle coincide, then the triangle is equilateral.

   (c) Prove that if the circumcenter and centroid of a triangle coincide, then the triangle is equilateral.

   (d) Prove that if the incenter and orthocenter of a triangle coincide, then the triangle is equilateral.

   (e) Prove that if the incenter and centroid of a triangle coincide, then the triangle is equilateral.

   (f) Prove that if the orthocenter and centroid of a triangle coincide, then the triangle is equilateral.

15. Use geometric software to produce the Euler line.

16. Use geometric software to produce the Nine-point circle.

17. Use geometric software to demonstrate that the center of the Nine-point circle lies on the Euler line.

18. Heron's formula states that the area of $\triangle ABC$ with sides of length a, b, and c is given by $Area(\triangle ABC) = \sqrt{s(s-a)(s-b)(s-c)}$, where s is the semiperimeter of the triangle, namely $s = \frac{1}{2}(a+b+c)$. Suppose $\triangle ABC$ has incenter D and inscribed circle with radius r, as shown in Figure 11.15. Label the sides of $\triangle ABC$ as $AB = c$, $AC = b$ and $BC = a$. Follow the steps outlined below to give a proof of Heron's Formula.

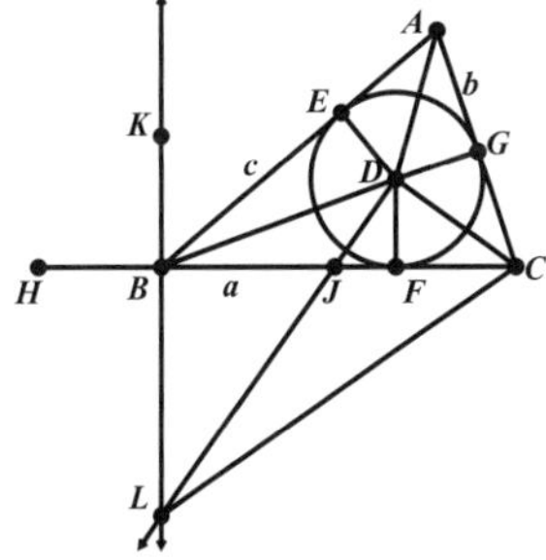

Figure 11.15. Exercise 11.2.18

   (a) Explain why the area of triangle $\triangle ABC$ is given by
$$Area(\triangle ABC) = r\left(\frac{a+b+c}{2}\right) = rs.$$

(b) Show that $\triangle BDE \cong \triangle BDF$, $\triangle CDF \cong \triangle CDG$ and $\triangle ADE \cong \triangle ADG$. Using the pairs of congruent triangles, explain why $\angle ADG + \angle BDC$ is equivalent to two right angles.

(c) Extend CB past B to a point H such that $BH = AG$. Show that $HC = s$.

(d) Construct quadrilateral $BDCL$ as follows: Construct $DJ \perp DC$ so that DJ meets BC at J. Construct $BK \perp HC$ at B, meeting DJ at L. Construct segment CL. Using the converse of Thale's Theorem (Exercise 4), explain why quadrilateral $BDCL$ is cyclic. [*Hint: The midpoint of LC is the circumcenter of the circle on which B, D, C and L lie.*]

(e) Show that $\angle CLB \cong \angle ADG$, and therefore, $\triangle CLB \sim \triangle ADG$. Using these similar triangles, explain why $\dfrac{BC}{BH} = \dfrac{BL}{r}$.

(f) Show that $\triangle JBL \sim \triangle JFD$. Using these similar triangles, explain why $\dfrac{BC}{BH} = \dfrac{JB}{JF}$, and therefore, $\dfrac{CH}{BH} = \dfrac{BF}{JF}$. Explain why this gives

$$\frac{CH^2}{BH \cdot CH} = \frac{BF \cdot FC}{JF \cdot FC}.$$

(g) Show that $\triangle JFD \sim \triangle DFC$. Using these similar triangles, explain why $JF \cdot FC = r^2$, and hence, $r^2 s^2 = s \cdot BH \cdot BF \cdot FC$. Finish the proof by showing that $BH = s - a$, $BF = s - b$ and $FC = s - c$.

11.3　Circles & squares

Quadrilaterals are the natural successor to triangles, and this is precisely what Euclid tackles next. While it is possible to inscribe a circle in and circumscribe a circle about any triangle, this is not the case for every quadrilateral. Looking at the diagram, if we consider $ABDx$, where x is the fourth vertex of the quadrilateral, then any potential circumcircle about quadrilateral $ABDx$ is completely determined by only three of its points, say A, B and D. This implies that it is not possible to circumscribe a circle about quadrilateral $ABDx$ whenever x is not on the circle determined by A, B and D. For example, we cannot circumscribe a circle about the quadrilateral $ABDF$ shown in the diagram. From Proposition III.22, we know that the opposite angles of a cyclic quadrilateral must sum to two right angles. As an exercise, we will see that this condition is sufficient for a quadrilateral to be cyclic. Not surprisingly, there are also quadrilaterals that do not have incircles. For example, while a rectangle has a circumcircle, it will not have an incircle, unless it is also a square. Quadrilaterals with incircles are called **tangential**. (Think of an example of a quadrilateral that is tangential, but not cyclic.) A convex quadrilateral with both an incircle and a circumcircle, as demonstrated in the diagram, is both cyclic and tangential and is called a **bicentric quadrilateral**. Swiss mathematician Nicolaus Fuss (1755–1826), student, friend and secretary to Euler during the last decade of his life, made significant early contributions to the study of bicentric quadrilaterals, as did a mathematician who discovered the Nine-point circle, Poncelet. We will explore both cyclic and tangential quadrilaterals in the exercises.

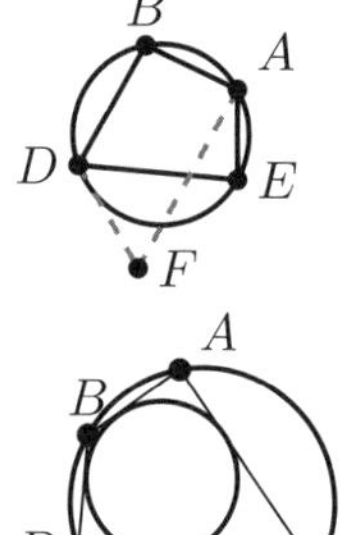

A square is an example of a bicentric quadrilateral, and these are the only cyclic quadrilaterals with an inscribed circle that Euclid considered. Furthermore, with the exception of Proposition IV.10, Euclid restricts his constructions to those related to regular n-gons for the remainder of Book IV. For the propositions involving squares, IV.6 through IV.9, we will state the theorem and give the construction, but leave the verification step of the proof to the reader.

Proposition IV.6. *In a given circle to inscribe a square.*

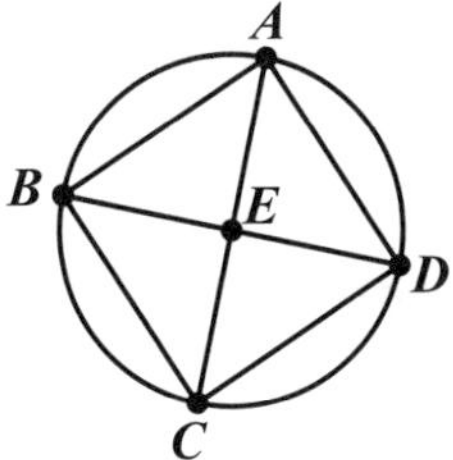

Figure 11.16. Proposition IV.6

Proof. Given a circle, let E be its center. Construct diameters AC and BD such that $AC \perp BD$. Join AB, BC, CD, DA. Claim: $ABCD$ is the inscribed square. The verification is left to the reader. □

Proposition IV.7. *About a given circle to circumscribe a square.*

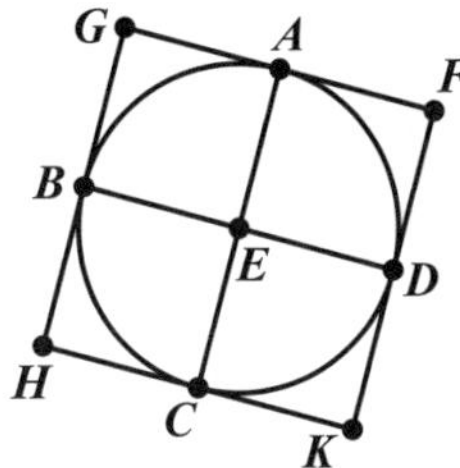

Figure 11.17. Proposition IV.7

Proof. Given a circle, let E be its center. Construct diameters AC and BD such that $AC \perp BD$. Construct perpendicular lines to AC through A and C. Similarly, construct perpendicular lines to BD through B and D. Let F be the intersection of the lines through A and D, G the intersection of the lines through A and B, H the intersection of the lines through B and C and K the intersection of the lines through C and D. Claim: $FGHK$ is the circumscribed square. The verification is left to the reader. □

Proposition IV.8. *In a given square to inscribe a circle.*

Proof. Consider the square □$ABCD$. Bisect AD at E and bisect AB at F. Construct a line through E that is parallel to AB, and let its intersection with BC be H. Construct a line through F that is parallel to AD, and let its intersection with DC be K. Let G be the intersection of EH and FK. Claim: The circle with center G and radius GE is the inscribed circle. The verification is left to the reader. □

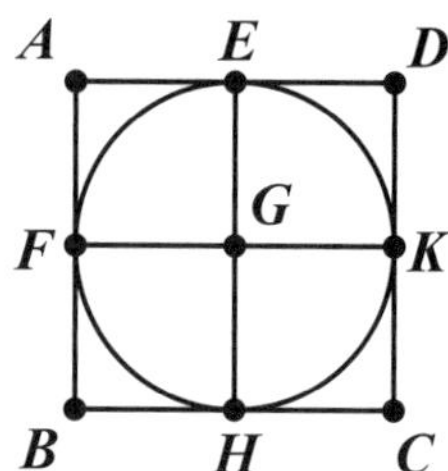

Figure 11.18. Proposition IV.8

Proposition IV.9. *About a given square to circumscribe a circle.*

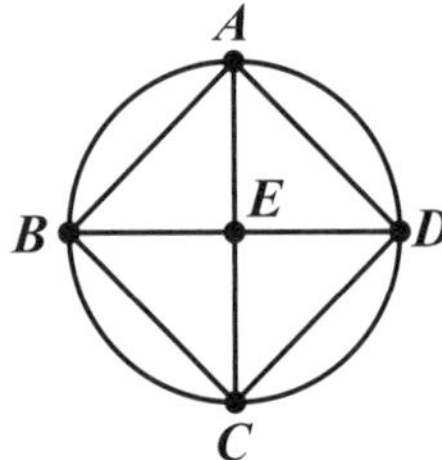

Figure 11.19. Proposition IV.9

Proof.   Consider the square $\square ABCD$. Draw AC and BD and let E be their point of intersection. Claim: The circle with center E and radius AE is the circumscribed circle. The verification is left to the reader. $\square$

Exercises 11.3

1. Prove that the construction given in each proposition works.

  (a) Proposition IV.6   (c) Proposition IV.8

  (b) Proposition IV.7   (d) Proposition IV.9

2. Give a different construction for Proposition IV.8. Prove that it works.

3. Let r_1 be the radius of the circumscribed circle for a square, and let r_2 be the radius of the inscribed circle for the same square. Find the ratio r_1/r_2. See Figure 11.20.

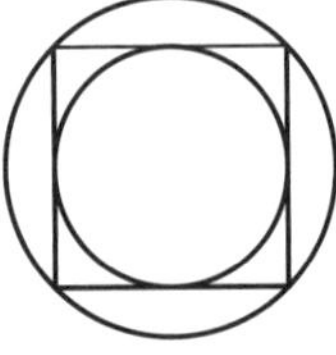

Figure 11.20. Exercise 11.3.3

4. In Proposition III.22, Euclid shows that the opposite angles of a cyclic quadrilateral
   are supplementary. Consider its converse: *If the opposite angles of a quadrilateral
   are supplementary, then the quadrilateral is cyclic.* Prove that this also holds. [*Hint:
   Try a proof by contradiction.*]

5. Prove the following theorem: *A quadrilateral is cyclic if and only if the perpendicular
   bisectors for all of its sides are concurrent.*

6. A trapezoid $ABCD$ with exactly one pair of parallel sides, $AB \parallel CD$, is **isosceles** if
   $AD = BC$. Prove that an isosceles trapezoid is cyclic. [*Hint: Use Exercise* 4.]

7. Prove that if the quadrilateral $ABCD$ is tangential, then $AB + CD = AD + BC$.

8. Let $ABCD$ be a convex quadrilateral. Prove that if $AB + CD = AD + BC$, then $ABCD$
   is tangential. [*Hint: Assume that ABCD is not tangential. Construct a circle c that is
   tangent to AB, BC and AD. Construct the tangent from C to c, and let its intersection
   with $\overrightarrow{AD}$ be E. One possible case is illustrated in Figure* 11.21. *Find an appropriate
   contradiction.*]

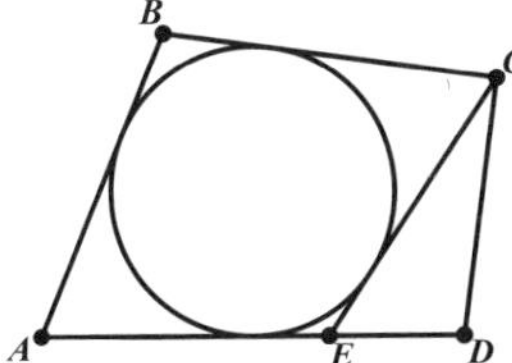

Figure 11.21. Exercise 11.3.8

9. Assume that quadrilateral $ABCD$ is tangential with an inscribed circle of radius r.
   Prove that the area of $ABCD$ is given by $Area = r \cdot s$, where s is the **semiperimeter**,
   $$s = \frac{AB + BC + CD + AD}{2}.$$

10. Let $ABCD$ be a cyclic quadrilateral, and let E be the intersection of its diagonals
    AC and BD. Construct perpendiculars from E to AB, BC, CD and AD, and let the
    intersections be F, G, H and J, respectively, as illustrated in Figure 11.22. Prove that
    quadrilateral $FGHJ$ is tangential. [*Hint: Quadrilaterals AFEJ, DHEJ, BFEG and
    CHEG are all cyclic.*]

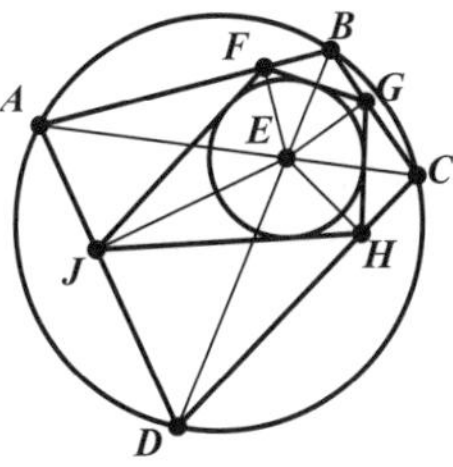

Figure 11.22. Exercise 11.3.10

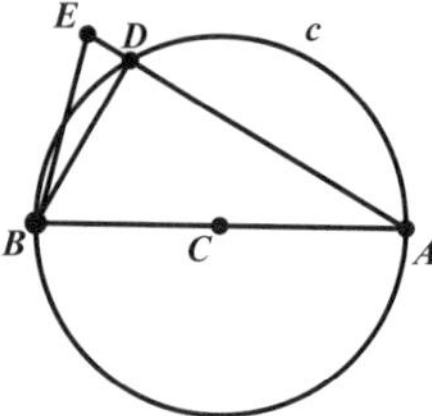

Figure 11.23. Exercise 11.3.11

11. Let AB be the diameter of a circle c with center C. Construct D on c such that $BD = BC$. Extend the chord AD to a point E such that $AE = AB$. Consider $\triangle BDE$, as illustrated in Figure 11.23. Prove that $\angle BED = 75°$ and $\angle DBE = 15°$.

12. In his work entitled *A Book on the Geometrical Constructions Necessary to the Artisan*, Abū al-Wafū'al Būzjānī included the following construction: *To inscribe an equilateral triangle in a square so that its angles touch its sides* [**78**]. Given a square $\square ABCD$, the construction proceeds as follows:

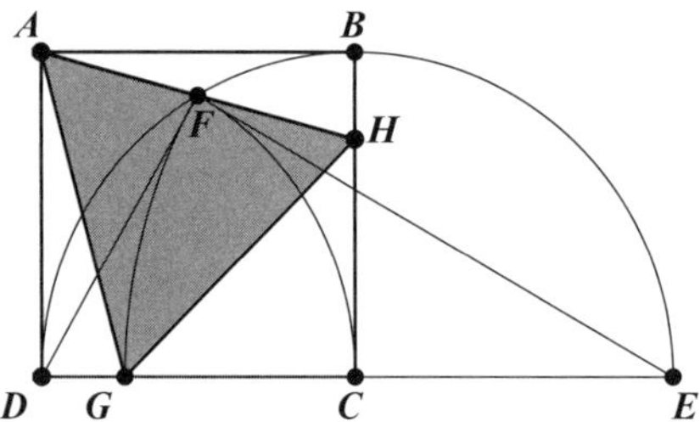

Figure 11.24. Exercise 11.3.12

- Extend DC to a point E such that $DC = EC$. Construct the circle with center C and radius DC.
- Construct a circle with center D and radius DC. Let the intersection of the two circles be F.
- Construct a circle with center E and radius EF. Let the intersection with DE be G.
- Construct H on BC such that $BH = DG$.
- Triangle $\triangle AGH$ is the desired equilateral triangle.

Use Exercise 11 to prove that $\triangle AGH$ is equiangular, and hence, equilateral. [*Hint: Show that $\triangle ADG$ is a $15° - 75° - 90°$ triangle.*]

13. In her 1912 book *A Source Book of Problems for Geometry, Based upon Industrial Design and Architectural Ornament*, Mabel Sykes presents the following constructions. Using dynamic software, such as Geometer's Sketchpad or GeoGebra, recreate each construction.
    (a) The ring of tangent circles illustrated
        in Figure 11.1.

(b) A cusped quadrilateral.

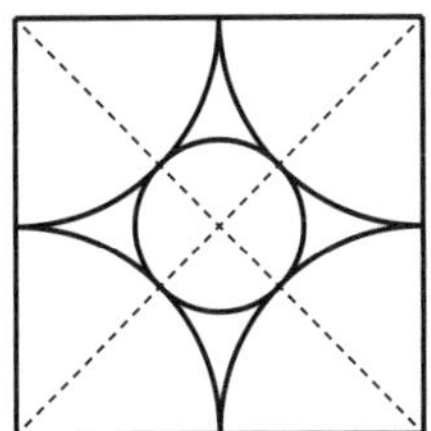

(e) A rounded quadrifoil.

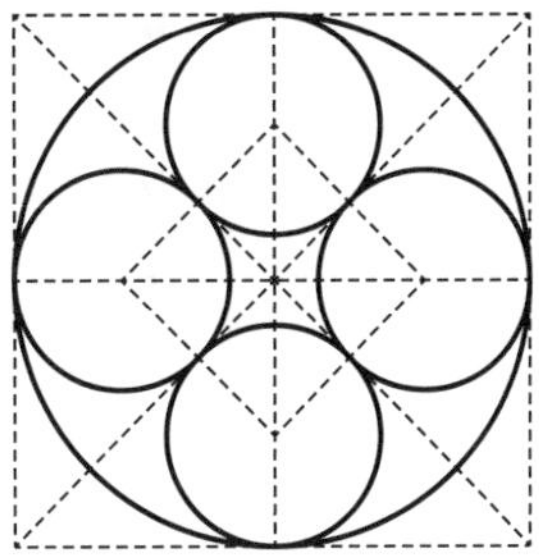

(c) Another cusped quadrilateral. [*Hint: To obtain C, construct G on AB so that GF = DE, then let C be the intersection of AB with the perpendicular bisector of DG.*]

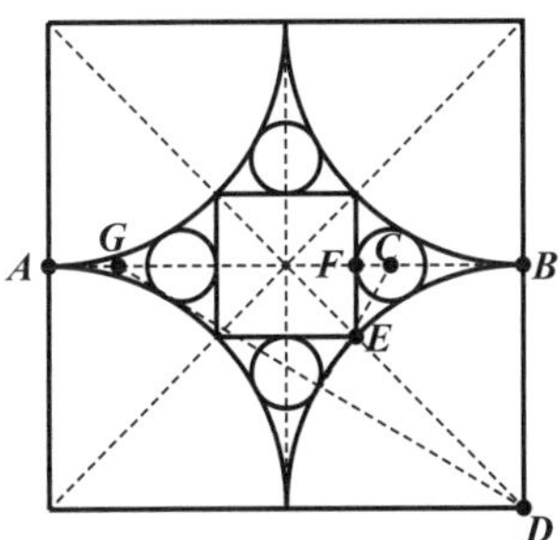

(f) Another rounded quadrifoil. [*Hint: Modify the previous quadrifoil.*]

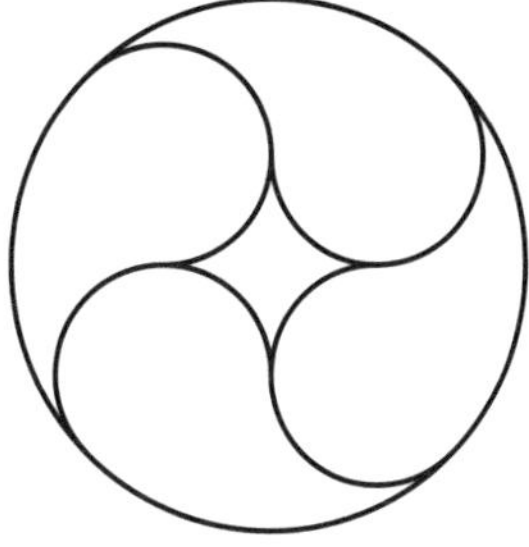

(d) A rounded trefoil.

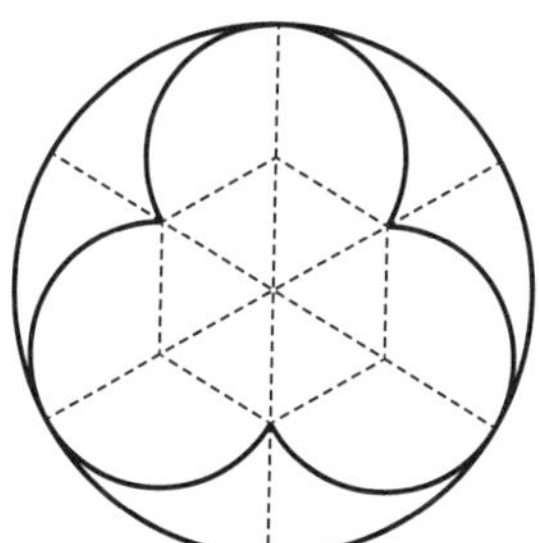

(g) A rounded multifoil. [*Hint: To create the regular octagon, start with a circle inscribed in a square, and use the diagonals of the square to determine the other four points on the circle that determine the octagon.*]

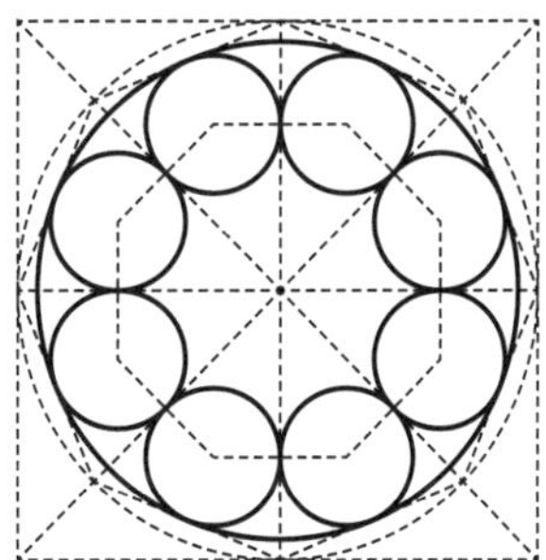

11.4 Circles & pentagons

In Proposition IV.10, Euclid constructs a $36°-72°-72°$ triangle that he needs for the construction of a regular pentagon. To do this, he constructs an isosceles triangle where each of the equal angles is twice the remaining angle. If we let α be the measure of the smallest angle, then the other two angles are each 2α. Thus, $\alpha + 2\alpha + 2\alpha = 180°$. Hence, $\alpha = 36°$ and the other two angles are each $72°$. Any such triangle is called a *golden triangle*, because as we shall learn in the construction, the ratio of the lengths of the equal sides to the base is the golden ratio. In his construction and subsequent proof, Euclid uses Propositions II.11, III.32 and III.37. We give a slightly revised proof.

Proposition IV.10. *To construct an isosceles triangle having each of the angles at the base double of the remaining one.*

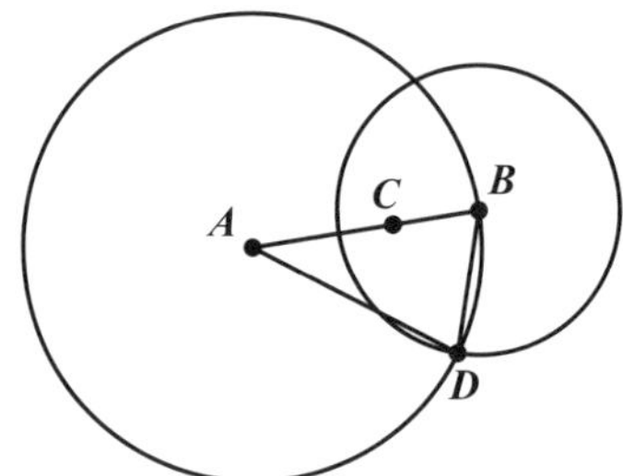

Figure 11.25. Euclid's construction for Proposition IV.10

Euclid's Construction: Consider segment AB. Using II.11, construct a point C on AB such that $AB \cdot CB = AC^2$. Construct a circle with center A and radius AB. Construct a circle with center B and radius AC. Let D be an intersection of the two circles. Join BD and AD.

Claim: Triangle $\triangle ABD$ is the desired isosceles triangle.

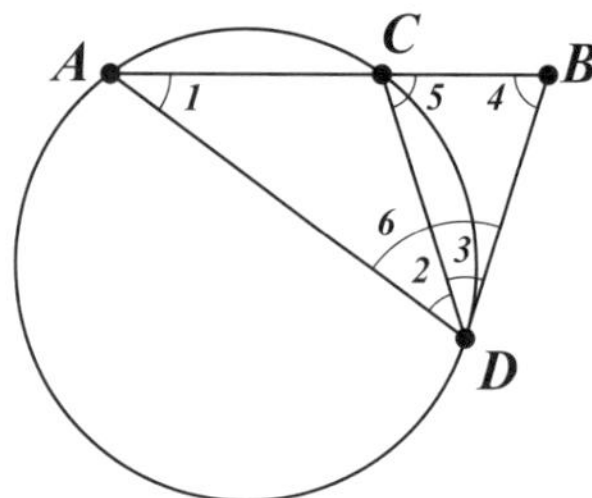

Figure 11.26. Justification for Proposition IV.10

Justification: Join CD. Using IV.5, construct the circle that circumscribes triangle $\triangle ACD$. Note that B lies outside of this circle, BD intersects the circle at D, and BCA is a line that intersects the circle at points C and A. Recall, by construction, $AC^2 = AB \cdot CB$ and $BD = AC$. Substituting, we have $BD^2 = AB \cdot CB$. Thus, by III.37, BD is tangent to circle $\bigcirc ACD$ at point D. But then, $\angle BDC \cong \angle BAD$ (or $\angle 3 = \angle 1$, as shown in Figure 11.26) by III.32.

By I.32, $\angle 5 = \angle 1 + \angle 2 = \angle 3 + \angle 2 = \angle 6$. Since $AB = AD$, $\triangle ABD$ is isosceles and $\angle 4 = \angle 6$ by I.5. Thus, $\angle 5 = \angle 4$, and $CD = BD$ by I.6 applied to $\triangle BCD$. Since $BD = AC$ by construction, we have $AC = CD$. Therefore, $\triangle ACD$ is isosceles and $\angle 1 = \angle 2$ by I.5. Putting this all together, we have

$$2 \cdot (\angle 1) = \angle 1 + \angle 2 = \angle 5 = \angle 4 = \angle 6,$$

and $\triangle ABD$ is as desired. $\square$

In Proposition IV.11, Euclid inscribes a regular pentagon in a circle, thereby completing his construction of a regular pentagon. In addition to Euclid's construction, we give two other constructions. The first is German artist Albrecht Dürer's construction, though it is originally attributed to Ptolemy's *Almagest*. The second, which follows after Propositions IV.13 and IV.14, is a construction using 72° angles.

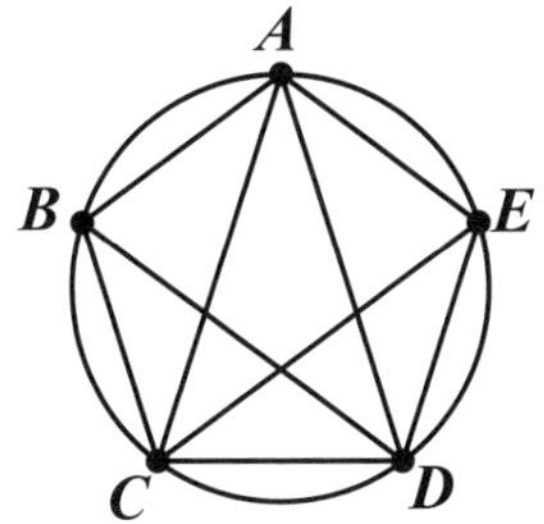

Figure 11.27. Proposition IV.11

Proposition IV.11. *In a given circle to inscribe an equilateral and equiangular pentagon.*

Euclid's Construction: Given a circle, inscribe an isosceles triangle $\triangle ACD$ where $AC = AD$ and $\angle ACD = \angle ADC = 2 \cdot (\angle CAD)$ by IV.10 and IV.2. Bisect angles $\angle ACD$ and $\angle ADC$, and let E and B be the intersections of the angle bisectors with the circle, respectively. Join AB, BC, DE and AD.

Claim: The pentagon $\bigcirc ABCDE$ is a regular pentagon.

Justification: By construction, the inscribed angles are equal, namely $\angle BDA = \angle CDB = \angle CAD = \angle ECD = \angle ACE$. By the Circle Equivalence Theorem, their associated chords are equal, namely $AB = BC = CD = DE = AE$. Thus, $\bigcirc ABCDE$ is equilateral.

Since $AB = DE$, $\overset{\frown}{AB} = \overset{\frown}{DE}$. Adding $\overset{\frown}{BCD}$ to both arcs, we have $\overset{\frown}{ABD} = \overset{\frown}{BDE}$. Therefore, their corresponding inscribed angles, $\angle AED$ and $\angle BAE$, are equal. Similarly, it can be shown that $\angle AED = \angle CDE = \angle BCD = \angle ABC$. Thus $\bigcirc ABCDE$ is equiangular. $\square$

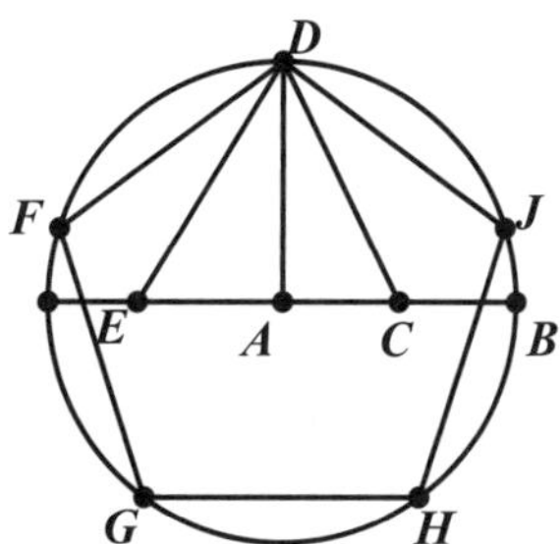

Figure 11.28. Dürer's regular pentagon

Dürer's Construction of a Regular Pentagon: Given a circle with center A, let B be a point on the circle. Bisect the segment AB at a point C. Construct a perpendicular to AB through A and let D be its intersection with the circle. Join CD and construct a circle with center C and radius CD. Let E be its intersection with $\overrightarrow{BA}$. Claim: The length of a side of a regular pentagon inscribed in the circle is equal to DE. The verification is left to the reader. $\square$

We will skip Proposition IV.12 (*About a given circle to circumscribe an equilateral and equiangular pentagon*), to construct the incircle for a regular pentagon in IV.13. We will see that Euclid's method for inscribing the circle can be generalized to a regular n-gon.

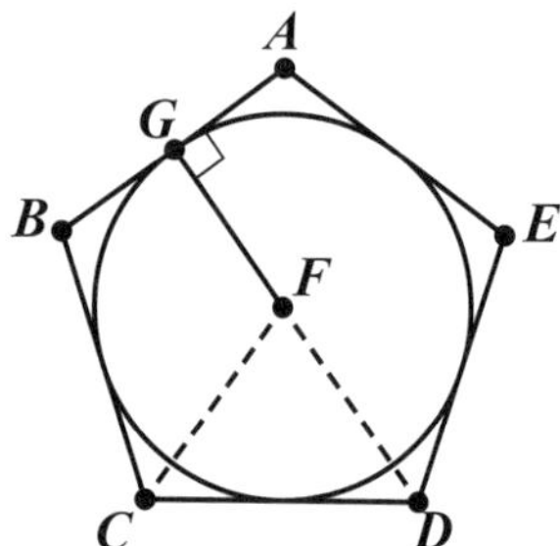

Figure 11.29. Proposition IV.13

Proposition IV.13. *In a given pentagon, which is equilateral and equiangular, to inscribe a circle.*

Euclid's Construction: Consider the regular pentagon $\bigcirc ABCDE$. Bisect angles $\angle BCD$ and $\angle CDE$, and let F be the intersection of the angle bisectors. Construct a perpendicular from F to AB and let G be the point of intersection with AB.

Claim: The circle with center F and radius FG is the inscribed circle. The verification is left to the reader. $\square$

For his proof of IV.13, Euclid joins F with each of the other vertices and shows that these segments bisect all of the other angles. He then proves that the distance from F to each side is the same. Clearly, the construction that Euclid gives will work in general for a regular n-gon. Furthermore, his proof can also be generalized. We leave it to the reader to show this.

As expected, Euclid follows this by constructing the circumcircle for a regular pentagon in IV.14. For his proof, he once again bisects adjacent angles in the given pentagon and argues that the intersection of the angle bisectors is the center of the desired circle. This time he uses the segment joining the point of intersection with one of the vertices as the radius of the circumscribed circle. Again, this construction clearly generalizes to any regular n-gon. The proof of Proposition IV.14 follows directly from the proof of Proposition IV.13.

Proposition IV.14. *About a given pentagon, which is equilateral and equiangular, to circumscribe a circle.*

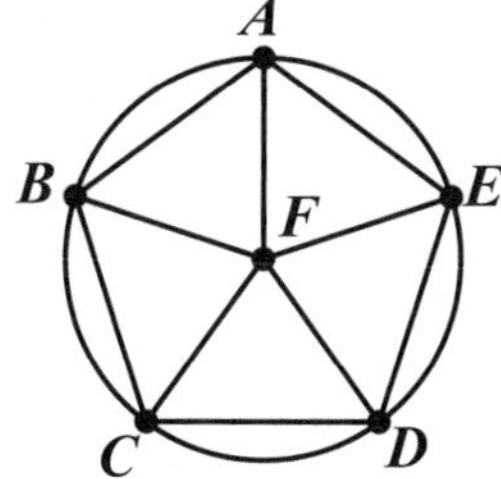

Figure 11.30. Proposition IV.14

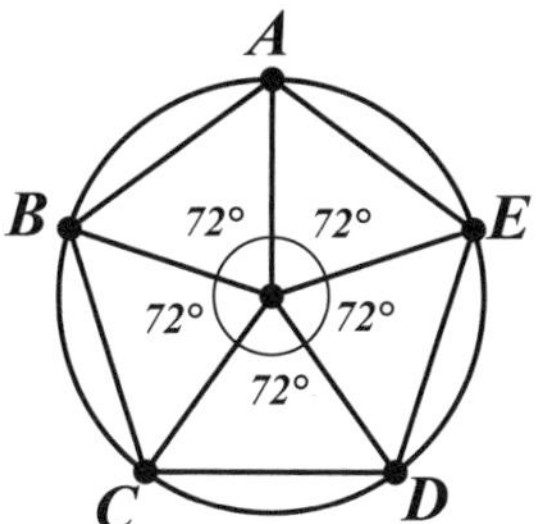

Figure 11.31. Central angles of a pentagon

Notice that in the previous two constructions, the center for the inscribed and the circumscribed circle is the same. We will call this point the *center* of a regular pentagon. When we join the center to each of the vertices, we have five congruent central angles, each of measure $\frac{360°}{5} = 72°$. This method gives an alternative construction of a regular pentagon.

Construction of Regular Pentagon Using Central Angles: Construct a circle and, at its center, copy five nonoverlapping central angles of 72°. Label the successive intersections of rays of the angle with the circle A, B, C, D, E. Then $\bigcirc ABCDE$ is a regular pentagon. $\square$

Since the last two propositions can be generalized to any regular n-gon, and as both circles will have the same center, we have the following definition.

Definition 11.12. The **center** of a regular n-gon is the center, O, of its circumcircle (or incircle). If A and B are adjacent vertices of the regular n-gon, then angle $\angle AOB$ is a **central angle of the regular n-gon.**

It is clear that a central angle of a regular n-gon is of measure $\frac{360°}{n}$. (Why?). The Circle Equivalence Theorem (page 204) specifies that, in equal circles, two inscribed angles are equal if and only if their corresponding chords and arcs are also equal. We state the following lemma, leaving its proof to the reader.

Lemma 11.13. *A regular n-gon is constructible if and only if its central angle $\frac{360°}{n}$ is constructible.*

Exercises 11.4

1. Use geometric software to produce Dürer's construction of the regular pentagon.

2. The following set of instructions is for constructing a pentagon given a side AB of length d.

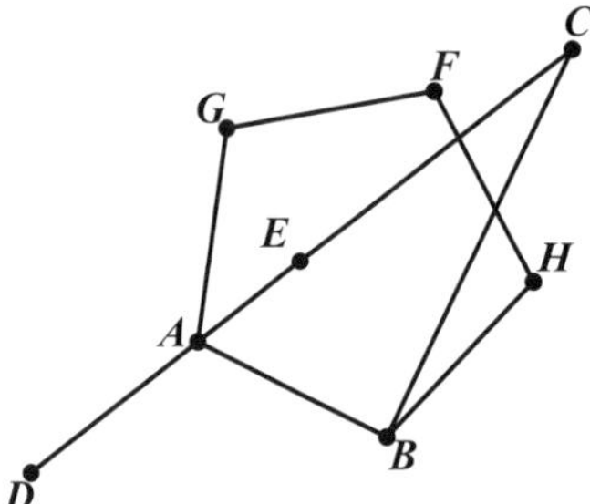

Figure 11.32. Exercise 11.4.2: Constructing a regular pentagon given side AB

*   Construct CB perpendicular to AB such that $CB = 2d$.
*   Construct segment CA and extend ray CA to a point D such that $AD = d$, where A lies between C and D.
*   Bisect CD at point E.
*   Let F be the intersection of the two circles centered at A and B, each with radius CE.
*   Let G be the intersection of circles with centers A and F and radius AB, and let H be the intersection of circles with centers B and F and radius AB.
*   Form pentagon $\bigcirc ABHFG$ by connecting vertices A, B, H, F and G.

(a) Use geometric software to produce this construction of the regular pentagon.

(b) Prove that if $AB = 1$, then $CE = \dfrac{1 + \sqrt{5}}{2}$, and hence, $\triangle FAB$ is a golden triangle.

(c) Use part (b) to prove that $\bigcirc ABHFG$ is a regular pentagon.

3. Prove that the construction given in Proposition IV.13 works for any regular n-gon.

4. Prove that the construction given in Proposition IV.14 works for any regular n-gon.

5. Prove that the construction given above using central angles of 72° produces a regular pentagon.

6. Prove Lemma 11.13.

11.5 Constructing regular polygons

Euclid constructed both the equilateral triangle and the square in Book I. We had to wait until the eleventh proposition of the fourth book for the construction of the next regular polygon, the more complex regular pentagon. Its construction and proof relies on results from all three of the previous books. In the two remaining propositions,

Euclid constructs a regular hexagon and then finishes Book IV with a regular 15-gon. In addition to these two constructions, we consider the more general question of which regular n-gons *can* be constructed using only an unmarked straightedge and compass.

In Proposition IV.15, Euclid inscribes a regular hexagon in a circle. We give a modified version of his construction and leave the verification step to the reader.

Proposition IV.15. *In a given circle to inscribe an equilateral and equiangular hexagon.*

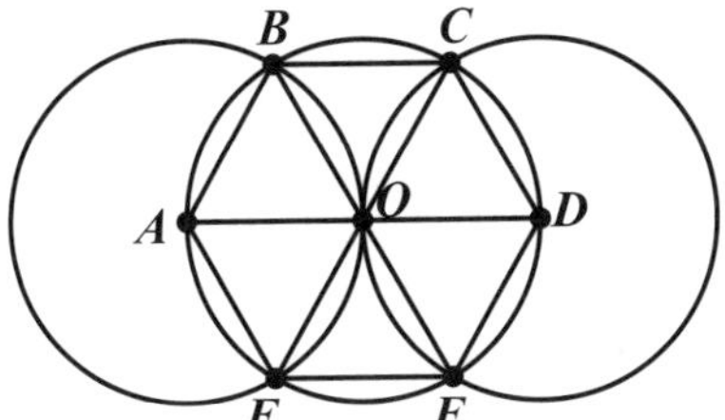

Figure 11.33. Regular hexagon

Construction: Given a circle with center O and point A, construct the diameter AD. Construct two circles, one with center A, the other with center D, and both with radius AO. Label the intersections of these circles with our original circle B, C, E and F so that the letters are in order from A either clockwise or counter-clockwise.

Claim: The hexagon $\bigcirc ABCDEF$ is a regular hexagon. The justification is left to the reader.   □

Another way to construct a regular hexagon is to notice that it has twice as many sides as an equilateral triangle, which is itself constructible. So, if we circumscribe a circle about the equilateral triangle, bisect each of the central angles and mark the intersection of the angle bisectors with the circle, then we'll have the six vertices of a regular hexagon marked on the circle. We will exploit this method to construct a regular $2n$-gon from any constructible n-gon: start with a constructible regular n-gon, circumscribe a circle around it, bisect each of the central angles and add the points where the angle bisectors meet the circle to the original set of vertices. Alternatively, we could bisect each of the n arcs into which the regular n-gon divides the circle. Either way, we now have $2n$ vertices which, when connected, produce a regular $2n$-gon. While Euclid did not include the following proposition in Book IV, his far more complicated constructions leads us to believe that he chose not to include it. We leave the proof to the reader.

Theorem 11.14. *If a regular n-gon is constructible, then so is a regular $2n$-gon.*

Using this doubling method on an inscribed equilateral triangle, square or regular pentagon, we are thus able to construct regular n-gons for $n = 6, 8, 10, 12, 16, 20, 32, 40, \ldots$.

For the last proposition in Book IV, Euclid constructs a regular 15-gon, also called a regular **pentadecagon**. We give a slightly revised version of his construction.

Proposition IV.16. *In a given circle to inscribe a fifteen-angled figure which shall be both equilateral and equiangular.*

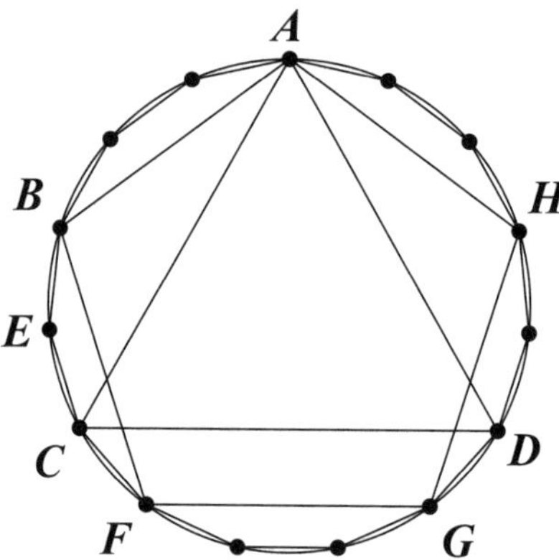

Figure 11.34. Proposition IV.16

Construction: Given a circle with a point A on it, inscribe an equilateral triangle $\triangle ACD$ with vertex A. Also inscribe a regular pentagon $\bigcirc ABFGH$. The triangle divides the circumference into three arcs, each one-third of the whole, while the pentagon divides the circumference into five arcs, each one-fifth of the whole. The arc that lies between B and C must be two-fifteenths of the whole since $\frac{1}{3} - \frac{1}{5} = \frac{2}{15}$. If we bisect this arc at E, then $\overset{\frown}{BE}$ will be exactly one-fifteenth of the whole. Copying this arc around the circle, we construct a regular 15-gon, as desired. $\square$

At first glance, the construction technique employed in the proof appears fruitful, as if it would yield other new regular n-gons from those that we already know to be constructible. Alas, it does not. (Why not? What makes the regular 3- and 5-gons special?) Euclid ends his regular polygon constructions with the 15-gon, but the hunt for other constructible regular polygons (besides those obtained by doubling) continued for nearly 2000 years until a 19-year-old finally found one. In 1796, a young Gauss constructed the **heptadecagon**, or 17-gon. Better still, he gave the formula to completely classify which regular n-gons are constructible. The reason the resolution to this problem took so long is that it required tools outside of geometry, specifically, tools from the new mathematical areas of abstract algebra and complex numbers. With those tools, Gauss completely classified the constructibility of regular n-gons as given in Theorem 11.15.

Theorem 11.15. [Gauss's Theorem]. *A regular n-gon is constructible if and only if* $n = 2^k p_1 p_2 p_3 \ldots p_m$ *where $n \geq 3$, $k \geq 0$, and p_i are distinct primes of the form $2^{2^j} + 1$.*

Primes of the form $2^{2^j} + 1$ are called **Fermat primes**, named for Pierre de Fermat (1601–1665). The known Fermat primes are:

$$2^{2^0} + 1 = 3$$
$$2^{2^1} + 1 = 5$$
$$2^{2^2} + 1 = 17$$
$$2^{2^3} + 1 = 257$$
$$2^{2^4} + 1 = 65537.$$

Figure 11.35. Close-up of 17-point star on monument to Gauss, Brunswick, Germany

Euler famously showed that $2^{2^5} + 1 = 4,294,967,297$ is not prime. It is known that $2^{2^j} + 1$ is not prime for $5 \leq j \leq 32$ and a smattering of other values for larger j. However, it is **unknown** whether there are only five primes of this form.[1] Therefore, as of this edition, we have the following:

A regular n-gon is constructible if $n = 2^k \cdot 3^{b_1} \cdot 5^{b_2} \cdot 17^{b_3} \cdot 257^{b_4} \cdot 65537^{b_5}$ where $n \geq 3$, $k \geq 0$, and $b_i \in \{0,1\}$ for all i.

If one day it is shown that there are no other Fermat primes, then this implication becomes an "if and only if."

As for the constructions of the regular 257-gon and the regular 65537-gon, Gauss left them to two other German mathematicians. Friedrich Julius Richelot (1808–1875) constructed the regular 257-gon in 1831, and Johann Gustav Hermes (1846–1912) spent ten years on the regular 65537-gon, finishing in 1894. The manuscript detailing this impossible-to-imagine construction sits in a large box at the University of Göttingen. For now, we are left to wonder why it is impossible to construct a regular n-gon when n does not take this form. For example, why can't we construct a 7-gon or a 9-gon? We will discuss this, as well as the three famous construction problems of antiquity, in Chapter 16.

Surely owing to its personal significance, Gauss remarked that the heptadecagon was a result worthy of being on his tombstone. His desire to memorialize his great accomplishment puts him in fine company. The greatest mathematician of antiquity and the subject of the following section, Archimedes of Syracuse (287–212 BCE), requested a representation of his most significant accomplishment, a cylinder circumscribing a sphere, on his tomb. The Roman philosopher Cicero (106–43 BCE) writes of finding a figure of a sphere in a cylinder when in search of Archimedes' grave [**24**]. On Gauss's simple grave in Göttingen there is no such polygon, but a 17-point star does grace the bottom of the bronze statue that sits in his honor in his native town of Brunswick. In Figure 11.36 we include a similar homage to Gauss in the form of a 17-point star made by Swiss artist Eugen Jost and appearing in a lovely book on art and mathematics, *Beautiful Geometry*.

---

[1]Current information on Fermat numbers can be found at *www.fermatsearch.org/news.html*.

Figure 11.36. Homage to Carl Friedrich Gauss by Eugen Jost (2014) [**86**]

Exercises 11.5

1. Prove that the construction given for Proposition IV.15 is valid.

2. Construct a regular $2n$-gon given a regular n-gon. Prove that this construction is valid.

3. Given a regular m-gon that is constructible and a regular n-gon that is constructible, prove that if m and n are relatively prime ($\gcd(m, n) = 1$), then a regular mn-gon is also constructible.

4. Find all n such that $3 \leq n \leq 100$ and the regular n-gon is constructible.

5. Find all n such that $101 \leq n \leq 200$ and the regular n-gon is constructible.

6. Use geometric software to produce the following "quadruple quadrisection" construction of a 17-gon due to Herbert W. Richmond (1863–1948), King's College,

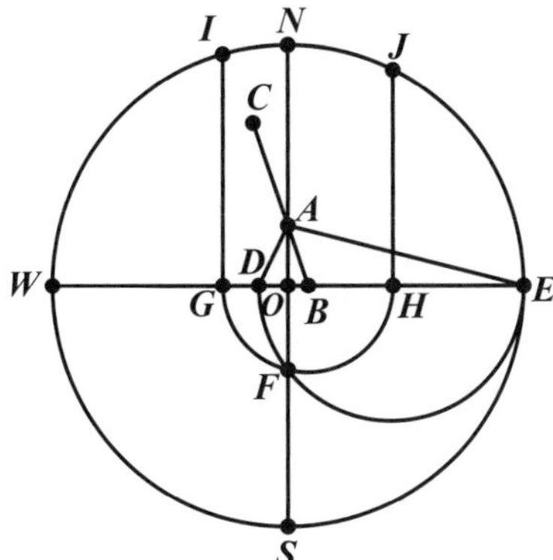

Figure 11.37. Exercise 11.5.6: Richmond's construction of a 17-gon

Cambridge, who gave it in the *Quarterly Journal of Pure and Applied Mathematics* in 1893. To **quadrisect** is to divide something into four equal parts. (Hence, to bisect into halves, and then, bisect each half.)

- Quadrisect the perimeter of the circle of center O by points N, E, S, W.
- Quadrisect the radius ON by the point A where A is closest to O.
- Quadrisect the angle $\angle OAE$ by the line AB (where B lies on OE and is closest to O). See Figure 11.37.
- Form line BAC.
- Quadrisect the straight angle BAC in the direction of W by the line AD, where D is on radius OW, and angle $\angle BAD$ is half of a right angle.
- Construct semicircle DFE, cutting OS at F.
- Construct semicircle GFH, centered at B, cutting OW at G and OE at H.
- Construct $GI \perp WE$ where I is on circle WNE in the 2nd quadrant.
- Construct $HJ \perp WE$ where J is on circle WNE in the 1st quadrant.
- Since arc $\widehat{IJ}$ is two seventeenths of the circumference of the circle, bisecting arc $\widehat{IJ}$ at K produces vertices I, K and J of the regular heptadecagon. Copy arc $\widehat{IK}$ around circle $ENWS$ to construct the other vertices of the heptadecagon.

11.6 The area of a circle & π

At the end of Chapter 10, we gave a decidedly modern proof using limits to show that the ratio of any circle's circumference to its diameter is a constant, and noted that this constant was named π in the eighteenth century. Euclid does not include a proof that this ratio is constant in the *Elements*, which is curious since there is evidence that it was known long before Euclid took brush to papyrus. There are estimates for π in the Bible, in ancient Egyptian papyri, and in ancient Chinese books, all of which predate Euclid. Certainly, a calculation of an estimate for π would have been out of place in the *Elements* for it is not a book that shows any concern for absolute measurement. Most of its propositions are relationships between geometric objects. So, while the lack of an estimate is unsurprising, we are left to wonder why he does not show that all circles satisfy this ratio. On the other hand, Euclid *does* present a different ratio that is satisfied by all circles. With the second proposition of his twelfth book, he proves the following: *Circles are to one another as the squares on the diameters.* Here, given any two circles with areas, *Area(circle₁)* and *Area(circle₂)*, and diameters, *diameter₁* and *diameter₂*, respectively, we have

$$\text{XII.2} \qquad \frac{Area(circle_1)}{Area(circle_2)} = \frac{diameter_1^2}{diameter_2^2}.$$

His proof relies on inscribing a sequence of regular polygons of an increasing number of sides. We will not prove this particular result, but we will revisit it after looking at a related proposition from Archimedes, a mathematician who followed closely on the heels of Euclid and is widely regarded as the greatest of antiquity. We present two results from Archimedes' treatise, *Measurement of the Circle.* The first is equivalent to our modern formula for the area of the circle, and the second is an estimate for π. Both of Archimedes' results rely on the same method Euclid uses in XII.2, inscribing and circumscribing regular polygons in and around a circle.

> The most important document containing the works of Archimedes was only rediscovered in Constantinople in 1906. Called the *Archimedes Palimpsest*, it contains seven treatises, including *Measurement of the Circle*, which contains the two propositions given in this section, and, notably, the only surviving copy *On Floating Bodies*. A palimpsest is a manuscript that has been erased so that the precious writing material, in this case goatskin parchment, can be reused. Disguised as a book of prayers since the thirteenth century, it laid undiscovered for hundreds of years in a monastery library. It was sold to a private collector in the 1920s, and resold at auction in 1998 at Christie's Auction House in New York City, where it sold to an anonymous buyer for over two million dollars. It has been an object of study at the Walters Art Museum in Baltimore, Maryland, since 1999.

Archimedes' Proposition 1. *The area of a circle is equal to the area of a right triangle in which one of the legs is equal to the radius and the other to the circumference.*

Restating this as a formula, we have $Area = \frac{1}{2}r \cdot Circumference$, where r is the radius of the circle. Since $Circumference = 2\pi r$, we have the familiar formula for area of a circle, $Area = \pi r^2$. Archimedes' proof (as does Euclid's XII.2) employs the **method of exhaustion**, a technique attributed to Eudoxus (408–355 BCE), a Greek mathematician who predates Euclid. Anyone familiar with integral calculus will recognize this method, which entails exhausting the difference between a known area and an unknown area. Here, the area of the circle is approximated by systematically inscribing regular polygons of an increasing number of sides in the circle. As the number of sides increases, the easily computed area of the polygon approaches the area of the circle from below, so that the difference between the unknown area of the circle and the known area of the polygon is less than any given quantity, no matter how small. At heart, this is the same limit method we saw at play in the proof that the ratio of any circle's circumference to its diameter is constant in Theorem 10.3, but here we are working with the two-dimensional area of the inscribed polygon instead of the one-dimensional perimeter. It is the idea that this process can continue forever, theoretically, and thus, that the difference between the unknown and the known can be exhausted, that assuredly places Greek mathematics as a forerunner to calculus.

As a visual aid, Figure 11.38 shows a systematic progression of regular n-gons inscribed in a circle. At this size, it is difficult to distinguish the circle from the regular 24-gon. Clearly, the area of the regular n-gon is a good approximation for the area of the circle when n is large.

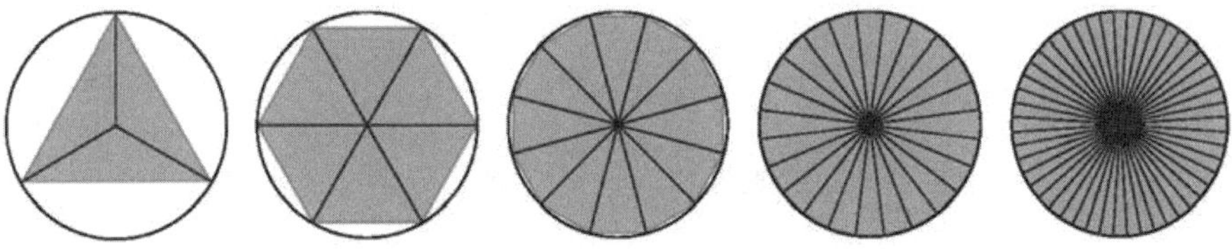

Figure 11.38. Inscribed regular 3-gon, 6-gon, 12-gon, 24-gon and 48-gon

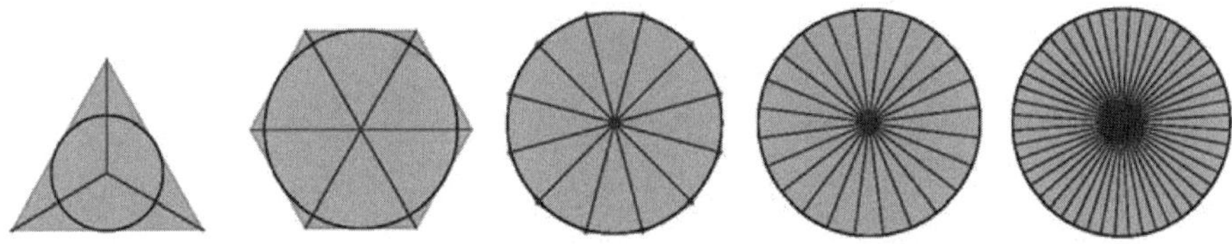

Figure 11.39. Circumscribed regular 3-gon, 6-gon, 12-gon, 24-gon and 48-gon

Archimedes takes Eudoxus' method of exhaustion one step further by approaching the area of the circle from above with circumscribed regular polygons, knowing that the actual area of the circle lies between these bounds. In fact, he uses a double reductio ad absurdum argument to show that the circle and triangle have the equal area, arguing that the other possibilities, when either is bigger than the other, lead to a contradiction. In approaching the area from above and below, those familiar with the work of Isaac Newton (1642–1727) and Gottfried Leibniz (1646–1716) will no doubt see the seeds of integral calculus in this method and want to turn this into a limit problem. We do just that, giving an updated proof of Archimedes' first proposition which relies on the limit concept. First, we need a new definition and a formula for the area of a regular n-gon.

Definition 11.16. The **apothem** of a regular polygon is a line segment from the center to the midpoint of one of its sides.

Lemma 11.17. *A regular polygon with perimeter P and apothem h has an area of $\frac{1}{2}h \cdot P$.*

Proof of Lemma. Suppose the regular polygon has n sides, each of length s. Join each vertex to the center, O. The polygon is composed of n congruent triangles, each with area $\frac{1}{2}hs$. Hence, the area of the polygon is $n \cdot (\frac{1}{2}hs) = \frac{1}{2}h \cdot (ns) = \frac{1}{2}h \cdot P$, as desired. $\square$

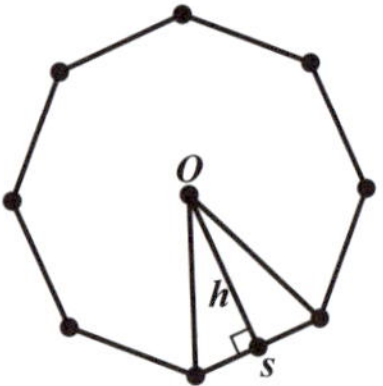

Figure 11.40. Area of a regular polygon

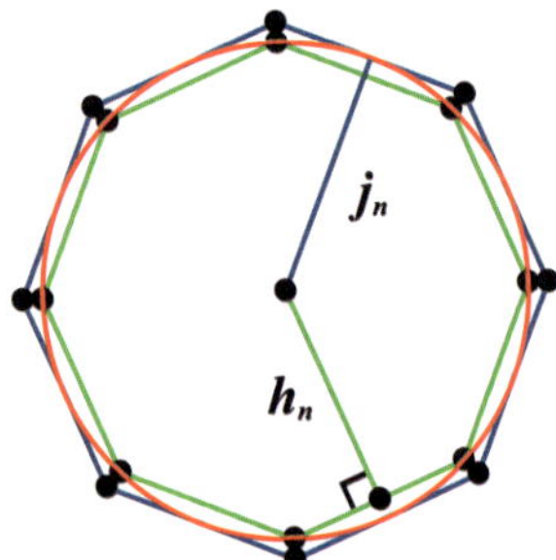

Figure 11.41. Archimedes' Proposition 1: Area of a circle

Proof of Archimedes' Proposition 1.   Let *Area(circle)* represent the area of the given circle, B_n be the area of an inscribed regular polygon with n sides, and D_n be the area of a circumscribed regular polygon with n sides. Clearly, $B_n \leq$ *Area(circle)* $\leq D_n$ for all $n \geq 3$. Let h_n be the apothem of B_n, and P_n be the perimeter of B_n. Likewise, let j_n be the apothem and Q_n the perimeter of D_n. By Lemma 11.17, we have

$$\frac{1}{2}h_n \cdot P_n \ \leq \ \textit{Area(circle)} \ \leq \ \frac{1}{2}j_n \cdot Q_n.$$

Taking the limit as n approaches infinity, we have

$$\lim_{n \to \infty} \frac{1}{2}h_n \cdot P_n \ \leq \ \textit{Area(circle)} \ \leq \ \lim_{n \to \infty} \frac{1}{2}j_n \cdot Q_n.$$

But clearly, $\lim_{n \to \infty} h_n = \lim_{n \to \infty} j_n = r$, and $\lim_{n \to \infty} P_n = \lim_{n \to \infty} Q_n = $ *Circumference*. Thus,

$$\frac{1}{2}r \cdot \textit{Circumference} \ \leq \ \textit{Area(circle)} \ \leq \ \frac{1}{2}r \cdot \textit{Circumference},$$

and the area of the circle is *Area(circle)* $= \frac{1}{2}r \cdot$ *Circumference*.   □

In light of this proposition, let's revisit Euclid's Proposition XII.2. A notable difference between these results is that Euclid relates geometric figures of the same dimension, whereas Archimedes relates figures of differing dimensions. Consequently, in combination, these two results tell us something very interesting. Neither result relies on the constancy of the ratio of any circle's circumference to its diameter, and together, they provide a proof of this result.

$$\frac{Circumference_1}{Circumference_2} = \frac{\frac{2}{r_1} \cdot Area_1}{\frac{2}{r_2} \cdot Area_2} \quad \text{by Archimedes Prop. 1}$$

$$= \frac{2r_2}{2r_1} \cdot \frac{Area_1}{Area_2} \quad \text{by rearranging}$$

$$= \frac{d_2}{d_1} \cdot \frac{d_1^2}{d_2^2} \quad \text{by Euclid XII.2}$$

$$= \frac{d_1}{d_2}.$$

Therefore, we have another proof that the ratio of any circle's circumference to its diameter is constant, that is

$$\frac{Circumference_1}{diameter_1} = \frac{Circumference_2}{diameter_2}.$$

For our second result, we consider Archimedes' third proposition in which he provides an estimate for this constant ratio, π.

Archimedes' Proposition 3. *The ratio of the circumference of any circle to its diameter is less than* $3\frac{1}{7}$ *but greater than* $3\frac{10}{71}$.

Archimedes method entails bounding this quantity between two fractions with the use of an inscribed and circumscribed regular 96-gon. He begins by inscribing a regular hexagon in the circle. He then proceeds to double the number of sides, successively producing an inscribed regular 12, 24, 48 and 96-gon. He relates the circumference of the inscribed $2n$-gon to that of the n-gon, and thus, generates a recursive formula to estimate a lower bound for π. Of course, he does all of this without the benefit of algebra or the ease of calculation afforded by our decimal number system. He then circumscribes a regular hexagon around the circle and, once again, doubles the number of sides, successively producing a circumscribed regular 12, 24, 48 and 96-gon. This gives him an upper bound for π. We present a revised argument, and where Archimedes used fractions to estimate square roots, we take full advantage of modern technology when giving our lower and upper bounds.

We begin by inscribing a regular hexagon in a circle of radius 1. Notice that the perimeter of the regular 6-sided polygon, $P_6 = 6$, is an underestimate for the circumference, 2π. More generally, if a_n denotes the length of the side of a regular n-sided polygon inscribed in a circle of radius 1, then its perimeter, P_n, is given by $P_n = n \cdot a_n$. Since the perimeter of any polygon inscribed in the circle is less than the circumference of the circle, that is $P_n < 2\pi$, we have

$$\pi \geq \frac{n \cdot a_n}{2}.$$

We now give the following lemma.

Lemma 11.18. *If a_n denotes the length of the side of a regular n-sided polygon inscribed in a circle of radius 1, then*

$$a_{2n} = \sqrt{2 - \sqrt{4 - a_n^2}}.$$

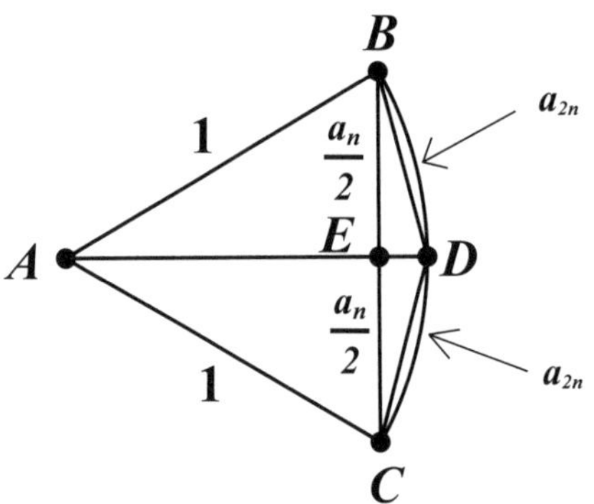

Figure 11.42. Sides of n-gon and $2n$-gon

Proof of Lemma.   Let A be the center of the given circle of radius 1, $BC = a_n$ and $CD = a_{2n}$, as shown in Figure 11.42. Note that AD bisects arc $\widehat{BC}$, meeting BC at E. Since angle $\angle AEB$ is a right angle (by proof of Theorem 11.14), then by the Pythagorean Theorem we have $1 = \left(\dfrac{a_n}{2}\right)^2 + AE^2$, and $a_{2n}^2 = \left(\dfrac{a_n}{2}\right)^2 + DE^2$. Since $DE = 1 - AE$, this gives

$$
\begin{aligned}
a_{2n}^2 &= \left(\tfrac{a_n}{2}\right)^2 + (1 - AE)^2 \\[2mm]
&= \left(\tfrac{a_n}{2}\right)^2 + \left(1 - \sqrt{1 - \left(\tfrac{a_n}{2}\right)^2}\right)^2 \\[2mm]
&= \left(\tfrac{a_n}{2}\right)^2 + 1 - 2\sqrt{1 - \left(\tfrac{a_n}{2}\right)^2} + 1 - \left(\tfrac{a_n}{2}\right)^2 \\[2mm]
&= 2 - \sqrt{4 - a_n^2}. \qquad\qquad \square
\end{aligned}
$$

Substituting the appropriate values for n gives the following:

$$
\begin{aligned}
a_{12} &= \sqrt{2 - \sqrt{3}} \approx 0.5176381 \\[2mm]
a_{24} &= \sqrt{2 - \sqrt{2 + \sqrt{3}}} \approx 0.2610524 \\[2mm]
a_{48} &= \sqrt{2 - \sqrt{2 + \sqrt{2 + \sqrt{3}}}} \approx 0.1308063 \\[2mm]
a_{96} &= \sqrt{2 - \sqrt{2 + \sqrt{2 + \sqrt{2 + \sqrt{3}}}}} \approx 0.0654382.
\end{aligned}
$$

Thus,

$$
\pi \geq 48 a_{96} \approx 3.14103 \geq 3\tfrac{10}{71}.
$$

Let's consider the circumscribed polygons that produce the upper bound for π. Suppose b_n denotes the length of the side of a regular n-sided polygon circumscribed about a circle of radius 1. We have two possible plans of attack for the circumscribed polygons. The first is to use the fact that $b_6 = \dfrac{2}{\sqrt{3}}$, and then prove the following recurrence relationship.

Lemma 11.19. *If b_n denotes the length of the side of a regular n-gon circumscribed about a circle of radius 1, then*

$$b_{2n} = \frac{2\left(\sqrt{4 + b_n^2} - 2\right)}{b_n}.$$

An alternative choice is to establish the following relationship between a_n and b_n.

Lemma 11.20. *If a_n and b_n denote the lengths of the sides of regular n-gons inscribed in and circumscribed about a circle of radius 1, respectively, then*

$$b_n = \frac{2a_n}{\sqrt{4 - a_n^2}}.$$

A representative diagram for regular pentagons is shown in Figure 11.43. We leave both proofs to the reader, but note that $b_{96} \approx 0.0654732$. Hence,

$$\pi \leq 48b_{96} \approx 3.142715 \leq 3\frac{1}{7}.$$

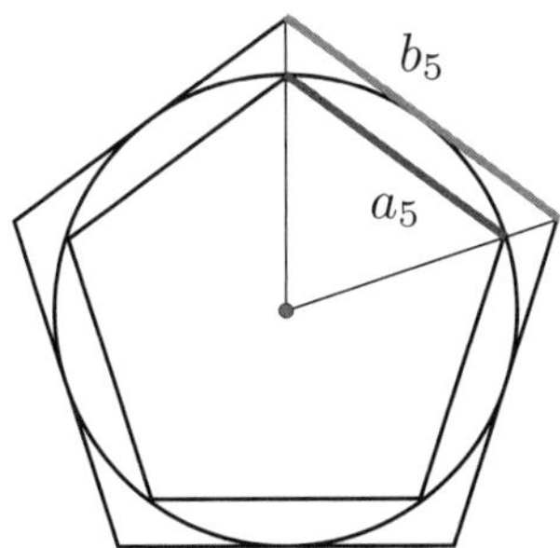

Figure 11.43. Sides of inscribed and circumscribed regular pentagon

Exercises 11.6

1. Prove Lemma 11.19.

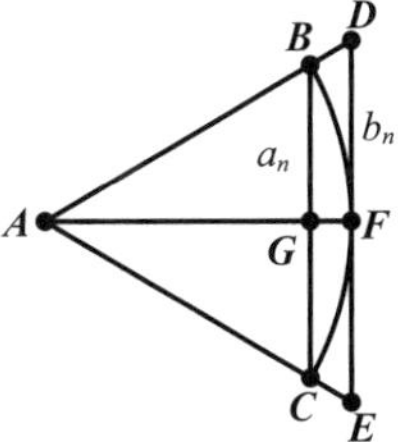

Figure 11.44. Exercise 11.6.2: Sides of inscribed and circumscribed regular *n*-gons

2. Using Figure 11.44 as a guide, prove Lemma 11.20. Note that the circle of radius 1 is centered at A, $AB = AC = 1, BC = a_n, DE = b_n$, and DE is tangent to the circle at F. Be sure to justify why $AG \perp BC$, $BG = \dfrac{a_n}{2}$ and $DF = \dfrac{b_n}{2}$.

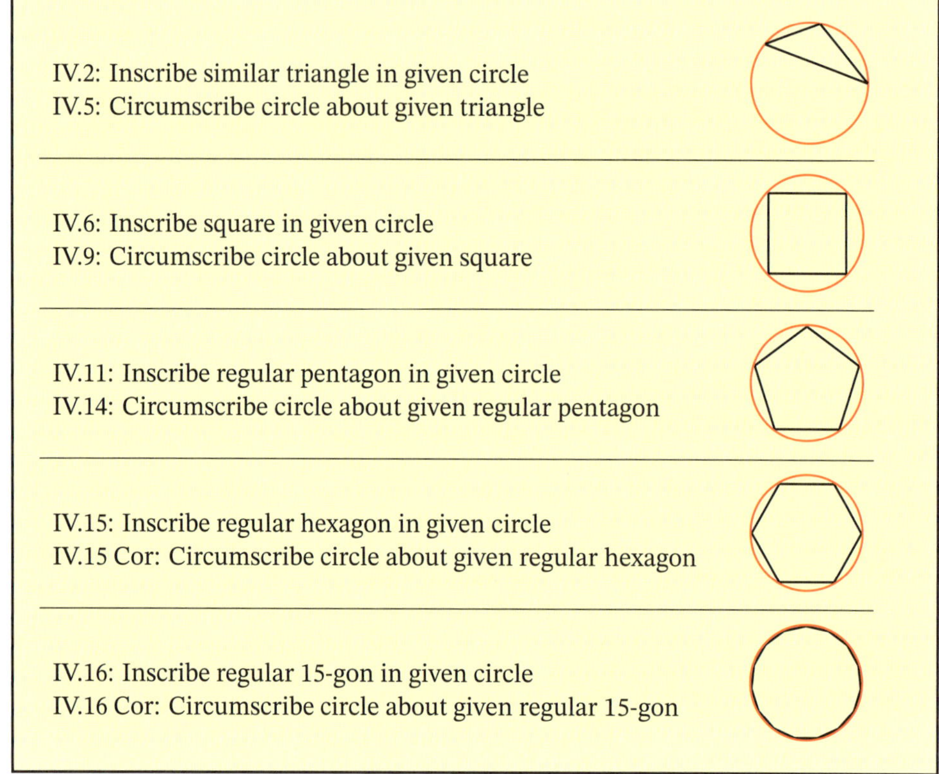

Figure 11.45. Inscribed polygon or circumscribed circle propositions in Book IV

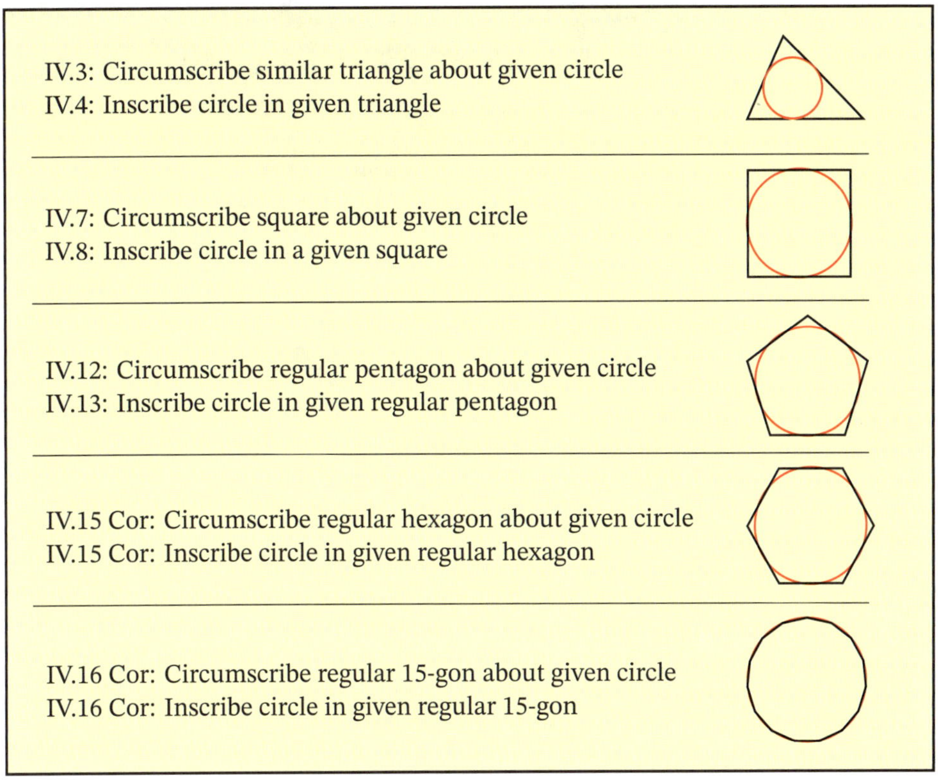

IV.3: Circumscribe similar triangle about given circle
IV.4: Inscribe circle in given triangle

IV.7: Circumscribe square about given circle
IV.8: Inscribe circle in a given square

IV.12: Circumscribe regular pentagon about given circle
IV.13: Inscribe circle in given regular pentagon

IV.15 Cor: Circumscribe regular hexagon about given circle
IV.15 Cor: Inscribe circle in given regular hexagon

IV.16 Cor: Circumscribe regular 15-gon about given circle
IV.16 Cor: Inscribe circle in given regular 15-gon

Figure 11.46. Circumscribed polygon or inscribed circle propositions in Book IV

12

Models for the Hyperbolic Plane

Figure 12.1. M.C. Escher's *Circle Limit III* (1959)

For over 2000 years, mathematicians were convinced that Euclid's fifth postulate, the Parallel Postulate, was actually a theorem. A simple review of the five postulates as given in Chapter 2 is sufficient to understand the source of this conjecture. Not only is Euclid's last postulate not immediately self-evident, its description requires more words than the other four postulates combined, and its complicated nature necessitates the assistance of a diagram. Consequently, there were many unsuccessful attempts at furnishing its proof. Along the way, geometers found a number of statements about parallel lines that are logically equivalent to the fifth postulate, under the assumption of Neutral geometry. In Theorem 7.1 of Chapter 7, we proved that the Parallel Postulate is equivalent to two of these well-known statements, Proclus' Axiom and Playfair's Axiom. The latter tends to be the parallel axiom most remembered by high-school geometry students, and we restate it here: *Through a point not on a given straight line, there exists at most one straight line that is parallel to the given line.* The eventual conclusion to this quest was the surprising realization that this postulate (or any of its equivalents) cannot be proven from the other axioms. Therefore, it can be replaced by a different, independent axiom. In this chapter, we will see what happens when we replace it with the *Characteristic Axiom*.

Characteristic Axiom. *Through a point not on a given straight line, there exist at least two straight lines that are parallel to the given line.*

By making this substitution, while keeping all of Hilbert's other axioms the same, we create a completely new geometry that is as valid and consistent as Euclidean geometry, but with more than one parallel, is clearly *not* Euclidean geometry. In fact, the geometry resulting from this axiom swap was the first of the new **non-Euclidean geometries** to be explored, and it came to be known as *Hyperbolic geometry*. In this chapter and the next, we delve into this strange new world.

12.1 Historical overview

12.1.1 Girolamo Saccheri. One of the first geometers to explore this new, non-Euclidean world was a Jesuit Professor at the University of Pavia in northern Italy, Girolamo Saccheri (1667–1733). In truth, he did not set out to explore a new geometry, but rather, he attempted to prove Euclid's fifth postulate in his 1733 book *Euclides ab omni naevo vindicatus (Euclid Freed of Every Fleck)*. Before discussing his findings, we should pause to examine the influence this book had on future geometers. Though Saccheri's book was mentioned in a doctoral thesis in the 1760s, it was only brought to the attention of the greater mathematical community 150 years after its publication when it was rediscovered in 1889 by fellow Jesuit, Angelo Manganotti. He relayed the book to Italian mathematician Eugenio Beltrami (1835–1900), who in the same year, published a paper detailing Saccheri's results [2]. Mathematicians were intrigued, and it wasn't long before Saccheri's book was translated into German, Italian and English, though it appears that the appreciation of Saccheri's contribution may have been further delayed by a cataloging error. In the preface to its English translation, American mathematician George Bruce Halsted (1853–1922) explains that Sir Thomas Heath, whose three-volume edition of *Euclid's Elements* is still considered the definitive English translation (and the one on which our text relies), had "listed Saccheri's *Euclides vindicatus* as a Latin edition of Euclid, calling Saccheri its *editor*, a mistake so gross it could never have been made by any one who had been privileged to see Saccheri's diadem." Though demoted from author to editor/translator by Heath, Saccheri clearly anticipates fundamental theorems of Hyperbolic geometry even though he is attempting to prove the Parallel Postulate.

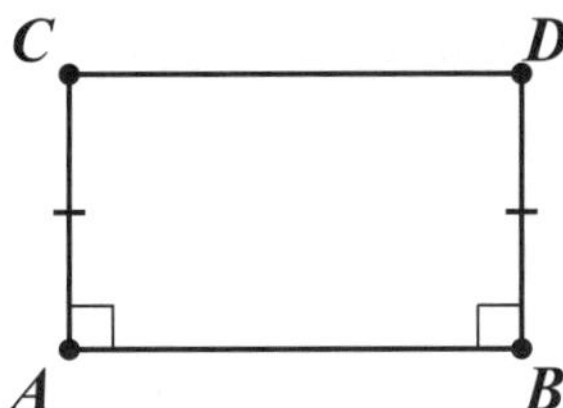

Figure 12.2. A Saccheri quadrilateral

In his book, Saccheri begins by considering a quadrilateral $ABDC$ where $AC \perp AB$, $BD \perp AB$ and $AC = BD$. Using Neutral geometry, Saccheri proves that the **summit angles**, those at C and D, must be congruent. To be successful in his attempt to prove the Parallel Postulate, he must show that these summit angles are right. He takes a double reductio ad absurdum approach, first assuming the summit angles to be obtuse, then acute. He calls the first **the hypothesis of the obtuse angle (HOA)**, the second **the hypothesis of the acute angle (HAA)**, and he sets out to find a contradiction in each case. By Proposition XIV of Saccheri's book, he proves that the hypothesis of the obtuse angle is false by finding a contradiction with Neutral geometry, specifically Euclid's Proposition I.17.[1] Saccheri states his finding in a dramatic fashion.

Proposition XIV. *The hypothesis of obtuse angle is absolutely false, because it destroys itself.*

Now he only needs to eliminate the acute case to prove Euclid's fifth postulate. This turns out to be the significantly more difficult case. After deriving strange results under *HAA* in eighteen more propositions spanning fifty-six pages, he comes to the following conclusion in Proposition XXXIII, though his "proof" is unconvincing as it finds no contradiction with Neutral geometry.

Proposition XXXIII. *The hypothesis of the acute angle is absolutely false; because repugnant to the nature of the straight line.*

Given our modern understanding of axiomatic systems and undefined terms, this proposition may make us smile. However, while Saccheri had the wrong conclusion, we will find that the tools he used and the results he derived will be helpful for us in the next chapter as we prove fundamental theorems in Hyperbolic geometry.

12.1.2 Gauss, Bolyai and Lobachevsky.

Though Saccheri anticipated fundamental results of Hyperbolic geometry, ultimately he is not credited with its discovery since he rejected its existence. Nearly one hundred years later, the three mathematicians who *are* credited with discovering the first non-Euclidean geometry were all working on the problem independently, and at roughly the same time in the early part

---

[1] HOA is valid in a different non-Euclidean geometry, Elliptic geometry, where, as we would expect, some Neutral geometry propositions, such as Euclid's Propositions I.16 and I.17, do not hold.

of the 1800s. These geometric explorers are Gauss, arguably the best mathematician of his day, and two relatively unknown younger mathematicians, Hungarian János Bolyai (1802–1860) and Russian Nikolai Ivanovich Lobachevsky (1793–1856). While Bolyai and Lobachevsky eagerly published their work, the more famous of the trio kept his strange new findings to himself.

Through notes and correspondence, we know that by 1813 Gauss had made little progress, writing, "In the theory of parallels we are now no further than Euclid was" [**58**]. However, by 1817, Gauss had become convinced that the fifth postulate was independent of the other four postulates. He began to work out the consequences of a geometry in which more than one line can be drawn through a given point parallel to a given line. In a letter dated November 8, 1824, he writes:

> The assumption that the sum of the three angles (of a triangle) is less than 180°
> leads to a special geometry, quite different from ours (Euclidean), which is ab-
> solutely consistent and which I have developed quite satisfactorily for myself....
> [**38**]

Realizing that his results were extremely controversial, and fearing "the clamor of the Boeotians," Gauss refused to publish anything on this subject during his lifetime[2] [**58**].

Independent of Gauss, János Bolyai also unearthed fundamental properties of this non-Euclidean geometry. János' father, Farkas (Wolfgang) Bolyai (1775–1856), and Gauss had become friends when they were students at Göttingen. In 1804, Farkas was appointed professor of mathematics and physics at a university in Romania. Once there, he picked up his own work that he had started on the Parallel Postulate in 1798 and continued, without success, to work on it over the next two decades. János followed in his father's footsteps, graduating from college in 1817, and entering the Academy of Military Engineers in Vienna in 1818. In 1820, the younger Bolyai began to think that perhaps a geometry without the Parallel Postulate was possible and wrote to his father about his plans to pursue this idea. Farkas tried to dissuade his son from an academic path that was, in his view, all-consuming and, in his experience, fruitless. János persisted with success, and his results appeared as an appendix to an 1832 work published by Farkas. We will introduce János's *angle of parallelism* in the next chapter.

The third mathematician to discover non-Euclidean geometry, Lobachevsky, was a professor at a Russian university, the University of Kasan. In 1829 and 1830, he published several articles in the Kasan *Messenger* describing an alternative to Euclid's geometry. Later he published a series of memoirs from 1835 to 1838. Unfortunately, he was not a compelling writer and his work was not well-reviewed. Around 1840, Gauss received a copy of Lobachevsky's work. In a letter from 1844, Gauss wrote that the memoirs were like "a confused forest through which it is difficult to find a passage and perspective, without having first gotten acquainted with all the trees individually" [**38**]. Criticism aside, Gauss ultimately praised Lobachevsky's work and later wrote that he derived "exquisite enjoyment" from reading it" [**38**]. As proof of this, Gauss recommended him for membership to the Göttingen Academy of Sciences, one of the few ways Lobachevsky was recognized for his achievement in his lifetime.

---

[2]To be thrown in with the Boeotians is akin to being called unintelligent and uncultured.

> **The time for Hyperbolic geometry had come**
>
> When the younger Bolyai wrote his father of his plans to pursue a geometry without the Parallel Postulate, his father replied in alarm:
>
> > You must not attempt this approach to parallels. I know this way to the very end. I have traversed this bottomless night, which extinguished all light and joy in my life. I entreat you, leave the science of the parallels alone.... [62]
>
> Undeterred, János continued his work, and as late as 1829, father and son continued to argue. After János found success, Farkas published his son's work, and János explained why his father thought it was important to publish as soon as possible.
>
> > He advised me that... certain things ripen at the same time and then appear in different places in the manner of violets coming to light in early spring. ... for here pre-eminence comes to him who is first. [62]
>
> Farkas was correct in predicting that his son's ideas could "ripen" elsewhere. When Farkas sent a copy of János' work to his friend, Gauss, he replied:
>
> > If I commenced by saying that I am unable to praise this work, you would certainly be surprised for a moment. But I cannot say otherwise. To praise it, would be to praise myself. Indeed the whole contents of the work, the path taken by your son, the results to which he is led, coincide almost entirely with my meditations, which have occupied my mind partly for the last thirty or thirty-five years. [48]
>
> While Farkas was happy to have Gauss validate his son's work, János was convinced that his father had shared his work without his permission, and that Gauss was claiming these ideas as his own. Father and son did not speak for many years, but reconciled before Farkas' death in 1856.

Though Saccheri published in 1733, Bolyai in 1832, and Lobachevsky as early as 1829, their work was largely ignored until 1868 when Beltrami revisited Saccheri's work. Two impediments to the proliferation of these ideas were the obscurity of some of the publications and the languages in which they were written. Furthermore, and most importantly, it took decades for the mathematical community to develop a general understanding of this strange new non-Euclidean world. The natural question to ask, then, is what does it take to understand a non-Euclidean plane?

As we discussed in Section 6.5, a model of a geometry is an interpretation of its primitive terms, or essentially, a way to visualize the geometry. Our model of Euclidean geometry is so natural that it seems innately human. As nineteenth-century mathematicians began to explore the strange new world of non-Euclidean geometry, they came to realize that Euclidean geometry only *seems* natural since it mimics our local experience. In other words, to the human eye the physical world appears to follow Euclidean geometry, and thus, the world within arms reach makes for a very convenient model for Euclidean geometry. The strange new world created by Gauss, Bolyai and Lobachevsky only existed on paper, and its acceptance and proliferation were severely hampered by the lack of a model, or a way to envision the elements of the geometry.

Throughout the nineteenth century, the hunt was on for a way to model this new geometry, and simultaneously, the genie was out of the bottle where the taboo of altering Euclidean axioms was concerned. By the turn of the century there was not only a model for this geometry, but a few more non-Euclidean geometries.

It would be decades before mathematicians and scientists understood the first applications of what initially appeared to be a purely intellectual pursuit. One of these geometries went on to become a critical component of Einstein's theory of general relativity. It is interesting to note that while we envision our local environment as Euclidean, in the greater universe, Einstein found space to be curved, thus non-Euclidean. This application illustrates the sentiment expressed by mathematician Henri Poincaré (1854–1912) in 1902: "One geometry cannot be more true than another; it can only be more convenient." It was in the latter part of the 1800s that Beltrami, Klein and Poincaré developed three models that gained traction within the mathematical community. We consider them in the following section.

12.2 Models of the hyperbolic plane

A map of the earth's surface is a two-dimensional representation of a three-dimensional space. The map is called a **projection** as it projects from 3-d to 2-d. There are many ways to create a projection, each with its own compromise, as all projections distort some of the properties of lines, angles, areas or shapes of objects. For example, one projection might preserve angles but distort distances, and another projection might preserve areas but distort shapes. One of the most common maps of the earth is the Mercator projection, as shown in Figure 12.3.

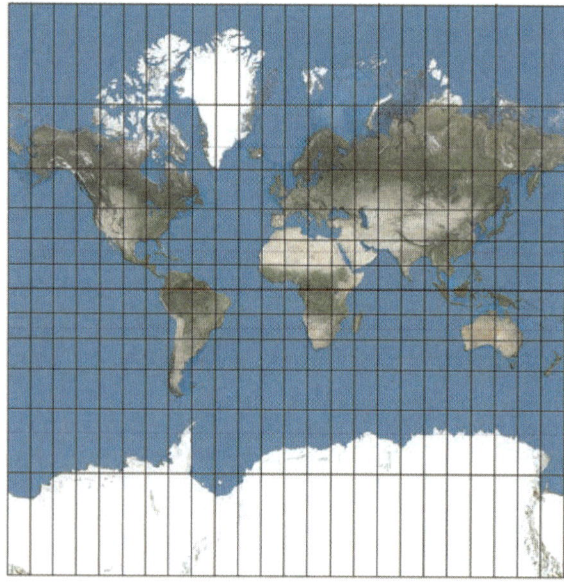

Figure 12.3. Mercator projection

This map greatly distorts both areas and distances. For example, if someone asked you which is larger, the United States or Greenland, upon glancing at Figure 12.3 your answer would be Greenland. However, the United States (3.72 million square miles) is actually more than four times larger than Greenland (0.8 million square miles). Worse yet, Antarctica (5.1 million square miles) is only 1.4 times larger than the United States, but on the map, it looks almost 10 times larger. Notice that the entire bottom boundary of the map represents just the South Pole! Since this standard map distorts areas so terribly, you would be correct to suspect that it must represent some other property quite accurately: It represents angle measure correctly. A sailor or pilot could draw a line between her current position and any distant point, and know the compass heading that will take her there. This was very important for navigational purposes for the Belgian, Gerardus Mercator (1512–1594), during a time of intense nautical exploration. Alternatively, the Mollweide projection, named after Karl Mollweide (1774–1825) and shown in Figure 12.4, sacrifices angle measure for accuracy in area. Take a moment to compare the relative sizes of the areas of the United States, Antarctica and Greenland in the Mollweide versus the Mercator projection.

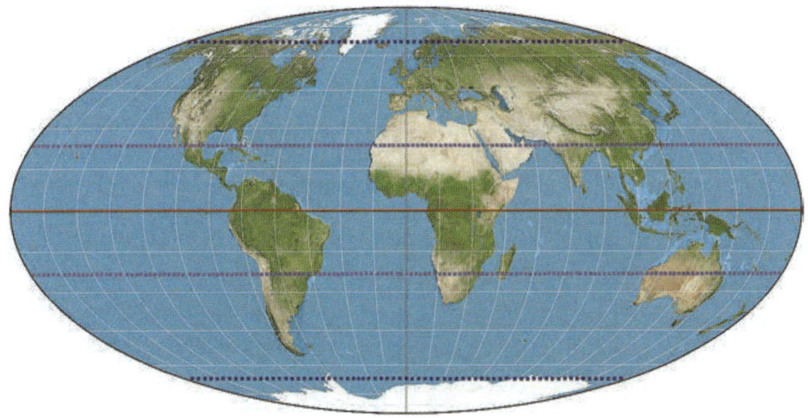

Figure 12.4. Mollweide projection

Just as Mercator's map helped sailors navigate their way around the globe, three mathematicians helped geometers negotiate the strange new world of non-Euclidean geometry by providing models. As there are many map projections of our globe, so there were many models of the new geometry. One may represent a particular feature of the geometry very well, but fail miserably in some other. We will choose which feature to compromise depending on our specific requirements. All of these models, however, give life to this new geometry as they help us build intuition for working in this strange new world.

Beltrami, and others, tried to construct a physical, rigid three-dimensional model for Hyperbolic geometry, but all attempts failed. It wasn't until 1901 that Hilbert proved that it is impossible to do so. (It is, however, possible to construct a model that has no rigid embedding in Euclidean space [**70**].) On the other hand, they *were* successful at providing two-dimensional models, though it should not be surprising that we encounter distortion problems when we try to represent the hyperbolic plane inside of the Euclidean plane. We will consider three such models for Hyperbolic geometry.

(1) Beltrami-Klein Disc model (1868, 1871)

(2) Poincaré Disc model (1882)

(3) Poincaré Half-plane model (1882)

Unlike our models for Euclidean geometry, which are unbounded, some models of hyperbolic space are bounded (e.g. Beltrami-Klein and Poincaré Disc models), while others (e.g. Poincaré Half-plane) are unbounded structures embedded in Euclidean space. Regardless of the model employed, a few of the amazing properties of Hyperbolic geometry that we prove in the next chapter are listed below.

- The angles in any triangle sum to less than 180°.

- Rectangles (and thus, squares) do not exist.

- Parallel lines are not everywhere equidistant.

- The summit angles of a Saccheri quadrilateral are acute.

Let's take a close look at each of these models.

12.2.1 The Beltrami-Klein model. Based on the work of Bernhard Riemann, Eugenio Beltrami created the first model of non-Euclidean geometry in 1868. In 1871, German mathematician Felix Klein (1849–1925) reinterpreted this model in the context of projective geometry which allowed him to explicitly calculate distances between points. Thus, this model is attributed to both mathematicians.

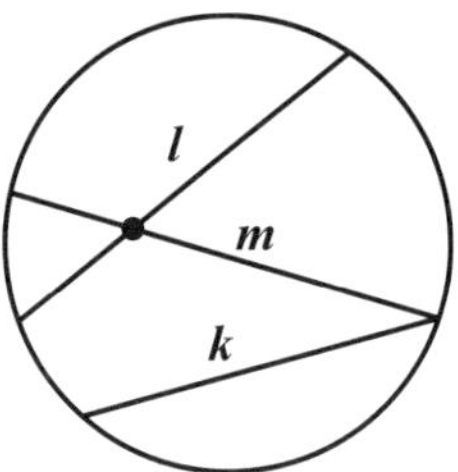

Figure 12.5. Lines in the Beltrami-Klein disc

In this model, the infinite hyperbolic plane is represented by a disc. The *points* in the model are points that lie *inside* the disc. Points that lie on the boundary circle are excluded from the model. The *lines* in the model are *open chords* in the circle (chords without their endpoints). The boundary of this finite structure plays an essential role in understanding parallel lines in hyperbolic space. There are two distinct types of parallel lines, and these definitions apply to all of our models for Hyperbolic geometry.

Definition 12.1. Two lines that intersect on the model's boundary, but have no common points within the model, are called **sensed parallel** (or asymptotically parallel).

Definition 12.2. Two lines that share no common points either within the model or on the boundary are called **ultraparallel** (or divergently parallel).

Definition 12.3. The points on the boundary, while not part of the model, are called **ideal points**.

In the example shown in Figure 12.5, lines l and m are not parallel since they intersect in the disc, lines k and l are ultraparallel, and lines k and m are sensed parallel. Is the Characteristic Axiom satisfied in this model? Through a point A not on line l, can you find at least two lines parallel to l? The example in Figure 12.6 demonstrates that it is particularly easy to illustrate the Characteristic Axiom in this model.

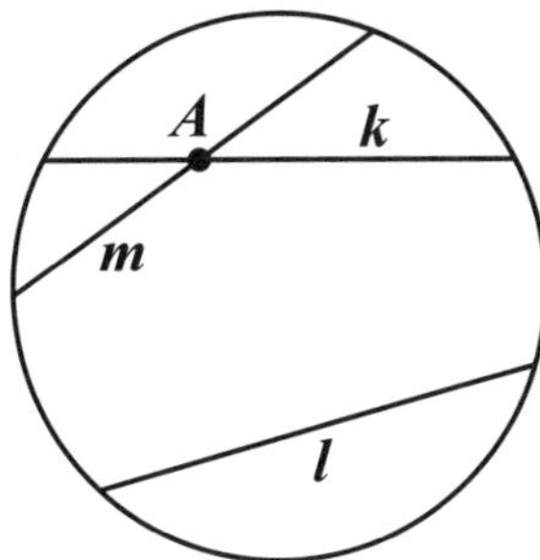

Figure 12.6. Characteristic Axiom for the Beltrami-Klein disc

Calculating lengths and angles, however, is difficult in this model because, as is the case in the Mercator map of the globe, distances are distorted. Two lengths that appear identical to an external viewer may actually be quite different due to the distortion concealed by the model. This is similar to what we see when we look at the Mercator projection of the earth where, to an external viewer, it may appear that Greenland is larger than the United States, but in actuality, it is smaller. In the Beltrami-Klein model, segments near the center of the disc are actually much shorter than segments located closer to the boundary that appear to have the same length. As as example, AB

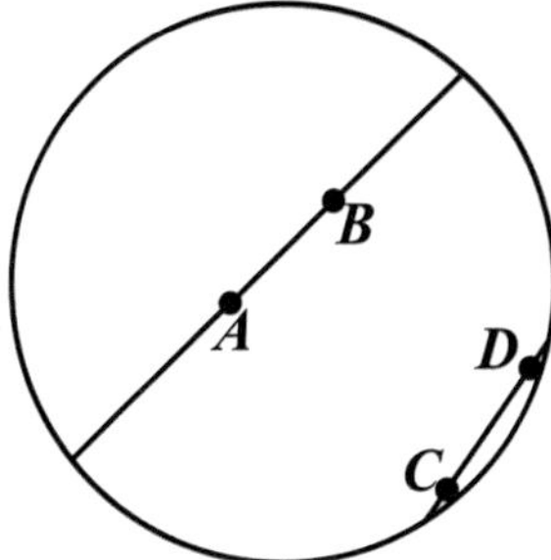

Figure 12.7. Distortion of distance: CD is much longer than AB

and CD in Figure 12.7 share the same Euclidean length (Check it with a ruler!); however, in the hyperbolic model, CD is much longer than AB. Using the same notation for distance between points introduced in Chapter 5, we will write $d_H(P, Q)$ to represent the hyperbolic distance between points P and Q. When we refer to $d_E(P, Q)$, we mean the distance that we see as an external Euclidean viewer of this model. Thus, in this example, we have $d_E(A, B) = d_E(C, D)$, but $d_H(A, B) < d_H(C, D)$. Keep in mind that dwellers who live within the model do not recognize any distortion since they have no preconceived notion that the length of the path from A to B should have any relation to the length of the path from C to D. It is important to note that every model for Hyperbolic geometry encounters this distortion near the boundary. We discuss the distortion at greater length within the context of the Poincaré Half-plane model in Section 12.2.3, and we calculate hyperbolic lengths in Section 12.3.5 with the help of calculus. In the Beltrami-Klein model, angle measure is also not as it appears. These compromises with both lengths and angles are too dear a price to pay, and thus, we will not spend more time working within this particular model.

Figure 12.8. M.C. Escher's *Circle Limit IV* (1960)

12.2.2 Poincaré Disc model. In 1882, Poincaré kept the disc and the points from the Beltrami-Klein model, but changed the representation of the lines. Once again, a *point* in the model is any point inside the disc, points that lie on the boundary circle are not in the model, and boundary points are ideal points. A *line* in the model is either an open diameter of the boundary circle (diameter without its endpoints) or an open arc of a circle that is *orthogonal* to the boundary circle of the disc. To fully understand this second type of line, we must define what it means for circles to be orthogonal.

Definition 12.4. Two intersecting circles are **orthogonal** if their radii at a point of intersection are perpendicular.

The two circles shown in Figure 12.9 are orthogonal, and thus, ℓ is a line in the model. Suppose r is the radius and C is the center of the dashed circle in this figure.

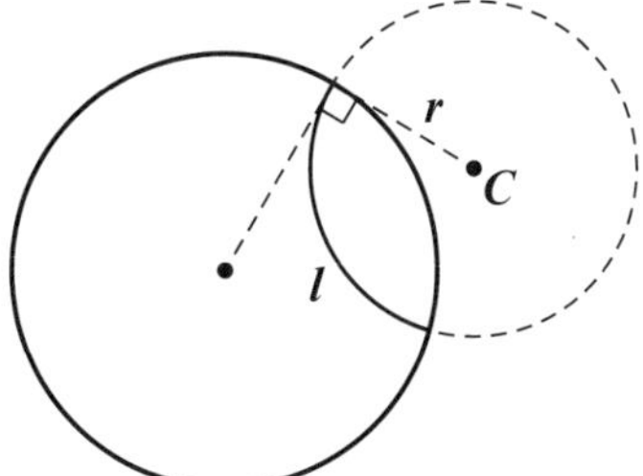

Figure 12.9. Line l: an arc of orthogonal circles in the Poincaré Disc model

With dynamic geometry software, we can drag C to change the size of the dashed circle while maintaining the orthogonality of the two circles. When doing this, the shape of ℓ changes as follows: ℓ approaches a semicircle as the length of r decreases, and ℓ approaches a diameter of the boundary circle as r increases. (Try it!) The definitions of ultraparallel and sensed parallel remain the same in this model. For example, in Figure 12.10, lines k and l are ultraparallel, k and m are sensed parallel, and l and m intersect. As shown in Figure 12.11, the Characteristic Axiom is easily illustrated in this model

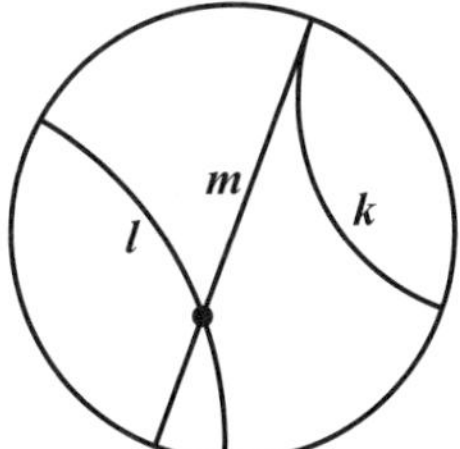

Figure 12.10. Lines in the Poincaré disc

by considering the number of possible lines through point A that are parallel to line l.

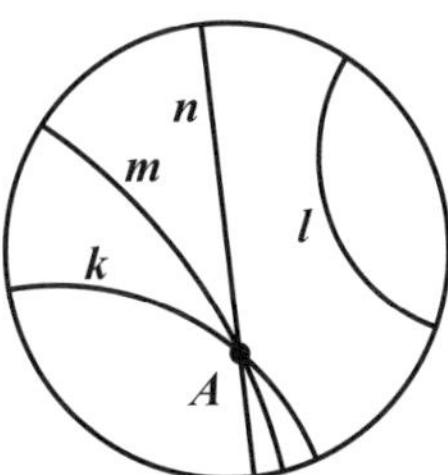

Figure 12.11. Characteristic Axiom for the Poincaré disc

While lengths are still fairly difficult to calculate, the advantage of this model over the Beltrami-Klein model is that angles are as they appear. That is, the angle between two hyperbolic lines is the Euclidean angle between their tangent lines at the point of intersection. (Note: As discussed in Section 10.2, nonrectilineal angles are calculated by constructing tangent lines to curves.) Since the angles here are measured as they are in Euclidean space, this model is called **conformal**. Triangle $\triangle ABC$ in Figure 12.12 was created using Geometer's Sketchpad®, and a measurement of the angles at the vertices gives the surprising and unexpected result that the angle sum for this triangle is roughly 112°.

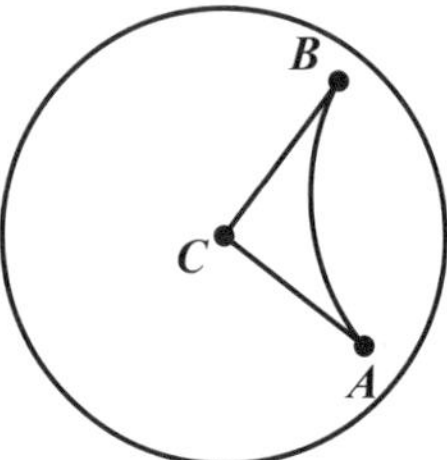

Figure 12.12. A triangle in the Poincaré disc

Distances, however, remain distorted in the model, with an arc (or segment) near the center of the disc being much shorter than an arc (or segment) of the same Euclidean length that is close to the boundary. This is the same type of distortion that we discussed in relation to Figure 12.7 in the Beltrami-Klein model. M.C. Escher (1898–1972) explores the concept of the extrinsic view of the Disc model in Figures 12.1 and 12.8. Consider the fish in Escher's *Circle Limit III*, or the angels and demons in his *Circle Limit IV*. The nearer the boundary, the smaller the creatures appear to the external viewer. To a fish living in this hyperbolic space created by Escher, every other fish in the space is exactly the same size as itself. The same is true of the angels or the demons. That is, all of the figures that appear similar are, in fact, congruent in Hyperbolic geometry.

Visualization software

The orthogonality requirement of this model makes for difficult construction of lines by hand, but software is an extremely helpful visual aid. Non-Euclid is a free interactive Java program that can be used for creating compass and unmarked straightedge constructions, and measuring angles and lengths in both the Poincaré Disc model as well as the Poincaré Half-plane model that we consider in the following section. It is available at

www.cs.unm.edu/~joel/NonEuclid/NonEuclid.html.

The online Advanced Sketch Gallery for Geometer's Sketchpad® has downloadable files for exploring both the Poincaré Disc model and the Poincaré Half-plane model that we consider in the following section. These resources can be found under the General Resources tab at *www.dynamicgeometry.com.*

12.2.3 Poincaré Half-plane model. In our final model, the Poincaré Half-plane model, the infinite hyperbolic plane is represented using the Euclidean upper half-plane with the x-axis boundary excluded. A *point* is any point of the Euclidean upper half-plane; that is, a point is an ordered pair (x, y) where x and y are real numbers with $y > 0$. The *lines* in the model are either vertical rays starting at the x-axis, or semicircles centered at a point on the x-axis. Hence, lines either have the equation $x = k$, where $y > 0$, or $y = \sqrt{r^2 - (x - c)^2}$, where $(c, 0)$ is the center, and r is the radius of the semicircle. In Figure 12.13, lines l and m intersect, lines k and l are ultraparallel,

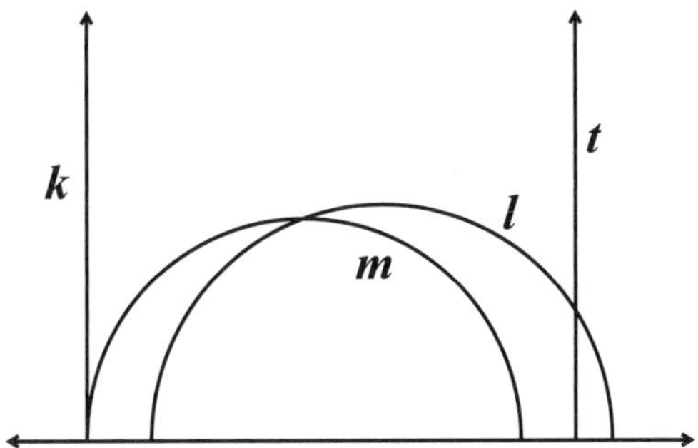

Figure 12.13. Lines in the Poincaré Half-plane

and lines k and m are sensed parallel. For reasons that will be made clear in Chapter 13, any two distinct vertical lines are sensed parallel. So, lines k and t in Figure 12.13 are sensed parallel. We leave it to the reader to illustrate the Characteristic Axiom in this model.

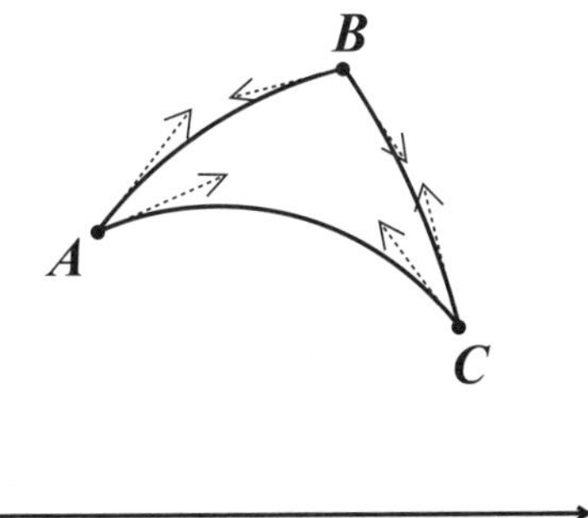

Figure 12.14. Triangle in the Poincaré Half-plane

As in the Poincaré Disc model, angles are as they appear, hence the model is conformal. To elaborate, the angle between two hyperbolic lines is the Euclidean angle between their tangent lines at the point of intersection. Triangle $\triangle ABC$ shown in Figure 12.14 was created using Geometer's Sketchpad®, and a measurement of the angles at the vertices (using the dashed tangent rays) produces an angle sum of approximately 163°. Notice that this is a different angle sum from the triangle illustrated in the Poincaré Disc model in Figure 12.12. In Hyperbolic geometry, unlike Euclidean geometry, triangles do not share the same angle sum.

Once again, distances are distorted to an external viewer. We will develop a specific formula for calculating the length of these paths in Section 12.3, but for now, let's concentrate on developing a general understanding of the difference between the intrinsic and extrinsic views. To help us, let's bring back the mechanical ant from Chapter 4 as a resident of Poincaré's Half-plane model. Just as the creatures in Escher's sketches appear to shrink as they get closer to the boundary of the disc, we can imagine, as an external viewer, that the size of the ant would appear to shrink as it moves closer to the boundary line at the x-axis. Specifically, consider an ant traveling along the vertical line $x = 1$ in the Half-plane model. First, since this is a line in the model, the ant will *feel* as if it is traveling on a straight line. The distance the ant travels from $(1, 2)$

to $(1, 1)$ is over seven times longer than the distance it travels from $(1, 11)$ to $(1, 10)$. The external viewer sees the ant continue to shrink as it travels towards the x-axis, its legs getting shorter and shorter, making it impossible for the ant to ever reach the unattainable boundary, no matter how far it travels. By visualizing an ant shrinking as it approaches the boundary, and growing as it travels farther away, an external viewer can make sense of the distortion within the model. It's similar to a person appearing smaller as they walk away from us. Keep in mind that, intrinsically, just like the fish, angels and demons in Figures 12.1 and 12.8, the ant living within the model does not recognize any distortion and does not actually change in size.

Even though distances are distorted to an external viewer, they are, fortunately, relatively easy to calculate. While it is fun to imagine observing creatures living within a hyperbolic plane, and while this may help us visualize a strange world, it will be reassuring to support these intuitive ideas with some actual calculations. In fact, we will put our considerable analytic geometry skills to use as we spend the remainder of this chapter investigating lines and distances in the Half-plane model. First, we show that there is a unique hyperbolic line joining any two points in our model; then we discuss how to find the equation of this line.

Theorem 12.5. *Between any two points, there exists a unique hyperbolic line.*

Proof. Suppose we are given points $A = (x_1, y_1)$ and $B = (x_2, y_2)$.

Case 1. If $x_1 = x_2 = k$, then no semicircle centered at the x-axis can pass through these points. Therefore, $x = k$ is the only hyperbolic line that passes through A and B.

Case 2. If $x_1 \neq x_2$, then the unique Euclidean line joining A and B is not vertical. Thus, we need to show that there is a unique Euclidean semicircle centered on the x-axis that passes through both A and B.

Any Euclidean circle passing through both A and B must have a center which lies on the perpendicular bisector of AB. Let m be the slope of the Euclidean line AB, and $m_\perp$ the slope of its perpendicular bisector. Since AB is not vertical, we know m is defined. If $m \neq 0$, then $m_\perp = -\dfrac{1}{m} \neq 0$. If $m = 0$, then the perpendicular bisector is a vertical line. In either case, the perpendicular bisector of AB meets the x-axis at a unique point $C = (c, 0)$. Thus, the Euclidean semicircle centered at C of radius AC is the unique hyperbolic line passing through A and B. $\square$

Now that we have seen the general strategy for finding the hyperbolic line joining two points, let's determine an algebraic formula for these lines in terms of the coordinates for a given $A = (x_1, y_1)$ and $B = (x_2, y_2)$.

Case 1. If $x_1 = x_2 = k$, then the hyperbolic line has equation $x = k$.

Case 2. If $x_1 \neq x_2$ and $y_1 \neq y_2$, then AB has slope $m = \dfrac{y_2 - y_1}{x_2 - x_1} \neq 0$. The slope of the perpendicular bisector of AB is $m_\perp = -\dfrac{1}{m} = -\left(\dfrac{x_2 - x_1}{y_2 - y_1}\right)$, and the midpoint of AB is $M = (x_M, y_M) = \left(\dfrac{x_1 + x_2}{2}, \dfrac{y_1 + y_2}{2}\right)$. Thus, the equation of the perpendicular bisector is

$$y - y_M = m_\perp(x - x_M).$$

The center, $(c, 0)$, occurs where this line meets the x-axis. Setting $y = 0$ and solving for c, we get:

$$
\begin{aligned}
0 - y_M &= -\frac{1}{m}(c - x_M) \\
-y_M &= -\frac{1}{m}c + \frac{1}{m}x_M \\
\frac{1}{m}c &= \frac{1}{m}x_M + y_M \\
c &= x_M + my_M.
\end{aligned}
\tag{12.1}
$$

Next, we need to find the radius, r, of the semicircle on which A and B reside by calculating a Euclidean distance. Specifically, $r = d_E(A, C) = d_E(B, C)$ where $C = (c, 0)$ is the center of the semicircle. This gives

$$
r = \sqrt{(x_i - c)^2 + (y_i - 0)^2} \text{ for } i = 1, 2.
\tag{12.2}
$$

Thus, the equation of the hyperbolic line between A and B is

$$
y = \sqrt{r^2 - (x - c)^2},
\tag{12.3}
$$

where c is given by Equation (12.1), and r by Equation (12.2).

Case 3. If $x_1 \neq x_2$, but $y_1 = y_2$, then the perpendicular bisector for AB is the vertical line $x = \dfrac{x_1 + x_2}{2}$. Therefore, the equation of the hyperbolic line between A and B is given by Equation (12.3), where $c = \dfrac{x_1 + x_2}{2}$ and

$$
r = \sqrt{\frac{(x_2 - x_1)^2}{4} + y_1^2}.
$$

Example 12.6. *(To illustrate Case 2)* Consider points $A = (1, 1)$ and $B = (2, 4)$. Then $m = 3$, and $(x_M, y_M) = \left(\dfrac{3}{2}, \dfrac{5}{2}\right)$. So, $c = x_M + my_M = \dfrac{3}{2} + 3 \cdot \dfrac{5}{2} = 9$, and $r = \sqrt{(x_1 - c)^2 + (y_1 - 0)^2} = \sqrt{(1 - 9)^2 + (1 - 0)^2} = \sqrt{65}$. Thus, the equation of the line joining A and B is given by $y = \sqrt{65 - (x - 9)^2}$.

Example 12.7. *(To illustrate Case 3)* Consider points $A = (1, 10)$ and $B = (11, 10)$. Then $c = 6$, and $r = \sqrt{125}$. Thus, the equation of the line joining A and B is given by $y = \sqrt{125 - (x - 6)^2}$.

While it is reassuring that Hilbert's first two Axioms of Incidence hold in the Half-plane model, a careful reader may wonder about the properties inherent to these two particular curves that allow them to play the role of a line in the Half-plane model. What does a vertical ray or a semicircle have that, say, a parabola or horizontal line in the upper half-plane does not? To get a better understanding of this, we need to learn about how arc lengths, and hence, distances, are calculated in this model.

> **The line and the circle in Hyperbolic geometry**
> We have discussed two bounded models and one unbounded model for Hyperbolic geometry. Though a circle marks the boundary in the Disc models, our main character is on the outside looking in since the boundary is not included in

> the space. On the other hand, lines appear familiar in the Beltrami-Klein model
> where they are represented by chords. The same is true of the diameter lines in
> the Poincaré Disc model, and the vertical lines in the Half-plane model. The clos-
> est the circle has come to playing a part in these models is in a minor capacity,
> really only a partial role, since arcs of circles are masquerading as lines in both
> Poincaré models. The line is front and center in the starring role, and though
> we have not laid eyes on a hyperbolic circle, we will have more to say about our
> missing character at the end of the chapter.

Exercises 12.2

1. Using geometric software, create a triangle in the Poincaré Disc model and measure
   its angle sum.

2. Given two points A and B in the Poincaré Disc model, explain how to construct the
   hyperbolic line joining them. There should be two cases: one where A and B lie on
   a common diameter and one where they do not.

   The following problems are for the Poincaré Half-plane model.

3. Illustrate the Characteristic Axiom in the Poincaré Half-plane model.

4. Find the equation of the hyperbolic line joining each pair of points.

   (a) $A = (1, 1)$ and $B = (3, 5)$
   (b) $A = (-3, 2)$ and $B = (3, 4)$
   (c) $A = (1, 1)$ and $B = (5, 1)$
   (d) $A = (1, 10)$ and $B = (5, 10)$

5. Consider the point $(-1, 7)$ and the hyperbolic line $y = \sqrt{16 - (x + 1)^2}$.

   (a) Give the equations of two hyperbolic lines passing through $(-1, 7)$ that do not
       intersect the hyperbolic line $y = \sqrt{16 - (x + 1)^2}$.
   (b) Graph all three hyperbolic lines.

6. Consider the two hyperbolic lines $y = \sqrt{36 - x^2}$ and $y = \sqrt{9 - (x - 3)^2}$.

   (a) Find their point of intersection in the Euclidean plane.
   (b) Are these hyperbolic lines parallel or do they intersect? Explain.
   (c) What are these two lines called?

7. Consider points $A = (x_1, y_1)$ and $B = (x_2, y_2)$, where $x_1 \neq x_2$. Let $C = (c, 0)$ be the
   center of the Euclidean semicircle that passes through A and B. Show that

$$c = \frac{(x_2^2 + y_2^2) - (x_1^2 + y_1^2)}{2(x_2 - x_1)}.$$

8. In Chapter 13, we will prove that any pair of ultraparallel lines has a unique common
   perpendicular. Find this unique common perpendicular for the pair of hyperbolic
   lines $y = \sqrt{4 - x^2}$ and $x = 5$. Be sure to graph the hyperbolic lines as a first step.
   [*Hint: The common perpendicular cannot be a Euclidean horizontal line since this is
   not a hyperbolic line. Recall that two intersecting circles are orthogonal if their radii
   are perpendicular at a point of intersection.*]

12.3 Arc length & distance in the half-plane model

12.3.1 Arc length in the Euclidean plane.
As a warm-up exercise for calculating arc length and distances in the Half-plane model, we first revisit how to calculate arc length in the analytic model of the Euclidean plane. Recall from Chapter 5, the Euclidean distance between points $A = (x_1, y_1)$ and $B = (x_2, y_2)$ is given by the familiar formula:

$$d_E(A, B) = \sqrt{(x_1 - x_2)^2 + (y_1 - y_2)^2}.$$

This is the same as the length of the unique line segment joining these points. When a path between two points is not necessarily a straight line, then we need to use calculus in order to find the length of the path. Recall, in the Euclidean plane, the **arc length** of a curve given parametrically as $x = f(t)$ and $y = g(t)$, where $a \leq t \leq b$, is given by the following integral:

$$s = \int_a^b \sqrt{\left(\frac{dx}{dt}\right)^2 + \left(\frac{dy}{dt}\right)^2}\, dt. \tag{12.4}$$

As an example, let's consider three different paths from point $A = (1, 1)$ to point $B = (2, 4)$, and use the corresponding integrals to calculate the length of each.

Example 12.8. The linear path, shown as the dashed line in Figure 12.15, is given parametrically by $x = 1 + t$ and $y = 1 + 3t$ for $0 \leq t \leq 1$. Using Equation (12.4) to determine the length of the path, we have

$$s = \int_0^1 \sqrt{(1)^2 + (3)^2}\, dt = t\sqrt{10}\,\Big|_0^1 = \sqrt{10} \approx 3.162.$$

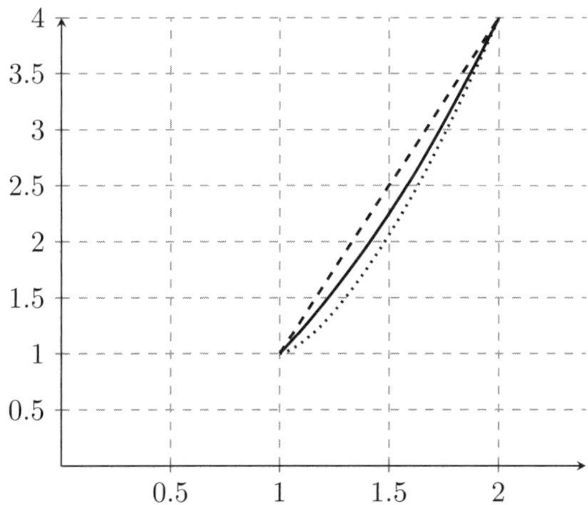

Figure 12.15. Three paths from $(1, 1)$ to $(2, 4)$

Example 12.9. The lower curved path, shown as the dotted curve in Figure 12.15, is given parametrically by $x = 1 + t$ and $y = 3t^{\frac{3}{2}} + 1$ for $0 \le t \le 1$. Using Equation (12.4) to determine the length of the path, we have

$$
\begin{aligned}
s = \int_0^1 \sqrt{(1)^2 + \left(3 \cdot \frac{3}{2}t^{\frac{1}{2}}\right)^2}\, dt \;\; &= \;\; \int_0^1 \sqrt{\left(1 + \frac{81}{4}t\right)}\, dt \\[2mm]
&= \;\; \frac{4}{81} \cdot \frac{2}{3}\left(1 + \frac{81}{4}t\right)^{\frac{3}{2}}\Bigg|_0^1 \\[2mm]
&= \;\; \frac{-8 + 85\sqrt{85}}{243} \approx 3.192.
\end{aligned}
$$

Example 12.10. The middle curved path, shown as the solid black curve in Figure 12.15, is given parametrically by $x = t$ and $y = t^2$ for $1 \le t \le 2$. Using Equation (12.4), trigonometric substitution and integration by parts to determine the length of the path, we have

$$
\begin{aligned}
s = \int_1^2 \sqrt{(1)^2 + (2t)^2}\, dt \;\; &= \;\; \frac{1}{2}\int_{t=1}^{t=2} \sec^3\theta\, d\theta \\[2mm]
&= \;\; \frac{1}{2}\sec\theta\tan\theta\Bigg|_{t=1}^{t=2} - \frac{1}{2}\int_{t=1}^{t=2}\sec\theta\tan^2\theta\, d\theta \\[2mm]
&= \;\; \frac{1}{2}t\sqrt{1 + 4t^2} + \frac{1}{4}\ln\left(2t + \sqrt{1 + 4t^2}\right)\Bigg|_1^2 \\[2mm]
&\approx \;\; 3.168.
\end{aligned}
$$

Not surprisingly, of the three paths, the Euclidean straight line given in Example 12.8 is the shortest path.

12.3.2 Arc length in the Poincaré Half-plane.

To calculate the length of a curve in the Half-plane model, we will modify our integral to give the hyperbolic length. The arc length of a curve defined parametrically by $x = f(t)$ and $y = g(t)$, where $a \le t \le b$, is given by the following integral:

$$
s_H = \int_a^b \frac{1}{y}\sqrt{\left(\frac{dx}{dt}\right)^2 + \left(\frac{dy}{dt}\right)^2}\, dt. \tag{12.5}
$$

This formula is very similar to the Euclidean arc length formula given in Equation (12.4), with the only difference between these integral formulas the division by y in the integrand for the Hyperbolic arc length. This produces the desired effect of increasing the length of the path near the x-axis boundary of our model, and decreasing the length of the path when far from the boundary. That is, when y is small (close to the x-axis), division by y produces a larger integrand than when y is large, and thus, farther from the boundary. [Exercise 12.3.1 offers the reader a good way to observe the effect of $\frac{1}{y}$ in the integrand.] As an example, let's reconsider the first two parameterizations of paths from $A = (1, 1)$ to $B = (2, 4)$ given in Examples 12.8 and 12.9.

Example 12.11. For the path given parametrically by $x = 1 + t$ and $y = 1 + 3t$ for $0 \leq t \leq 1$, using Equation (12.5) to determine the length of the path in the hyperbolic model, we have

$$
\begin{aligned}
s_H &= \int_0^1 \frac{1}{(1 + 3t)} \sqrt{(1)^2 + (3)^2}\, dt \\
&= \frac{\sqrt{10}}{3} \ln(1 + 3t)\Big|_0^1 = \frac{\sqrt{10}}{3} \ln 4 \approx 1.4613.
\end{aligned}
$$

Example 12.12. For the path given parametrically by $x = 1 + t$ and $y = 3t^{\frac{3}{2}} + 1$ for $0 \leq t \leq 1$, using Equation (12.5) to determine the length of the path in the hyperbolic model, we have

$$
\begin{aligned}
s_H &= \int_0^1 \frac{1}{3t^{\frac{3}{2}} + 1} \sqrt{(1)^2 + \left(3 \cdot \frac{3}{2} t^{\frac{1}{2}}\right)^2}\, dt \\
&= \int_0^1 \frac{1}{3t^{\frac{3}{2}} + 1} \sqrt{1 + \frac{81}{4} t}\, dt \\
&\approx 1.5114.
\end{aligned}
$$

What is the shortest possible path between these two points in the hyperbolic plane? How do we determine it? Here, calculus will help us. As discussed in Chapters 3, 4 and 5, the distance between two points is given by the length of the shortest path joining them. In Taxicab geometry, the Euclidean line segment joining two points is the shortest path, but there are other paths that share this length. Hence, the shortest path is not unique. In Spherical, there is a unique shortest path (the minor arc) between two nonantipodal points, but for antipodal points, once again, the shortest path is not unique. In Euclidean geometry, the line segment joining two points *is* the unique shortest path between the points, and thus, the distance between two points is the length of this line segment. The story in Hyperbolic geometry is no different than in Euclidean: there is a unique shortest path joining any two points. Moreover, we call this unique path a hyperbolic line segment, and its length gives the distance between its endpoints. For brevity, the unique shortest path joining two points in either Euclidean or Hyperbolic geometry is called the **geodesic** between the two points. Using the arc length formula for the Poincaré Half-plane model, we will show that the geodesics in this model are either segments of vertical rays or arcs of semicircles centered on the x-axis, and thereby, support the claim that vertical rays and semicircles *are* the *hyperbolic lines* in this model. We will also show that a parabolic path, or a horizontal path, or any other possible path, is not a geodesic, thus demonstrating that these other curves are *not* lines in the model.

We recall that, given two points in the Half-plane model, there is a unique vertical segment or arc of a semicircle centered on the x-axis that joins them. Our strategy will be to derive a general formula for the length of a vertical segment or an arc of a semicircle, then show that these particular paths are the shortest. In the course of doing so, we will produce the formula for distance between any two points in the Half-plane model.

12.3.3 Length of hyperbolic segments. In order to derive a function for the length of the hyperbolic segment joining points $A = (x_1, y_1)$ and $B = (x_2, y_2)$, we consider two cases.

Case 1. Assume $x_1 = x_2$.

In this case, the hyperbolic segment joining A and B is a vertical line segment. We may use the hyperbolic arc length formula given in Equation (12.5) once we have a parameterization of this vertical line segment. If we assume that $y_2 > y_1$, then one possible parameterization is given by $x(t) = x_1$ and $y(t) = t$, where $y_1 \le t \le y_2$. The resulting integral for the arc length is given by

$$s_H = \int_{y_1}^{y_2} \frac{1}{t}\sqrt{(0)^2 + (1)^2}\, dt = \ln|t|\Big|_{y_1}^{y_2} = \ln|y_2| - \ln|y_1| = \ln\left|\frac{y_2}{y_1}\right|.$$

Since every y value is positive in the upper half-plane, we are free to remove the absolute values. Finally, since $\ln\frac{p}{q} = -\ln\frac{q}{p}$, we can remove the assumption that $y_2 > y_1$, and allow for traveling the path in either direction (from A to B, or vice versa), as long as we take the absolute value of the natural logarithm. This produces the formula for the length of the hyperbolic segment joining A and B in the case where A and B share an x-coordinate:

$$s_H = \left|\ln\left(\frac{y_2}{y_1}\right)\right|.$$

Case 2. Assume $x_1 \ne x_2$. WLOG, assume $x_1 > x_2$.

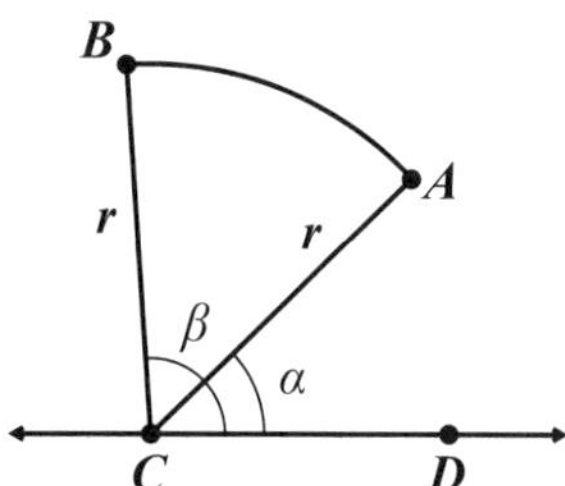

Figure 12.16. Parameterization for hyperbolic line segment

In this case, the hyperbolic segment joining A and B is a circular arc. Consider the unique circle with center $C = (c, 0)$, and radius r, that passes through A and B. To parameterize the circular arc joining A and B, we use a modified version of polar coordinates, centered at $(c, 0)$, rather than the origin. We leave it to the reader to verify that a parameterization of this circle is given by $x = c + r\cos t$ and $y = r\sin t$. We must specify limits for the angles, t, which fully sweep out the arc from A to B. To describe the circular arc joining A and B, let D be a point on the x-axis to the right of C, as shown in Figure 12.16. Define α as angle $\angle ACD$, and β as angle $\angle BCD$. Since $x_1 > x_2$, we know $\alpha < \beta$. Therefore, the arc from A to B is swept out by angles, t, where $\alpha \le t \le \beta$.

The resulting integral for the arc length is given by

$$
\begin{aligned}
s_H &= \int_{\alpha}^{\beta} \frac{1}{r \sin t} \sqrt{(-r \sin t)^2 + (r \cos t)^2}\, dt \\
&= \int_{\alpha}^{\beta} \frac{1}{r \sin t} \sqrt{r^2(\sin^2 t + \cos^2 t)}\, dt \\
&= \int_{\alpha}^{\beta} \frac{1}{r \sin t} \sqrt{r^2}\, dt \\
&= \int_{\alpha}^{\beta} \csc t\, dt \\
&= \ln|\csc t - \cot t|\Big|_{\alpha}^{\beta} \\
&= \ln\left|\frac{\csc \beta - \cot \beta}{\csc \alpha - \cot \alpha}\right|.
\end{aligned}
$$

To rewrite $\csc \alpha$ and $\cot \alpha$ in terms of x_1, y_1, r and c, consider Figure 12.17 which illustrates the case when $0 < \alpha < 90°$. For this triangle, we have

$$
\csc \alpha = \frac{r}{y_1} \quad \text{and} \quad \cot \alpha = \frac{x_1 - c}{y_1}.
$$

We leave it to the reader to verify that the same formulas hold for the alternative cases, where $90° < \alpha < 180°$ or $\alpha = 90°$. By a similar argument, the formulas for angle β are

$$
\csc \beta = \frac{r}{y_2} \quad \text{and} \quad \cot \beta = \frac{x_2 - c}{y_2}.
$$

Continuing with our work above, we have

$$
\ln\left|\frac{\csc \beta - \cot \beta}{\csc \alpha - \cot \alpha}\right| = \ln\left|\frac{\frac{r}{y_2} - \frac{x_2 - c}{y_2}}{\frac{r}{y_1} - \frac{x_1 - c}{y_1}}\right| = \ln\left|\frac{\frac{r - x_2 + c}{y_2}}{\frac{r - x_1 + c}{y_1}}\right| = \ln\left|\frac{y_1(c + r - x_2)}{y_2(c + r - x_1)}\right|.
$$

To see why the absolute value is unnecessary, we note that every x value on the open semicircle in the upper half-plane lies strictly between $c - r$ and $c + r$. That is, $c - r < x < c + r$. Rearranging gives $c + r - x > 0$. Combining this with the fact that every y value in the upper half-plane is positive, we have

$$
\frac{y_1(c + r - x_2)}{y_2(c + r - x_1)} > 0.
$$

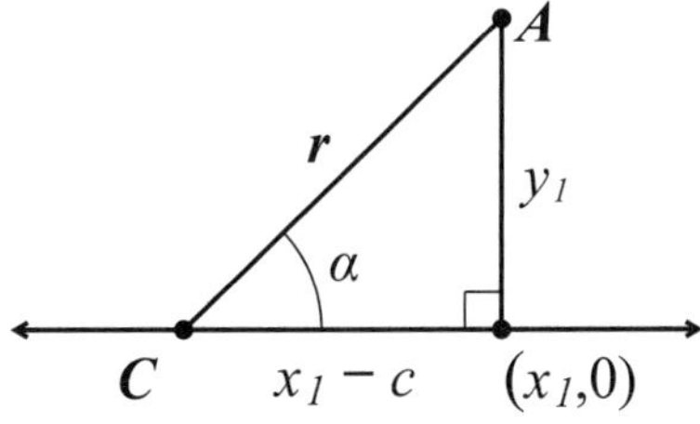

Figure 12.17. Representative triangle for $0 < \alpha < 90°$

Finally, since $\ln \frac{p}{q} = -\ln \frac{q}{p}$, we can remove the assumption that $x_1 > x_2$, and allow for travelling the path in either direction (from A to B, or vice versa), as long as we take the absolute value of the natural logarithm. This produces the formula for the length of the hyperbolic segment joining A and B in the case where A and B do not share an x-coordinate:

$$s_H = \left| \ln \left(\frac{y_1(c + r - x_2)}{y_2(c + r - x_1)} \right) \right| = \left| \ln \left(\frac{y_2(c + r - x_1)}{y_1(c + r - x_2)} \right) \right|.$$

To summarize, there are two cases to consider when calculating the length of the segment joining two points $A = (x_1, y_1)$ and $B = (x_2, y_2)$ in the Poincaré Half-plane.

Case 1. If $x_1 = x_2$, then the length of the hyperbolic segment between A and B is given by

$$\left| \ln \left(\frac{y_2}{y_1} \right) \right|. \tag{12.6}$$

Case 2. If $x_1 \neq x_2$, then the length of the hyperbolic segment between A and B is given by

$$\left| \ln \left(\frac{y_2(c + r - x_1)}{y_1(c + r - x_2)} \right) \right|, \tag{12.7}$$

where $(c, 0)$ is the center of the semicircle passing through A and B and r is its radius.

Our next step is to prove that any other path from A to B will be longer than the path along the vertical ray or circular arc, thus showing that hyperbolic line segments in the form of vertical rays or arcs of circles are geodesics in the Half-plane model.

12.3.4 Hyperbolic segments are geodesics.

Theorem 12.13. *In the Poincaré Half-plane, the hyperbolic segment joining two points is the unique geodesic between them.*

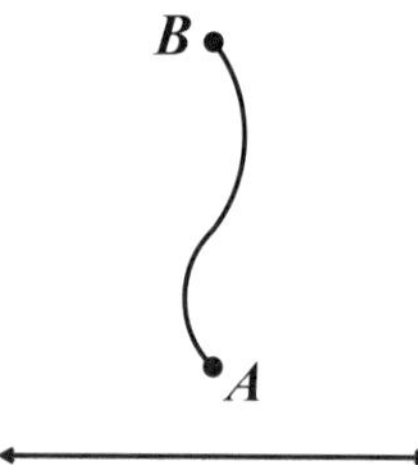

Figure 12.18. Possible path from A to B when $x_1 = x_2$

Proof. Let $A = (x_1, y_1)$ and $B = (x_2, y_2)$ be two points in the upper half-plane.

Case 1. Assume that $x_1 = x_2$ and, WLOG, let $y_2 > y_1$.

Any path from A to B can be parameterized by $x = x(t)$ and $y = t$, where $y_1 \leq t \leq y_2$. Clearly, $(x'(t))^2 \geq 0$ and $y'(t) = 1$. Therefore

$$\left(\frac{dx}{dt}\right)^2 + \left(\frac{dy}{dt}\right)^2 = \left(\frac{dx}{dt}\right)^2 + 1 \geq 0 + 1.$$

Thus,

$$s_H = \int_{y_1}^{y_2} \frac{1}{y}\sqrt{\left(\frac{dx}{dt}\right)^2 + \left(\frac{dy}{dt}\right)^2}\, dt \geq \int_{y_1}^{y_2} \frac{1}{t}\sqrt{1}\, dt = \ln\left(\frac{y_2}{y_1}\right).$$

Since $y_2 > y_1$, this is equivalent to $\left|\ln\left(\frac{y_2}{y_1}\right)\right|$. Note that $x'(t) = 0$ if and only if $x(t) = x_1$, a constant function. Hence, the vertical line segment is the only path that produces the lower bound, $\left|\ln\left(\frac{y_2}{y_1}\right)\right|$, for s_H. Therefore, the hyperbolic line segment *is* the shortest path between A and B, and any other path is longer.

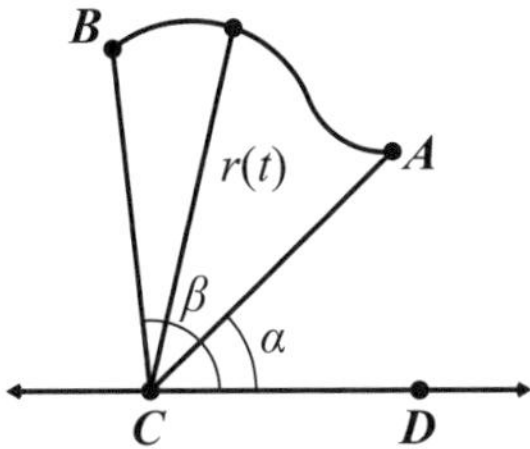

Figure 12.19. Possible path from A to B when $x_1 \neq x_2$

Case 2. Assume that $x_1 \neq x_2$ and, WLOG, let $x_1 > x_2$. Consider the unique circle with center $C = (c, 0)$ and radius r that passes through A and B. Let D be a point on the x-axis to the right of C. Define α as angle $\angle ACD$, and β as angle $\angle BCD$. Once again, $\alpha < \beta$ since $x_1 > x_2$, and a path from A to B can be parameterized by $x = c + r(t)\cos(t)$ and $y = r(t)\sin(t)$, where $\alpha \leq t \leq \beta$. Here, we assume that the radius, $r(t)$, is a function of t. We leave it to the reader to show that

$$\left(\frac{dx}{dt}\right)^2 + \left(\frac{dy}{dt}\right)^2 = (r'(t))^2 + (r(t))^2.$$

Thus,

$$
\begin{aligned}
s_H &= \int_{y_1}^{y_2} \frac{1}{y} \sqrt{\left(\frac{dx}{dt}\right)^2 + \left(\frac{dy}{dt}\right)^2}\, dt \\[2mm]
&= \int_{\alpha}^{\beta} \frac{1}{r(t)\sin t} \sqrt{(r'(t))^2 + (r(t))^2}\, dt \\[2mm]
&\geq \int_{\alpha}^{\beta} \frac{1}{r(t)\sin t} \sqrt{0 + (r(t))^2}\, dt \quad \text{since } (r'(t))^2 \geq 0 \\[2mm]
&= \int_{\alpha}^{\beta} \frac{1}{r(t)\sin t} |r(t)|\, dt \\[2mm]
&= \int_{\alpha}^{\beta} \frac{1}{r(t)\sin t} r(t)\, dt \quad \text{since radius } r(t) \geq 0 \\[2mm]
&= \int_{\alpha}^{\beta} \frac{1}{\sin t}\, dt \\[2mm]
&= \ln\left| \frac{\csc \beta - \cot \beta}{\csc \alpha - \cot \alpha} \right|.
\end{aligned}
$$

We recognize this last function as the length of the hyperbolic segment joining A and B. Once again, notice that $r'(t) = 0$ if and only if r is a constant. Hence, the semicircle is the only path that produces the lower bound, $\ln\left|\frac{\csc \beta - \cot \beta}{\csc \alpha - \cot \alpha}\right|$, for s_H. Therefore, the hyperbolic line segment *is* the shortest path between A and B, and any other path is longer. $\qquad\qquad\qquad\qquad\qquad\qquad\qquad\qquad\qquad\qquad\qquad\qquad\qquad\qquad\square$

We have done it! We have shown that the geodesics in the Half-plane model are either segments of vertical rays or arcs of semicircles centered on the x-axis. In doing so, we have also calculated the length of these hyperbolic segments, and thus, we now have a formula for calculating the distance between any two points in our model.

12.3.5 Hyperbolic distance. Let's summarize our findings. The formula for calculating the distance between points $A = (x_1, y_1)$ and $B = (x_2, y_2)$ in the Poincaré Half-plane model is given by the following two equations, depending upon whether the points are aligned vertically.

Case 1. If $x_1 = x_2$, then the hyperbolic distance between A and B is given by

$$
d_H(A, B) = \left| \ln\left(\frac{y_2}{y_1}\right) \right|. \tag{12.8}
$$

Case 2. If $x_1 \neq x_2$, then the hyperbolic distance between A and B is given by

$$
d_H(A, B) = \left| \ln\left(\frac{y_2(c + r - x_1)}{y_1(c + r - x_2)}\right) \right|, \tag{12.9}
$$

where $(c, 0)$ is the center of the semicircle passing through A and B, and r is its radius. (Note that c may be calculated with Equation (12.1), and r with Equation (12.2). Furthermore, while we acknowledge that c and r have no meaning in the case where $x_1 = x_2$, Equation (12.9) *does* simplify algebraically to Equation (12.8) when $x_1 = x_2$.)

Example 12.14. Using Equation (12.8), the hyperbolic distance between points $P = (0, 2)$ and $Q = (0, 20)$ is $d_H(P, Q) = \ln 10 \approx 2.3026$.

Example 12.15. To calculate the distance between $A = (1, 1)$ and $B = (2, 4)$, recall from Example 12.6 that $c = 9$ and $r = \sqrt{65}$. Applying the formula in Equation (12.9), we have

$$d_H(A, B) = \left| \ln \left(\frac{4(9 + \sqrt{65} - 1)}{1(9 + \sqrt{65} - 2)} \right) \right| = \ln \left(\frac{4(8 + \sqrt{65})}{7 + \sqrt{65}} \right) \approx 1.4506.$$

Notice that in Example 12.15, the distance between points A and B is less than the lengths of the two alternative paths between these points as calculated in Examples 12.11 and 12.12, which were 1.4613 and 1.5114, respectively. This makes sense since the shortest path between A and B in the hyperbolic plane *should* be along a hyperbolic line.

With a model to visualize the basic objects of Hyperbolic geometry, we are ready to take an axiomatic approach to the development of this geometry in the next chapter.

Where's the circle?

Though we have worked with "circles" in all three models presented in this chapter, they were not **hyperbolic circles**. The boundary disc, while circular, is *not* a figure within our geometry, and the arc of a circle is merely a line. So, what does our neglected main character, the circle, look like in the Poincaré Half-plane model? Every Euclidean circle that can be drawn entirely in the upper half-plane is also a hyperbolic circle. Moreover, if a circle has Euclidean center (h, k), and Euclidean radius r, then it has hyperbolic center $(h, \sqrt{k^2 - r^2})$, and hyperbolic radius R, where $R = \frac{1}{2} \ln \frac{k+r}{k-r}$. Conversely, every hyperbolic circle is also a Euclidean circle. (For a proof, see [**110**].) Thus, circles appear the same, but their centers and radii change. The center of a hyperbolic circle in this model lies closer to the x-axis than its Euclidean counterpart, (h, k), since its y-coordinate, $\sqrt{k^2 - r^2}$, is less than k. For example, the diagram shown here is a hyperbolic circle of radius $r = \ln 3$ with center $A = (5, 3)$. Note that $AB = AC = AD$ since each is a radius of the circle.

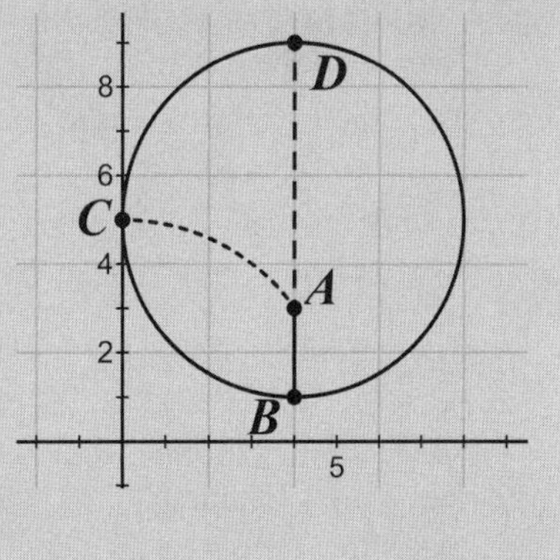

Exercises 12.3

The following problems are for the Poincaré Half-plane model.

1. Calculate the hyperbolic distance between each pair of points on the vertical ray $x = 1$.

   (a) $(1, 1)$ and $(1, 2)$

   (b) $(1, 10)$ and $(1, 11)$

(c) $(1, 10)$ and $(1, 20)$

(d) $(1, 0.1)$ and $(1, 1.1)$

2. Calculate the hyperbolic distance between each pair of points. (Exercise 12.2.4 will help.)

(a) $(1, 1)$ and $(3, 5)$

(b) $(-3, 2)$ and $(3, 4)$

(c) $(1, 1)$ and $(5, 1)$

(d) $(1, 10)$ and $(5, 10)$

3.(a) Give an example of points A, B, C and D where $d_H(A, B) = d_H(C, D)$, but $d_E(A, B) \neq d_E(C, D)$.

(b) Give an example of points A, B, C and D where $d_E(A, B) = d_E(C, D)$, but $d_H(A, B) \neq d_H(C, D)$.

(c) Give an example of points A, B, C and D where $d_E(A, B) = d_E(C, D)$, and $d_H(A, B) = d_H(C, D)$.

4. Compute the hyperbolic arc length of the following curves joining points $A = (0, 1)$ and $B = (3, 2)$. For part (a), evaluate the integral directly, then estimate the final answer with a calculator. For parts (b) and (c), use Maple or WolframAlpha.com to estimate the arc length.

(a) $x = 3t$ and $y = t + 1$ for $0 \leq t \leq 1$

(b) $x = 3t$ and $y = t^2 + 1$ for $0 \leq t \leq 1$

(c) $x = 3t$ and $y = t^3 + 1$ for $0 \leq t \leq 1$

(d) Find the center, $C = (c, 0)$, and radius, r, of the semicircle through A and B. Then, give the equation of the semicircle.

(e) Use c and r from part (d) to compute the hyperbolic arc length of the semicircular path from A to B.

(f) How do your answers for parts (a), (b) and (c) compare to your answer for part (e)? In particular, which curve is the shortest? Which is the longest?

5. Compute the hyperbolic arc length of the following curves joining points $P = (0, 1)$ and $Q = (1, 3)$. In each case, determine the parameterization and set up the appropriate integral. For part (a), evaluate the integral directly, then estimate the final answer with a calculator. For parts (b) and (c), use Maple or WolframAlpha.com to estimate the arc length.

(a) $y = 2x + 1$

(b) $y = 2x^2 + 1$

(c) $y = 2\sqrt{x} + 1$

(d) Find the center, $C = (c, 0)$, and radius, r, of the semicircle through P and Q. Then, give the equation of the semicircle.

(e) Use c and r from part (d) to compute the hyperbolic arc length of the semicircular path from P to Q.

(f) How do your answers for parts (a), (b) and (c) compare to your answer for part (e)? In particular, which curve is the shortest? Which is the longest?

6. Show that $x = c + r\cos t$ and $y = r\sin t$ is a parameterization of the circle with center $(c, 0)$ and radius r.

7. Let $A = (x_1, y_1)$ and $B = (x_2, y_2)$ where $x_1 > x_2$. In the parameterization of the hyperbolic line segment in the case of a circular arc from A to B, as shown in Figure 12.16, show that equations

$$\csc \alpha = \frac{r}{y_1} \quad \text{and} \quad \cot \alpha = \frac{x_1 - c}{y_1}$$

hold in the each of the following alternative cases.

(a) $90° < \alpha < 180°$

(b) $\alpha = 90°$ [*Hint: In this case, $(x_1, y_1) = (c, r)$.*]

8. Given the parameterization $x = c + r(t)\cos(t)$ and $y = r(t)\sin(t)$, show that

$$\left(\frac{dx}{dt}\right)^2 + \left(\frac{dy}{dt}\right)^2 = (r'(t))^2 + (r(t))^2.$$

13

Axiomatic Hyperbolic Geometry

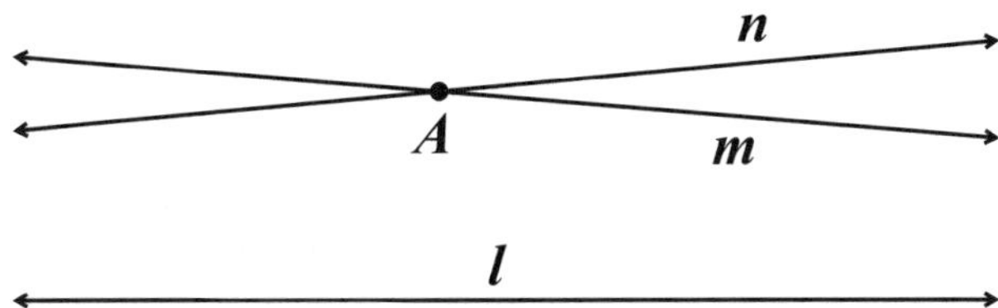

Figure 13.1. The Characteristic Axiom

Now that we have explored Hyperbolic geometry through the lens of several models, we step back and prove some amazing results about this strange new world using an axiomatic approach. We begin by assuming Hilbert's Axioms of Incidence (I.1–I.4), Axioms of Order (II.1–II.4), and Axioms of Congruence (III.1–III.5). We replace his Axiom of Parallels (commonly known as Playfair's Axiom) with the Characteristic Axiom.

Characteristic Axiom. *Through a point not on a given straight line, there exist at least two straight lines that are parallel to the given line.*

We will also replace his Axioms of Continuity (V.1: Archimedes' Axiom and V.2: Axiom of Line Completeness) with a logical equivalent, Dedekind's Axiom.

Dedekind's Axiom. *For every partition of the points on a line into two nonempty sets such that no point of either lies between two points of the other, there is a point in one set that lies between every other point of that set and every point in the other set.*

For a basic example that utilizes the real number line, let $A = (-\infty, 1]$ and $B = (1, \infty)$. These two nonempty sets form a partition of the real number line, and the number 1 satisfies Dedekind's Axiom since it is an element of A, and it lies between any element

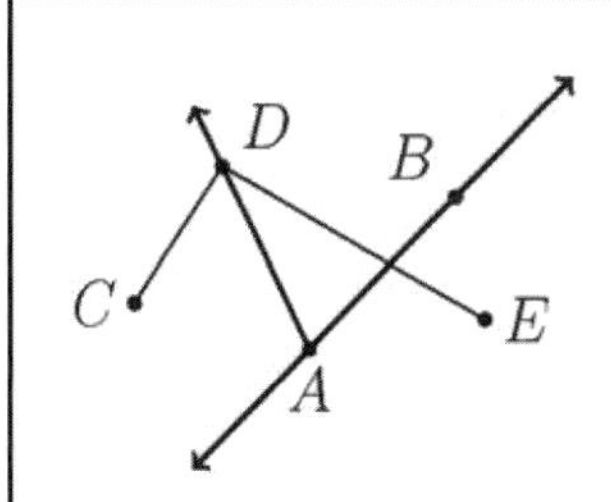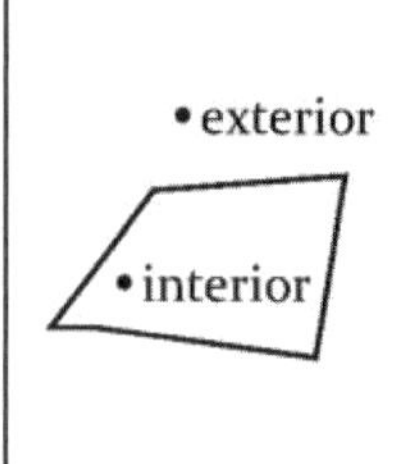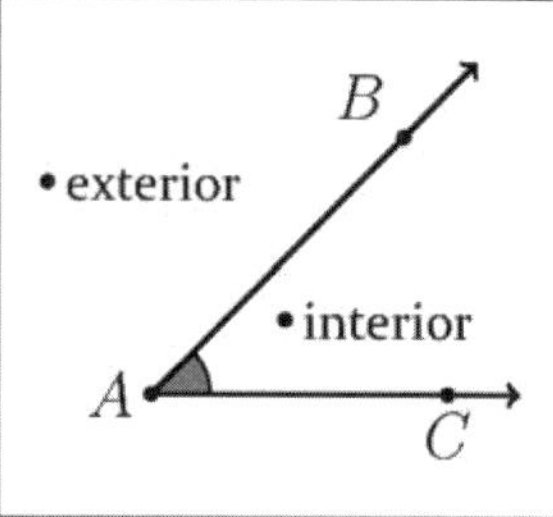

Figure 13.2. Hilbert's Plane Separation Theorems

of B and any element of A besides itself. Using this new set of axioms, some of the unusual theorems we will prove include the following:

- Parallel lines are not everywhere equidistant.

- The angles in any triangle sum to less than two right angles.

- Rectangles (and thus, squares) do not exist.

- AAA is a congruence scheme for triangles.

- The summit angles of a Saccheri quadrilateral are acute.

Finally, we develop the area function in Hyperbolic geometry and show how it relates to the area function in Spherical geometry as briefly discussed in Chapter 4.

Any of Euclid's propositions belonging to Neutral geometry will still hold in Hyperbolic geometry, and thus, we are free to use them. We may also use any results that Hilbert's Axioms (minus Playfair's Axiom) allow us to prove. Consequently, it would be helpful to review a few highlights from Section 6.3, in particular, Hilbert's separation theorems. His *Plane Separation by Lines* (Hilbert's Theorem 8 of Section 6.3) proves that any line separates the points of the plane that are not on the line into two regions, where a segment joining two points in the same region will not intersect the line, but one joining points in opposite regions will. As demonstrated in the first box of Figure 13.2, since C and D are in the same region as determined by line $\overleftrightarrow{AB}$, CD does not intersect the line, but DE does since E is not in the same region as D. This implies that all points, other than A, that lie on the ray $\overrightarrow{AD}$ lie on one side of $\overleftrightarrow{AB}$. By his *Plane Separation by Polygons* (Hilbert's Theorem 9 of Section 6.3), every polygon separates the points of the plane that are not on the polygon into two regions, the *interior* and the *exterior*. Just as in Euclidean geometry, angle $\angle BAC$ is formed by rays $\overrightarrow{AB}$ and $\overrightarrow{AC}$ at vertex A. It is also the case that any acute angle $\angle BAC$ divides the points of the plane that are not on rays $\overrightarrow{AB}$ and $\overrightarrow{AC}$ into two regions, the **interior** and **exterior**. These regions, as determined by either a polygon or an angle, are demonstrated in Figure 13.2. It is surprising that, unlike Euclidean geometry, it is possible for an entire line to lie in the interior of an angle. The reader is asked to demonstrate this in the Poincaré Half-plane model in the exercises. Lastly, we will need both Pasch's Axiom and the Crossbar Theorem (Theorem 6.12). We restate them here for easy reference.

Pasch's Axiom. *Let A, B and C be three distinct points that do not lie on a line, and let ℓ be a line that does not meet any of the points A, B or C. If line ℓ passes through a point of segment AB, then it also passes through a point of segment AC or a point of segment BC.*

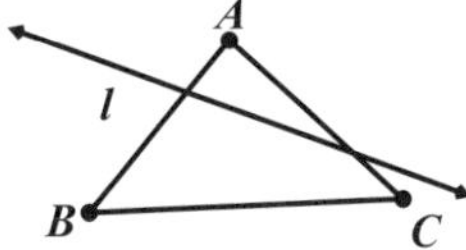

Figure 13.3. Pasch's Axiom

Crossbar Theorem (Theorem 6.12). *If D is in the interior of angle $\angle BAC$, then ray $\overrightarrow{AD}$ intersects segment BC.*

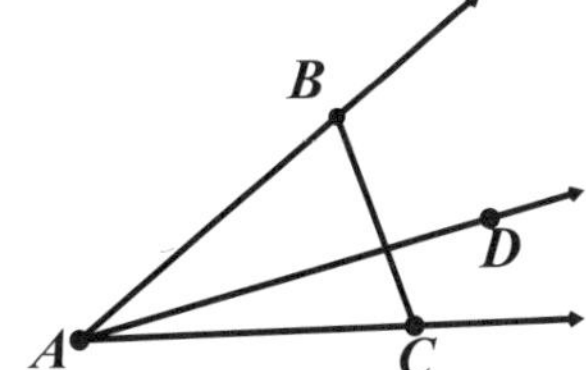

Figure 13.4. Crossbar Theorem

We start with an exploration of parallel lines.

Exercises 13.0
1. Give an example in the Poincaré Half-plane model where a line lies entirely within the interior of an angle.

13.1 Parallel lines

One of the surprising properties of Hyperbolic geometry is that lines can be parallel to each other in two different ways. In each model explored in Chapter 12, there were parallel lines that met on the boundary (called **sensed parallel**) and those that never met (called **ultraparallel**). To see why these two types of parallel lines exist regardless of the model, we start by proving that the assumption of at least two parallel lines opens the metaphorical floodgates to the existence of infinitely many such lines.

Theorem 13.1. *Through a point not on a given straight line, there are infinitely many straight lines parallel to the given line.*

Proof. Suppose that we given a line, ℓ, and a point, A, not on it. By Euclid I.12, we can construct a line through A that is perpendicular to ℓ. Let B be the intersection of this line with ℓ. By Euclid I.11, we can construct a line, m, through A that is perpendicular to $\overleftrightarrow{AB}$. Let C be another point on m. Using Hilbert's Axiom II.2, pick D on m such that A is between C and D. By Euclid I.28, we have $m \parallel \ell$. By the Characteristic Axiom, there is another line, n, through A, that is also parallel to ℓ. Consider triangle $\triangle BCD$ as illustrated in Figure 13.5. By Pasch's Axiom, since n intersects segment CD at A, it must also intersect either BC or BD. WLOG, assume n intersects BC at a point E.

Claim: If F is any point on the segment joining C and E, then line $\overleftrightarrow{AF}$ does not intersect ℓ. Why? To obtain a contradiction, suppose that it does intersect ℓ at a point G. Consider triangle $\triangle BFG$. Since n intersects BF at E, by Pasch's Axiom it must also

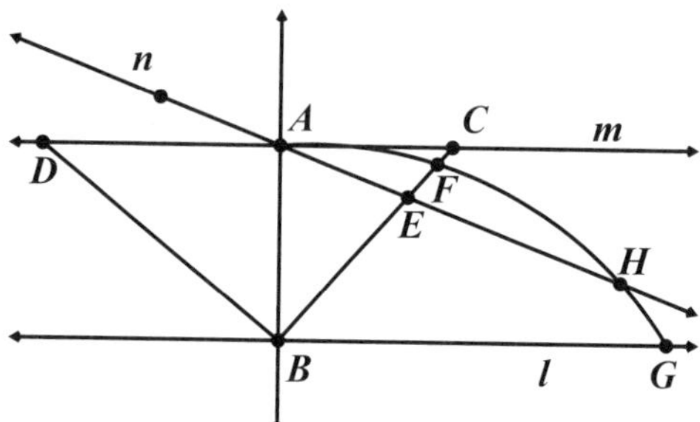

Figure 13.5. An infinite number of parallel lines

intersect either BG or FG. It cannot intersect BG as $\overleftrightarrow{BG} = \ell$, and $n \parallel \ell$. If we assume that n intersects FG at a point H, then n intersects $\overleftrightarrow{FG}$ at two distinct points, namely, A and H, which contradicts Hilbert I.2 as there can be at most one line containing any two given points. Thus $\overleftrightarrow{AF} \parallel \ell$.

Since there are infinitely many points on the segment joining C and E, we have infinitely many lines through A that are parallel to ℓ.[1]   □

In the proof of Theorem 13.1, we gain valuable insight into parallel lines. Once again, consider a line, ℓ, and a point, A, not on it. As in the proof, let's use Euclid I.12 to construct a line through A that is perpendicular to ℓ, and let B be the intersection of this line with ℓ. Again, by Euclid I.11, construct a line, m, through A that is perpendicular to $\overleftrightarrow{AB}$ and let C be another point on m. The line $\overleftrightarrow{AB}$ divides the plane into two regions which we will arbitrarily call the **right side** and the **left side**. Moreover, if we assume that C lies in the *right side*, then all points on the open ray $\overrightarrow{BC}$ will also be in that side, as illustrated in Figure 13.6. (Why? This follows from Plane Separation by Lines.)

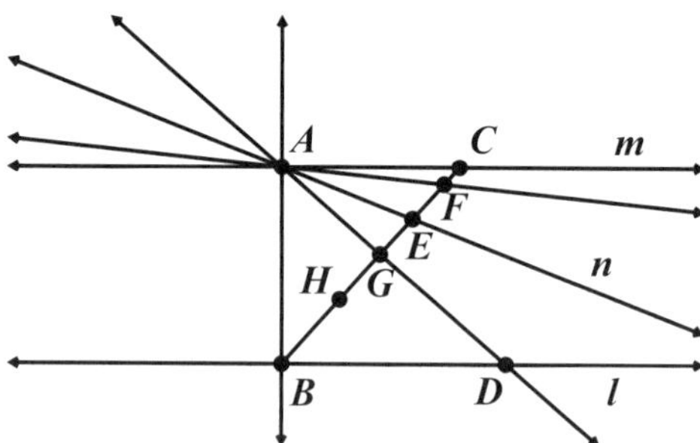

Figure 13.6. Partitioning BC

Consider the following sets that form a partition of the points on segment BC:

$$\mathcal{P} = \{F \mid F \text{ lies on } BC \text{ and } \overleftrightarrow{AF} \parallel \ell\}$$

and

$$\mathcal{J} = \{H \mid H \text{ lies on } BC \text{ and } \overleftrightarrow{AH} \nparallel \ell\}.$$

---

[1]We convince ourselves that a segment AB has infinitely many points by executing a sequence of bisections. First, bisecting segment AB produces point B_1 in AB. Next, bisecting AB_1 produces point B_2 in AB_1. Continuing to bisect in this manner produces an infinite sequence of points in AB.

Since $C \in \mathcal{P}$ and $B \in \mathcal{J}$, both sets are nonempty. Suppose that $G \in \mathcal{J}$. We leave it as an exercise for the reader to show that if H lies on segment BG, then $\overleftrightarrow{AH}$ must intersect ℓ. Combining this discussion with Theorem 13.1, if E is any point on BC such that $\overleftrightarrow{AE} \parallel \ell$, then for any point, F, between E and C, $\overleftrightarrow{AF} \parallel \ell$ and, if G is any point on BC such that $\overleftrightarrow{AG} \nparallel \ell$, then for any point, H, between B and G, $\overleftrightarrow{AH} \nparallel \ell$. Thus, no point of $\mathcal{P}$ can lie between any two points of $\mathcal{J}$, or vice versa. Thus, by Dedekind's Axiom, there is a point in one of these two sets that lies between every other point of that set and every point in the other set. In Figure 13.7, this point is labelled J. It acts as a "separating point" of segment BC, where, every point above J in the figure has a line through A parallel to ℓ, and every point below J has a line through A that intersects ℓ. We claim that the line joining A and J is parallel to ℓ, and thus, the separating point is an element of $\mathcal{P}$. Let's show why this is true.

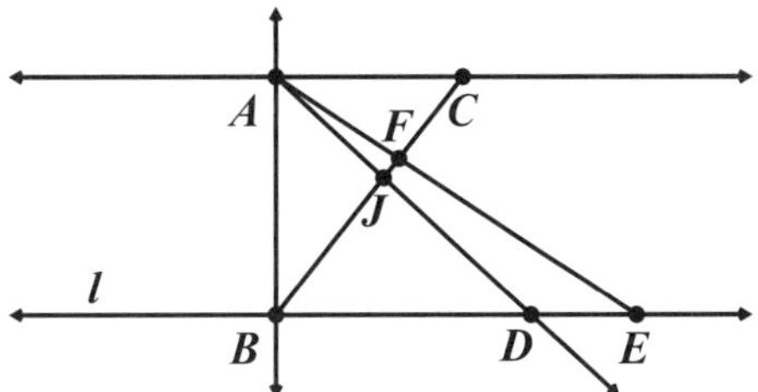

Figure 13.7. A separating point, J, on BC

Claim: $J \in \mathcal{P}$, that is, $\overleftrightarrow{AJ} \parallel \ell$. Why? To obtain a contradiction, we will assume that $\overleftrightarrow{AJ}$ intersects ℓ at a point D as demonstrated in Figure 13.7. Pick E on ℓ such that D lies between B and E. Since $\angle BAD < \angle BAE < \angle BAC$, ray $\overrightarrow{AE}$ will lie in the interior of angle $\angle JAC$. Thus, by the Crossbar Theorem, $\overrightarrow{AE}$ intersects JC at a point, F, on BC. Note that F lies between J and C, and AF intersects ℓ at E. This is a contradiction since every point between J and C has a line through A that is parallel to ℓ. $\qquad\square$

Now that we have established J as the "separating point" on BC, consider the angle $\angle BAJ$. Since $\angle BAC = 90°$, we have $\angle BAJ \leq 90°$. Given another point G on BC, if $\angle BAG > \angle BAJ$, $\overleftrightarrow{BG} \parallel \ell$, but if $\angle BAG < \angle BAJ$, $\overleftrightarrow{BG} \nparallel \ell$. Thus, as the angle $\angle BAG$ decreases, we can think of $\overleftrightarrow{AJ}$ as the "last" line on the right side of $\overleftrightarrow{AB}$ that goes through A and is parallel to ℓ. We define $\overleftrightarrow{AJ}$ as the **right-sensed parallel** to ℓ through A, and call $\angle BAJ$ the **angle of parallelism to ℓ through A on the right**. Clearly, we can reproduce this construction on the left side of $\overleftrightarrow{AB}$ to obtain a **left-sensed parallel** to ℓ through A, and the **angle of parallelism to ℓ through A on the left**. From the proof of Theorem 13.1, we know that at least one of these two angles of parallelism must be acute. Any other line through A that is parallel to ℓ will be called **ultraparallel** to ℓ through A.

Figure 13.8 illustrates these ideas using the Half-plane Model. We started with a point A not on a line ℓ, and then constructed the line $\overleftrightarrow{AB}$ perpendicular to ℓ. Line m is the right-sensed parallel to ℓ through A, with angle of parallelism on the right calculated by Geometer's Sketchpad® as $\angle BAC = 45°$. Line n is the left-sensed parallel to ℓ through A, with angle of parallelism on the left, $\angle BAD$, also equal to $45°$. The reader will use dynamic geometry software to find angles of parallelism in the exercises. The

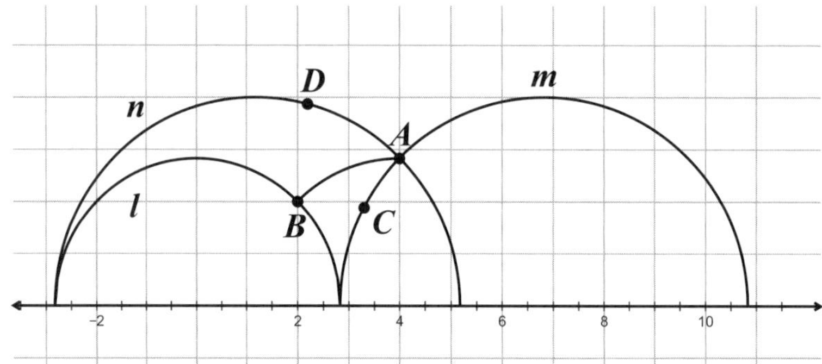

Figure 13.8. Angles of parallelism in the Half-plane model

equality of angles of parallelism on the right and left occurs in all cases, regardless of the point, line, or model, as the following theorem shows.

Theorem 13.2. *Given a point A not on a line ℓ, the angle of parallelism to ℓ through A on the left equals the angle of parallelism to ℓ through A on the right.*

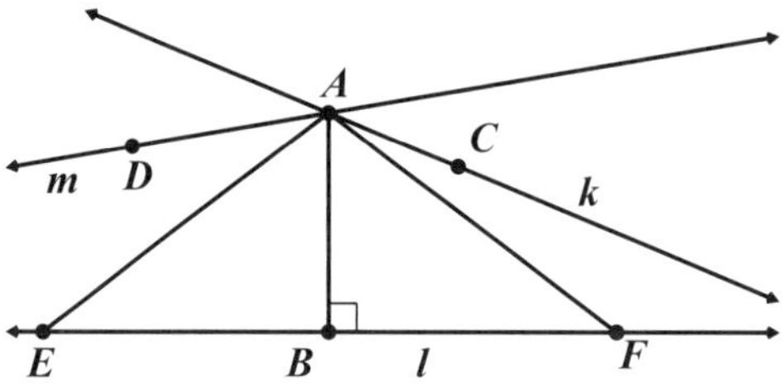

Figure 13.9. Left and right angles of parallelism to ℓ through A are equal

Proof.   Given a point A not on a line ℓ, once again, construct the perpendicular from A to ℓ, and let the intersection with ℓ be B. Suppose that k and m are the right- and left-sensed parallels to ℓ through A, respectively. Pick C on k that lies on the right side of $\overleftrightarrow{AB}$, and D on m that lies on the left side of $\overleftrightarrow{AB}$. We wish to show that $\angle BAC = \angle BAD$. To obtain a contradiction, suppose that they are not equal. WLOG, we will assume that $\angle BAD > \angle BAC$. Copy $\angle BAC$ onto the left side of $\overleftrightarrow{AB}$ such that E lies on ℓ and $\angle BAC = \angle BAE$ (Euclid I.23). (Note: Because $\angle BAD$ is the smallest angle such that AD does not intersect ℓ, we know that such an E must exist on ℓ.) Construct F on ℓ on the right side of $\overleftrightarrow{BA}$ such that $BE = BF$. Join AF. By SAS, we have $\triangle ABE \cong \triangle ABF$. Thus, $\angle BAE = \angle BAF$. But, by construction, $\angle BAE = \angle BAC$, so $\angle BAF = \angle BAC$. This is clearly an impossibility since $\overleftrightarrow{AF}$ intersects ℓ, but $\overleftrightarrow{AC}$ does not.   □

With this theorem, we no longer have to concern ourselves with the distinction between angles of parallelism on the right or left. Given a point A not on a line ℓ, we can simply define the **angle of parallelism** to ℓ through A as the common value of the angles of parallelism on the right and on the left. Also, note that the angle of parallelism must be acute.

Corollary 13.3. *Any angle of parallelism is acute.*

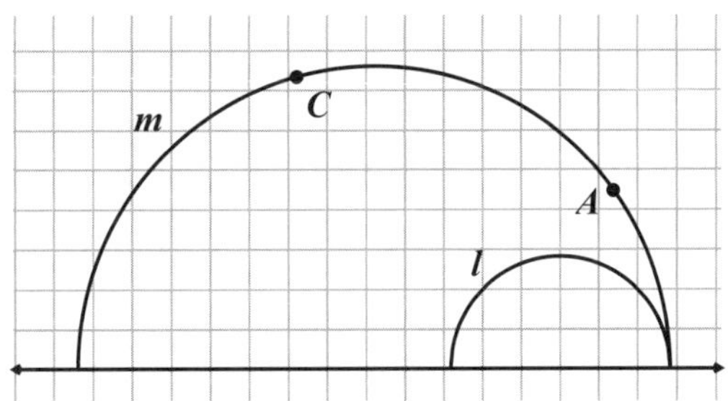

Figure 13.10. Right-sensed parallel in the Half-plane model

Next, let's show that the two types of parallels to a given line are mutually exclusive. Suppose that m and ℓ are parallel lines, and let A and C be distinct points on m, as shown in Figure 13.10. We would like to show that it is not possible for m to be simultaneously sensed parallel to ℓ through A, and ultraparallel to ℓ through C. Given a line ℓ, this will allow us to divide the set of lines parallel to ℓ into two distinct classes: **sensed parallels** to ℓ, and **ultraparallels** to ℓ. In particular, we will start by assuming that m is right-sensed parallel to ℓ through a point A. Then, given any other point C on m, we'd like to show that m is also the right-sensed parallel to ℓ through C. Figure 13.10 which illustrates this situation in the Half-plane model seems to suggest that this will be the case. Let's prove it.

Theorem 13.4. *If m is the right-sensed parallel to a line ℓ at a point A, then it is also the right-sensed parallel to ℓ for all other points that lie on m.*

Proof. Suppose that m is the right-sensed parallel to ℓ through a point, A, and let C be another point on m. We will show that m is also the right-sensed parallel to ℓ through C. Construct B and D on ℓ such that $AB \perp \ell$ and $CD \perp \ell$. Join BC. We will consider two cases depending on whether C lies on the right or left side of $\overleftrightarrow{AB}$.

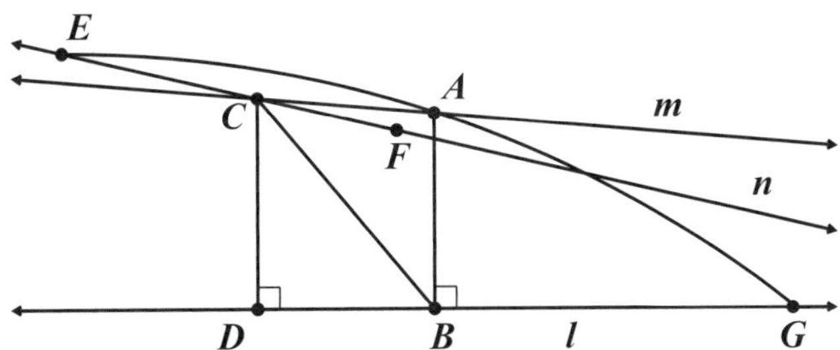

Figure 13.11. Case 1: C is to the left of A

Case 1. C lies to the left of $\overleftrightarrow{AB}$. To obtain a contradiction, we will assume that m is not the right-sensed parallel to ℓ at C. Suppose that, instead, n is. Pick points E and F on n such that F lies on the right side of $\overleftrightarrow{CD}$, and C is between E and F. Then E is on the opposite side of m as ℓ. Consider the line $\overleftrightarrow{EA}$. Since m is the right-sensed parallel to ℓ at A, $\overleftrightarrow{EA}$ must intersect ℓ at some point, G. Since n is the right-sensed parallel to ℓ at C, we have $\angle DCB < \angle DCF < \angle DCA$. Thus $\overrightarrow{CF}$ lies within the angle $\angle ACB$, and hence, by the Crossbar Theorem, must intersect AB (somewhere other than A or B). Consider triangle $\triangle ABG$. Since n intersects AB, by Pasch's Axiom it must intersect either AG or BG. Since n is parallel to ℓ, it cannot intersect BG, and thus, n must intersect $\overleftrightarrow{AG}$ at some new point H. The lines n and $\overleftrightarrow{AG}$ also intersect at E. This contradicts Hilbert I.2 since E and H may lie on at most one line. Thus we have obtained a contradiction and m must be the right-sensed parallel to ℓ at C.

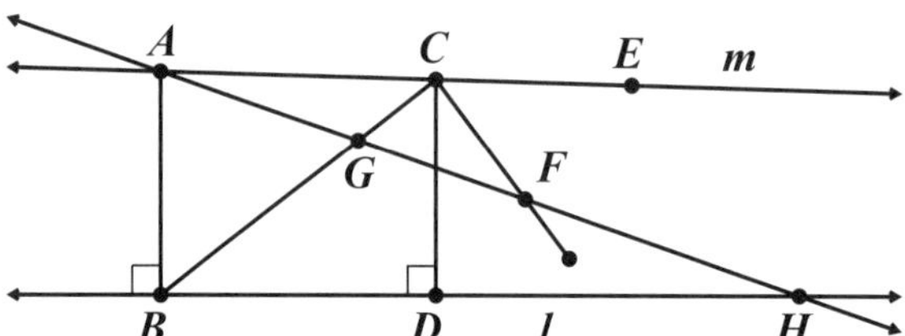

Figure 13.12. Case 2: C is to the right of A

Case 2. C lies to the right of $\overleftrightarrow{AB}$. Pick a point E on m such that C lies between A and E. Given a point F between the parallel lines m and ℓ and on the same side of $\overleftrightarrow{CD}$ as E, we need to show that $\overleftrightarrow{CF}$ intersects ℓ. Consider the ray $\overrightarrow{AF}$. Since $\angle BAF < \angle BAC$, by the Crossbar Theorem $\overrightarrow{AF}$ must intersect BC at a point G. As m is the right-sensed parallel to ℓ at A, AF must intersect ℓ at a point H, where F lies between G and H. Consider triangle $\triangle BGH$. Since $\overleftrightarrow{CF}$ intersects GH, by Pasch's Axiom it must intersect either BG or BH. Since $\overleftrightarrow{CF}$ already intersects $\overleftrightarrow{BG}$ at C, $\overleftrightarrow{CF}$ must therefore intersect BH (line ℓ) as desired. $\qquad\qquad\square$

By this theorem, any point on a right-sensed (or left-sensed) parallel to a line ℓ is as good as any other. This allows us to make the following definitions.

Definition 13.5. Let m and ℓ be distinct parallel lines. If m is right-sensed (or left-sensed) parallel to ℓ at some point A, then we can simply say that m is **right-sensed (or left-sensed) parallel** to ℓ. If m is not sensed parallel to ℓ, then it is **ultraparallel** to ℓ.

The distinction between sensed parallel to a line at a point versus simply being sensed parallel to the line is a subtle change in language, but a big change in our definition of what it means to be sensed parallel. We are no longer tethered to a specific point when discussing whether two lines are sensed parallel. Given a line ℓ, we are now free to discuss the set of all lines that are sensed parallel to ℓ.

Now that we have defined what it means for a line m to be right-sensed or left-sensed parallel to a line ℓ, we might wonder whether or not this relationship is *symmetric*. In other words, if m is right-sensed parallel to ℓ, is ℓ right-sensed parallel to m?

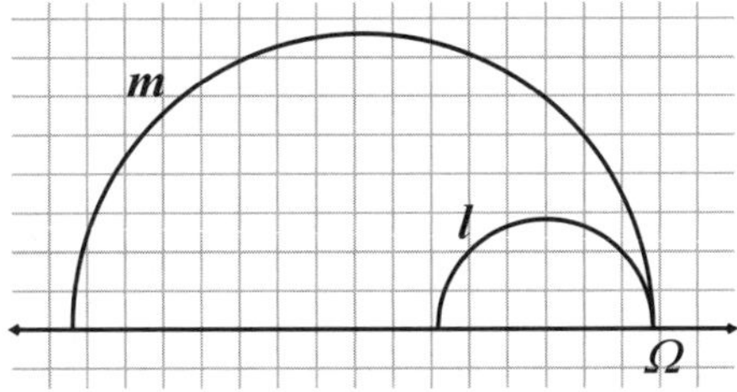

Figure 13.13. Two sensed parallels in the Half-plane model

The example in the Half-plane model shown in Figure 13.13 suggests that this is the case. It turns out that sensed parallelism is symmetric and is stated in the following theorem. For a proof, see [23] or [87].

Theorem 13.6 [Sensed parallelism is symmetric]. *If m is right-sensed (left-sensed) parallel to ℓ, then ℓ is right-sensed (left-sensed) parallel to m.*

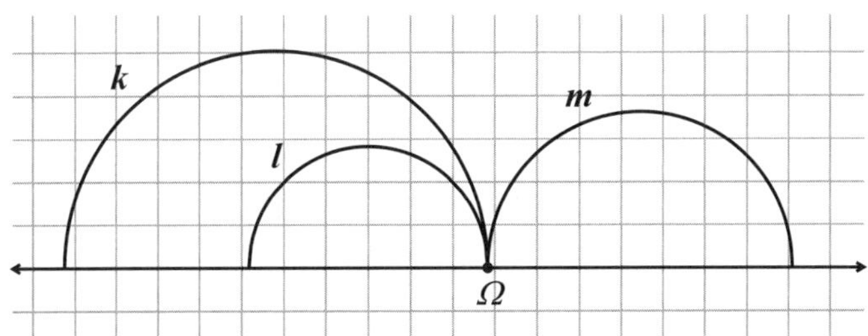

Figure 13.14. Three sensed parallels in the Half-plane model

Because sensed parallelism is symmetric, if m is right-sensed parallel to ℓ, then we know that ℓ is right-sensed parallel to m, and thus, we may simply say that m and ℓ are right-sensed parallel. Similarly, if m and ℓ are parallel, but not sensed parallel, we say they are ultraparallel. We leave the proof of the following theorem as an exercise for the reader.

Theorem 13.7. *Two lines that have a common perpendicular must be ultraparallel.*

The converse is also true, but its proof requires tools that we do not yet have. We will see the converse as Theorem 13.34 in Section 13.3.

The Characteristic Axiom illustrates why parallelism, in general, spectacularly fails to be transitive in Hyperbolic geometry. Figure 13.1, for example, shows point A not on ℓ, $m \parallel \ell$ and $\ell \parallel n$, but clearly $m \nparallel n$ since they share point A. However, right-sensed (or left-sensed) parallelism is transitive as suggested by Figure 13.14. For a proof, see [23] or [87].

Theorem 13.8 [Sensed parallelism is transitive]. *If k and ℓ are right-sensed (left-sensed) parallel, and ℓ and m are right-sensed (left-sensed) parallel, then k and m are right-sensed (left-sensed) parallel.*

At this point, we recall that while the boundary points in our three models of Hyperbolic geometry in Chapter 12 are not points in the hyperbolic plane, they nonetheless have a special designation as the meeting point of sensed parallels. In Chapter 12, we called them **ideal points**. To see how this fits in with our axiomatic development of Hyperbolic geometry, we need Theorems 13.6 and 13.8. Since *right-sensed (left-sensed) parallelism* is both symmetric and transitive, it has two of the three properties necessary to be an equivalence relation. When the definition of *right-sensed (left-sensed) parallelism* is expanded to allow a line to be sensed parallel to itself, then it is clear that right-sensed (left-sensed) parallelism forms an equivalence relation on the set of lines in the plane.

Every equivalence relation on a set creates a partition of the set into disjoint subsets called equivalence classes. Within any equivalence class, any two elements are related to each other, but two elements from two different equivalence classes are not. For example, in Euclidean geometry similarity forms an equivalence relation on the set of all triangles where each equivalence class consists of all triangles with the same set of three angles. The set of all $30° - 60° - 90°$ triangles, in particular, is one equivalence class of this equivalence relation. Likewise for sensed parallelism here, all lines that are right-sensed parallel to each other belong to the same equivalence class, and thus, share the same **ideal point**. The same is true for all lines that are left-sensed parallel to each other. We will use capital Greek letters, such as Ω or Λ, to represent our ideal points. For example, in Figure 13.13, lines ℓ and m share the ideal point Ω. Each line, ℓ, will have exactly two ideal points, one for the set of lines that are right-sensed parallel to ℓ, and another for the set of lines that are left-sensed parallel to ℓ.

Now that we have introduced ideal points, we can additionally refine our nomenclature for sensed parallels. The convention of describing a sensed parallel with the adjectives *right* and *left* is completely arbitrary. We could just as easily have called them *green-sensed parallel* and *red-sensed parallel*. Furthermore, given a particular model, such as the Poincaré Half-plane, the terms left-sensed and right-sensed can be confusing if we attribute the everyday meanings of the words left and right to these parallels. For example, compare the sensed parallels ℓ and m sharing ideal point Ω shown in Figure 13.13 with those given the same labels in Figure 13.14. Both pairs *can* be called right-sensed parallels even though their orientation to each other is not the same in these figures. Even further, both pairs *can* be called left-sensed parallels. So, do we say that ℓ and m are right-sensed or left-sensed parallel? As it turns out, while the adjectives were useful in developing the fundamentals of sensed parallels, they have finished serving their purpose. Instead, we will say ℓ and m are **sensed parallel with ideal point** Ω, and we note again that every line in Hyperbolic geometry has exactly two ideal points.

Let's interpret these terms within the Poincaré Half-plane model. For example, in Figure 13.15, k and ℓ are sensed parallel with ideal point Λ, ℓ, m and v are all sensed parallel with ideal point Ω, k and m are sensed parallel with ideal point Σ. (It's interesting to note that lines k, ℓ, and m create a strange geometric object known as an **ideal triangle**, a triangle with three ideal points for "vertices.") Note that lines k, ℓ and m clearly have two ideal points. What about v and w? The unbounded nature of the Half-plane model results in vertical lines that appear to have only one ideal point, a problem that does not arise with lines in the bounded disc models. To address this, we introduce the **ideal point at infinity** that is shared by all vertical lines in the Half-plane model. Thus, vertical lines in the Half-plane model have two ideal points, one on the x-axis and the other at infinity. For example, as shown in Figure 13.15, line v has its ideal points at Ω and infinity. Recall our claim from Section 12.2.3 that distinct vertical lines are sensed parallel. While we cannot fully support this claim until Theorem 13.34 in Section 13.3, this new ideal point is consistent with this claim. In the figure, for example, v and w are sensed parallel since they share the ideal point at infinity.

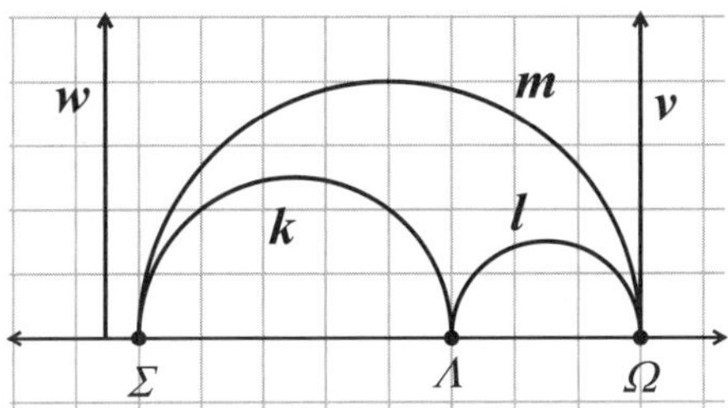

Figure 13.15. Sensed parallels with different ideal points

Using this language, Theorem 13.8 can be rewritten as follows.

Theorem 13.8 Revisited. (Sensed parallelism is transitive) *If k and ℓ are sensed parallel with ideal point Ω, and ℓ and m are sensed parallel with ideal point Ω, then k and m are sensed parallel with ideal point Ω.*

Due to the nature of lines and ideal points, rays from a point in the hyperbolic plane to an ideal point of a line ℓ are sensed parallel to ℓ. This allows for the following definition of sensed parallel rays.

Definition 13.9. Consider a line, ℓ, with ideal points Ω and Λ, and a point, A, not on it. There exist two rays with endpoint A that are sensed parallel to ℓ, one with ideal point Ω and the other with ideal point Λ, denoted as $\overrightarrow{A\Omega}$ and $\overrightarrow{A\Lambda}$, and called **sensed parallel rays** to ℓ with endpoint A.

The axiomatic approach to Hyperbolic geometry taken in this section necessitated a long and detailed refinement of nomenclature. The scaffolding employed to build the fundamental properties of the hyperbolic plane allows us to fully understand the origin of the two types of parallel lines, the angle of parallelism and ideal points. We can pull away the scaffolding for the remainder of the chapter as we will have no need to make such fine distinctions. We *do* need a basic familiarity with the concepts of sensed parallel, ultraparallel, angle of parallelism and ideal points. So, with an understanding of the nature of the fundamental elements of Hyperbolic geometry, we are ready to proceed to omega triangles and Saccheri quadrilaterals.

> **Angle of Parallelism in the Poincaré Half-plane model** Non-Euclid is a free interactive java program that can be used for creating compass and unmarked straightedge constructions, and measuring angles and lengths in both the Poincaré Disc model as well as the Poincaré Half-plane model. It is available at
> *www.cs.unm.edu/~joel/NonEuclid/NonEuclid.html*

where the java file NonEuclid.jar can be downloaded for exploring these models. Below, we give an exercise designed to build intuition for the measure of an angle of parallelism in the Half-plane model. After downloading and then running the java program, follow the steps as outlined. At the end of the walk-through, use your new skills to create more examples in the Half-plane or disc model.

(1) Under VIEW in the menu at the top of the window, choose HYPERBOLIC MODEL, then choose UPPER HALF-PLANE.

(2) Under FILE, choose NEW. The bottom of the visible black window is the x-axis boundary of the Half-plane.

(3) Under CONSTRUCTIONS, choose PLOT POINT, then use the mouse to right-click at a spot as close as possible to the bottom of the window. The program automatically labels this first point as A and will continue in alphabetical order with all newly constructed points.

(4) Right-click in the black window above and to the right of A to construct another point B.

(5) Under CONSTRUCTIONS, choose DRAW LINE, then use the mouse to right-click on point A and then right-click on point B. The program will construct line $\overleftrightarrow{AB}$.

(6) Under CONSTRUCTIONS, choose PLOT POINT, then use the mouse to right-click at another spot above the line $\overleftrightarrow{AB}$ to construct new point C.

(7) Under CONSTRUCTIONS, choose DRAW LINE, then use the mouse to right-click on point A and then right-click on point C to construct line $\overleftrightarrow{AC}$.

(8) Under CONSTRUCTIONS, choose PLOT POINT ON OBJECT, then right-click while highlighting line $\overleftrightarrow{AC}$ to construct point D on $\overleftrightarrow{AC}$.

(9) Under CONSTRUCTIONS, choose DRAW PERPENDICULAR. Right-click on point D when it is highlighted. Right-click on $\overleftrightarrow{AB}$ when it is highlighted. Now we have a perpendicular line from D to $\overleftrightarrow{AB}$ which intersects $\overleftrightarrow{AB}$ at E.

(10) Under EDIT, choose MOVE POINT. Right-click and hold D, then drag point D until E is not near any other point on $\overleftrightarrow{AB}$.

(11) Under MEASUREMENTS, choose MEASURE ANGLE, then right-click in this order: E, then D then A. The measurement of $\angle EDA$ will appear in the grey box at left. As long as point A is very close to the bottom of the black window, this is roughly the angle of parallelism to line $\overleftrightarrow{AB}$ at point D. Use Move Point to drag A closer to the bottom of the window if necessary.

> (12) Drag D back and forth on line $\overleftrightarrow{AC}$ and watch how the angle of parallelism at D to line $\overleftrightarrow{AB}$ changes.
>
> Question: As D moves away from the "nearly ideal point" A, does the angle of parallelism become bigger or smaller?
>
> Repeat this process with other points on other lines in the Half-plane.

Exercises 13.1

1. Consider a line, ℓ, and a point, A, not on it. Construct B on ℓ such that $AB \perp \ell$. Construct a line, m, through A that is perpendicular to $\overleftrightarrow{AB}$, and let C be another point on m, as illustrated in Figure 13.6. Consider the following partition of the points on segment BC:

$$\mathcal{P} = \{F \mid F \text{ lies on } BC \text{ and } \overleftrightarrow{AF} \parallel \ell\}$$

and

$$\mathcal{J} = \{H \mid H \text{ lies on } BC \text{ and } \overleftrightarrow{AH} \nparallel \ell\}.$$

Suppose that $G \in \mathcal{J}$. Show that if H lies on segment BG, then $\overleftrightarrow{AH}$ must intersect ℓ.

2. Consider the line, ℓ, in the Poincaré Half-plane model given by the equation $y = \sqrt{4 - x^2}$. This line has ideal points $(\pm 2, 0)$.

   (a) Find the equation for the sensed parallel to ℓ through point $A = (0, 5)$ that shares the ideal point $(2, 0)$.

   (b) Find a point B on ℓ such that $AB \perp \ell$. Using dynamic geometry software such as Geometer's Sketchpad®, determine the angle of parallelism to ℓ through A. Be sure to include your sketch.

3. Consider the line, ℓ, in the Poincaré Half-plane model given by equation $x = 4$. This line has $(4, 0)$ and the point at infinity as its ideal points.

   (a) Find the equation for the sensed parallel to ℓ through point $A = (-1, 3)$ that shares the ideal point at infinity.

   (b) Find the equation for the sensed parallel to ℓ through A that shares the ideal point $(4, 0)$.

   (c) Find a point B on ℓ such that $AB \perp \ell$. Using dynamic geometry software such as Geometer's Sketchpad®, determine the angle of parallelism to ℓ through A using each of your sensed parallels. Be sure to include your sketch.

4. Prove Theorem 13.7: Two lines that have a common perpendicular must be ultra-parallel.

5. Prove that any two vertical lines in the Half-plane model cannot have a common perpendicular.

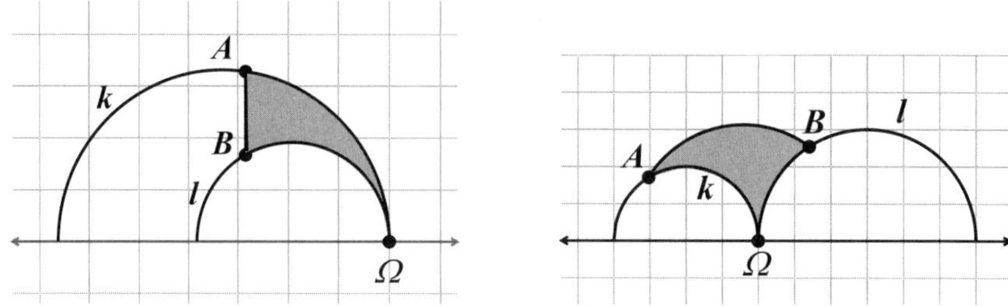

Figure 13.16. Omega triangles, $\triangle AB\Omega$, in the Half-plane model

13.2 Omega triangles

In this section we use sensed parallels and ideal points to define a new mathematical object called an *omega triangle* . Two examples of omega triangles in the Half-plane model are shown in Figure 13.16.

Definition 13.10. Consider two sensed parallels k and ℓ with ideal point Ω. Let A be a point on k, and B a point on ℓ. The figure consisting of the segment AB along with the two sensed parallel rays $\overrightarrow{A\Omega}$ and $\overrightarrow{B\Omega}$ is **omega triangle** $\triangle AB\Omega$. If a point lies between the parallel lines $\overleftrightarrow{A\Omega}$ and $\overleftrightarrow{B\Omega}$ and on the same side of $\overleftrightarrow{AB}$ as the sensed rays, then the point is in the **interior** of $\triangle AB\Omega$.

We use $\angle AB\Omega$ to represent the angle at B formed by the ray $\overrightarrow{BA}$ and the sensed parallel ray $\overrightarrow{B\Omega}$. Notice that, as demonstrated by Figure 13.16, an omega triangle only has one measurable side, AB, and two measurable angles, $\angle AB\Omega$ and $\angle BA\Omega$. While we call them "triangles" since they bear a resemblance to a triangle in any of our models for Hyperbolic geometry, it is important to keep in mind that they are not actually hyperbolic triangles. An omega "triangle" only has two vertices since the ideal point, Ω, is not a point in the hyperbolic plane. Moreover, two of the sides are rays instead of segments. By contrast, a hyperbolic triangle has three vertices in the plane and three hyperbolic line segments for sides, as demonstrated in Figure 13.17.

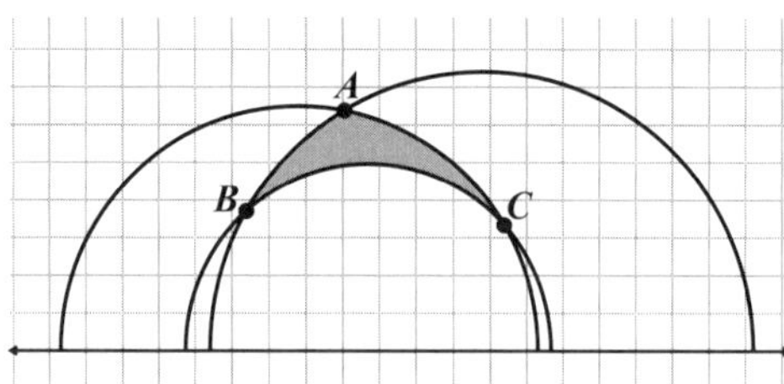

Figure 13.17. Triangle $\triangle ABC$ in the Half-plane model

Though they are not triangles, for ease of language and notation we still refer to them as **omega triangles**, and denote them by $\triangle AB\Omega$, for example. Since we adhere to the notational convention of capital Greek letters for ideal points, the presence or absence of a Greek letter will be sufficient to indicate an omega triangle or a hyperbolic triangle, respectively. These new figures will prove to be quite useful tools. In this section, as we build a toolbox for working with omega triangles, we establish a number of omega triangle properties analogous to familiar Euclidean results such as the Crossbar Theorem, the Exterior Angle Theorem, and triangle congruence schemes.

Theorem 13.11 [Crossbar Theorem for Omega Triangles]. *If C is in the interior of omega triangle $\triangle AB\Omega$, then*

- *ray $\overrightarrow{AC}$ intersects ray $\overrightarrow{B\Omega}$,*
- *ray $\overrightarrow{BC}$ intersects ray $\overrightarrow{A\Omega}$, and*
- *line $\overleftrightarrow{C\Omega}$ intersects segment AB.*

Proof. Consider omega triangle $\triangle AB\Omega$ with interior point C. We first show that $\overrightarrow{AC}$ intersects ray $\overrightarrow{B\Omega}$. Construct D on $\overleftrightarrow{B\Omega}$ such that $AD \perp \overleftrightarrow{B\Omega}$.

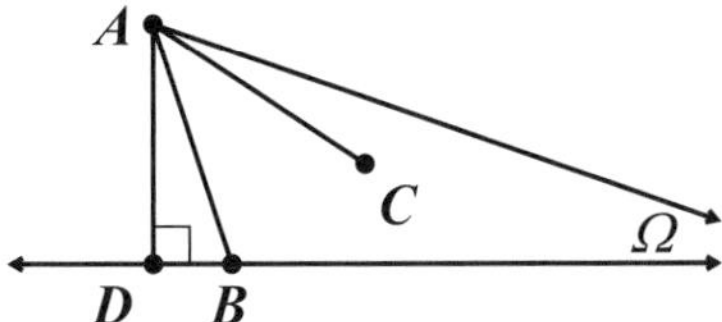

Figure 13.18. Crossbar Theorem for Omega Triangles: Case 1

Case 1. B is to the right of D (see Figure 13.18). Notice that $\angle DA\Omega$ is the angle of parallelism to $\overleftrightarrow{B\Omega}$ at A. Since $\angle DAC < \angle DA\Omega$, and $\overrightarrow{A\Omega}$ is sensed parallel to $\overleftrightarrow{B\Omega}$, $\overrightarrow{AC}$ must intersect $\overrightarrow{D\Omega}$. Since C is an interior point of $\triangle AB\Omega$, $\overrightarrow{AC}$ cannot intersect DB. Therefore, $\overrightarrow{AC}$ must intersect $\overrightarrow{B\Omega}$.

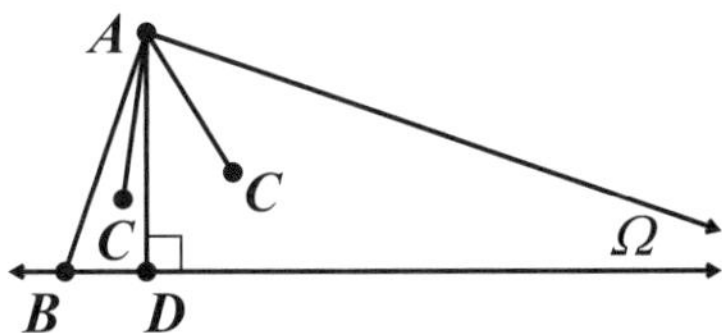

Figure 13.19. Possible positions for C in Case 2

Case 2. B is to the left of D (see Figure 13.19). We leave the proof of this case as an exercise for the reader.

The proof that $\overrightarrow{BC}$ intersects the ray $\overrightarrow{A\Omega}$ is nearly identical.

Finally, let's show that $\overleftrightarrow{C\Omega}$ must intersect AB. Consider line $\overleftrightarrow{C\Omega}$. As proven above, $\overrightarrow{AC}$ will intersect $\overrightarrow{B\Omega}$ at a point, say E, as shown in Figure 13.20. Consider triangle $\triangle ABE$. Since $\overleftrightarrow{C\Omega}$ intersects AE at C, by Pasch's Axiom, the line must intersect either AB or BE. However, $\overleftrightarrow{C\Omega} \parallel \overrightarrow{B\Omega}$, so it must intersect AB. $\qquad\square$

Our next theorem is a version of Euclid I.16, the *Exterior Angle Theorem*, for omega triangles.

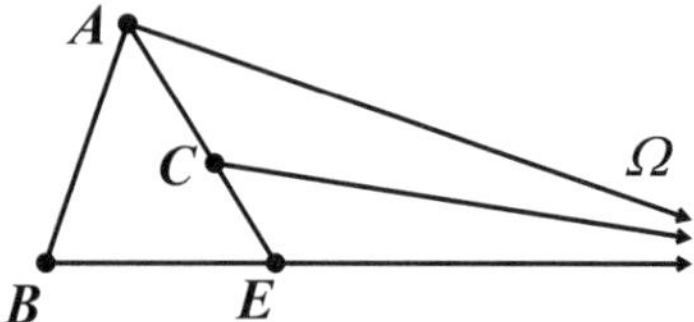

Figure 13.20. The line $\overleftrightarrow{C\Omega}$ intersects AB

Theorem 13.12 [I.16$_\Omega$: Exterior Angle Theorem for Omega Triangles]. *The exterior angle of an omega triangle is greater than the opposite interior angle.*

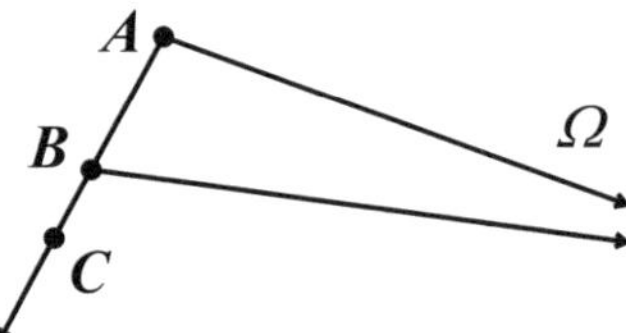

Figure 13.21. I.16$_\Omega$: Euclid's Proposition I.16 for omega triangles

Proof. Let $\triangle AB\Omega$ be an omega triangle. Consider points C on line $\overleftrightarrow{AB}$ such that B is between A and C. We wish to show that $\angle CB\Omega > \angle BA\Omega$. We will assume that this is not the case and obtain a contradiction.

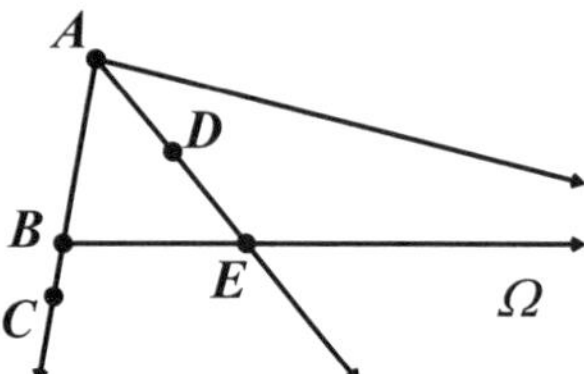

Figure 13.22. Exterior Angle Theorem: Case 1

Case 1. Assume that $\angle CB\Omega < \angle BA\Omega$. Construct a point D in the interior of $\triangle AB\Omega$ such that $\angle CB\Omega = \angle BAD$ (Euclid I.23). By the Crossbar Theorem for Omega Triangles, we have $\overrightarrow{AD}$ must intersect $\overrightarrow{B\Omega}$ at a point, E. Consider $\triangle ABE$. By Euclid I.16 applied to $\triangle ABE$, exterior angle $\angle CBE$ must be greater than interior angle $\angle BAE$. This is a contradiction since $\angle CB\Omega$ equals the opposite interior angle $\angle BAE$ by construction. Therefore, this case is not possible.

Case 2. Assume that $\angle CB\Omega = \angle BA\Omega$. We will show that this would imply that there is a point where the angle of parallelism to $\overleftrightarrow{B\Omega}$ is right, contradicting Corollary 13.3.

Let D be the midpoint of AB. Construct E on $\overleftrightarrow{B\Omega}$ such that $DE \perp \overleftrightarrow{B\Omega}$. If $E = B$, then the angle of parallelism to $\overleftrightarrow{B\Omega}$ from A would be right (Why?) which is not possible,

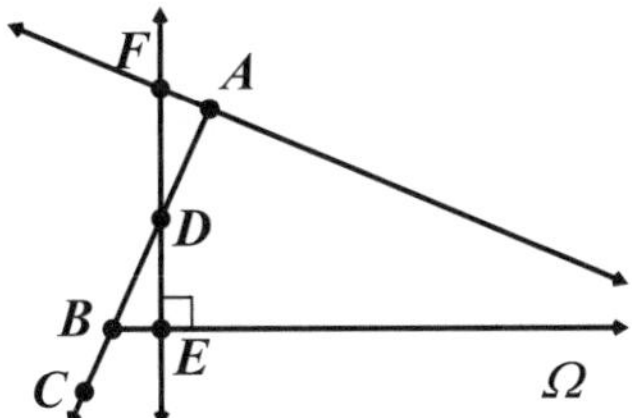

Figure 13.23. Exterior Angle Theorem: Case 2

so $E \neq B$. Note that E may be to the right or left of B, and the case where E is to the right of B is shown in Figure 13.23. Construct F on $\overleftrightarrow{A\Omega}$ so that F is on the opposite side of $\overleftrightarrow{AB}$ as E, and $AF = BE$. Join DF, and consider the triangles $\triangle DBE$ and $\triangle DAF$. We leave it as an exercise for the reader to show that $\triangle DBE \cong \triangle DAF$ regardless of the position of E. Thus, $\angle BDE \cong \angle ADF$, which implies ED and DF lie in a straight line (Euclid I.13 and I.14). Additionally, $\angle DFA \cong \angle DEB$, and hence, both are right angles. Thus, the angle of parallelism to $\overleftrightarrow{B\Omega}$ from F is right, which is not possible.

Since both cases, $\angle CB\Omega < \angle BA\Omega$ and $\angle CB\Omega = \angle BA\Omega$, are impossible, we must have $\angle CB\Omega > \angle BA\Omega$, as desired. $\qquad\qquad\square$

When we introduced omega triangles, we noted that omega triangle $\triangle AB\Omega$ has exactly three measurable elements, namely the finite segment AB and the two angles $\angle AB\Omega$ and $\angle BA\Omega$. Two omega triangles are congruent when they have equivalent corresponding measures.

Definition 13.13. Two omega triangles $\triangle AB\Omega$ and $\triangle CD\Lambda$ are said to be **congruent** if $\angle AB\Omega = \angle CD\Lambda, \angle BA\Omega = \angle DC\Lambda$ and $AB = CD$. This congruence will be denoted by $\triangle AB\Omega \cong \triangle CD\Lambda$.

Somewhat surprising is the fact that any two of the three measurable quantities of an omega triangle uniquely determines the third. This is presented as two congruence schemes for omega triangles, *angle-side* (AS_Ω) and *angle-angle* (AA_Ω).

Theorem 13.14 [AS_Ω: Angle-Side Congruence for Omega Triangles]. *Given omega triangles $\triangle AB\Omega$ and $\triangle CD\Lambda$, if $\angle AB\Omega = \angle CD\Lambda$ and $AB = CD$, then $\angle BA\Omega = \angle DC\Lambda$.*

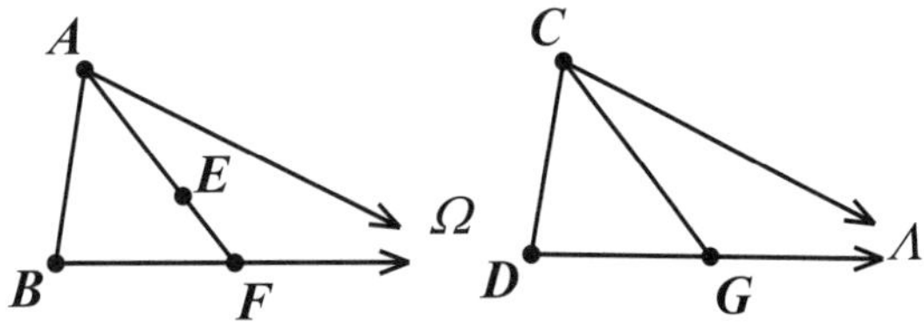

Figure 13.24. AS_Ω: *Angle-Side* Congruence for Omega Triangles

Proof. Let $\triangle AB\Omega$ and $\triangle CD\Lambda$ be omega triangles with $\angle AB\Omega = \angle CD\Lambda$ and $AB = CD$. In order to obtain a contradiction, we will assume $\angle BA\Omega \neq \angle DC\Lambda$. WLOG, assume $\angle BA\Omega > \angle DC\Lambda$. Construct a point E in the interior of $\triangle AB\Omega$ such that $\angle BAE = \angle DC\Lambda$ (Euclid I.23), as shown in Figure 13.24. By the Crossbar Theorem for Omega Triangles, we have $\overrightarrow{AE}$ must intersect $\overrightarrow{B\Omega}$ at a point, F. Construct G on $\overrightarrow{D\Lambda}$ such that $BF = DG$. By SAS, $\triangle ABF \cong \triangle CDG$. Thus, $\angle DCG = \angle BAE = \angle DC\Lambda$, which is clearly a contradiction. Therefore, $\angle BA\Omega = \angle DC\Lambda$, as desired. $\square$

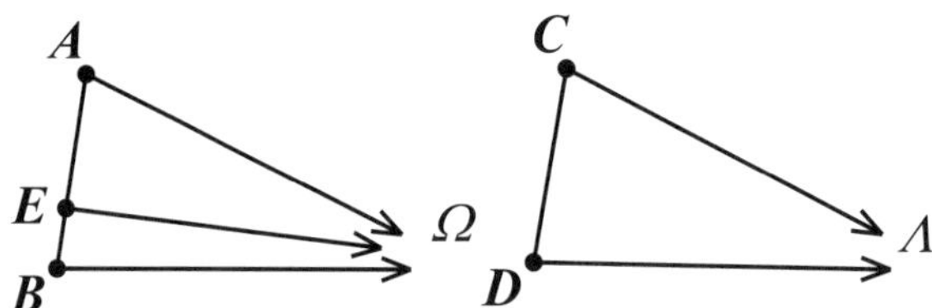

Figure 13.25. AA_Ω: *Angle-Angle* Congruence for Omega Triangles

Theorem 13.15 [AA_Ω: Angle-Angle Congruence for Omega Triangles]. *Given omega triangles $\triangle AB\Omega$ and $\triangle CD\Lambda$, if $\angle AB\Omega = \angle CD\Lambda$ and $\angle BA\Omega = \angle DC\Lambda$, then $AB = CD$.*

Proof. Let $\triangle AB\Omega$ and $\triangle CD\Lambda$ be omega triangles with $\angle AB\Omega = \angle CD\Lambda$ and $\angle BA\Omega = \angle DC\Lambda$. In order to obtain a contradiction, we will assume $AB \neq CD$. WLOG, assume $AB > CD$. Construct a point E on AB such that $AE = CD$. Let $\overrightarrow{E\Omega}$ be the sensed parallel to $\overrightarrow{B\Omega}$ through E with ideal point Ω, as shown in Figure 13.25. By AS_Ω, $\triangle AE\Omega \cong \triangle CD\Lambda$. Therefore $\angle AE\Omega = \angle CD\Lambda = \angle AB\Omega$. This contradicts the $I.16_\Omega$ as $\angle AE\Omega$ is the exterior angle to omega triangle $\triangle EB\Omega$, and thus, cannot equal the opposite interior angle $\angle AB\Omega$. $\square$

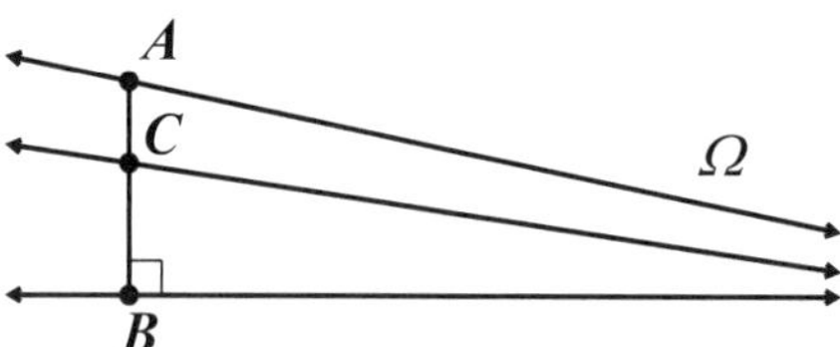

Figure 13.26. Comparing angles of parallelism

As a tool, omega triangles provide another means to understand the relative measure of angles of parallelism. Consider a line, ℓ, with ideal points Ω and Λ, and a point A not on it. Construct B on ℓ such that $AB \perp \ell$, and let C lie on $\overleftrightarrow{AB}$ between A and B. The angle of parallelism to ℓ through C must be larger than that through A. (Why?)

When the two measurable angles in an omega triangle are equal, then we say we have an *isosceles omega triangle*. We end this section with this definition and an isosceles omega triangle theorem.

Definition 13.16. An omega triangle $\triangle AB\Omega$ is **isosceles** if $\angle AB\Omega = \angle BA\Omega$.

We leave the proofs of the following lemma and theorem to the reader.

Lemma 13.17. *Consider an omega triangle $\triangle AB\Omega$ where C is the midpoint of AB. Omega triangle $\triangle AB\Omega$ is isosceles if and only if $\angle AC\Omega$ is a right angle.*

Theorem 13.18. *If $\triangle AB\Omega$ and $\triangle CD\Lambda$ are two isosceles omega triangles with $AB = CD$, then $\triangle AB\Omega \cong \triangle CD\Lambda$.*

Exercises 13.2

1. Prove Case 2 in Theorem 13.11.

2. Explain why, in the proof of Theorem 13.12, we have $\triangle DBE \cong \triangle DAF$. Consider both cases, the first where B lies to the right of E (as shown in Figure 13.23), and the second where B lies to the left of E. Be sure to illustrate this second case with an appropriate diagram.

3. Prove I.17$_\Omega$: *The two angles of an omega triangle taken together are less than two right angles.*

4. Prove Lemma 13.17.

5. Prove Theorem 13.18.

6. Let $\overleftrightarrow{AC}$ and $\overleftrightarrow{BD}$ be sensed parallels with ideal point Ω, $\overline{AB} \perp \overleftrightarrow{BD}$ and $\overline{CD} \perp \overleftrightarrow{BD}$. Suppose that $CD < AB$. Show that the angle sum of the quadrilateral $ABDC$ is less than four right angles.

7. Answer the parenthetical "Why?", posed at the end of the paragraph following the proof of Theorem 13.15, in order to justify the following claim: The angle of parallelism to ℓ through C must be larger than the angle of parallelism through A.

13.3 Saccheri quadrilaterals

We are now ready to revisit Saccheri and his quadrilaterals. In his 1733 book *Euclides ab omni naevo vindicatus (Euclid Freed of Every Fleck)*, Saccheri begins by considering a quadrilateral $ABDC$ where $AC \perp AB$, $BD \perp AB$ and $AC = BD$. In his honor, such a quadrilateral is now commonly called a *Saccheri quadrilateral*.

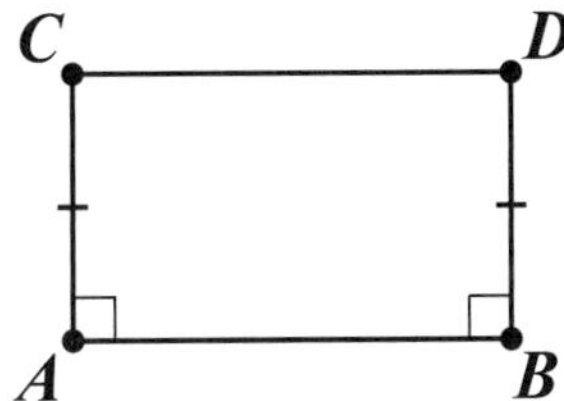

Figure 13.27. A Saccheri quadrilateral

Definition 13.19. A quadrilateral $ABDC$ where $AC \perp AB$, $BD \perp AB$ and $AC = BD$ is called a **Saccheri quadrilateral**. The segment AB is the **base**, CD is the **summit**, and angles $\angle ACD$ and $\angle BDC$ are the **summit angles.**

For his first proposition, Saccheri proves that the summit angles are congruent. We state and prove this using updated notation.

Proposition 13.20 [Saccheri's Proposition I]. *The summit angles of a Saccheri quadrilateral are equal.*

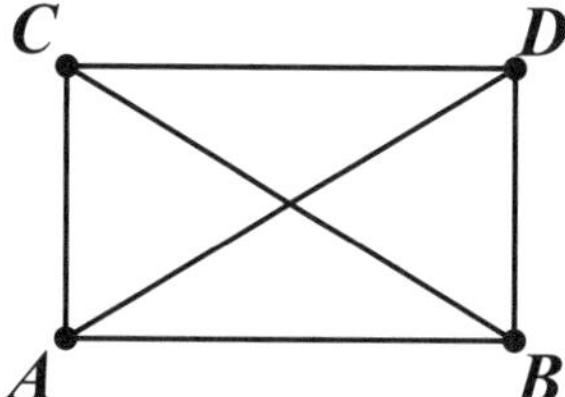

Figure 13.28. Proof of Proposition 13.20 (Saccheri's Proposition I)

Proof. Assume $ABCD$ is a Saccheri quadrilateral where $AC \perp AB$, $BD \perp AB$ and $AC = BD$. Construct AD and BC. Then by SAS, $\triangle CAB \cong \triangle DBA$. Therefore, $BC = AD$. Triangles $\triangle ACD$ and $\triangle BDC$ are congruent by SSS. Thus, $\angle ACD = \angle BDC$, as desired. $\qquad\square$

For his second proposition, Saccheri proves that the line joining the midpoints of the base and summit is a common perpendicular to both. Again, we state and prove this using updated notation.

Proposition 13.21 [Saccheri's Proposition II]. *The line joining the midpoints of the base and summit of a Saccheri quadrilateral is perpendicular to the base and the summit.*

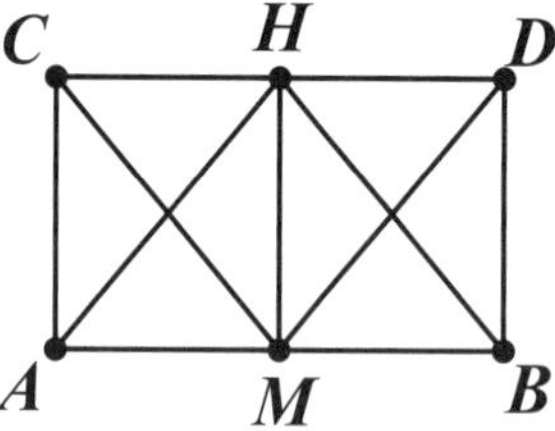

Figure 13.29. Proof of Proposition 13.21 (Saccheri's Proposition II)

Proof. Assume $ABCD$ is a Saccheri quadrilateral where $AC \perp AB$, $BD \perp AB$ and $AC = BD$. Let M and H be the midpoints of AB and CD, respectively. Construct MH, CM and DM. By SAS, we have $\triangle CAM \cong \triangle DBM$. Thus $CM = DM$. Triangles $\triangle CHM$ and $\triangle DHM$ are congruent by SSS. Thus, $\angle CHM = \angle DHM$ and therefore, by definition we have $HM \perp CD$. The proof that $HM \perp AB$ is similar. $\qquad\square$

A direct result of the perpendicularity of HM (the segment joining the midpoints) to both the base and summit is that the base and summit are ultraparallel. We leave the proof as an exercise for the reader.

Corollary 13.22. *The lines containing the base and summit of a Saccheri quadrilateral are ultraparallel.*

For his third proposition, Saccheri shows that, depending on whether or not the summit angles are right, obtuse or acute, we will have $CD = AB, CD < AB$ or $CD > AB$, respectively. We leave the proof as an exercise for the reader. We state Saccheri's proposition as it has been translated.

Proposition 13.23 [Saccheri's Proposition III]. *If two equal straights AC, BD, stand perpendicular to any straight AB: I say the join CD will be equal to, or less, or greater than AB, according as the angles at CD are right, or obtuse, or acute.*

Saccheri then goes on to prove that if the summit angles are right, obtuse, or acute for one of his quadrilaterals, then that must be the case for all of them. He called these three cases *the hypothesis of the right angle (HRA), the hypothesis of the obtuse angle (HOA)* and *the hypothesis of the acute angle (HAA)*, respectively. Since Saccheri was attempting to prove the Parallel Postulate, he wanted to show that HRA was the only valid hypothesis. Employing a double reductio ad absurdum technique, Saccheri hoped that he would reach an eventual contradiction under the assumptions of both HAA and HOA. While the hypothesis of the obtuse angle fell with relative ease, in the end Saccheri dismissed HAA solely because he could not reconcile his preconceived notions of straight lines with the bizarre results he proved under the assumption of the HAA hypothesis.

We are in a completely different position than Saccheri. Far from assuming HAA in the hopes of reaching a contradiction, we can prove Saccheri's hypothesis of the acute angle! Using our axioms, the mathematical objects we have defined, and the theorems we have already proven, we can now show that in Hyperbolic geometry the summit angles of a Saccheri quadrilateral are, in fact, acute.

Theorem 13.24. *The summit angles in a Saccheri quadrilateral are acute.*

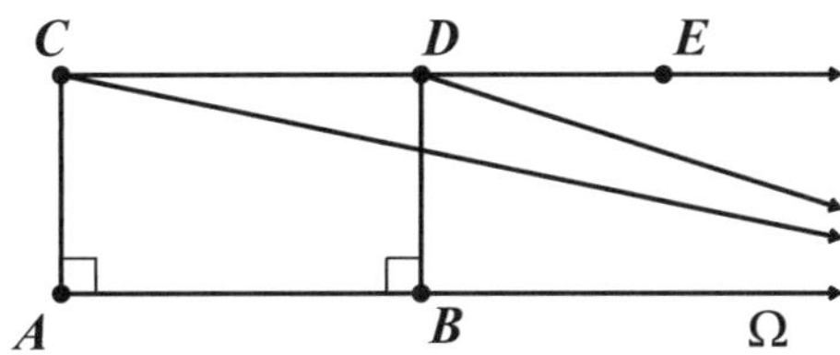

Figure 13.30. Summit angles in a Saccheri quadrilateral are acute

Proof. Assume $ABCD$ is a Saccheri quadrilateral where $AC \perp AB$, $BD \perp AB$ and $AC = BD$. Let E be a point on $\overleftrightarrow{CD}$ such that D is between C and E. Let $\ell = \overleftrightarrow{AB}$ have ideal point Ω on the side of $\overleftrightarrow{BD}$ that contains E. Consider the sensed parallel rays

$\overrightarrow{C\Omega}$ and $\overrightarrow{D\Omega}$. Since the summit angles of a Saccheri quadrilateral are congruent, it will suffice to show that one of them, say $\angle BDC$, is acute. Consider omega triangles $\triangle CA\Omega$ and $\triangle DB\Omega$. By AS_Ω, they are congruent. Thus, $\angle AC\Omega = \angle BD\Omega$. Next, consider omega triangle $\triangle CD\Omega$ with exterior angle $\angle ED\Omega$. By $I.16_\Omega$, we have $\angle ED\Omega > \angle DC\Omega$. Combining these results, we have

$$\begin{aligned}
\angle BDE &= \angle ED\Omega + \angle BD\Omega \\
&> \angle DC\Omega + \angle AC\Omega \\
&= \angle ACD \\
&= \angle BDC.
\end{aligned}$$

Thus, $\angle BDC < \angle BDE$. Since $\angle BDC + \angle BDE$ equals two right angles, $\angle BDC$ is acute, as desired. $\qquad\square$

Saccheri's second proposition reveals an interesting subfigure found within Figure 13.29. Notice that the angles of quadrilateral $AMHC$ consist of one acute angle and three right angles. Such a figure has come to be called a Lambert quadrilateral, named after Johann Lambert (1728–1777), who proved many of these same results in his own attempt to prove the Parallel Postulate.

Example 13.25. In Figure 13.31, which illustrates a Saccheri quadrilateral in the Half-plane model, the summit angles, $\angle ACD$ and $\angle BDC$, are approximately $61.5°$.

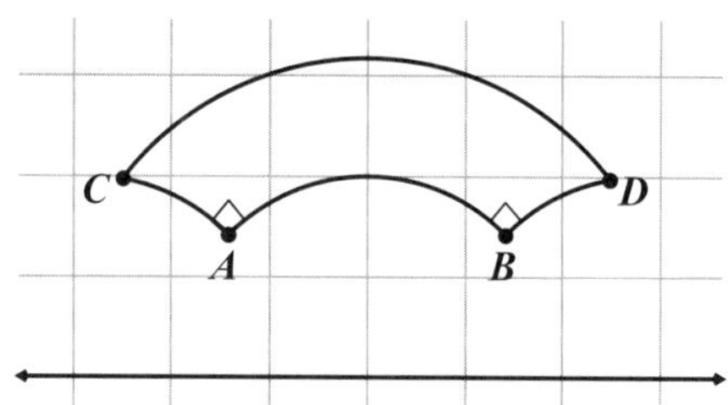

Figure 13.31. A Saccheri quadrilateral in the Half-plane model

A direct corollary to the acuteness of the summit angles is that parallel lines are not everywhere equidistant. As an example, take a look at side AC of the Saccheri quadrilateral shown in Figure 13.29. Since the summit and base are parallel, and both AC and HM are perpendicular to the base, our Euclidean intuition leads us to believe that HM and AC have equal length. This is the wrong conclusion in Hyperbolic geometry. As an exercise, the reader will show that, as given in the following corollary, the side of a Saccheri quadrilateral is longer than the segment joining the midpoints of its summit and base.

Corollary 13.26. *Consider a Saccheri quadrilateral $ABDC$ with $AC \perp AB$, $BD \perp AB$ and $AC = BD$. Let E and F be the midpoints of AB and CD, respectively. Prove that $EF < AC$, that is, prove that the distance from the midpoint of the summit to the base is less than the distance from a summit vertex to the base.*

We will have more to say about the distance between parallel lines in Hyperbolic geometry at the end of this section, but this discussion requires some results about triangles and quadrilaterals. First, let's use Saccheri quadrilaterals to prove the Hyperbolic analogue to Euclid I.32, that the angles in any hyperbolic triangle sum to less than two right angles. Appropriately, we refer to this theorem as $I.32_H$.

Theorem 13.27 [$I.32_H$]. *Given any triangle, the sum of the angles is less than two right angles.*

Proof.   Consider any triangle $\triangle ABC$. We start by constructing an associated Saccheri quadrilateral. To do this, bisect AB and AC at D and E, respectively, and construct the line $\overleftrightarrow{DE}$. Next, construct F, G and H on $\overleftrightarrow{DE}$ such that $AF \perp \overleftrightarrow{DE}$, $BG \perp \overleftrightarrow{DE}$, and $CH \perp \overleftrightarrow{DE}$.

Claim: Either D lies between F and G, or points D, F and G coincide.

We leave it to the reader to show this. Note: The same argument shows that either E lies between F and H, or all three of these points coincide.

There are essentially three different cases to consider depending upon where F lies. In each case, we will show that $BCHG$ is a Saccheri quadrilateral with base GH and summit BC, and that the angle sum for our triangle equals the sum of the summit angles in $BCHG$.

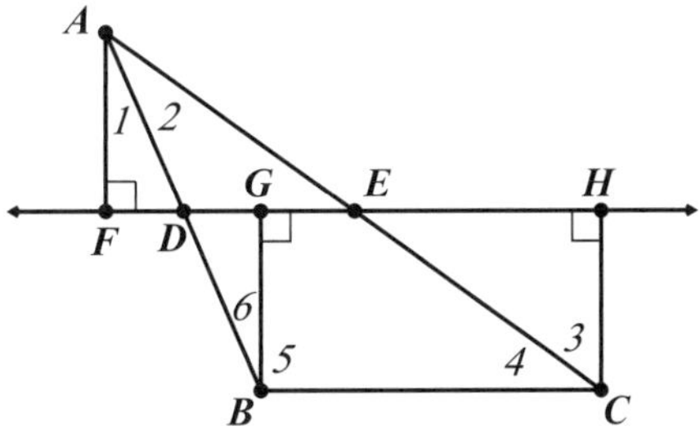

Figure 13.32. Case 1

Case 1. F lies outside of DE. There are two possibilities, either D lies between F and E, or E lies between D and F. WLOG, we assume that D lies between F and E. Recall, G must lie on the opposite side of D as F, and H must lie on the opposite side of E as F. See Figure 13.32 for an illustration.

Consider triangles $\triangle AFD$ and $\triangle BGD$. Angles $\angle AFD$ and $\angle BGD$ are both right angles. By construction, $AD = BD$. By Euclid I.15 (Vertical Angles), we have $\angle ADF = \angle BDG$. Thus, by AAS (Euclid I.26), $\triangle AFD \cong \triangle BGD$, $AF = BG$, and $\angle 1 = \angle 6$.

Next consider triangles $\triangle AFE$ and $\triangle CHE$. By construction, $\angle AFE$ and $\angle CHE$ are both right angles, and $AE = CE$. By Euclid I.15, $\angle AEF = \angle CEH$. Thus, by AAS, $\triangle AFE \cong \triangle CHE$, $AF = CH$, and $\angle 3 = \angle FAE = \angle 1 + \angle 2$.

Since $BG = AF = CH$, $BCHG$ is a Saccheri quadrilateral with base GH and summit BC. The angle sum of triangle $\triangle ABC$ equals the sum of the summit angles of $BCHG$ as follows:

$$\begin{aligned}
\angle BAC + \angle ACB + \angle ABC &= \angle 2 + \angle 4 + \angle 5 + \angle 6 \\
&= \angle 2 + \angle 4 + \angle 5 + \angle 1 \\
&= \angle 3 + \angle 4 + \angle 5 \\
&= \angle BCH + \angle CBG.
\end{aligned}$$

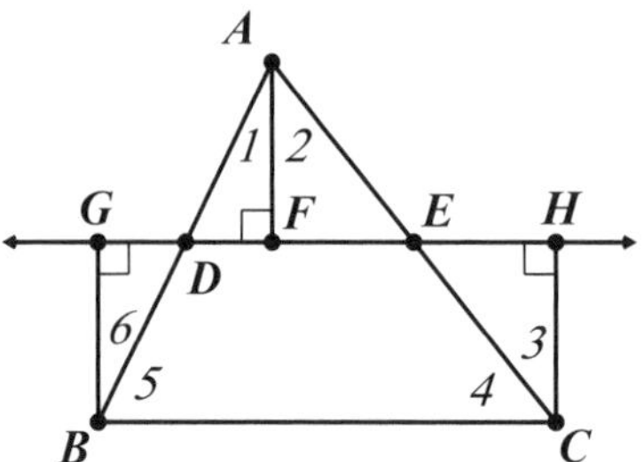

Figure 13.33. Case 2

Case 2. *F* lies between *D* and *E*. Recall, *G* must lie on the opposite side of *D* as *F*, and *H* must lie on the opposite side of *E* from *F*. See Figure 13.33 for an illustration. We leave it to the reader to show that $BG = CH$ (making *BCHG* a Saccheri quadrilateral with base *GH* and summit *BC*) and that

$$\angle BAC + \angle ACB + \angle ABC = \angle BCH + \angle CBG.$$

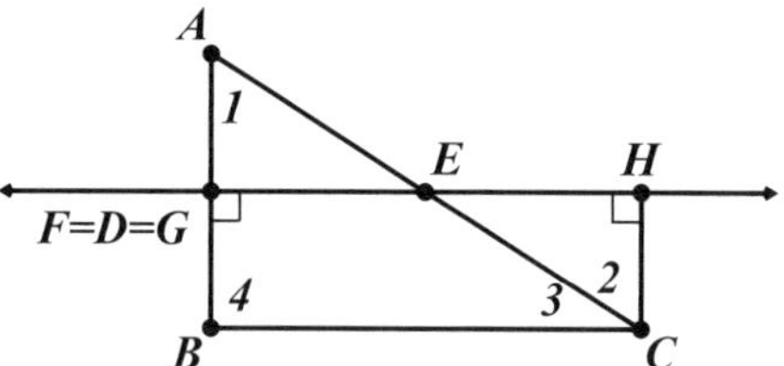

Figure 13.34. Case 3

Case 3. *F* coincides with either *D* or *E*. WLOG, assume that *F* coincides with *D*. Then it must also coincide with *G*. See Figure 13.34. A word of caution: it is often very difficult not to view these triangles through Euclidean goggles. While the angle $\angle ABC$ appears to be right, it is actually acute. (Why?). We leave it to the reader to show that $BG = CH$ and $\angle BAC + \angle ACB + \angle ABC = \angle BCH + \angle CBG$.

In all three cases, *BCHG* is a Saccheri quadrilateral whose summit angles equal the angle sum of triangle $\triangle ABC$. Since the summit angles are equal and acute, the angle sum of the triangle must be less than two right angles, as desired. $\square$

The proof of I.32$_H$ relies on the construction of a Saccheri quadrilateral associated to the given triangle by its angle sum. We imagine that Saccheri would be shocked to learn that his quadrilaterals allow us to prove a result that he would find, in the words of his thirty-third proposition, "repugnant to the nature of the" triangle. Despite any shock or revulsion on the part of the Jesuit professor, this is a very useful technique in Hyperbolic geometry, and as such, gives rise to the following definition.

Definition 13.28. Given triangle $\triangle ABC$, an **associated Saccheri quadrilateral** is constructed as follows: After bisecting *AB* at *D*, and *AC* at *E*, construct line $\overleftrightarrow{DE}$. Next, construct *G* and *H* on $\overleftrightarrow{DE}$ such that $BG \perp \overleftrightarrow{DE}$ and $CH \perp \overleftrightarrow{DE}$. Quadrilateral *BCHG* is the associated Saccheri quadrilateral with summit *BC*.

Note that, as shown in the proof of I.32$_H$, the sum of the summit angles of the Saccheri quadrilateral associated with $\triangle ABC$, namely $\angle B + \angle C$, is equal to the angle sum of $\triangle ABC$. Moreover, this result belongs to Neutral geometry.

With the upper bound for the sum of the angles in a triangle established, we are able to show a few other results of Hyperbolic geometry that offer a stark contrast to Euclidean geometry. After showing that the sum of the angles in a quadrilateral is less than 360°, we will see that we lose rectangles and squares, but gain AAA triangle congruence.

Corollary 13.29. *The angle sum of any quadrilateral is less than four right angles.*

Proof. Consider a quadrilateral $ABCD$. Construct diagonal AC. There are two cases to consider.

Case 1. $ABCD$ is the union of triangles $\triangle ABC$ and $\triangle ACD$.

Case 2. Both B and D lie on the same side of $\overleftrightarrow{AC}$. In this case, either D is in the interior of triangle $\triangle ABC$ or B is in the interior of $\triangle ACD$. Either way, $ABCD$ is the union of triangles $\triangle ABD$ and $\triangle CBD$.

We leave it to the reader to illustrate these two cases and to use I.32$_H$ to prove that the angle sum for $ABCD$ is less than four right angles. $\square$

The nonexistence of rectangles and squares follows immediately from the previous corollary.

Corollary 13.30. *Rectangles and squares do not exist in Hyperbolic geometry.*

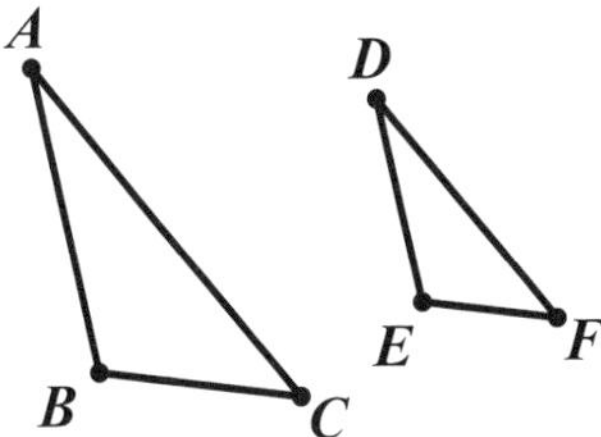

Figure 13.35. AAA$_H$ congruence scheme

Though we lose right-angled quadrilaterals, and, at best, have an upper bound for the angle sum in a hyperbolic triangle, like Spherical geometry, we gain a triangle congruence scheme. Any two triangles with corresponding angles must be congruent, that is, AAA$_H$ is a valid congruence scheme.

Theorem 13.31 [AAA$_H$]. *If the angles of one triangle are equal, respectively, to the angles of another triangle, then the triangles are congruent.*

Proof. Consider triangles $\triangle ABC$ and $\triangle DEF$, where $\angle A = \angle D$, $\angle B = \angle E$ and $\angle C = \angle F$. We wish to show that $\triangle ABC \cong \triangle DEF$. If we can show that $AB = DE$, then by ASA, we have $\triangle ABC \cong \triangle DEF$.

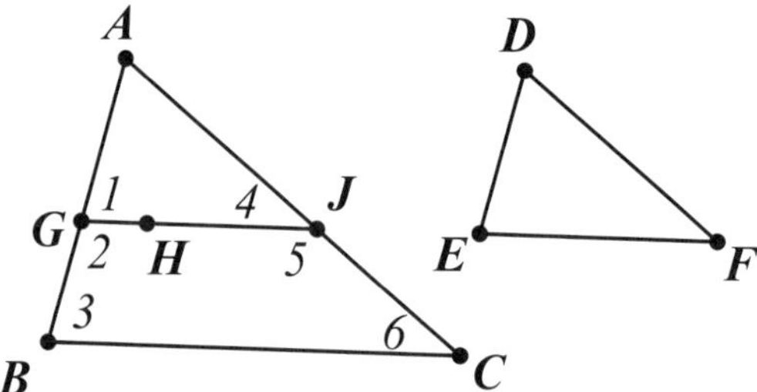

Figure 13.36. Assuming $AB \neq DE$

To obtain a contradiction, suppose $AB \neq DE$. WLOG, assume that $AB > DE$. Construct G on AB such that $AG = DE$. Construct H on the same side of $\overleftrightarrow{AB}$ as C, such that $\angle AGH = \angle DEF$. Since $\angle AGH = \angle ABC$, by Euclid I.28 we have $\overleftrightarrow{GH} \parallel \overleftrightarrow{BC}$. Thus, since $\overleftrightarrow{GH}$ intersects AB, by Pasch's Axiom $\overleftrightarrow{GH}$ must intersect AC at a point, J, distinct from C. Consider triangle $\triangle AGJ$ and the numbered angles as shown in Figure 13.36. By ASA, $\triangle AGJ \cong \triangle DEF$. Thus, $\angle 1 = \angle DEF = \angle 3$ and $\angle 4 = \angle DFE = \angle 6$. Consider the angle sum of quadrilateral $GBCJ$. We have $\angle 2 + \angle 3 + \angle 5 + \angle 6 = \angle 2 + \angle 1 + \angle 5 + \angle 4$. By Euclid I.13, $\angle 1 + \angle 2$ and $\angle 4 + \angle 5$ each sum to two right angles. Thus, the angle sum of quadrilateral $GBCJ$ is four right angles, which is impossible by Corollary 13.29. Thus, $AB = DE$ and $\triangle ABC \cong \triangle DEF$, as desired. $\qquad\square$

13.3.1 Distance between parallel lines. Equipped with a toolbox that includes omega triangles, Saccheri quadrilaterals, triangle angle sum, quadrilateral angle sum, and $\mathrm{AAA_H}$, we are ready to explore the distance between two parallel lines in Hyperbolic geometry. Recall, if ℓ is a line and A is a point not on it, then the distance from A to ℓ is the length of the segment AB, where B lies on ℓ, and $AB \perp \ell$. In Euclidean geometry, if ℓ and m are parallel lines with A and C any two points on m, then the distance from A to ℓ equals the distance from C to ℓ. This allows us to say that parallel lines are everywhere *equidistant*, and define the distance between two lines as the distance from one line to any point on the other line. Parallel lines in Hyperbolic geometry behave much differently than their Euclidean counterparts. For starters, in Corollary 13.26, we have already seen that the side of a Saccheri quadrilateral is longer than the segment joining the midpoint of its base and summit. In general, ultraparallel lines are not everywhere equidistant. Furthermore, sensed parallel lines are nowhere equidistant! We leave the proof of these results as exercises.

Theorem 13.32 [Sensed parallels are nowhere equidistant]. *If m and ℓ are sensed parallels with ideal point Ω, and A and C are distinct points on m, then the distance from A to ℓ does not equal the distance from C to ℓ.*

Actually, we can say a bit more about the relationship between the distances from these points to ℓ. If C lies on the sensed parallel ray $\overrightarrow{A\Omega}$, then the distance from C to ℓ is less than the distance from A to ℓ. Furthermore, if ℓ and m are sensed parallels, then for any segment BD there exists a point, E, on ℓ such that the distance from E to m is equal to the length of BD. We leave these verifications to the reader as Exercises 13.3.11 and 13.3.12.

As mentioned above, ultraparallel lines are not everywhere equidistant, and we leave the proof by contradiction for this theorem to the reader.

Theorem 13.33. *Let m and ℓ be ultraparallel lines. Given a point A on m, there is at most one other point on m that is the same distance from ℓ as A.*

This last theorem brings up a natural question: What is the shortest possible distance between two ultraparallel lines? To answer this question we will need the converse of Theorem 13.7. This particular proof is attributed to Hilbert.

Theorem 13.34. *If two lines are ultraparallel, then they have a unique common perpendicular.*

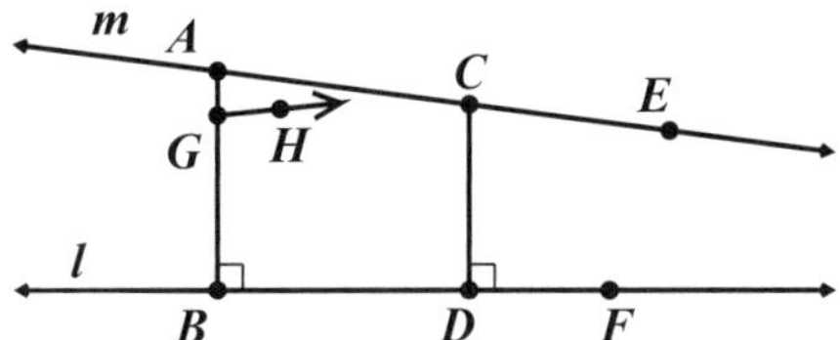

Figure 13.37. Ultraparallel lines have a unique common perpendicular: Case 2

Proof. Let m and ℓ be ultraparallel lines with A and C lying on m. Construct B and D on ℓ such that $AB \perp \ell$ and $CD \perp \ell$. Our goal is to construct a Saccheri quadrilateral whose summit lies on m and whose base lies on ℓ. We can then use the fact that the line joining the midpoints of the base and summit is perpendicular to both.

Case 1. If $AB = CD$, then $ABDC$ is the desired Saccheri quadrilateral, and hence, the line joining the midpoints of the base and summit is a common perpendicular to our two ultraparallel lines. We leave the proof by contradiction of the uniqueness of this perpendicular as an exercise for the reader.

Case 2. Suppose $AB \neq CD$. Then WLOG, assume $AB > CD$. Pick E on m such that C lies between A and E. Pick F on ℓ such that D lies between B and F. Construct G on AB such that $GB = CD$. Let $\overrightarrow{GH}$ be the unique ray on the same side of $\overleftrightarrow{AB}$ as C, such that $\angle BGH = \angle DCE$.

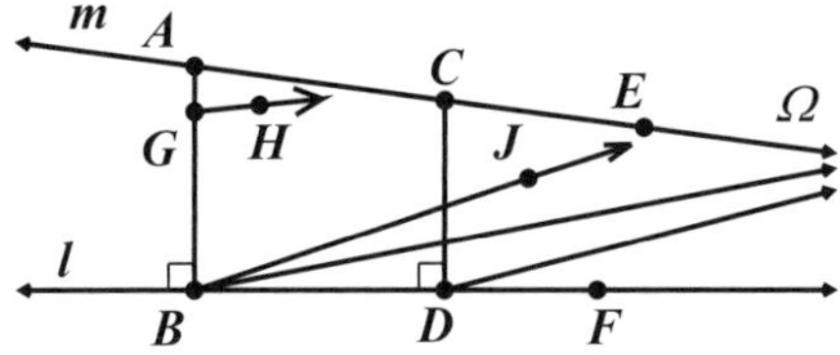

Figure 13.38. Ray $\overrightarrow{BJ}$ must intersect ray $\overrightarrow{AE}$

Claim: $\overrightarrow{GH}$ intersects $\overleftrightarrow{AE}$. Why? Pick J on the same side of $\overleftrightarrow{AE}$ as B so that $\overrightarrow{BJ}$ is sensed parallel to $\overrightarrow{GH}$ with ideal point Λ. Also, as shown in Figure 13.38, suppose $\overrightarrow{AE}$ has ideal point Ω, and thus, $\overrightarrow{B\Omega}$ and $\overrightarrow{D\Omega}$ are sensed parallel rays to $\overrightarrow{AE}$.

Omega triangles $\triangle GB\Lambda$ and $\triangle CD\Omega$ are congruent by AS_Ω. Thus, $\angle GB\Lambda = \angle CD\Omega$. Since $\angle ABD$ and $\angle CDF$ are both right, we have $\angle DB\Lambda = \angle FD\Omega$. Since $\angle FD\Omega$ is an exterior angle to omega triangle $\triangle BD\Omega$, $\angle FD\Omega > \angle DB\Omega$ by I.16$_\Omega$. Thus, $\angle DB\Lambda > \angle DB\Omega$, and J lies in the interior of omega triangle $\triangle AB\Omega$. By the Crossbar Theorem for Omega Triangles, $\overrightarrow{BJ}$ must intersect $\overrightarrow{AE}$ at a point, K.

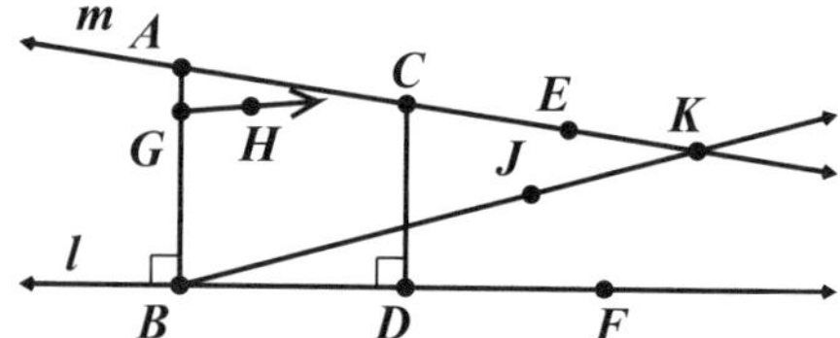

Figure 13.39. Ray $\overrightarrow{GH}$ must intersect ray $\overrightarrow{AE}$

Now consider triangle BAK as shown in Figure 13.39. Since $\overleftrightarrow{GH}$ intersects AB, then by Pasch's Axiom, it must intersect either AK or BK. Since $\overrightarrow{BJ}$ is sensed parallel to $\overrightarrow{GH}$, $\overleftrightarrow{GH}$ must intersect AK, or equivalently, $\overleftrightarrow{AE}$, as desired. This ends the proof of the claim.

Let the intersection of $\overrightarrow{GH}$ and $\overrightarrow{AE}$ be point L. We will now construct a Saccheri quadrilateral using L as one of the summit vertices. Construct N on m such that C lies between A and N, and $CN = GL$. Construct M and P on ℓ such that $LM \perp \ell$ and $NP \perp \ell$.

Claim: $LM = NP$ and thus $LMPN$ is a Saccheri quadrilateral with summit LN and base MP.

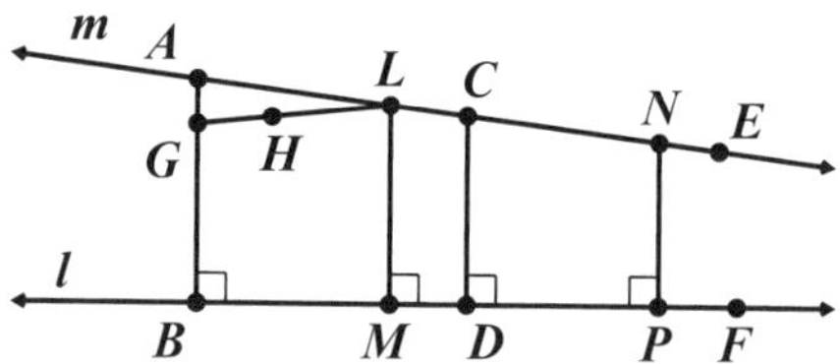

Figure 13.40. $LMPN$ is a Saccheri quadrilateral

We leave the proof of this claim and the fact that the common perpendicular is unique as exercises for the reader. $\square$

By combining Corollary 13.26 with Theorem 13.34, we can now prove that the unique shortest distance between two ultraparallel lines is the length of the segment that is the unique common perpendicular to both, which we state as follows.

Theorem 13.35. *Suppose that ℓ and m are ultraparallel with unique common perpendicular $\overleftrightarrow{EF}$, where E lies on m and F lies on ℓ. Let A be any other point on m. The distance from A to ℓ is greater than the length of EF.*

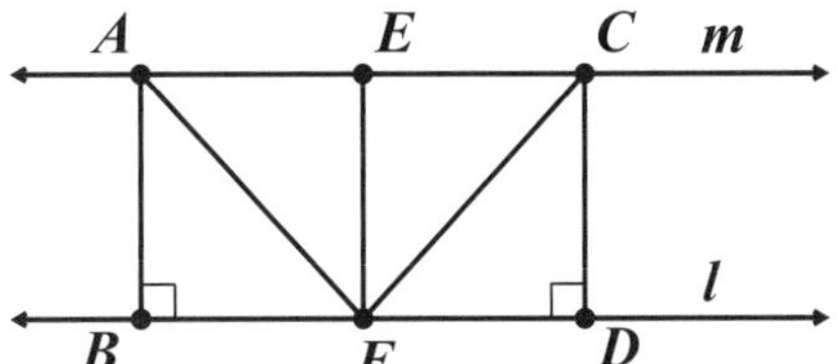

Figure 13.41. Theorem 13.35

Proof. Once again, our goal is to construct a Saccheri quadrilateral whose summit lies on m and whose base lies on ℓ. This time we will use A as one of the summit vertices and we will construct the quadrilateral with E as the midpoint of the summit and F as the midpoint of the base.

Construct C on m such that E lies between A and C, and $AE = EC$. Construct B and D on ℓ such that $AB \perp \ell$ and $CD \perp \ell$.

Claim: $AB = CD$ and $BF = FD$. We leave the proof of this to the reader.

Thus $ABDC$ is a Saccheri quadrilateral with summit AC and base BD, where E and F are the midpoints of AC and BD, respectively. Therefore, by Corollary 13.26, we have $EF < AB$, as desired. $\qquad\square$

One last item to address is related to vertical lines and the ideal point at infinity in the Half-plane model. Distinct vertical lines do not meet, and therefore, are either sensed parallel or ultraparallel. In Section 12.2.3, we claimed that these lines are sensed parallel, and earlier in this section we introduced the ideal point at infinity shared by all vertical lines in the Half-plane model. Now, we can show why vertical lines cannot be ultraparallel. From Exercise 13.1.5, we know that any two vertical lines in the Half-plane model cannot have a common perpendicular. This makes sense since any perpendicular to vertical line $x = k$ must be a semicircle with center at $(k, 0)$. With no common perpendicular, distinct vertical lines cannot be ultraparallel by Theorem 13.34. Hence, vertical lines in the Half-plane model must be sensed parallel, which justifies the existence of the ideal point at infinity in the model.

Circumscribing and inscribing a circle in Hyperbolic geometry

In Chapter 10, we saw that given any triangle in Euclidean geometry, we can both inscribe and circumscribe a circle using Propositions IV.4 and IV.5, respectively. One of these constructions was based on the incenter of the triangle, and the other, the circumcenter. As a reminder, the *incenter* is the concurrent intersection point of the three angle bisectors. The *circumcenter* is the concurrent intersection point of the perpendicular bisectors of the three sides. Since angle bisection and perpendicular bisector construction both belong to Neutral geometry,

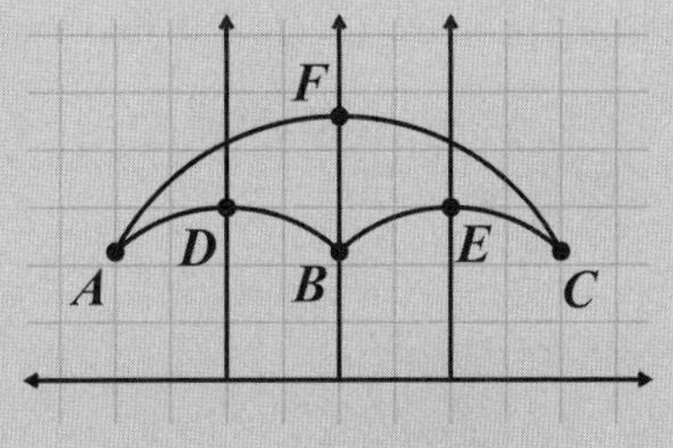

we might expect to be able to find the incenter and circumcenter of any triangle in Hyperbolic geometry. However, we are only guaranteed the existence of a
concurrent intersection point for *one* of these two constructions.

It turns out that while we will always be able to inscribe a circle in any triangle, we cannot always circumscribe one. As illustrated in the figure, $\triangle ABC$
cannot be circumscribed by a circle since its three perpendicular bisectors are
parallel. Hence, there is no point equidistant from all three vertices.

This means that the construction of the circumscribed circle cannot belong
to Neutral geometry. To satisfy your curiosity, revisit Euclid's Proposition IV.5.
You will find that Euclid I.29 is required to show that two perpendicular bisectors
must intersect.

Before we head off to discuss area in Hyperbolic geometry, let's take a moment to
reflect on what we've learned about this strange, new universe. Our main characters,
the line and the circle, as well as angles, triangles and some polygons, all still exist
and adhere to the properties of Neutral geometry. Yet, the nature of these objects has
changed dramatically. The distance between two parallel straight lines varies. There
are no similar triangles, and the angle sum of a triangle is no longer constant (much less
equal to two right angles). Rectangles and squares cannot exist. It's as if, in Hyperbolic
geometry, the geometrical objects we thought we knew became a bit less confined in
some ways, but more restricted in others.

Exercises 13.3

1. In Figure 13.29 which corresponds to Proposition 13.21, show that $HM \neq AC$.

2. Prove Corollary 13.22.

3. Consider Proposition 13.23, Saccheri's Proposition III.

   (a) Prove that if the summit angles are right, then $CD = AB$.
   (b) Prove that if the summit angles are obtuse, then $CD < AB$.
   (c) Prove that if the summit angles are acute, then $CD > AB$.

4. Prove Corollary 13.26.

5. In the proof of I.32$_H$, show that either D lies between F and G or all three of these
   points coincide. [*Hint: Use a proof by contradiction.*]

6. In Case 2 of the proof of I.32$_H$, prove that $BG = CH$ and $\angle BAC + \angle ACB + \angle ABC =$
   $\angle BCH + \angle CBG$.

7. In Case 3 of the proof of I.32$_H$, prove that $BG = CH$ and $\angle BAC + \angle ACB + \angle ABC =$
   $\angle BCH + \angle CBG$.

8. Prove the following corollary to I.32$_H$: *In a triangle, if one of the sides be produced,
   the exterior angle is greater than the two interior and opposite angles.*

9. Finish the proof of Corollary 13.29. That is, illustrate the two cases and use I.32$_H$ to
   prove that the angle sum for $ABCD$ is less than four right angles.

10. Prove Theorem 13.32.

11. Let m and ℓ be sensed parallels with ideal point Ω, and let A and C be distinct points on m where C lies on $\overrightarrow{A\Omega}$. Construct $AB \perp \ell$ and $CD \perp \ell$ where B and D are points on ℓ. Prove that $CD < AB$.

12. Prove the following theorem: *If ℓ and m are sensed parallels, then for any segment AB there exists a point, C, on ℓ such that the distance from C to m is equal to AB.*

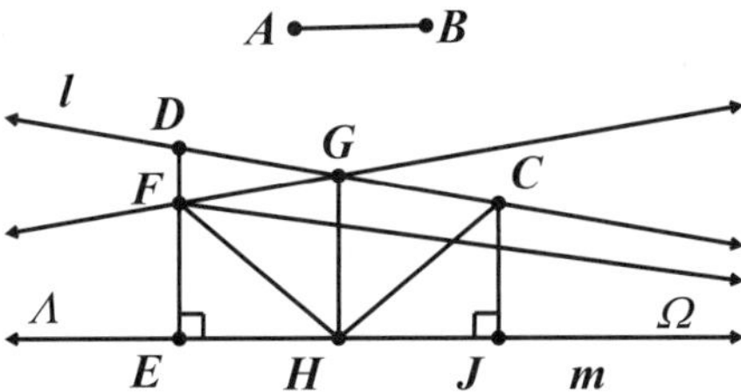

Figure 13.42. Exercise 13.3.12

Proof.   To begin, let AB be a given segment, and ℓ and m be sensed parallel lines with shared omega point Ω. Let Λ be the other omega point associated with line m. Pick a point D on ℓ and using Euclid I.12, construct E on m such that $DE \perp m$. If $DE = AB$, then we are finished. Otherwise, there are two cases to consider, one where $DE > AB$ and the other where $DE < AB$. Construct F on ray $\overrightarrow{ED}$ such that $FE = AB$. Consider the two sensed parallel rays to m through F, $\overrightarrow{F\Lambda}$ and $\overrightarrow{F\Omega}$. We will start with the case where $DE > AB$, as illustrated in Figure 13.42.

(a) Extend $\overrightarrow{F\Lambda}$ to the other side of DE, and show that this extension intersects ℓ at a point G.

(b) Construct C on ℓ such that $GF = GC$ and G lies between C and D. Using Euclid I.12, construct points H and J on m such that $GH \perp m$ and $CJ \perp m$. Join FH and CH. Show that $\triangle FGH \cong \triangle CGH$.

(c) Show that $\triangle FEH \cong \triangle CJH$, then use this result to finish the proof of the case where $DE > AB$.

(d) Modify the proof of the first case to show that the result holds when $DE < AB$, and include an appropriate diagram. $\square$

13. Prove Theorem 13.33.

14. Show that $LM = NP$ in the proof of Theorem 13.34.

15. Show that the common perpendicular found in both cases of the proof of Theorem 13.34 must be unique.

16. Show that $AB = CD$ and $BF = FD$ in the proof of Theorem 13.35.

17. Consider the lines, k and ℓ, given by the equations $y = \sqrt{4 - x^2}$ and $y = \sqrt{25 - x^2}$ in the Poincaré Half-plane model.

(a) Find the equation for the unique common perpendicular to k and ℓ.

(b) Determine the shortest distance between k and ℓ.

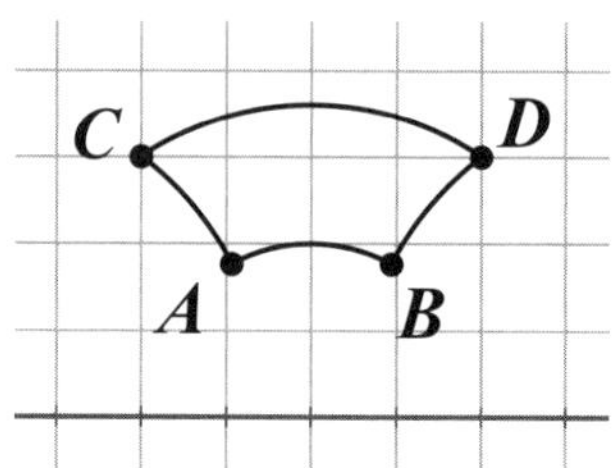

Figure 13.43. Exercise 13.3.18

18. Consider the Saccheri quadrilateral $ABDC$, with base AB and summit CD, where $A = \left(-\frac{16}{17}, \frac{30}{17}\right), B = \left(\frac{16}{17}, \frac{30}{17}\right), C = (-2, 3)$ and $D = (2, 3)$ are the vertices in the Poincaré Half-plane model, as illustrated in Figure 13.43.

  (a) Find the equations for lines $\overleftrightarrow{AB}$ and $\overleftrightarrow{CD}$.

  (b) Find the equations for lines $\overleftrightarrow{AC}$ and $\overleftrightarrow{BD}$.

  (c) Using dynamic geometry software, such as Geometer's Sketchpad®, graph the Saccheri quadrilateral $ABDC$. Construct the appropriate tangent lines and use them to calculate the summit angles. Be sure to include your sketch.

19. Consider triangles $\triangle ABC$ and $\triangle ABD$, where C lies on line segment BD, as illustrated in Figure 13.44. Which triangle has the larger angle sum? Prove your answer.

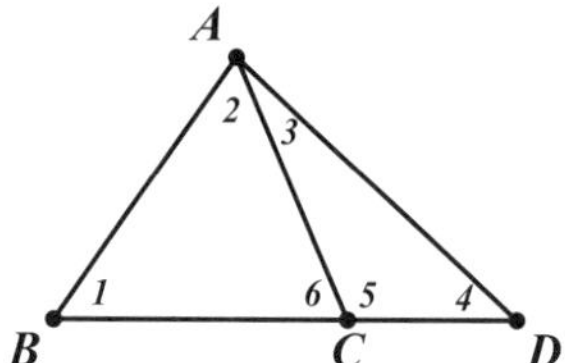

Figure 13.44. Exercise 13.3.19

13.4 Hyperbolic area

In this final section, we discuss the concept of area in Hyperbolic geometry. We start by recalling from Section 7.3 the axioms that any area function must satisfy, but remove the axiom concerning squares as these figures do not exist in Hyperbolic geometry.

A-1 Area is a function from the set of polygonal regions to the real numbers.

A-2 For every polygonal region R, $Area(R) > 0$.

A-3 *The Congruence Postulate.* If two triangular regions are congruent, then they have the same area.

A-4 *The Additivity Postulate.* If two polygonal regions intersect only in edges and vertices (or do not intersect at all), then the area of their union is the sum of their areas.

Any function that satisfies these four axioms in Hyperbolic geometry is an area function. Before proceeding, briefly think about why the formula for triangular area in Euclidean geometry is sufficient to calculate the area of irregular polygons, such as the irregular pentagon shown in Figure 13.45.

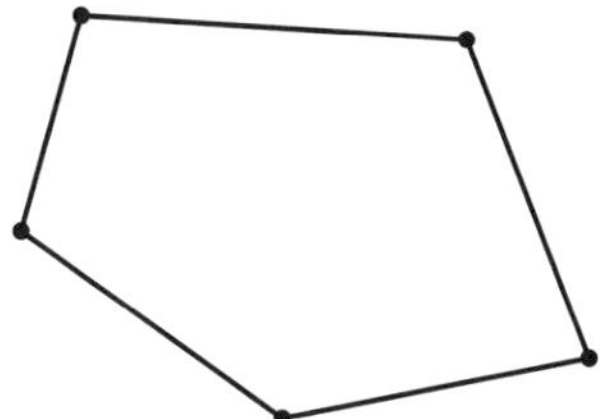

Figure 13.45. An irregular pentagon

One strategy is to divide the pentagon into triangles, calculate the area of each triangle and then sum the individual triangular areas to find the area of the pentagon. Dividing a polygon into nonoverlapping triangles is called *triangulating*, which we formally define here.

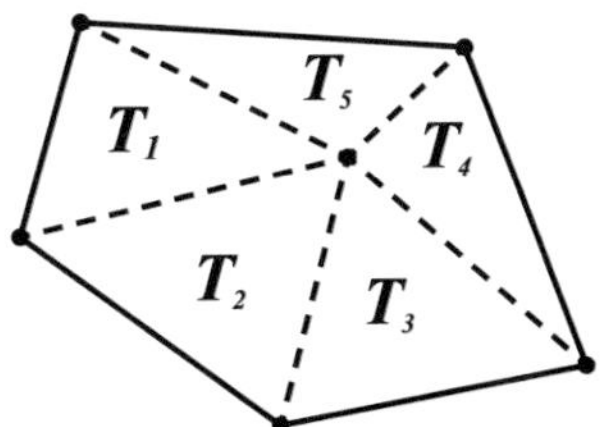

Figure 13.46. A star triangulation

Definition 13.36. A **triangulation** of a polygon, P, is a finite collection of triangles, $\mathcal{C} = \{T_1, T_2, \dots, T_n\}$, such that

(1) Distinct T_i and T_j may intersect only at edges and vertices.

(2) P is the union of the T_i.

Note that every polygon has infinitely many triangulations. Figure 13.46 shows an example of a **star triangulation**, whereas Figure 13.47 demonstrates a **border triangulation**. In order for an area function to be *well-defined*, we need ensure that, no matter how we triangulate our polygon, the resulting sum of the triangular areas will be the same. While this is indeed verifiable for the Euclidean area function, we will not prove it. (For a proof, see [87].)

With the idea of triangulation in mind, our strategy for developing a Hyperbolic area function follows the same path: define the area of a triangle and then extend the definition to all polygons. If our function satisfies the four area axioms, then we will have a valid area function. Perhaps not surprisingly, like its counterpart in Spherical geometry, the definition is based on angles instead of lengths.

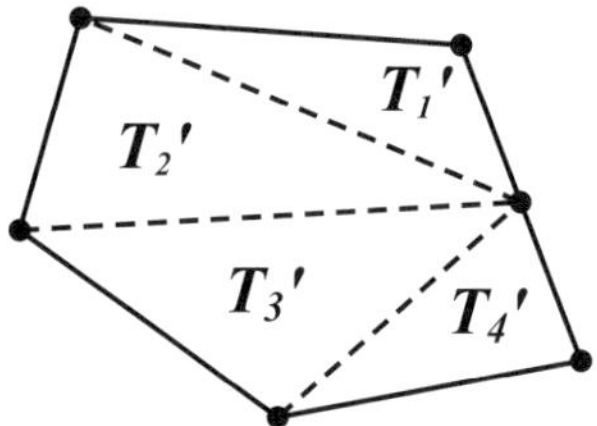

Figure 13.47. A border triangulation

Definition 13.37. The **defect of triangle** $\triangle ABC$ is defined as $\pi-(\alpha+\beta+\gamma)$, where α, β, and γ are the angles of the triangle given in radians. The triangular defect is denoted by $\delta(\triangle ABC)$.

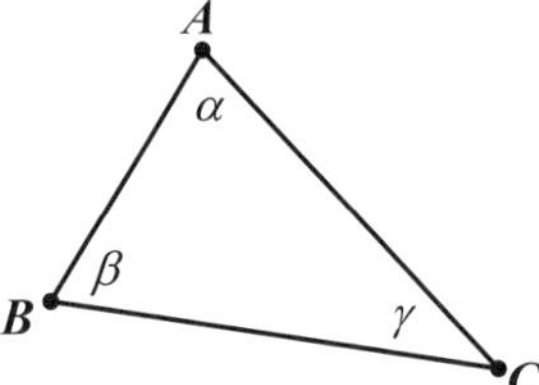

Figure 13.48. $\text{Area}(\triangle ABC) = \delta(\triangle ABC) = \pi - (\alpha + \beta + \gamma)$

Since a triangle's angle sum is less than π by I.32$_H$, we know that the defect of a triangle is a positive number. We now define the area of a triangle to be its defect.

Definition 13.38. The **area of a triangle** is its defect, that is

$$Area(\triangle ABC) = \delta(\triangle ABC).$$

This formula for triangular area is quite similar to Theorem 4.8 which gives the area of a Spherical triangle on a sphere of unit radius. Recall that the angle sum of a Spherical triangle exceeds π. Consequently, in Spherical geometry the difference is reversed so that triangular area is the angle sum minus π. Here, we would like extend the Hyperbolic definition of area to all polygons, and thus, we need to define the defect of a polygon. First, we define the defect of a given triangulation for a polygon.

Definition 13.39. Let P be any polygon and let $\mathcal{C} = \{T_1, T_2, \dots, T_n\}$ be a triangulation of P. The **defect of a triangulation** $\mathcal{C}$ is given by $\delta(\mathcal{C}) = \delta(T_1) + \delta(T_2) + \cdots + \delta(T_n)$.

We leave it to the reader to check that both the star triangulation in Figure 13.46 and the border triangulation in Figure 13.47 have the same defect. This is not a coincidence. The following theorem gives the rather surprising result that the defect of a polygon is independent of the triangulation. The proof is beyond the scope of this book, but can be found in [**87**].

Theorem 13.40. *If $\mathcal{C}_1$ and $\mathcal{C}_2$ are triangulations of polygon P, then $\delta(\mathcal{C}_1) = \delta(\mathcal{C}_2)$.*

Theorem 13.40 allows us to use any convenient triangulation of a polygon to find its defect, and assures us that it is well-defined.

Definition 13.41. The **defect of polygon** P, denoted $\delta(P)$, is equal to the defect of any of its triangulations.

Definition 13.42. The **area of polygon** P is its defect, that is, $Area(P) = \delta(P)$.

Let's check that the area function satisfies the four area axioms.

A-1 Since we have defined area for all polygons, the first axiom holds. ✓

A-2 Since the defect of a triangle is positive and the area of a polygon is the sum of defects of triangles, area is a positive function. Thus, the second axiom holds. ✓

A-3 Since congruent triangles have congruent angles, their defects are equal. Thus, the third axiom holds. ✓

A-4 If two polygons, P_1 and P_2, with respective triangulations, $\mathcal{C}_1$ and $\mathcal{C}_2$, intersect only in edges and vertices, then a triangulation of $P_1 \cup P_2$ would be $\mathcal{C} = \mathcal{C}_1 \cup \mathcal{C}_2$. The area of $P_1 \cup P_2$ is $Area(P_1 \cup P_2) = \delta(\mathcal{C}) = \delta(\mathcal{C}_1) + \delta(\mathcal{C}_2) = Area(P_1) + Area(P_2)$. The fourth axioms holds. ✓

Now that we have found one area function, you may wonder whether there are others. While we will not prove it, all other area functions in Hyperbolic geometry are simply constant multiples of this area function.

Given any convex polygon[2], we can simplify our calculation of its area. We leave it to the reader to prove the following result.

Theorem 13.43. *A convex polygon P with n sides and vertex angles α_1, α_2, ..., α_n, has area given by*

$$Area(P) = \pi(n - 2) - \sum_{i=1}^{n} \alpha_i.$$

With this theorem, the area of the convex pentagon from Figure 13.45 is easily computed as 3π minus its vertex angle sum. In comparison to its Euclidean counterpart, this hyperbolic area calculation is much easier. Theorem 13.43 highlights another unusual property of area in Hyperbolic geometry: no triangle has an area greater than or equal to π. A quadrilateral (convex or otherwise) cannot have an area of 2π or larger. If you lived in Hyperbolic geometry, then you could not enclose more area than $\pi(n - 2)$ with a convex n-gon.

We now introduce the concept of *equivalent by finite decomposition*, which is very closely related to area. This definition belongs to Neutral geometry, and therefore, exists in Euclidean as well as Hyperbolic geometry.

Definition 13.44. Let P and P' be polygons with triangulations $\mathcal{C} = \{T_1, T_2, ..., T_n\}$ and $\mathcal{C}' = \{T_1', T_2', ..., T_n'\}$, respectively, such that for each i, we have $T_i \cong T_i'$. We say that polygons P and P' are **equivalent by finite decomposition**, and we write $P \equiv P'$.

---

[2]A polygon is *convex* if the line segment joining any two points of the polygon lies in the interior of the polygon.

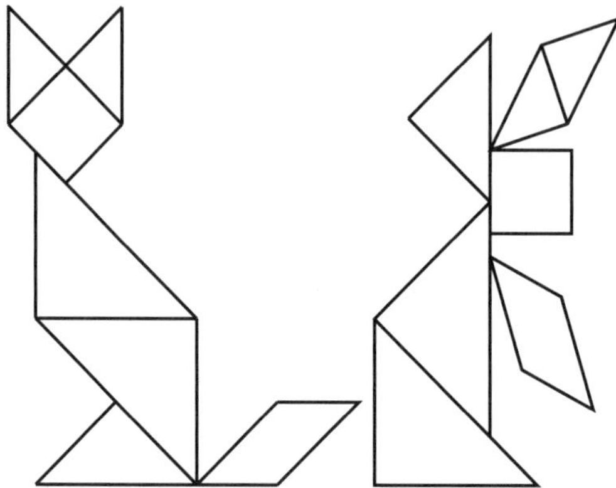

Figure 13.49. A cat and rabbit from a tangram set

Informally, we can think of two polygons as being equivalent if we can "cut" one of the polygons into a finite number of triangles, and then "rearrange" the triangles to form the other polygon. A tangram puzzle uses a similar concept: the seven tiles in a tangram set can be rearranged to form many different shapes. The cat and the rabbit shown in Figure 13.49 are two such examples. For an example in Hyperbolic geometry, we leave the proof of the following theorem as an exercise for the reader.

Theorem 13.45. *A triangle and its associated Saccheri quadrilateral are equivalent by finite decomposition.*

Let's consider whether the equivalence by finite decomposition relation is reflexive, symmetric or transitive.

Reflexive: Clearly, any polygon is equivalent to itself by finite decomposition. ✓

Symmetric: It is also apparent that if P_1 is equivalent to P_2, then P_2 is equivalent to P_1. ✓

Transitive: While it is true that if P_1 is equivalent to P_2, and P_2 is equivalent to P_3, then P_1 is equivalent to P_3, showing this requires some work. We start by considering the example illustrated in Figure 13.50. Note that, as shown in the upper left section of the figure, polygons P_1 and P_2 have triangulations $\mathcal{C}_1 = \{T_1, T_2\}$ and $\mathcal{C}_2 = \{T_1', T_2'\}$, respectively, such that for each i, we have $T_i \cong T_i'$. Furthermore, as shown in the lower left section of the figure, polygons P_2 and P_3 have triangulations $\mathcal{C}_2' = \{S_1, S_2\}$ and $\mathcal{C}_3 = \{S_1', S_2'\}$, respectively, such that for each i, we have $S_i \cong S_i'$. If we overlap the two triangulations for P_2, namely $\mathcal{C}_2$ and $\mathcal{C}_2'$, we can create a new triangulation $\mathcal{C}_2'' = \{U_1, U_2, U_3, U_4\}$ of P_2, as shown in the right section of the figure. Since each U_i lives entirely in one T_j' and in one S_k, this new triangulation of P_2 is a finer decomposition of P_2 than either $\mathcal{C}_2$ or $\mathcal{C}_2'$. Consequently, we can easily relate the components of one to another. For example, T_2' has a triangulation composed of U_1 and U_4, whereas S_1 has a triangulation composed of U_3 and U_4. Since $T_j' \cong T_j$ and $S_k \cong S_k'$ for each j and k, we can imagine rearranging the triangles of the new triangulation for P_2 to form either P_1 or P_3, as shown in the right section of the figure. More formally, we have $U_i \cong U_i' \cong U_i''$ for each i. Thus, P_1 is equivalent to P_3.

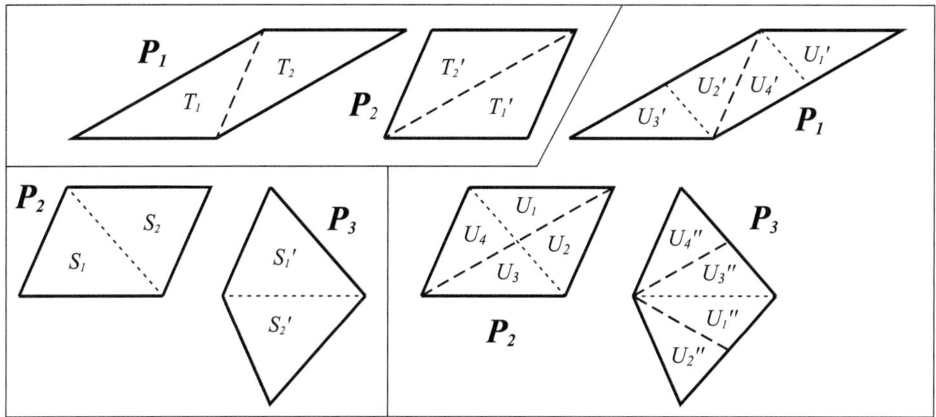

Figure 13.50. Transitivity of equivalence by finite decomposition

As noted in our example, each U_i lies completely in one T_j' and one S_k. For example, $U_4 \subseteq T_2'$ and $U_4 \subseteq S_1$. Any triangulation, $\mathcal{C}_2''$, of P_2 with this property is called a **common subdivision** of $\mathcal{C}_2$ and $\mathcal{C}_2'$. Do any two triangulations of the same polygonal region, P, always have a common subdivision? The answer is yes. To see this, we note that by overlapping two triangulations of P, we create a finite number of polygonal subregions. In our example, we created four triangular subregions. However, even if a particular subregion is not triangular, it has a triangulation. The finite union of the triangulations for each subregion is a triangulation of P. Since we can always create a common subdivision for the two triangulations of P_2, P_1 is equivalent to P_3. ✓

We summarize these three properties in the following theorem.

Theorem 13.46. *Equivalence by finite decomposition is an equivalence relation, that is, it satisfies the following three properties:*

(1) *For all polygons P, $P \equiv P$. (reflexive)*

(2) *If $P_1 \equiv P_2$, then $P_2 \equiv P_1$. (symmetric)*

(3) *If $P_1 \equiv P_2$ and $P_2 \equiv P_3$, then $P_1 \equiv P_3$. (transitive)*

By Theorem 13.46, the finite decomposition relation partitions the set of all polygons into equivalence classes, and each equivalence class consists of all polygons that can be cut up and rearranged to form each other. It is tempting to think of each equivalence class as the set of polygons formed by a specific set of tangram tiles (assuming any given set of tangram tiles cannot be formed from some other set). The fact that there are infinitely many polygons in each equivalence class makes this untrue. What we can say is that for each finite set of polygons in any given equivalence class, there is a finite set of triangles that we can use to show that any two of these triangles are equivalent. The following theorem relating finite decomposition and area holds in both Euclidean and Hyperbolic geometry.

Theorem 13.47. *If two polygons are equivalent by finite decomposition, then they have the same area.*

Proof.   Suppose P and P' are polygons such that $P \equiv P'$. By definition, there are triangulations $\mathcal{C} = \{T_1, T_2, \dots, T_n\}$ and $\mathcal{C}' = \{T_1', T_2', \dots, T_n'\}$ for P and P', respectively, such that $T_i \cong T_i'$ for all i. By [A-3], congruent triangles have the same area, so $Area(T_i) = Area(T_i')$ for all i. By [A-4], $Area(P) = \sum_i Area(T_i) = \sum_i Area(T_i') = Area(P')$. $\square$

Combining Theorems 13.45 and 13.47 gives the following corollary.

Corollary 13.48. *A triangle and its associated Saccheri quadrilateral have the same area.*

Perhaps more surprising is that, in both Euclidean and Hyperbolic geometry, the converse of Theorem 13.47 is also true.

Theorem 13.49 [Bolyai]. *If two polygons have the same area, then they are equivalent by finite decomposition.*

Thus, in both Euclidean and Hyperbolic geometry, having the same area and being equivalent by finite decomposition are really two sides of the same coin. Therefore each equivalence class determined by finite decomposition is specified by a positive real number, namely the area that is common to all polygons in the equivalence class. While we will not prove this general version of Bolyai's Theorem, we do prove the following special case.

Theorem 13.50 [Bolyai's Theorem for Hyperbolic Triangles]. *If two triangles have the same area, then they are equivalent by finite decomposition. That is, if $\delta(\triangle ABC) = \delta(\triangle DEF)$, then $\triangle ABC \equiv \triangle DEF$.*

To prove this theorem, we need the following two lemmas.

Lemma 13.51. *Two Saccheri quadrilaterals with congruent summits and congruent summit angles are congruent.*

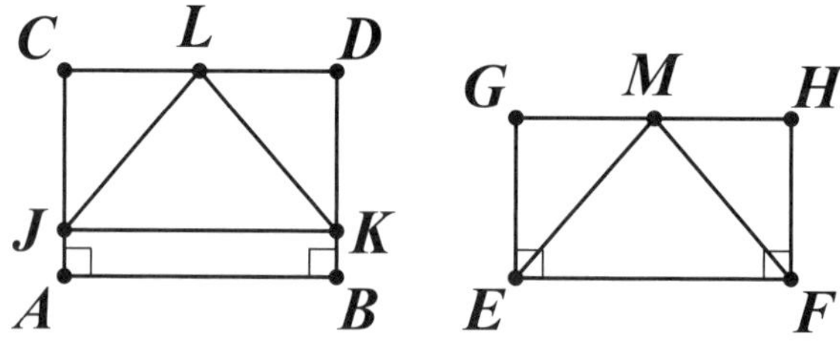

Figure 13.51. Saccheri quadrilaterals $ABDC$ and $EFHG$

Proof of Lemma.   Let $ABDC$ and $EFHG$ be Saccheri quadrilaterals with bases AB and EF, respectively. Assume that $CD = GH$, and $\angle C = \angle D = \angle G = \angle H$. To establish the congruence of these quadrilaterals, we must show that the corresponding sides and bases are congruent as they are the only remaining corresponding parts. Specifically, we need to show that $AB = EF$ and $AC = EG$. We use a proof by contradiction and suppose that $AC \neq EG$. WLOG, assume $AC > EG$.

Construct J on AC, and K on BD, such that $CJ = DK = GE = HF$. Construct JK. Let L and M be the midpoints of CD and GH, respectively, as illustrated in Figure 13.51. By SAS, we have $\triangle CJL \cong \triangle DKL \cong \triangle GEM \cong \triangle HFM$. Thus, $JL = KL = EM = FM$, $\angle CJL = \angle DKL = \angle GEM = \angle HFM$, and $\angle CLJ = \angle DLK = \angle GME = \angle HMF$. Subtracting equals from equals, we have $\angle JLK = \angle EMF$. Consider isosceles triangles $\triangle JLK$ and $\triangle EMF$. By SAS, they are congruent. Thus, $\angle LJK = \angle LKJ = \angle MEF = \angle MFE$. Adding equals to equals we have $\angle CJK = \angle DKJ = \angle GEF = \angle HFM$. Thus, $\angle CJK$ and $\angle DKJ$ are both right angles. But this implies that the quadrilateral $ABKJ$ has four right angles, which is impossible by Corollary 13.29. Thus, $AC = GE$, that is, the corresponding sides of these quadrilaterals are congruent.

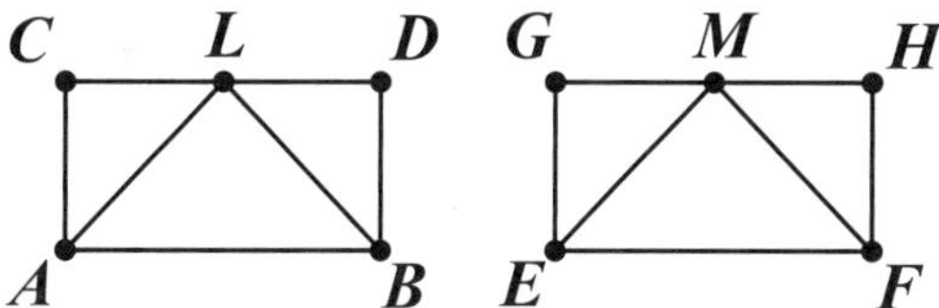

Figure 13.52. Saccheri quadrilaterals $ABDC$ and $EFHG$

To show that the corresponding bases are congruent, that is, $AB = EF$, we simply repeat the previous argument with AB replacing the unnecessary JK, as illustrated in Figure 13.52. Here, we have $\triangle CAL \cong \triangle DBL \cong \triangle GEM \cong \triangle HFM$, and hence, $\triangle ALB \cong \triangle EMF$. Thus, $AB = EF$, as desired. $\qquad\square$

Lemma 13.52. *Two triangles with the same defect and one pair of congruent sides are equivalent by finite decomposition.*

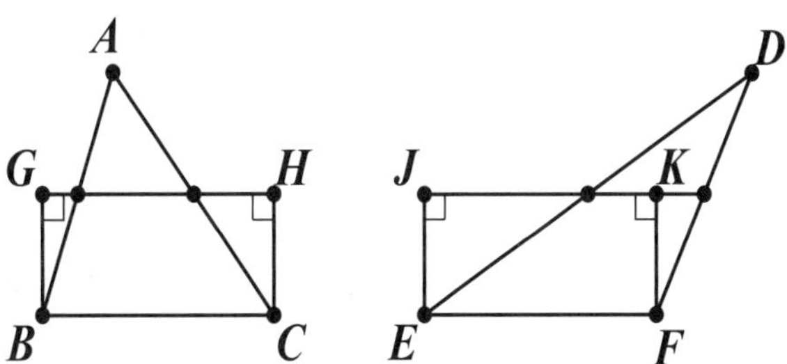

Figure 13.53. $\delta(\triangle ABC) = \delta(\triangle DEF)$ and $BC = EF$

Proof of Lemma. Let $\triangle ABC$ and $\triangle DEF$ be triangles with $BC = EF$ and $\delta(\triangle ABC) = \delta(\triangle DEF)$. Construct associated Saccheri quadrilaterals $BCHG$ and $EFKJ$, with summits BC and EF, for $\triangle ABC$ and $\triangle DEF$, respectively, as illustrated in Figure 13.53. Since a triangle and its associated Saccheri quadrilateral are equivalent by finite decomposition, $\triangle ABC \equiv \square BCHG$ and $\triangle DEF \equiv \square EFKJ$. By hypothesis, $\triangle ABC$ and $\triangle DEF$ have equal angle sum. Additionally, the angle sum of each triangle is equivalent to the sum of the summit angles of its associated Saccheri quadrilateral. Therefore, $\angle GBC + \angle HCB = \angle JEF + \angle KFE$. Since the summit angles of a Saccheri quadrilateral are congruent by Saccheri's Proposition 13.20, we have $\angle GBC = \angle JEF$.

By Lemma 13.51, the two Saccheri quadrilaterals are congruent. Since equivalence by decomposition is transitive, we have $\triangle ABC \equiv \triangle DEF$, as desired. $\square$

We are now ready for the proof of Theorem 13.50.

Proof of Theorem 13.50 *[Bolyai's Theorem for Hyperbolic Triangles].* Consider two triangles $\triangle ABC$ and $\triangle DEF$ with $\delta(\triangle ABC) = \delta(\triangle DEF)$. Our strategy will be to construct a third triangle that has the same angle sum as $\triangle ABC$ and $\triangle DEF$. Additionally, this new triangle will share one side with $\triangle ABC$ and another side with $\triangle DEF$. We can then use Lemma 13.52 twice to show that $\triangle ABC$ and $\triangle DEF$ are each equivalent by decomposition to this third triangle. Since equivalence by decomposition is transitive, the given triangles will, therefore, be equivalent.

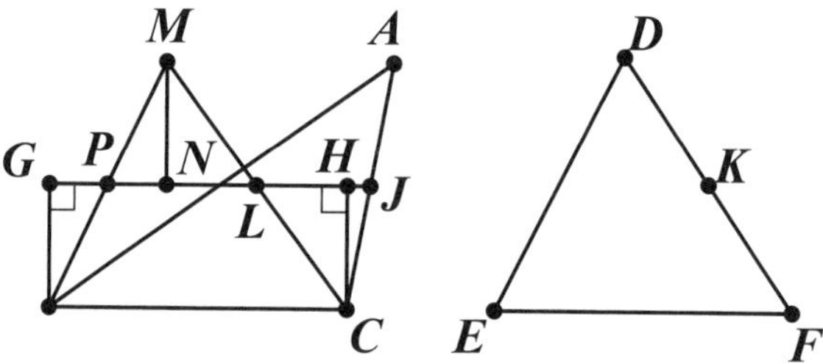

Figure 13.54. $\delta(\triangle ABC) = \delta(\triangle DEF)$ and $DF > AC$

Case 1. If $AC = DF$, then by Lemma 13.52 we have $\triangle ABC \equiv \triangle DEF$.

Case 2. If $AC \neq DF$, then WLOG, suppose $DF > AC$. We start by constructing the associated Saccheri quadrilateral $BCHG$ with summit BC for $\triangle ABC$. Let J and K be the midpoints of AC and DF, respectively. Construct a point, L, on ray $\overrightarrow{HG}$ such that $CL = FK$, as illustrated in Figure 13.54. (Such a point will exist since $FK > CJ \geq CH$, as the distance from C to $\overleftrightarrow{GH}$ is shortest at the perpendicular by Theorem 3.11.) Construct a point, M, on ray $\overrightarrow{CL}$ such that L is between C and M, and $ML = CL$. Note that M is on the opposite side of $\overleftrightarrow{GH}$ as B and C, and $CM = FD$. Construct N on $\overleftrightarrow{GH}$ such that $MN \perp GH$. Finally, construct MB and let P be the intersection of MB with $\overleftrightarrow{GH}$.

Notice that L must lie between N and H. (Why? See the box on page 329.) Consider the triangles $\triangle MNL$ and $\triangle CHL$. Since $ML = CL$, $\angle MNL = \angle CHL$, and $\angle MLN = \angle CLH$ (Vertical Angles), by AAS, $\triangle MNL \cong \triangle CHL$. Thus, $MN = CH = BG$.

Claim: Either P lies between G and N, or all three points coincide. (Why? See the box on page 329.)

In the case that all three coincide, then P is the midpoint of MB. Otherwise, consider triangles $\triangle MNP$ and $\triangle BGP$. Since $MN = BG$, $\angle BGP = \angle MNP$, and $\angle MPN = \angle BPG$ (Vertical Angles), by AAS, $\triangle MNP \cong \triangle BGP$. Thus, $MP = BP$, and again, P is the midpoint of MB.

Consider triangle $\triangle MBC$. Since P and L are the midpoints of MB and MC, respectively, and $\overleftrightarrow{PL} = \overleftrightarrow{GH}$, quadrilateral $BCHG$ is also the associated Saccheri quadrilateral for $\triangle MBC$. Thus, the angle sum of triangle $\triangle MBC$ equals the sum of the summit angles of $BCHG$, which equals the angle sum of triangle $\triangle ABC$ by construction. Since triangles $\triangle ABC$ and $\triangle MBC$ have the same angle sum and share BC, by Lemma 13.52, $\triangle ABC \equiv \triangle MBC$. Similarly, since $\triangle MBC$ and $\triangle DEF$ have the same angle sum and $MC = DF$, $\triangle MBC \equiv \triangle DEF$. By transitivity, $\triangle ABC \equiv \triangle DEF$, as desired. $\qquad\square$

How to answer several "Why?" questions in the chapter

After introducing Saccheri quadrilaterals, the proofs of I.32_H and Theorem 13.50 (and Exercise 13.3.5) include a step where the reader must verify a claim similar (or identical) to the following: Assume that triangle $\triangle ABC$ is given, and we have the following construction:

- D is the midpoint of AB,

- E is the midpoint of AC,

- F, G and H are points on $\overleftrightarrow{DE}$ such that $AF \perp \overleftrightarrow{DE}$, $BG \perp \overleftrightarrow{DE}$, $CH \perp \overleftrightarrow{DE}$,

- BC is the summit, and HG is the base of the quadrilateral $BCHG$.

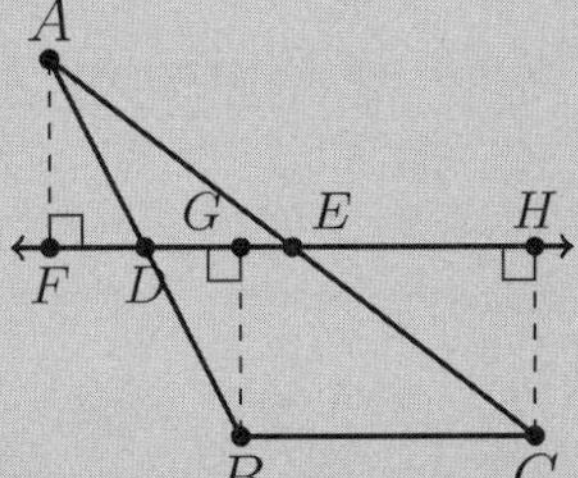

In order to prove that $BCHG$ is a Saccheri quadrilateral, we need verify the relative positions of D, F and G on $\overleftrightarrow{DE}$. The diagram shows one possible arrangement of these points, but we must verify that either D lies between F and G, or these three points coincide. We offer a representative proof that can be tailored to meet the conditions for each instance of this query.

Proof. Assume that the three points do not coincide, and show that D lies between F and G.

Case 1. Suppose exactly two of these points coincide. There are three sub-cases to consider, $D = F$, $D = G$ or $F = G$, as depicted below, where each marked angle is right by construction.

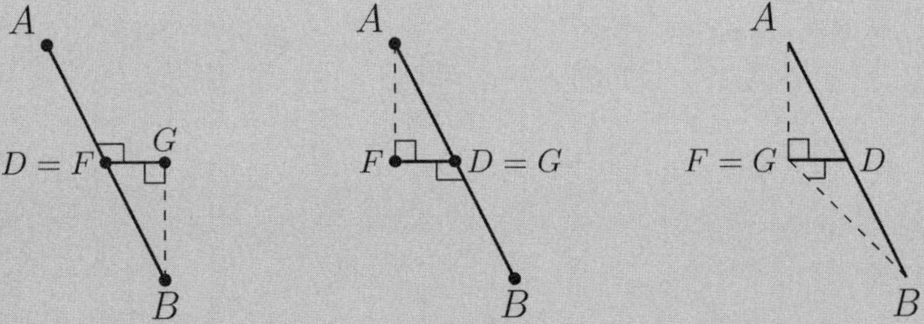

Each triangle shown in the first two sub-cases contradicts Euclid I.16. In the third sub-case, when $F = G$, there are two lines joining A and B, contradicting Hilbert I.2. Therefore, Case 1 is not possible.

> *Case 2.* The points D, F and G are distinct. We will show that D lies be-
> tween F and G by contradiction. While there are four possible arrangements
> of these points on line $\overleftrightarrow{DE}$, regardless of the arrangement, $\angle BDG$ is acute by
> Euclid I.17 when applied to $\triangle BDG$. By Euclid I.15, vertical angle $\angle ADX$ is
> equivalent to $\angle BDG$, and, is thus, acute. (Here, X is any point on $\overleftrightarrow{DE}$ such
> that D is between X and F.) Finally, $\triangle AFD$ contradicts Euclid I.16 as exte-
> rior angle $\angle ADX$ is less than interior right angle $\angle AFD$. Therefore, D must
> lie between F and G. □

Exercises 13.4

1. Show that $\delta(\triangle ABC) = \delta(\triangle ABD) + \delta(\triangle ADC)$ for the triangle shown in Figure 13.55.

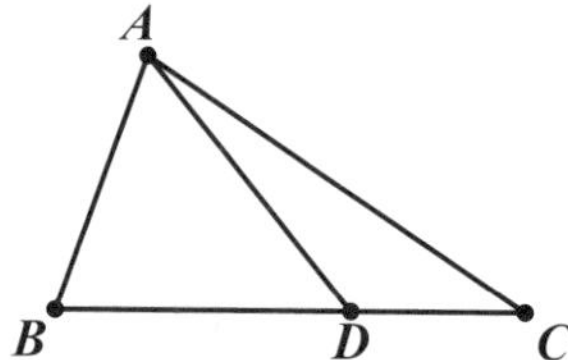

Figure 13.55. Exercise 13.4.1

2. Show that the star triangulation in Figure 13.46 and the border triangulation in Figure 13.47 have the same defect.

3. Using induction on integers $n \geq 3$, prove Theorem 13.43: A convex polygon P with n sides and vertex angles $\alpha_1, \alpha_2, \ldots, \alpha_n$, has area $Area(P) = \pi(n-2) - \sum_{i=1}^{n} \alpha_i$.

4. Prove Theorem 13.45: A triangle and its associated Saccheri quadrilateral are equivalent by finite decomposition, and thus, have the same area. [*Hint: Look at the three possible cases and use Exercises* 13.3.6 *and* 13.3.7.]

14

Finite Geometries

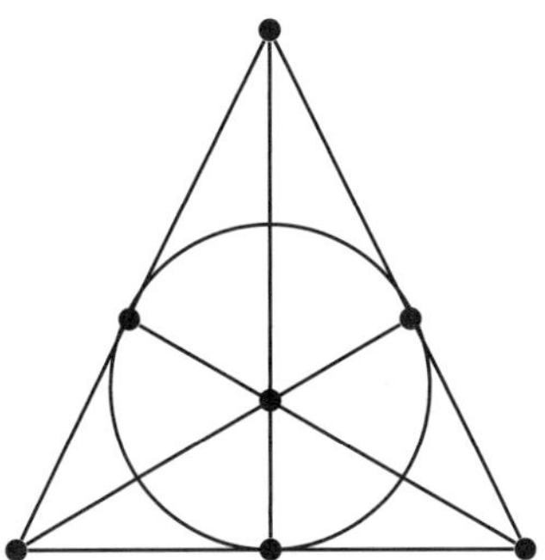

Figure 14.1. Fano's plane

The introduction of non-Euclidean geometries in the nineteenth century inspired a flurry of activity on the foundations of geometry in Germany and Italy in the late 1800s. It was only a matter of time before the unstated assumption of an infinite number of points was questioned, giving rise to geometries of finitely many points. Four Point geometry of Section 6.2 is one such example. Perhaps initially these small geometries were seen as a curiosity or cute mathematical plaything that belonged to the realm of the finite field theorists, but over the past century the study of finite geometries has blossomed into an exciting and thriving field of mathematics with an active research community connected to such diverse fields as statistics, combinatorics, quantum information theory, algebraic coding theory, graph theory and group theory.

In this chapter, we investigate two types of finite geometries, *projective planes* and *affine planes*. Before discussing these large families in general, we explore the smallest example of each type, starting where we left off in Chapter 6 with Four Point geometry. The reader should review this miniature geometry as presented in Section 6.2 before continuing with the current chapter.

14.1 Four Point geometry - Part 2

Let's continue our investigation of the small geometry introduced in Section 6.2, Four Point geometry. Two of its three axioms, which are restated below for ease of reference, describe the incidence structure; that is, the relation between points and lines. All of the finite geometries studied in this chapter are incidence geometries with axioms identical or similar in nature to Hilbert's Axioms of Incidence, but without any axioms of betweenness, congruence or continuity.

Four Point geometry

Undefined terms: *point, line* and *on*.

Axiom 4P-1. There exist exactly four points.

Axiom 4P-2. Any two distinct points are on exactly one line.

Axiom 4P-3. Each line is on exactly two points.

We will review our nomenclature for this geometry. As usual, a line with points A and B may be denoted by AB. Since every line consists of exactly two points, each line can be represented as a two-element set. For example, if line ℓ consists of points A and B, then ℓ can be written as $\{A, B\}$. For this reason, we also say that a line *contains* a point. Furthermore, to add variety to our language, we describe incidence in the following ways: a line *goes through* a point, a point *lies on* a line, or a point *is an element of* a line. When point P is on line ℓ, we also say that P is *incident* with ℓ, or ℓ is incident with P. We represent the geometry with diagrams such as those given in Figure 14.2, where a connecting segment between two points indicates that the points form a line.

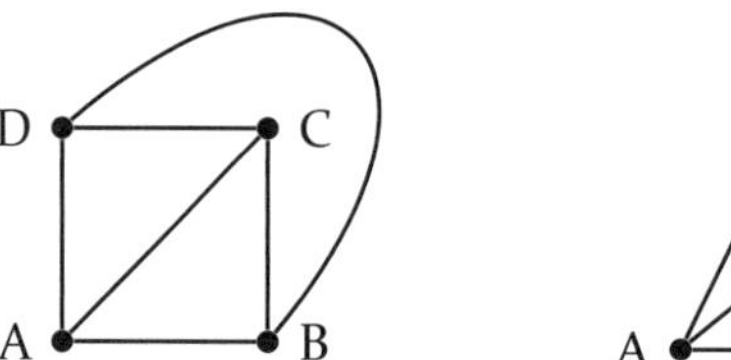

Figure 14.2. Diagrams to model Four Point geometry

By their very nature, finite geometries allow for the counting of points and lines, and consequently, we encounter many combinatorial results in this chapter. Our first example of a counting theorem is Theorem 6.8, stating that Four Point geometry has exactly six lines. Let's use the axioms to explore three more counting theorems for this miniature geometry. We start with a definition of intersecting lines before counting the number of times a pair of intersecting lines meet.

Definition 14.1. Two distinct lines on the same point P are said to be **intersecting** lines, and P is a **point of intersection**.

If a point P is on lines ℓ and m, we write ℓ intersects m at P, ℓ meets m at P, or $\ell \cap m = \{P\}$ when P is the only point of intersection.

Theorem 14.2. *In Four Point geometry, if two distinct lines intersect then they have exactly one point in common.*

Proof.   Assume lines ℓ and m meet at more than one point, say A and B. Thus, points A and B are on more than one line, contradicting Axiom 4P-2. $\square$

Next, we can count the number of lines through any point. Note that in each of our models, the point labeled A is on the following three lines: AB, AC, and AD. We leave it to the reader to use the axioms to prove that, regardless of the model, every point is on exactly three lines.

Theorem 14.3. *In Four Point geometry, each point is on exactly three lines.*

As is the case with all other geometries we have studied, distinct lines that do not intersect are called parallel.

Definition 14.4.   In a finite geometry, two distinct lines that do not have a point of intersection are **parallel**.

Each of our models in Figure 14.2 shows three pairs of lines that do not intersect, namely $\{AB, CD\}, \{AC, BD\}$, and $\{AD, BC\}$. Let's use the axioms to prove that, regardless of the model, every line has a unique parallel.

Theorem 14.5. *In Four Point geometry, to each line there is exactly one distinct line parallel to it.*

Proof.   Let ℓ be a line in the geometry. By Axiom 4P-3, ℓ has two points on it, say A and B. By Axiom 4P-1 there is a point C not on ℓ. By Theorem 14.3, C is on exactly three lines. By Axiom 4P-2, two of these lines intersect ℓ and the third does not intersect ℓ. Hence, there is at least one line distinct from and parallel to ℓ.

Assume there is a second line, m, parallel to ℓ. Then m is not on C since we have already accounted for the three lines on C. Since m is parallel to $\ell = AB$, m is not on A and m is not on B. Since m must be on two points by Axiom 4P-3, then there must be two more points in the geometry, giving a total of at least five points, which contradicts Axiom 4P-1. Therefore, each line in the geometry has exactly one distinct line parallel to it. $\square$

With Theorem 14.5, we see that Playfair's Axiom is valid in Four Point geometry, though certainly not independent.

Because this geometry is so small, it is easy to understand. Hence the proofs for the last two theorems may seem overly complicated. We ended Section 6.2 by noting that the set of axioms for Four Point geometry is categorical. Consequently, any statement about the undefined terms that holds true in one model will be true in all models. Thus for Four Point geometry, we could have invoked this fact to prove the previous two theorems. We will find, however, that this is not always possible for finite geometries, as some sets of axioms discussed in this chapter are not categorical.

On a final note regarding models for Four Point geometry, those given in Figure 14.2 live in the two-dimensional plane, but there are models for this geometry that are not planar. (See the box and associated problems, Exercises 14.1.1 and 14.1.2, for two examples.) Though we can interpret a model for a finite geometry as living in a higher dimension, we limit our consideration to models embedded in the two-dimensional plane for the remainder of the chapter.

Three-dimensional models for Four Point geometry

The models for Four Point geometry given in this section and Section 6.2 all live in a plane, but we are free to give a model that lives in a higher dimension since the axioms specify no such restriction. As shown on the left in Figure 14.3, a tetrahedron where each vertex is a point, and each edge is a line, is a model for Four Point geometry. An octahedron provides another interesting model where we identify the six vertices of the octahedron with the six lines of Four Point geometry. Here, each point is identified with a face of the octahedron. Since there are eight faces of an octahedron but only four points in our geometry, we may only use four faces. In order to satisfy Axiom 4P-3, each of the four points must be identified with a face from two sets of opposing faces of the octahedron. For example, we can identify the four points with the four shaded faces of the octahedron shown on the right in Figure 14.3, where each darker-hued face is opposite its lighter-hued counterpart. (We cannot choose any four faces. For example, the four faces in the upper half of the octahedron cannot be identified with the four points since the axioms would not hold. Why not?) It is left as an exercise for the reader to verify that the axioms are satisfied by these three-dimensional models.

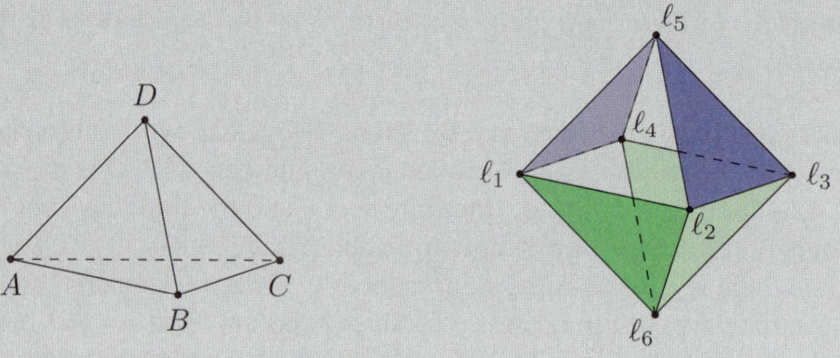

Figure 14.3. Nonplanar models of Four Point geometry

In this chapter we explore two types of finite geometries, affine and projective. When we explore affine planes in Section 14.4, we will see that Four Point geometry is the smallest example of a *finite affine plane*. Before doing this, we introduce another small geometry with a different set of axioms, Fano's Plane.

Exercises 14.1

1. Verify the three axioms of Four Point geometry in each of the following models:

   (a) the set model where lines are given by subsets of points

   (b) one of the diagrams in Figure 14.2

   (c) the tetrahedron in Figure 14.3 where the four points are the four vertices of the tetrahedron, and lines are the edges of the tetrahedron

   (d) the octahedron in Figure 14.3 where the four points are the four shaded opposite faces of the octahedron, and lines are the vertices of the octahedron

2. Give an isomorphism between one of the diagrams in Figure 14.2 and the octahedron in Figure 14.3 where the four points are the four shaded opposite faces of the octahedron, and lines are the vertices of the octahedron.

3. Using only the axioms of Four Point geometry, prove Theorem 14.3 : *In Four Point geometry, each point is on exactly three lines.*

4. Consider these axioms for a Three Point geometry given previously in Exercise 6.2.3.

   Axiom 3P-1. There exist exactly three points.

   Axiom 3P-2. Any two distinct points are on exactly one line.

   Axiom 3P-3. Two distinct lines are on at least one point.

   Axiom 3P-4. Not all points of the geometry are on the same line.

   (a) Construct a diagram to represent this geometry.
   (b) Prove that there are exactly three lines.
   (c) Prove that any two distinct lines are on exactly one point.
   (d) Prove that each line is on exactly two points.

5. Consider these axioms for a Four Line geometry.

   Axiom 4L-1. There exist exactly four lines.

   Axiom 4L-2. Any two distinct lines have exactly one point in common.

   Axiom 4L-3. Each point is on exactly two lines.

   (a) Construct a diagram to represent this geometry.
   (b) Prove that there are exactly six points.
   (c) Prove that each line has exactly three points.

14.2 Fano's plane

Our next example of a finite geometry is a little larger than Four Point geometry and has the distinction of being the first to appear in print. Given its historical significance, let's take a moment to learn about the mathematician who created it, Italian mathematician Gino Fano (1871–1952).

At seventeen, Fano began his studies at the University of Turin, home to several major figures in the Italian school of geometry who had taken up the axiomatic development of the field. (See page 110 for more on the efforts of the Italians and Germans to create a sound logical foundation for geometry in the second half of the 1800s.) Peano, Pieri, and Fano's advisor, Corrado Segre (1863–1924), formed a young, active research group whose focus of interest was projective geometry, a geometry we consider in the following section. At nineteen, Fano produced the first translation of what was to become one of the most influential mathematics papers of the nineteenth century, Klein's *Erlangen Programm*, which described how geometry could be studied through algebraic group theory. At twenty-two, Fano published original work in which he proposed a minimal set of axioms for projective geometry.

As we learned in Chapter 6, the measure of an axiomatic system is in its consistency, independence, and completeness, and a model is a powerful tool that can be used to determine whether these desired properties hold. (For example, see the small models used to demonstrate independence of axioms in Exercises 6.2.1 and 6.2.3.) Consequently, Fano used a model to verify the independence of one of his axioms. This technique was no different from that employed by his contemporaries, other Italian and German mathematicians who were proposing their own axiomatic systems during this highly active period of work in the foundations of geometry. The distinguishing feature of Fano's model was that it had a finite number of points and lines. For Fano, this tiny model solved his problem of establishing independence, but did not open up a new avenue of research for him in 1892. He worked in projective and algebraic geometry for the rest of his career, but did not develop any results in finite geometry. In fact, finite geometries did not gain prominence until 1906 when Americans Oswald Veblen (1880–1960) and W.H. Bussey (1879–1962) gave a systematic study of finite projective geometries [**122**].

Gino Fano came from a wealthy Jewish family in Mantua, Italy. After receiving his degree at the University of Turin in 1892, he took positions in Rome and Messina (Sicily) before returning to Turin in 1901 to an appointed professorship. He married and had two sons in Turin, spending the remainder of his academic life at the University until the fascist regime enacted racial laws against the Jews in 1938. He was barred from holding a position at any Italian academy or scientific institution. He and his wife fled to Switzerland while his sons both went to the United States to pursue their doctorates.

In 2002, on the occasion of the 50th anniversary of his death, Fano's younger son Robert shared some remembrances of his father for an international conference hosted by the University of Turin. "The 1938 dismissal from his professorship, his having to seek refuge in Switzerland and the dispersion of his family were very traumatic for my father because they amounted to the collapse of the three pillars of his life: his family, his Country and his profession. Those events also caused the only serious disagreement between my parents: my mother wanted to follow her children to the United States, while my father, as he told me before my departure, would never go to a country likely to be at war with Italy" [**47**].

At the end of the war, his position was reinstated, though he was 74 and soon retired. Until his death in 1952, he summered in Italy and wintered in the United States, lecturing in both countries.

The finite geometry presented by Fano in 1892 was three-dimensional, consisting of 15 points, 35 lines and 15 planes. When we restrict the scope of our consideration to one of the planes of this three-dimensional geometry, we have the two-dimensional geometry that has come to be known as the Fano plane. The following set of axioms completely characterizes this small geometry. We will see that this set of axioms has a concrete model, and it can be shown that the set is consistent, independent, and categorical, hence complete. Here, our undefined terms are the usual suspects, *point*, *line* and *on*.

Fano Plane

Axiom F-1. There exists at least one line.

Axiom F-2. There are exactly three points on every line.

Axiom F-3. Any two distinct points are on exactly one line.

Axiom F-4. Any two distinct lines are on exactly one common point.

Axiom F-5. Not all points are on the same line.

The wording for Axiom F-4 may sound a bit odd, but it is equivalent to stating that two distinct lines intersect in exactly one point. Recall that in Four Point geometry, we were given the number of points, but used the axioms to determine the number of lines. Here, we can use the axioms to determine the number of points *and* the number of lines in the plane. First, Axiom F-1 provides a line ℓ, then by Axiom F-2, there must be at least three points, A, B, and C, on ℓ. Axiom F-5 prescribes a fourth point, D, not on line ℓ. In order to keep track of these results as we derive them, we give a diagram of each step, shown in Figures 14.4 and 14.5. The configuration of points thus far is shown in Figure 14.4a.

Continuing our analysis, by Axiom F-3 there must be a line m through A and D. By Axiom F-2, there must be another point on line m. By Axiom F-3, this point cannot be B or C. If we label this new point E, then line m is $\{A, D, E\}$ by Axiom F-2, as shown in Figure 14.4b. By the same reasoning, we must have point F and line $\{B, D, F\}$, as shown in Figure 14.4c. By Axiom F-3, there is a line containing points A and F, and by Axiom F-4, the third point on this line cannot be B, C, D or E. Therefore, there must be a seventh point, G, forming the line $\{A, F, G\}$, as shown in Figure 14.4d.

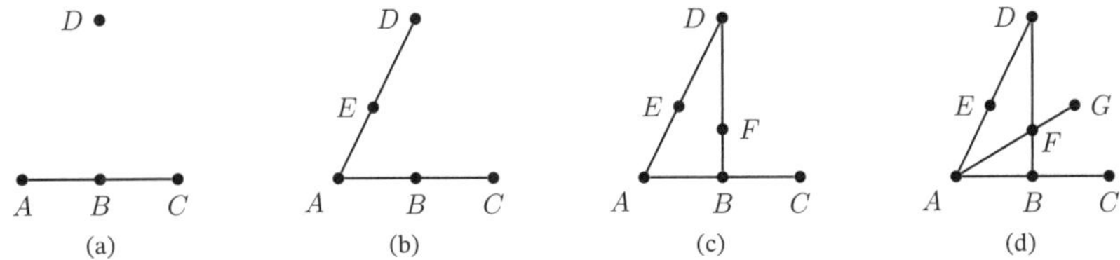

Figure 14.4. First set of steps to construct the Fano plane

At this stage of our analysis, we have at least 7 points and 4 lines. By Axiom F-3, we must consider the possible lines $\{B, E, \_\}, \{B, G, \_\}, \{C, D, \_\}, \{C, E, \_\}, \{C, F, \_\}, \{C, G, \_\}, \{D, G, \_\}, \{E, F, \_\}$ and $\{E, G, \_\}$, where the underscore is a placeholder for the unknown third point on the line. Let's start with $\{B, E, \_\}$. Suppose that the third point is a new point, say H. Then by Axiom F-4, $\{B, E, H\}$ and $\{A, F, G\}$ must have exactly one point in common, which is clearly not the case. This means that the third point on line BE must be one of the five existing points. Again, by Axiom F-4, the only point that can complete line BE is G, as shown in Figure 14.5a. This completed line, $\{B, E, G\}$, eliminates $\{B, G, \_\}$ and $\{E, G, \_\}$ from our list of possible lines. Note that while we connected points B, E and G with a circle, this is not an appearance of our main character as it is merely a representation of a set of points forming a line in this geometry, and really, any curve connecting these three points, but not meeting any of the other four points, would suffice.

By similar reasoning, the only point in our set of available points that can complete line $\{C, D, \_\}$ is also G, as shown in Figure 14.5b. This completed line, $\{C, D, G\}$, eliminates $\{C, G, \_\}$ and $\{D, G, \_\}$ from our list of possible lines. Likewise, the only point in our set of available points that can complete line $\{C, E, \_\}$ is F. This completed line, $\{C, E, F\}$, eliminates $\{C, F, \_\}$ and $\{E, F, \_\}$ from our list of possible lines.

Thus, there are at least seven points and at least seven lines in Fano's plane. To show that there cannot be more than seven points, suppose H is another point in the geometry. By Axioms F-2 and F-3, there is a line $t = \{A, H, \_\}$. By Axiom F-4, line t must meet line $\{B, D, F\}$. However, if B is on line t, then Axiom F-3 is violated since A

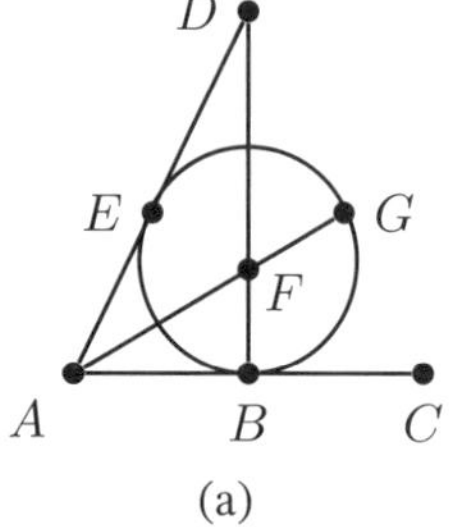
(a)

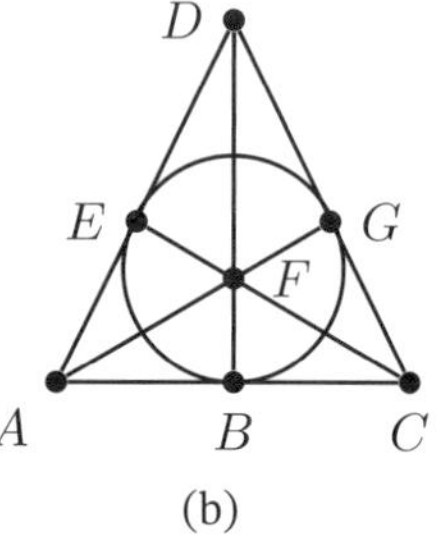
(b)

Figure 14.5. Second set of steps to construct the Fano plane

would be on two lines with B. Neither D nor F can be on line t for the same reason. Thus, there would have to be an extra point on line $\{B, D, F\}$, contradicting Axiom F-2. Therefore, we cannot have more than seven points. With no new points, we cannot add more lines, and thus, we have established the following theorem.

Theorem 14.6. *There are exactly seven points and seven lines in Fano's plane.*

At the end of the previous section, we noted that the focus of this chapter is finite affine and projective geometries, and that Four Point geometry is an example of a finite affine plane. Similarly, Fano's miniature geometry is an example of a *finite projective plane*. Having discussed a small example of each type, we are poised to study either of the large general families of *affine planes* or *projective planes*. Though it is the less familiar of the two, we start with projective geometry since the historical development of this unusual geometry is an interesting journey that takes us back to the Renaissance.

> **Does the circle have a part to play in the script of this geometry?**
>
> In Four Point geometry, as in any finite geometry, a line is a finite set of points, and "connecting the dots" in the diagram simply indicates a set of points comprising a line. Consequently, there is no part for the circle to play in Four Point geometry, or, for that matter, in *any* of the finite geometries under consideration in this chapter. Here, the line is a solo performer. The fact that line $\{B, E, G\}$ appears as a circle in Figure 14.5b, for example, does raise the interesting notion that the representation of a line in our model is not necessarily "straight." Any curve can represent a finite set of points forming a line in these geometries. With this change in perspective, we encourage the reader to revisit their essay in answer to the question, "What is a straight line?," in Chapter 1.

Exercises 14.2

1. Rewrite the axioms for the Fano plane, replacing the undefined terms point and line with *player* and *team*, respectively.

2. Using the axioms, prove that each point lies on exactly three lines in Fano's plane.

3. We can produce another small geometry by modifying one of Fano's axioms. Keep Axioms F-1, F-2, F-3 and F-5, but replace Axiom F-4 with Axiom Y-4 as follows:

If point P is not on line ℓ, then there exists a unique line on P that has no points in common with ℓ.

The resulting plane of nine points and twelve lines, known as **Young's geometry**, is named after American mathematician John Wesley Young (1879–1932).[1]

(a) Construct a diagram to represent this geometry.

(b) Prove that there are at least nine points.

(c) Prove that every point is on at least four lines.

(d) Prove that two lines parallel to a third line are parallel to each other.

(e) Prove that each line has at least two lines parallel to it.

4. For Fano's geometry:

(a) Use a model to verify that Axiom F-1 is independent.

(b) Use a model to verify that Axiom F-2 is independent.

(c) Use a model to verify that Axiom F-3 is independent.

(d) Use a model to verify that Axiom F-4 is independent. [Exercise 3 will be helpful.]

(e) Use a model to verify that Axiom F-5 is independent.

14.3 Projective geometry

14.3.1 History: Geometry of perspective develops into projective

geometry. In Chapter 12, we discussed the problem of making a two-dimensional map of the earth's surface, and how it became vitally important to produce accurate maps as Europeans ventured far and wide in search of trade routes in the 1500s. Painting a faithful representation of a three-dimensional world on a two-dimensional canvas presents a similar challenge. In this section, we take a brief historical tour to explain how a cartographer's need for accuracy and an artist's desire for realism led to a new geometry, projective geometry.

Spanning nearly 500 years, the history of projective geometry is a long and winding road, and it may help to first summarize the distinct phases of its development in order to provide a road map for the tour. Each stage of its evolution has an accompanying set of brackets with the names of contributing artists or mathematicians mentioned in the section. Initially, projective geometry grew from the art of perspective drawing practiced by artists, architects, cartographers and engineers who created illustrations of three-dimensional objects using their intuition and artistic skill. [Masaccio, Brunelleshi, da Vinci] The next phase is the codification of the methods used in practicing the craft of perspective drawing. The techniques were recorded formally, like recipes, and in some cases, accompanied by illustrations of tools of the trade designed to aid the artist. [Alberti, Dürer] When the subject proved sufficiently interesting, mathematicians modelled the problem, presenting solutions that could be incorporated into the practice of the craftsperson, and other solutions that were simply theoretically intriguing. [Kepler, Desargues, Pascal, Monge] Finally, a systematic study produced a

---

[1]Young was a charter member of the Mathematical Association of America who later served as its president. In his 1911 textbook, Young gave this small geometry as an example to explain the properties of consistency, independence and categoricalness of axiomatic systems [**127**].

theoretical framework encompassing a body of theorems for which a solid logical foundation was provided at a later date. [Poncelet, Carnot, Brianchon, Chasles, Gergonne, von Staudt, Steiner, Plücker, Klein, Möbius] With this road map in hand, let's take our tour of the history of the development of this geometry.

Our story goes back to the Renaissance, when artists developed perspective drawing techniques to depict realistic three-dimensional scenes. The basic idea employed by these artists was inspired by what our eyes see when we look off into the distance as we stand on a set of railroad tracks. Even though we know that the two tracks do not intersect, these parallel lines appear to meet at a point on the horizon which artists refer to as a **vanishing point**. Even though a vanishing point in a two-dimensional perspective drawing has no corresponding point in the three-dimensional world, it matches the image captured by our eyes. Furthermore, each family of parallel lines has its own vanishing point, and the set of all vanishing points forms the artists' **horizon line**. Developed in early Renaissance Italy in the 1400s, Masaccio (ca. 1401–1428) and Fillipo Brunelleshi (1377–1446) were among the first to use these principles in their paintings, and Leon Battista Alberti (1404–1472) was the first to publish the rules in a painting manual in 1435. In Albrecht Dürer's book on geometry, *Underweysung der Messung mit dem Zirkel und Richtscheyt* (Numberberg, 1525), he illustrates several mechanisms, including the one shown in Figure 14.6, designed to help an artist draw in perspective. Let's take a look at the information conveyed by the illustration.

Figure 14.6. Dürer's illustration of a mechanism for drawing with perspective

Dürer's practical solution to the problem is a one-point perspective method akin to a system for tracing an image, albeit from 3-d to 2-d. The idea is to create a one-to-one correspondence between the visible points of the three-dimensional subject and the points of the two-dimensional canvas by employing a fixed point in space marking the eye of the viewer. As shown in the figure, the frame of the canvas is placed between the subject (a lute) and the viewing position (a hook on the wall at right), and with the help of an assistant, a string connects the viewing position to points on the subject, passing through the canvas's frame along the way. The artist undergoes the tedious process of measuring the position of the string in the frame in order to transfer the corresponding point to the canvas. (Today, we could replicate this method by replacing the string with a laser pointer and the canvas with a piece of glass.) Take a moment to convince yourself that a visible line of the subject, a lute string, for example, becomes a line on the canvas. Like the railroad tracks in the photograph, if the strings of the lute are parallel, then the artist's depiction of the strings will appear to converge on the canvas. To help visualize the resulting image of the lute, Figure 14.7 shows a close-up of a lute depicted in 1533 by German artist Hans Holbein the Younger (1497–1543). The positioning of the lute is so similar to Dürer's illustration that it appears to be a homage to his countryman.

Figure 14.7. Detail of Holbein's *The Ambassadors* (1533)

Dürer's one-point perspective method is theoretically sound, but ultimately time-consuming and impractical. As expected, more practical perspective techniques developed to meet artists' needs. When drawing in one-point perspective, an artist could simply start by choosing the location of the horizon line and the desired vanishing point on the canvas, then ensure that all parallel lines meet at this point. For example, among the most famous perspective paintings is Leonardo da Vinci's (1452-1519) *The Last Supper*, which has a horizon line at the eye level of the table guests, and a single vanishing point at the right eye of Jesus in the center of the painting where the lines defined by the walls, tapestries and ceiling converge. Regardless of the mechanism employed, the device of Dürer or the chosen point and concurrent lines of da Vinci, the illusion of depth created by these new perspective techniques resulted in paintings that looked significantly more realistic.

Many interesting mathematical questions arise when we analyze the perspective techniques that aim to recreate the image of a subject as seen by our eye. The first

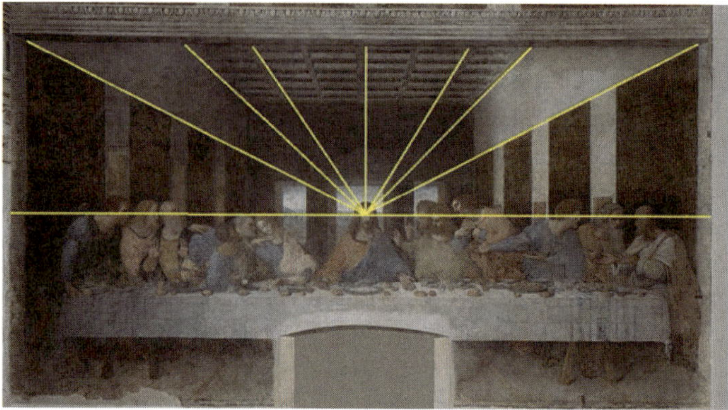

Figure 14.8. Horizon line and vanishing point in da Vinci's *The Last Supper* (1498)

solutions to these problems were developed within the confines of three-dimensional Euclidean geometry by architects, engineers, mathematicians and astronomers who produced noteworthy early results. German astronomer Johannes Kepler (1571–1630), French architect and engineer Girard Desargues (1591–1661), and Desargues' precocious student, French mathematician and philosopher Blaise Pascal (1623–1662), are among the notable contributors to the early development of the subject, though a few results appear as early as about 340 AD in Pappus of Alexandria's (ca. 290–350) *Synagoge*. Both Kepler and Desargues wrote of parallel lines having a common point at an infinite distance, a notion that may seem outrageous, but for now, consider it as the theoretical equivalent of a vanishing point in a perspective drawing, not unlike the meeting point at the horizon of a pair of parallel railroad tracks as they recede into the theoretically infinite distance.

Desargues' strange geometric results would eventually find a home in projective geometry, but at this stage of its development, there was no proper field of projective geometry. His first treatise, published in 1639, did not spark a flurry of new mathematics. Relatively few copies were printed, probably for his inner social circle, an impressive set of friends that included the major French mathematicians of his day, Mersenne, Descartes, Fermat, Roberval and Pascal. His book earned scathing reviews and never enjoyed a wide audience. His dense style of writing, lack of symbolism, and unusual nomenclature, with an introduction of seventy new terms such as "palm," "tree," "stump" and "trunk," made for a book that mathematicians and engineers alike found abstruse. As Descartes wrote to Mersenne, "Between you and me, I can hardly imagine what he may have written concerning conics" [**32**]. In addition to the inaccessibility of his writing, the timing of his publication was unfortunate, coming only two years after Descartes' development of analytic geometry in *La Géométrie*. As a result, Desargues' work was cast aside by all but his student, Pascal, and his significant early contribution to the field went unrecognized until its rediscovery over two hundred years later, in 1845.

It was another 140 years until the next contribution by French mathematician Gaspard Monge (1746–1818), who, notably, served as a draftsman at an engineering school in his early career. In this role, Monge applied his considerable geometry skills to the task of creating technical drawings. In 1765, he invented a **parallel projection** method for producing a two-dimensional representation of an object that consisted of two orthogonal projections of the object onto planes that are also orthogonal to each other. An **orthogonal projection** differs from the perspective techniques described above in that it eliminates the fixed and finite viewpoint. Thus, Dürer's string affixed to the hook on the wall has no application here. With this method, the canvas is called a projection plane, and orthogonal means meeting at a right angle. As demonstrated in Figure 14.9, a two-dimensional orthogonal projection of an object onto the "canvas" is produced by establishing a one-to-one correspondence between points of the object (visible points of a 3-d surface) and points of the "canvas" such that the lines joining corresponding points are orthogonal to the "canvas." (For those who have studied vector calculus or linear algebra, this is the same notion as the orthogonal projection of a vector.)

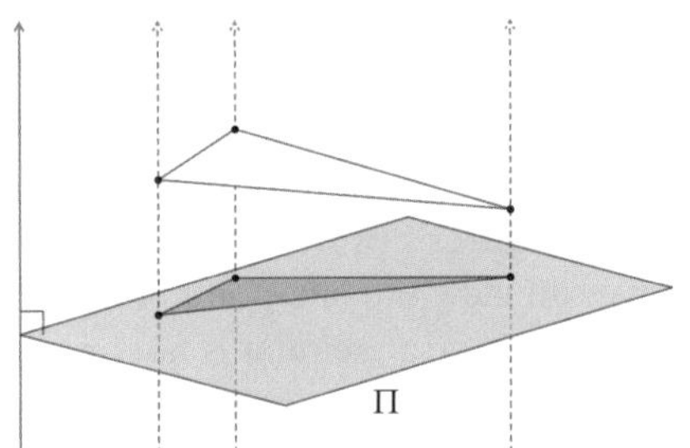

<table>
<tr><td>

Figure 14.9. Orthogonal projection of triangle onto plane Π

</td><td>

Figure 14.10. Orthogonal projection of $\triangle ABC$ onto $\triangle A'B'C'$ in plane Π

</td></tr>
</table>

In the more detailed version of the projection shown in Figure 14.10, segment DE has been added where DE is parallel to AB. Notice that the projections of these segments, $D'E'$ and $A'B'$, are parallel. Since an orthogonal projection respects parallelism, this method does not replicate the image as seen by our eyes. For example, if railroad tracks were floating above the "canvas" in Figure 14.9, then no matter how the tracks were positioned in space, the orthogonal projection of the tracks would neither meet at a point nor appear to meet at a point. Monge's concerns were that of an engineer rather than an artist: to produce precise technical plans for accurate reconstruction of an object by an engineer in a distant location. This is why his method required two orthogonal projections. As demonstrated by the two floating triangles in Figure 14.11, any triangles with vertices on the dashed orthogonal lines will produce the same orthogonal projection on the plane below. The second projection onto a plane orthogonal to the first plane, as demonstrated in Figure 14.12, serves to differentiate the two triangles. Monge called his projection techniques, including a few additional steps that we do not detail here, **descriptive geometry**. As France was in a period of war during this time, engineers in the French military studied Monge's geometry, but the methods remained a military secret until 1799 when *Géométrie descriptive* was published from Monge's lecture notes.

Let's compare the methods of Dürer and Monge. While orthogonal projection and one-point perspective both result in a two-dimensional representation of a three-dimensional object, the resulting 2-d images are as different as the goals of the painter

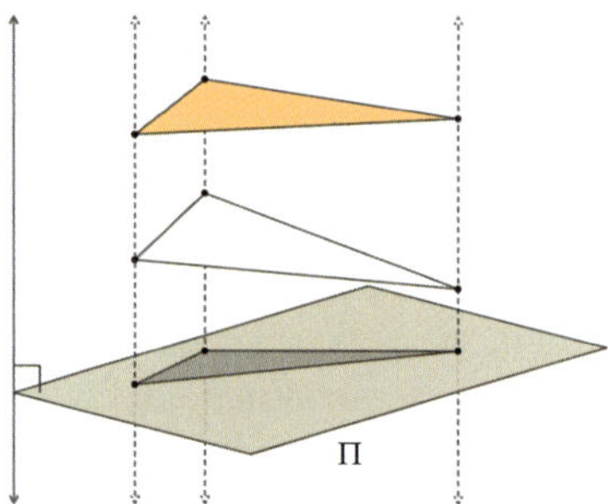

Figure 14.11. Orthogonal projection of two triangles onto plane Π

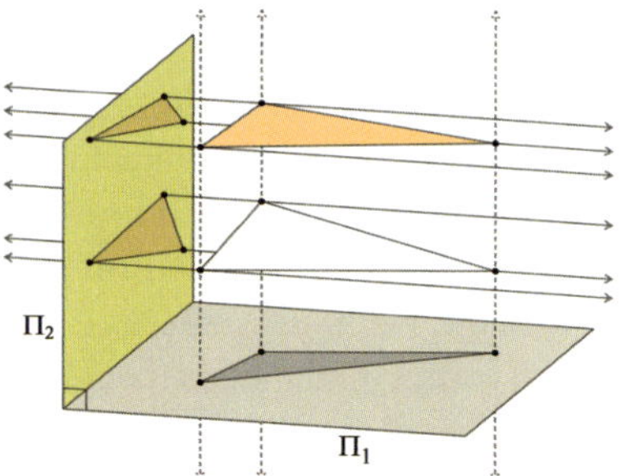

Figure 14.12. Orthogonal projections onto orthogonal planes Π_1 and Π_2

and draftsperson; the artist seeks to replicate an image as seen by our eyes, while the engineer aims to produce technical plans for 3-d modelling. Despite this difference, there is a way to view one-point perspective as a projection, and a projection as one-point perspective. When we transpose the position of Dürer's canvas (plane) and subject (triangle), as demonstrated in Figure 14.13, we can view the one-point perspective method as a **central projection** from a point. While this arrangement is not practical for an artist, it makes perfect sense for a mathematician. Alternatively, if we admit the possibility of Desargues' seemingly outrageous idea that parallel lines meet at a point at an infinite distance, then we can interpret parallel projection as a perspective drawing where the viewpoint is at an infinite distance from the projection plane. This idea was not so shocking to the generation of students that followed Monge, for they arrived at the same concept of a point at infinity without any knowledge of the work of Desargues.

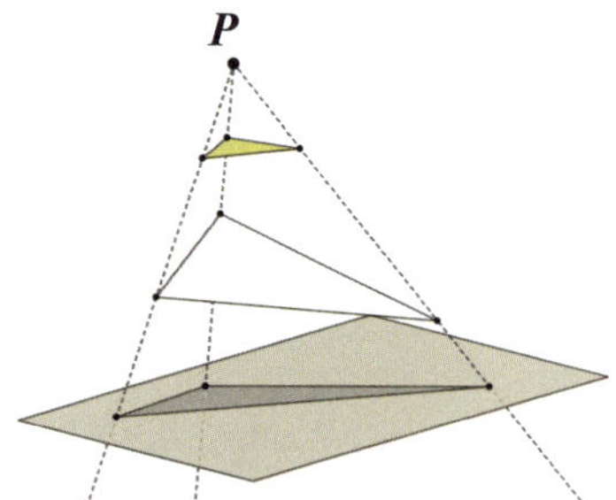

Figure 14.13. Perspective drawing (central projection)

By the beginning of the nineteenth century, French geometers were unknowingly reinventing Desargues' results before pushing the subject into uncharted territory. It was Monge's student, Jean Victor Poncelet (1788–1867), whose 1822 book is widely regarded as the foundation of modern projective geometry as a separate discipline [**97**]. Schooled in descriptive geometry by Monge, Poncelet and his French contemporaries, Lazare Carnot (1753–1823), Charles-Julien Brianchon (1783–1864), Charles Dupin (1784–1873), Michel Chasles (1793–1880), and Joseph Gergonne (1771–1859), produced foundational work in projective geometry. As was the custom for men of their class, these mathematicians served in the French military. Of particular note, after being captured as a prisoner of war in 1813 during Napolean's disastrous Russian campaign, Poncelet spent a year in a Russian prison where he first set out to reconstruct from memory all that Monge had taught, then went on to prove original results. Poncelet's 1822 seminal book on projective geometry was formed from the notes he compiled during his imprisonment.

A group of mathematicians working in Germany, including Karl von Staudt (1798–1867), Jakob Steiner (1796–1863), Julius Plücker (1801–1868), Felix Klein, and August Möbius (1790–1868), were the next group to make significant early contributions to the field. Twenty-five years after Poncelet's book, von Staudt provided the first synthetic development of projective geometry. In the latter half of the nineteenth century, projective geometry was a highly active research area, with many universities of the time hosting named chairs and professorships of projective geometry. Postulational developments of projective geometry were investigated alongside those of Euclidean geometry during the push for rigor in the foundations of geometry in the late 1800s. As mentioned in the previous section, it was during this period of time that Fano produced the first finite geometry when devising his own set of axioms for projective geometry.

While we have discussed the one-point perspective method of Dürer, the orthogonal projection method of Monge, and the mathematicians credited with foundational work in projective geometry, we have not actually described this new geometry. With that in mind, let's consider a natural question that arises with Dürer's method, and explain how the answer led to this geometry. First, recall that Dürer's one-point perspective method creates a one-to-one correspondence between visible points on a 3-d object and points on the 2-d plane of the canvas. Moreover, a line of the 3-d object is mapped to a line on the canvas plane. For example, as demonstrated by Holbein's homage to Dürer in Figure 14.7, the image of a lute string is a line. Here's why this is true, in general: If ℓ is the line in space determined by the lute string, and V is the fixed viewpoint in space, then there is a unique plane through ℓ and V (as long as ℓ doesn't pass through V). The 2-d image of the lute string is found at the intersection of this plane with the plane of the canvas, and these planes intersect in a line. Therefore, the one-point perspective method maps points to points, and lines to lines. This mapping process does not, in general, preserve properties of measure such as lengths, areas, and angles, and, it does not necessarily preserve parallelism. The image of two parallel lines that extend away from the canvas, like the lute strings, for example, will appear to meet, just as the railroad tracks do in the photograph. Though these inherent properties of an object, which seem rather important, are not found in the two-dimensional image, there is no denying that, when drawn well, the identity of the original object is perfectly obvious. So, even though these properties are lost, there must be a sufficient amount of information retained by this process, and the properties that remain must be significant. The natural question arises: What properties of the original are retained by this process?

Let's describe the process a bit more formally. The one-point perspective method is an example of a **perspective transformation**. At it simplest, a perspective transformation is a relation from one set of points $\mathscr{P}$ (e.g. visible points of lute) to another set of points $\mathscr{P}'$ (e.g. painted image of lute) by a *perspectivity* about a viewpoint. There is a pyramid formed by the viewpoint (e.g. hook on the wall) and the rectangular plane acting as the backdrop to the subject (e.g. wall beyond the lute), where the viewpoint forms the apex of the pyramid, the backdrop plane forms its base, and the subject is within the pyramid's interior. Every line joining the apex of the pyramid and a visible point on the subject, if extended, will meet the pyramid's base. Notice that each of the lines emanating from the apex is the theoretical equivalent of Dürer's string joining the hook on the wall to a point on the subject as it passes through the frame. These lines create the one-to-one correspondence between the visible points of the subject and the points of the canvas. Here, the canvas is merely a cross-section of the pyramid separating its apex from the subject. Since the frame may be placed anywhere between the subject and the viewpoint, the image created by Dürer's method can be any such cross-section of the pyramid. Thus, a perspective transformation of the subject is any of the infinitely many cross-sections of the pyramid that exist between the subject and the apex. Some are closer to the subject while others are closer to the viewer's eye, and some are parallel to the backdrop plane while others are situated obliquely. Mathematicians wanted to know what properties were shared by all such perspective transformations of the same object. The characteristics which are unchanged by this process, and thus, appear in all perspective transformations, are called **invariants**. If this question seems daunting, then consider this: Poncelet allowed for any number of perspective transformations of perspective transformations (akin to composition of functions, or a photograph of a photograph), even from different viewpoints, when determining the answer to this question.

A sequence of perspective transformations is called a **projective transformation**. Projective geometry began as a formal area of study to determine which properties of the original figure remain unchanged by this transformation process. Poncelet found that the characteristics unchanged by projection are the same as those demonstrated by the, admittedly, simplified example of the lute. It is precisely the incidence properties that are invariant under projection, namely, collinearity of points and concurrency of lines. Poncelet called these **projective properties** since they are invariant under projection. Consequently, it should be no surprise that these invariant incidence properties form the foundation of our new geometry, serving as axioms for the projective plane given at the start of the next section. The only other basic projective property comes from the loss of parallelism: there are no parallel lines in projective geometry. This certainly meshes with our experience of the image of railroad tracks, and more generally, with our understanding that parallelism is not preserved by every perspective transformation. What may be surprising is that our set of axioms for projective geometry requires only three axioms! It is also interesting to note that the point in projective geometry corresponding to the artist's vanishing point will be of no concern to us until Section 14.5.

On a final note, even though the genesis of this geometry is found in the artist's desire to represent three-dimensional space, our consideration of projective geometry will be limited to planes. So, how much mathematics can we do in a plane with only three axioms? As H.S.M. Coxeter wrote at the start of his text, *Projective Geometry*, "what happens in a single plane is sufficiently exciting to occupy our attention for a long time" [**29**].

14.3.2 Projective planes.

One should not be overwhelmed by Hilbert's 20 axioms. The Columbia Encyclopedia of Sports lists 43 axioms for baseball, Hoyle lists 37 for bridge. Hence, geometry is easier than baseball and bridge. [**125**]

– Albert Wilansky (1921–2017)

Following the logic set out by Wilansky, projective plane geometry should be nearly seven times easier than Euclidean geometry since it only requires three axioms. As usual, the undefined terms in our axioms are *point*, *line* and *on*. A **projective plane** is a set of points, $\mathscr{P}$, and a set of lines, $\mathscr{L}$, which satisfy the following three axioms:

Projective Plane

Axiom P-1. Any two distinct points are on exactly one line.

Axiom P-2. Any two distinct lines are on exactly one common point.

Axiom P-3. There are at least four distinct points, no three of which are on the same line.

First, notice that the projective plane provides a symmetry that does not exist in the Euclidean plane where two points determine a unique line but two lines may not determine a unique point. In the projective plane, this disparity in the roles played by points and lines is eliminated by Axioms P-1 and P-2. Next, let's consider a model for these axioms. Note that Euclidean, Taxicab, Spherical and Hyperbolic geometries all fail to satisfy Axiom P-2. Thus, we need to look elsewhere. It is clear that Fano's plane shares Axioms P-1 and P-2. Take a minute to verify that Fano's plane also satisfies Axiom P-3. Therefore, this axiomatic system is consistent. The reader will also verify that the set of axioms is independent in Exercise 14.3.1. When the set of points is finite, as in Fano's plane, then we have a **finite projective plane**. Nearly all of the work in projective geometry in the nineteenth century assumed an infinite set of points. Elliptic geometry, briefly mentioned in Chapter 4, is an example of a geometry with an infinite number of points that satisfies these axioms. Our primary focus is on the finite case as it produces a geometry far different than all others in this book, and takes us into a highly active area of mathematical research in the twentieth century.[2]

Once again, the unique line on points A and B, as given by Axiom P-1, will be denoted by AB. In the finite case, AB is simply shorthand for the unique set of points comprising the line on points A and B. Hence, if points A, B and C are on a line we can also denote the line by ABC. Since betweenness has no meaning here (and sets are not ordered), this is equivalent to writing the points in any order, e.g. ACB or BAC. As we found when working with Four Point geometry and Fano's plane, finite planes have combinatorial properties that are easily verified. We can count the number of points and lines, as well as the number of points on each line, and the number of lines through each point. We spend the rest of this section proving classic, fundamental results about these planes, as well as introducing the new concept of *duality*.

As a start, let's see what we can say about the number of points and the number of lines in a finite projective plane. Axiom P-3 provides at least four distinct points, no three of which are on the same line. Since there are six ways to pair these four points, Axiom P-1 guarantees the existence of at least six distinct lines. We can, however, prove that there are more points and more lines than this.

---

[2]As a side note, we give a model for the infinite real projective plane on page 368.

Theorem 14.7. *There are at least seven lines and at least seven points in any projective plane.*

Proof.   Let A, B, C and D be the four points guaranteed by Axiom P-3. Axiom P-1 guarantees the existence of six lines: AB, AC, AD, BC, BD and CD. By Axiom P-3, these lines must be distinct since no three of the points A, B, C and D are on the same line. This is illustrated in Figure 14.14a.

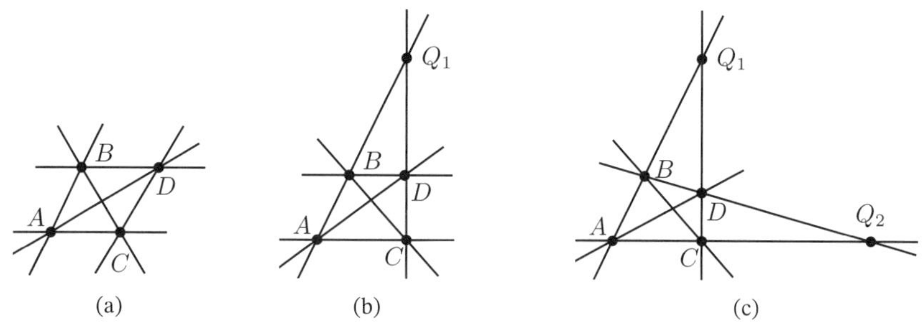

Figure 14.14. First steps to determine minimum number of lines

By Axiom P-2, all fifteen pairs of these six lines must intersect at exactly one point. For twelve of the pairs, the point of intersection clearly lies in the set $\{A, B, C, D\}$, but for three of the pairs, it cannot. Specifically, AB and CD must meet, AC and BD must meet, and AD and BC must meet, and by Axiom P-3, none of these points of intersection can lie in $\{A, B, C, D\}$. Thus, AB and CD meet at a new point, let's say Q_1. We now have lines ABQ_1 and CDQ_1 in our geometry, as demonstrated in Figure 14.14b.

Could AC and BD meet at Q_1? This is not possible since it would violate Axiom P-2 as line ABQ_1 would meet AC at two points, namely A and Q_1. Therefore, AC and BD meet at new point, Q_2, as demonstrated in Figure 14.14c. By a similar argument, AD and BC meet at third new point, Q_3, as illustrated in Figure 14.15a.

We now have the required seven points, but still only six distinct lines, namely ABQ_1, ACQ_2, ADQ_3, BCQ_3, BDQ_2, and CDQ_1. We will show that the newly added points, Q_1, Q_2, and Q_3, will generate at least one new line. By Axiom P-1, there exists line Q_1Q_2. Line Q_1Q_2 must be distinct from lines ABQ_1 and ACQ_2, else Axiom P-2 is violated as these lines would meet in more than one point. Line Q_1Q_2 must be distinct from lines BDQ_2 and CDQ_1 for the same reason. There are only two more lines to check. If Q_1Q_2 were identical to ADQ_3, then ADQ_3 meets ABQ_1 at more than one point, violating Axiom P-2. Similarly, Q_1Q_2 must be distinct from BCQ_3. Therefore, Q_1Q_2 is our seventh line.   □

In the exercises, the reader will verify that Q_3 is on Q_1Q_2. As demonstrated in Figure 14.15b, this arrangement of seven points and seven lines is easily shown to be equivalent to Fano's plane. Thus, a copy of Fano's plane can be found in every projective plane. This implies that Fano's plane is the smallest finite projective plane.

You may have noticed that Axioms P-1 and P-2 have an interesting symmetry. Aside from the word 'common', which is only added for clarity, exchanging the words *point* and *line* in one of these axioms produces the other. For this reason, we say that Axioms P-1 and P-2 are **dual** statements.   In general, any mathematical statement about projective planes has a corresponding dual statement that can be produced by

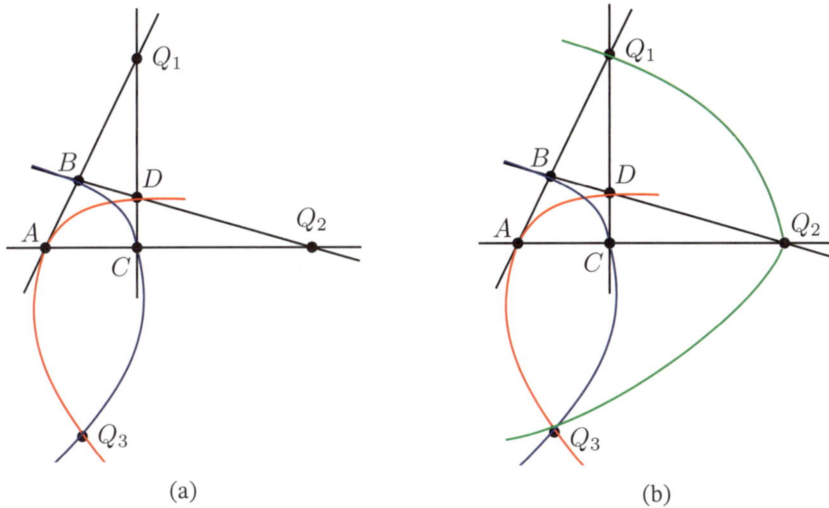

Figure 14.15. More steps to determine minimum number of lines and points

interchanging these words. The dual of Axiom P-3 is given below as Theorem 14.8 since it, too, is true for projective planes.

Theorem 14.8. *There are at least four distinct lines, no three of which are on the same point.*

Proof. By Axiom P-3, we have points A, B, C and D, and four distinct lines AB, BC, CD and AD, as explained in the previous proof. We will show by contradiction that no three of these lines pass through a common point. Without loss of generality, suppose AB, BC and CD meet at an unknown point X. If X were B, then B is on line CD. Thus, distinct lines BC and CD have two points, B and C, in common, violating Axiom P-2. If X is not B, then lines AB and BC have points B and X in common, violating Axiom P-2. $\square$

Theorem 14.8 puts us in a powerful position, the likes of which we have not encountered before in our geometric travels. Since Axioms P-1 and P-2 are duals, and Axiom P-3 and Theorem 14.8 are duals, this means that any mathematical statement that we can prove from these building blocks has a dual counterpart that is also true. It's like a Buy One Get One Free deal, but really it's a Prove One Get One Free deal! Once a statement has been proved, the proof of its dual counterpart could be produced by replacing each line in the proof, step by step, with its dual counterpart. This property is known as the *Principle of Duality* and is given as the following theorem.

Theorem 14.9 [Principle of Duality]. *In projective geometry, if a statement is true, then its dual is also true.*

For example, we could now revisit Theorem 14.7, knowing that we only need to prove the existence of at least seven points, let's say, since duality provides the justification for the existence of at least seven lines.

Tables and chairs, pizzas and toppings

The elegance of the duality property of projective geometry must have gone a long way towards convincing nineteenth-century mathematicians that the words used for the undefined terms were irrelevant, and it was only the relationships between the terms, as prescribed by the axioms, that mattered. This is an appropriate time to think back to Chapter 6 to recall that it was Hilbert who said that we could replace the undefined terms *point* and *line* with *chair* and *table*, as long as the axioms are satisfied. Many modern geometry texts give the axioms for finite geometries such as Four Point geometry or Fano's plane using substitutes for *point* and *line* such as *bead* and *wire*, *book* and *shelf*, *abba* and *dabba*, *x* and *y*, *politician* and *committee*, *Fe* and *Fo*, or as we did in Exercises 6.2.3 and 14.2.1, *topping* and *pizza*, or *player* and *team*. These nongeometric words highlight the fact that the axioms alone must prescribe the rules for playing with the undefined terms in our system, and that any preconceived notion we have of an undefined term is prejudicial. (Perhaps Desargues' writing style, utilizing nonmathematical words like palm, tree, stump and trunk, was simply too far ahead of its time to be fully appreciated.) On the other hand, without some interpretation of the undefined terms it can be quite difficult to build intuition and formulate proofs. So, we continue to encourage our readers to do so, and to draw diagrams in order to facilitate comprehension, but be cautious when drawing conclusions from these aids. Use only the axioms or previously proven results to justify the steps of a proof, and try to avoid hidden assumptions generated by an interpretation of the undefined terms.

Unlike the axiom sets for the Fano plane or Four Point geometry, the projective plane axioms do not indicate the minimum number of points on a line or the minimum number of lines on each point. Nevertheless, we can prove the following two theorems.

Theorem 14.10. *In a projective plane, every line is on at least three points.*

Proof.   Let ℓ be a line, and A, B, C and D be the points, as given by Axiom P-3.

Case 1.   If ℓ is one of the six distinct lines AB, AC, AD, BC, BD or CD, then from the proof of Theorem 14.7, ℓ is on at least three points.

Case 2.   On the other hand, suppose ℓ is distinct from these six lines. Then either A or B is not on ℓ (else ℓ is AB). Without loss of generality, suppose A is not on ℓ. By Axiom P-2, ℓ meets AB at some W_1, ℓ meets AC at some W_2, and ℓ meets AD at some W_3, as demonstrated in Figure 14.16. Notice that for each i, W_i cannot be A since A is not on ℓ. Also, $W_1 \neq W_2$ by Axiom P-2, else AB and AC meet at two points (A and $W_1 = W_2$). Likewise, $W_2 \neq W_3$ and $W_1 \neq W_3$. Therefore, ℓ is on at least three points. $\square$

By the Principle of Duality, we automatically have the following dual theorem.

Theorem 14.11. *In a projective plane, every point is on at least three lines.*

In Fano's plane, it is clear that every line is on exactly three points and every point is on exactly three lines. Up to this stage of our analysis, every projective plane theorem

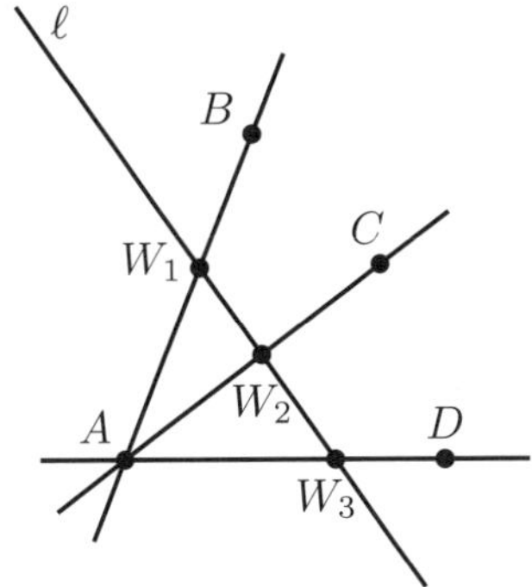

Figure 14.16. There are at least three points on any line.

holds regardless of whether or not there are finitely many points. For the rest of the section, we focus our attention on *finite projective planes*. With only finitely many points, it follows that the number of points on any line is, necessarily, finite. Additionally, we will show that all lines have the same number of points (and hence, by the Principle of Duality, all points are on the same number of lines). Therefore, again by the Principle of Duality, the number of points on a line must be the same as the number of lines on a point. In order to prove this, we need a preliminary lemma whose proof is left to the reader in the exercises. The lemma holds for all projective planes.

Lemma 14.12. *In a projective plane, given two distinct lines, there exists a point which is not on either of the given lines.*

With this lemma, we are ready to show that any two lines have the same number of points. The idea for the proof is to pair the points of distinct lines. With two lines ℓ and m and a point Q not on either, we can join Q to every point on ℓ. Each of these lines must meet m, thus pairing the points on ℓ and m.

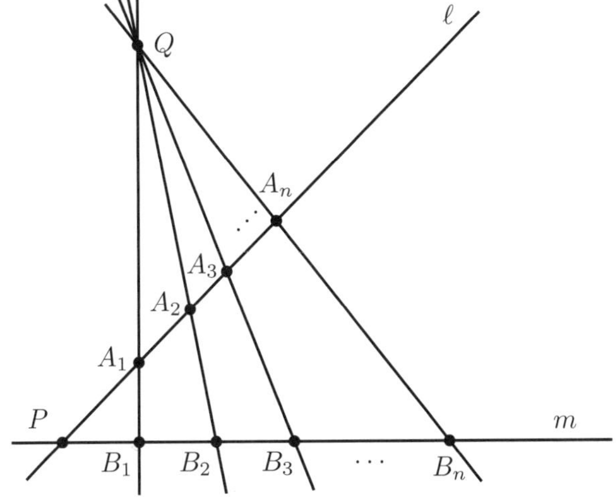

Figure 14.17. All lines have the same number of points.

Theorem 14.13. *In a finite projective plane, every line is on the same number of points and every point lies on the same number of lines. Furthermore, the number of points on a line equals the number of lines on a point.*

Proof. We will show that every line is on the same number of points, and appeal to duality for the other two statements.

Our strategy is to show that, given distinct lines ℓ and m, we can find a one-to-one map from the set of points on ℓ to the set of points on m. Let ℓ and m be distinct lines as guaranteed by Theorem 14.7. By Axiom P-2, these lines meet at a unique point, say P. By Lemma 14.12, there exists a point, say Q, not on ℓ and not on m. By Theorem 14.10, ℓ is on at least three points. For any point A_1 on ℓ which is distinct from P, there exists line $A_1 Q$ by Axiom P-1. By Axiom P-2, $A_1 Q$ meets m in exactly one point, say B_1. Notice that B_1 is necessarily distinct from P, else $A_1 Q$ meets ℓ at two points, violating Axiom P-2. If A_2 is another point on ℓ distinct from P and A_1, then similarly, $A_2 Q$ meets m at exactly one point, say B_2. Notice that B_2 is distinct from P for the same reason as given above for B_1. We also claim that B_2 is distinct from B_1, for if they were not, then $A_1 Q$ and $A_2 Q$ would meet at both Q and B_1, violating Axiom P-2.

This process can be repeated for each point on line ℓ, ensuring that every point on ℓ is paired with a new, distinct point on m, as demonstrated in Figure 14.17. This assures that m has at least as many points as ℓ. To show that ℓ has at least as many points as m, we can repeat this argument by starting with a point B_i on m, then finding the intersection of line $B_i Q$ with ℓ.

Therefore, any two lines must have the same number of points. By duality, any two points are on the same number of lines. Furthermore, again by duality, if any line is on exactly n points, then any point is on exactly n lines. $\square$

The uniformity of the number of points on a line in a finite projective plane allows us to define the order of the plane. If there are $n + 1$ points on every line (and $n + 1$ distinct lines through each point) of a finite projective plane, then we say that the *order* of the plane is n.

Definition 14.14. The **order** of a finite projective plane is one less than the number of points on any line in the plane. A projective plane of order n is denoted by Π_n.

For example, Fano's plane is a projective plane of order 2 since there are exactly three points on every line. The reason that the order of the plane and the number of points on a line differs by one will become clear after we discuss the connection between projective planes and affine planes in Section 14.5. With the ability to count the number of points on each line and lines on each point, we can now count the number of distinct lines in any finite projective plane.

Theorem 14.15. *In a finite projective plane of order n, there are exactly $n^2 + n + 1$ lines.*

Proof. Let ℓ be any line of the plane. By definition, there are $n + 1$ distinct points on ℓ. Each of these points is on n distinct lines other than ℓ, as demonstrated by Figure 14.18. Axiom P-1 guarantees that a line other than ℓ through point P_i is necessarily distinct from a line other than ℓ through P_j, where $i \neq j$. Since all of these lines are distinct, there are n lines at each of the $n + 1$ points, giving $n(n + 1)$ lines, not including line ℓ. Once we account for line ℓ, we see that there are $n^2 + n + 1$ lines as required. $\square$

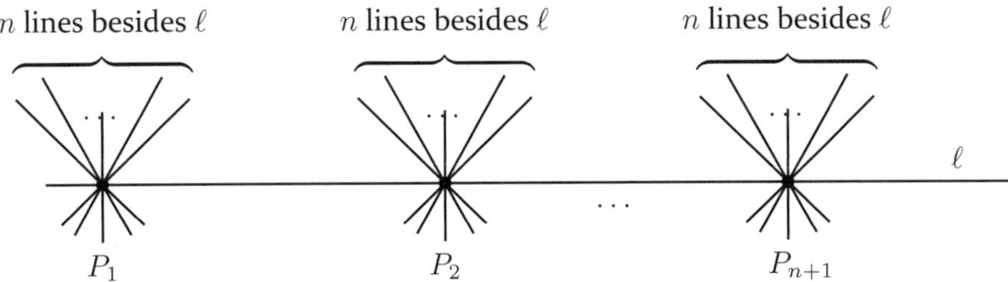

Figure 14.18. Number of lines on a projective plane of order n

The Principle of Duality gives the following theorem for the number of points in a projective plane.

Theorem 14.16. *In a finite projective plane of order n, there are exactly $n^2 + n + 1$ points.*

These theorems specifying combinatorial information about a finite projective plane of order n are true whenever one of these planes exists. It is important to note that there is nothing in our work thus far to guarantee that such a plane exists for any given order n. In fact, the search for which orders exist is a problem that has been fascinating and frustrating mathematicians for the past 75 years. We'll discuss the current status of the answer to this question in Section 14.6. Since we've only seen one example of a finite projective plane, you might be wondering if there are others. For now, we'll just say that there are infinitely many such planes, but the diagrams we've been working with become unwieldy as the order grows larger. See Exercise 14.3.7 for a diagram of the projective plane of order 3.

Exercises 14.3

1. For each part, give an example of a model to verify the independence of the specified projective plane axiom.

    (a) Axiom P-1

    (b) Axiom P-2

    (c) Axiom P-3

2. Consider a set of five points given by $\mathscr{P} = \{A, B, C, D, E\}$ and a set of 10 lines given by $\mathscr{L} = \Big\{\{A, B\}, \{A, C\}, \{A, D\}, \{A, E\}, \{B, C\}, \{B, D\}, \{B, E\}, \{C, D\}, \{C, E\}, \{D, E\}\Big\}$.

    (a) Draw a diagram to represent the given points and lines.

    (b) Use the axioms to determine if this is a model for a projective plane.

3. Show that point Q_3 of Theorem 14.7 is on line $Q_1 Q_2$. This is demonstrated in Figure 14.15b. Now verify that the seven points and seven lines of Theorem 14.7 is equivalent to Fano's plane.

4. Prove Lemma 14.12.

5. Prove the following: *In a projective plane, for any point P, there is at least one line that is not on P.*

6. Assuming that there is a projective plane of order 5, how many points and lines would it have?

7. A diagram for the projective plane of order 3 is shown in Figure 14.19.

   (a) Verify that the axioms are true for this model.
   (b) How many points are on each line? How many lines are on each point?
   (c) How many points are there? How many lines?
   (d) Label the vertices in alphabetical order starting with A at the 12 o'clock position and proceeding clockwise. Give the set of points, $\mathscr{P}$, and the set of lines, $\mathscr{L}$, for this plane.
   (e) Is the set of projective plane axioms categorical? Justify your answer.

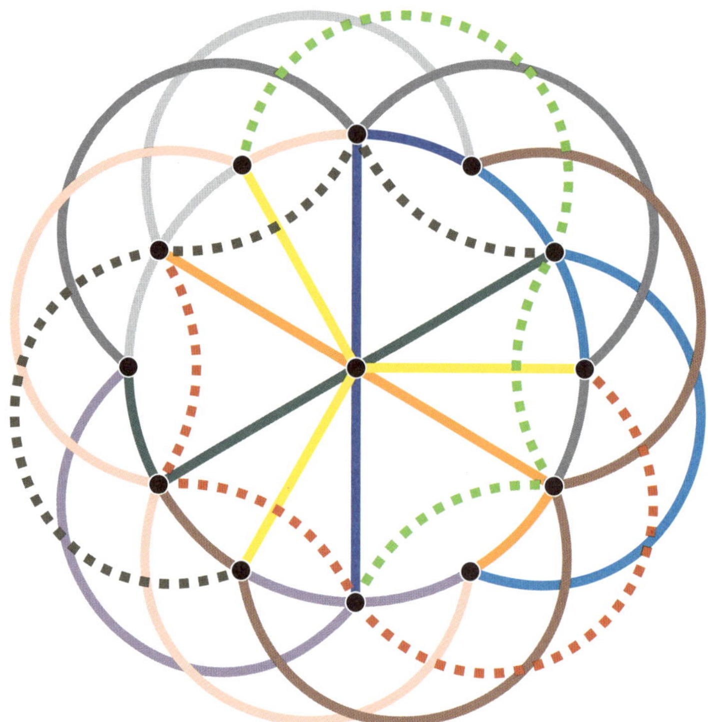

Figure 14.19. Exercise 14.3.7: Projective plane of order 3

The best approach to geometry: analytic or synthetic?

As detailed in the history of the development of projective geometry in Section 14.3.1, despite the early work of Desargues, the field did not enjoy a large following until the nineteenth century. Desargues' dissemination of his groundbreaking projective geometry results was doubly thwarted by his puzzling style of exposition and his misfortune of publishing within the shadow created by Descartes' revolutionary work, *La Géométrie.* The attention of mathematicians during the seventeenth and eighteenth centuries was singularly focused on the power and utility of the analytic geometry of Descartes and Fermat, which soon gave way to the calculus of Newton and Leibniz. These breakthroughs lead to an explosion of new mathematics and, simultaneously, an abandonment of the old ways of synthetic geometry.

Bucking the fashion of the day, the early French contributors to projective geometry, with the exception of Gergonne, went against the grain of popular opinion and were determined to champion the supremacy of the lost art of synthetic geometry. With his 1822 book on projective geometry, Poncelet intended to free his geometry of algebraic analysis, though the debate among early nineteenth-century geometers over the best approach to their field, synthetic or analytic, raged for decades. Carnot and Poncelet were staunch supporters of the synthetic approach, and Carnot wished to "free geometry from the hieroglyphics of analysis." Another later disciple of the synthetic camp referred to analytic geometry as the "clatter of the coordinate mill" [81]. The geometers working in Germany were not immune to the debate. Steiner, a staunch syntheticist, vowed to stop writing for the premiere research journal of the day if its editors continued publishing the analytic works of Plücker.

It wasn't until the late 1800s that geometers came to realize the advantage to keeping both methods at the ready, since some problems gave way easily to the synthetic methods and others to the analytic. Present-day mathematicians are scarcely aware that there was ever any such controversy within their ranks. Our presentation of projective geometry falls squarely within the synthetic camp since we focus on the combinatorial aspects of finite projective planes without introducing coordinates to the points, but we hope our short excursion into this unusual geometry motivates the curious reader to explore the analytic side. On a final historical note, very late in his life, the diehard syntheticist Poncelet published the notes he had compiled while in prison nearly fifty years earlier. Amusingly, these notes reveal that he, too, relied on coordinate analysis as a means of checking his synthetic results.

14.4 Affine planes

We obtain an affine plane by changing just one of the three axioms for a projective plane. In doing so, we regain parallel lines and return to a much more familiar world. As we will soon see, Euclidean, Taxicab and Four Point geometry are all examples of *affine* geometries.

Recall, two distinct lines, ℓ and m, are **parallel** when they share no common points, denoted $\ell \parallel m$. In the language of sets, this means $\ell \cap m = \emptyset$. In a projective plane, Axiom P-2 guarantees that any two distinct lines meet in a point. There are, therefore, no parallel lines in projective geometry. In the axioms for an affine plane, we assume a stronger version of Playfair's Axiom guaranteeing a unique parallel. As before, our undefined terms are *point, line* and *on*. An **affine plane** is a set of points, $\mathcal{P}$, and a set of lines, $\mathcal{L}$, which satisfy the following three axioms:

Affine Plane

Axiom A-1. Any two distinct points are on exactly one line.

Axiom A-2. If point P is not on line ℓ, then there exists a unique line on P that has no points in common with ℓ.

Axiom A-3. There are at least four distinct points, no three of which are on the same line.

Notice that Axioms A-1 and A-3 are identical to Axioms P-1 and P-3, respectively, of projective geometry. Axiom A-2 is the only axiom that has changed, and thus, produces the differences between projective and affine geometry. Whereas a projective plane has no parallel lines, Axiom A-2 guarantees that pairs of parallel lines exist in an affine plane. Hence, at least as far as parallels are concerned, this geometry appears to have a Euclidean nature, and we may expect a certain degree of familiarity when working with it. The first axiom combines Hilbert's first two axioms of incidence, asserting the uniqueness of a line determined by two points. The third axiom eliminates trivial one-point or one-line geometries. Clearly, the Euclidean plane satisfies these three axioms, and hence, Euclidean geometry is affine. (If we were to add the appropriate axioms to the system, namely betweenness, congruence and continuity axioms, we could produce all of Euclidean geometry.) The converse, however, does not hold. For example, Taxicab geometry also satisfies the three affine plane axioms.

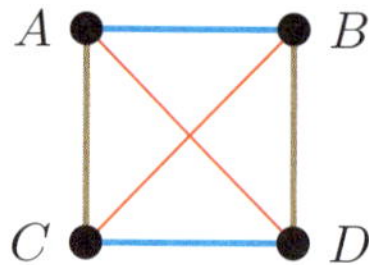

Figure 14.20. The smallest affine plane

Since our focus in this chapter is finite geometry, we'll start by constructing the smallest affine plane. With pen and paper, let's produce a diagram to model the axioms. Start with the four points given by Axiom A-3, then make sure there is a line between any two points to satisfy Axiom A-1. Next, verify that your diagram is equivalent to one of the models for Four Point geometry given in Figure 14.2 or Figure 14.20. Clearly, Four Point geometry satisfies Axioms A-1 and A-3. Take a moment to verify that the parallel axiom, Axiom A-2, is satisfied. Any of our two-dimensional models for Four Point geometry is a model for the axioms of an affine plane, thus verifying that this axiomatic system is consistent. (See Exercise 14.4.1.) The reader will verify that the set of axioms is independent in Exercise 14.4.5.

We formalize the above discussion with the following theorem.

Theorem 14.17. *There are at least six lines in any affine plane.*

Proof.   Let points A, B, C and D be the four points given by Axiom A-3. By Axiom A-1, there exist lines AB and AC, AD, BC, BD and CD. A pairwise grouping of any of these lines shows that these lines are distinct, otherwise three of these four points are on the same line, violating Axiom A-3.   □

This configuration of four points and six lines is equivalent to our Four Point geometry. Thus, a copy of Four Point geometry can be found in every affine plane. This implies that the smallest affine plane is the Four Point plane.

Since parallelism is a significant property in an affine plane, let's carefully consider the lines in the Four Point plane, highlighting sets of parallels. The six lines found in the graph are shown individually in Figure 14.21. Notice that the three pairs grouped by color are parallel lines since they share no common points. We will say that each set of parallel lines forms a *parallel class*. This gives rise to the following definition.

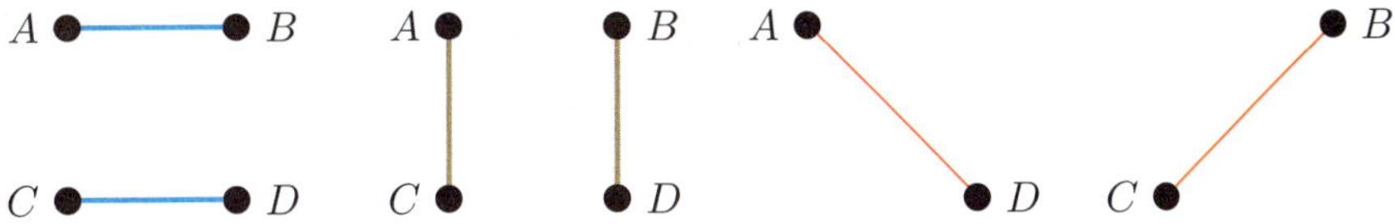

Figure 14.21. Parallel classes for the smallest affine plane

Definition 14.18. In an affine plane, a **parallel class** is a set consisting of a line together with all the lines parallel to it.

As shown in Figure 14.21, there are exactly three parallel classes in the smallest affine plane. In general, we can count the number of parallel classes and the number of lines in each class for any finite affine plane. To help determine some patterns, we encourage the reader to produce a larger finite affine plane. Start by adding another point to line AB, then follow the axioms, introducing as few new points and lines as possible. This is not an easy construction, but nonetheless, a worthwhile exercise. Give it an honest effort before peeking at the next figure.

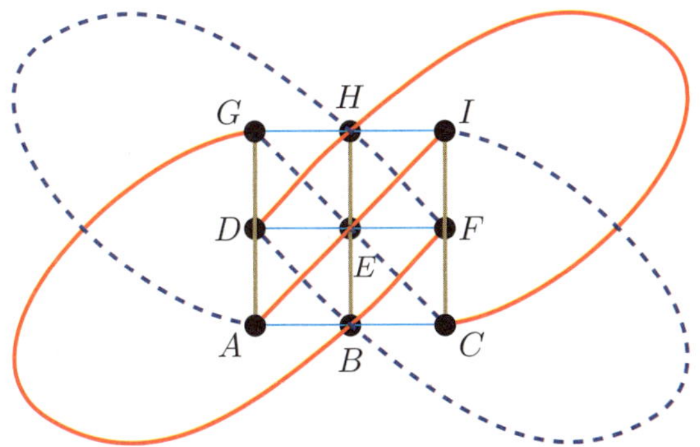

Figure 14.22. An affine plane with three points on every line

The next smallest affine plane is shown in Figure 14.22. There are 9 points and 12 lines. Listing the points and lines as sets, we have

$$\mathscr{P} = \{A, B, C, D, E, F, G, H, I\}$$

and
$$\mathcal{L} = \{\,\{A,B,C\},\ \{D,E,F\},\ \{G,H,I\}, \qquad\qquad \text{(light blue)}$$
$$\{A,D,G\},\ \{B,E,H\},\ \{C,F,I\}, \qquad\qquad \text{(brown)}$$
$$\{A,E,I\},\ \{A,H,F\},\ \{D,B,I\}, \qquad\qquad\ \text{(red)}$$
$$\{C,E,G\},\ \{G,B,F\},\ \{C,H,D\}\,\}. \qquad\ \text{(blue dashed)}$$

Notice that each line has three points and each point is on four lines. We recognize this as a model for Young's geometry from Exercise 14.2.3. Thus, Young's geometry is an affine plane. Let's consider the parallel classes for this affine plane given in Figure 14.22. In Figure 14.23, we see the four sets of mutually parallel lines grouped by color, where each set consists of three lines.

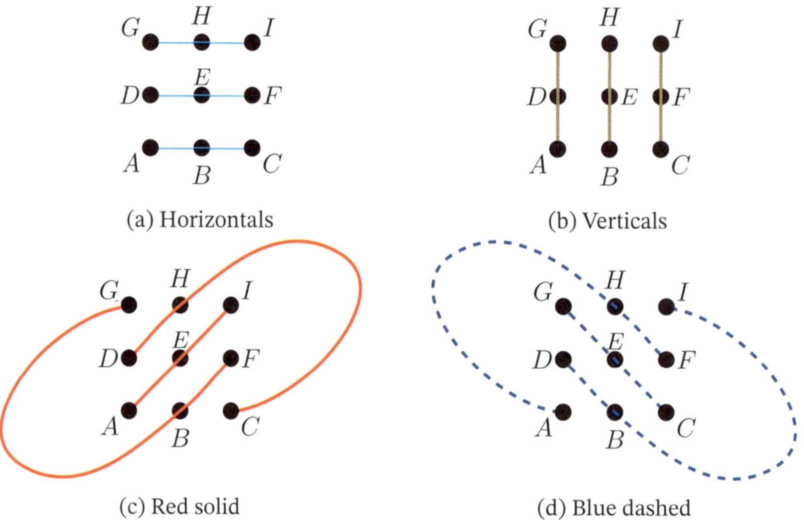

Figure 14.23. Sets of mutually parallel lines in the affine plane with 12 lines

Depending upon the ease with which you produced a figure equivalent to that of Figure 14.22, you might be wary of adding another point to line ABC and following the axioms once again. As you can surely imagine, these constructions become much more difficult as the number of points increase. It will be far easier to establish the general properties of these planes using the same techniques as in the previous section.

Let's begin by proving some familiar properties of the Euclidean plane. Our first set of theorems hold for any affine plane, regardless of whether it is finite or infinite. Be sure to provide your own diagram to accompany these proofs when one is not provided.

Theorem 14.19. *In an affine plane, two distinct lines can meet in at most one point.*

Proof. Let ℓ and m be distinct lines in an affine plane. If $\ell \parallel m$, then they have no points in common and the proof is completed. Suppose $\ell \nparallel m$. For the sake of reaching a contradiction assume that they meet at more than one point, say P and Q. This implies that points P and Q are on two lines, violating Axiom A-1. $\qquad\square$

Theorem 14.20. *There are at least two points on every line in any affine plane.*

Proof. Let ℓ be a line in an affine plane. Let points A, B, C and D be the four points given by Axiom A-3. If any two of these points are on ℓ, then the proof is complete. Otherwise, at least three of these points are not on ℓ. Without loss of generality, suppose A is not on ℓ. By Axiom A-2, there is a unique line through A that is parallel to ℓ. So, at most one of the distinct lines AB, AC or AD (given in Theorem 14.17) is parallel to ℓ. This means that at least two of these lines must intersect ℓ. Since these lines already meet at point A, Theorem 14.19 guarantees that these lines intersect ℓ at two different points. $\qquad\square$

Theorem 14.21. *In an affine plane, for any line ℓ there is a point not on ℓ.*

Proof. Let ℓ be an arbitrary line. By Theorem 14.17, let m be a line distinct from ℓ. By Theorem 14.19, lines ℓ and m have at most one point in common. By Theorem 14.20, there is at least one point on m that is not on ℓ. $\qquad\square$

Recall from the previous section that the dual of a statement can be produced by interchanging the words *point* and *line*. What is the dual statement for each of the three previous theorems? The Principle of Duality states that if a theorem is provable, then so is its dual. We leave it as an exercise for the reader to explain why duality is not a property of an affine plane. There are, however, still theorems with true dual statements. Theorems 14.20 and 14.21 are two examples, and the reader will prove their duals in Exercises 14.4.9 and 14.4.10.

The next two theorems are equivalent to Proclus' Axiom and Euclid's I.30. Notice that, not surprisingly, these proofs are very similar to those given for Theorem 7.1.

Theorem 14.22 [Proclus' Axiom]. *In an affine plane, if a line (distinct from two different parallel lines) meets one of the two parallel lines, then it meets the other.*

Proof. Suppose there are three distinct lines, ℓ, m and k, such that $m \parallel k$, and ℓ meets m at point P. We must show that ℓ meets k. For the sake of an eventual contradiction, suppose $\ell \parallel k$. Thus, there are two lines through P that are both parallel to line k. This contradicts Axiom A-2, thus ℓ must intersect k. $\qquad\square$

Theorem 14.23 [Euclid I.30]. *In an affine plane, two distinct lines, each parallel to a third line, must be parallel to each other.*

Proof. Suppose there are three distinct lines, ℓ, m and k, such that $\ell \parallel k$ and $m \parallel k$. We must show that $\ell \parallel m$. For the sake of an eventual contradiction, suppose ℓ meets m at point P. Thus, there are two lines through P that are parallel to k, contradicting Axiom A-2. Therefore, $\ell \parallel m$. $\qquad\square$

Before we begin investigating properties of a general *finite* affine plane, it would be wise to take a moment to identify any patterns evident from Figures 14.20 and 14.22. Write down a list of properties for each of these finite affine planes, noting features that are true of every point or every line. For example, note the number of points on a line, the number of lines on a point, and the number of parallel classes. If you recognize any patterns, then speculate as to what may be true in a finite affine plane by jotting down a conjecture. We suspect that some of your conjectures will coincide with the theorems to come. After considering Figures 14.20 and 14.22, a likely conjecture is that, just as

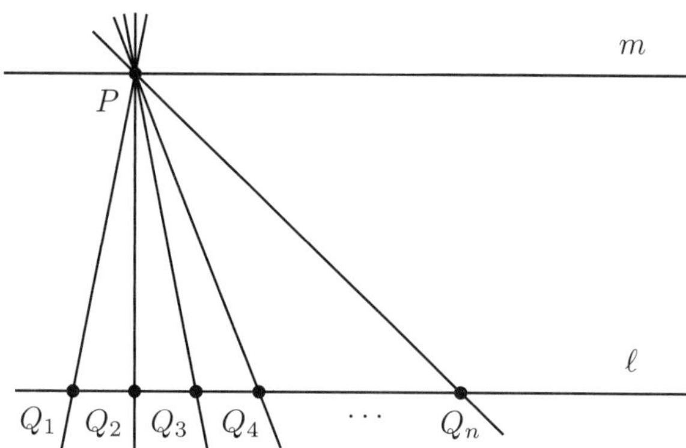

Figure 14.24. Number of lines on a point

was the case in finite projective plane, every line must have the same number of points. This is indeed true, and we establish this result after the following lemma.

Lemma 14.24. *In a finite affine plane, if point P is not on line ℓ, then the number of lines on point P is one more than the number of points on ℓ.*

Proof. By Axiom A-2, there is exactly one line on point P, say m, that is parallel to ℓ. Therefore, every line on P other than m must meet ℓ. Combining this with Theorem 14.19, every line on P other than m meets ℓ in exactly one point. By Axiom A-1, there must be a line between P and each point on ℓ. This establishes a one-to-one correspondence between the points of ℓ and the lines on P other than m. This is illustrated in Figure 14.24. Therefore, the number of lines on P is one more than the number of points on ℓ. $\qquad\Box$

Theorem 14.25. *In a finite affine plane, each line has the same number of points.*

Proof. Let points A, B, C and D be the four points given by Axiom A-3. By Axiom A-3, A is not on the distinct lines BC or CD. By Lemma 14.24, the number of lines on A is one more than the number of points on BC. Likewise, the number of lines on A is one more than the number of points on CD. Therefore, the number of points on BC equals the number of points on CD.

Using the same technique, we can show that the six distinct lines determined by points A, B, C and D have the same number of points. If these are the only lines in the plane, then the proof is complete. Otherwise, suppose ℓ is some other line in the plane. Since ℓ is distinct from the six lines determined by A, B, C and D, by Axiom A-1 at most one of these points is on ℓ. Without loss of generality, suppose A, B and C are not on ℓ. Since C is not on either of the distinct lines ℓ or AB, we can employ Lemma 14.24 using the same argument as given earlier in the proof to show that ℓ has the same number of points as AB. $\qquad\Box$

This theorem allows us to define the order of a finite affine plane in a manner similar to that of finite projective planes.

Definition 14.26. In a finite affine plane, the number of points on each line is called the **order** of the plane. An affine plane of order n is denoted by π_n.

Now we see that Four Point geometry is π_2, the smallest possible affine plane. The diagram in Figure 14.22 is an affine plane of order 3. More generally, the previous results can be combined to show that every point in an affine plane of order n is on $n + 1$ lines. The reader will verify the details of this proof in the exercises.

Corollary 14.27. *In an affine plane of order n, every point is on n + 1 lines.*

As they did for finite projective planes, these combinatorial results allow us to count the number of points and lines in any finite affine plane. In the following theorem, we show that there are n^2 points in any affine plane of order n.

Theorem 14.28. *In a finite affine plane of order n, there are n^2 points.*

Proof.   Let ℓ be a line in the plane of order n. Since the order of the plane is n, ℓ consists of exactly n distinct points, say $B_1, B_2, \ldots, B_n$. By Theorem 14.21, let P be a point not on ℓ. By Axiom A-1, there exists unique line B_1P. Notice that distinct points $B_2, B_3, \ldots, B_n$ are not on B_1P, else ℓ would meet B_1P in more than one point, violating Theorem 14.19. By Axiom A-2, for each point $B_2, B_3, \ldots, B_n$, there exists a unique parallel line $m_2, m_3, \ldots, m_n$ such that B_i is on m_i, and $m_i \parallel B_1P$ for all $i = 2, 3, \ldots, n$, as demonstrated in Figure 14.25.

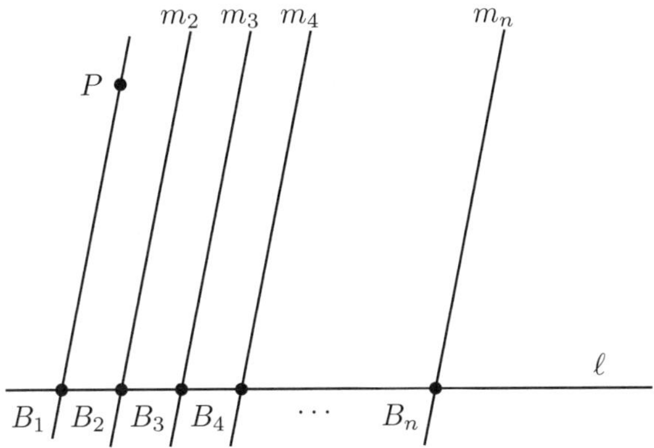

Figure 14.25. Set of parallel lines

Claim: Every point in the finite affine plane is on one of the lines in this set of parallel lines $B_1P, m_2, m_3, \ldots, m_n$. Why? Let Q be any point in the plane. If Q is on B_1P then our claim is true. If Q is not on B_1P, then by Axiom A-2 there exists a unique line m_Q such that Q is on m_Q and $m_Q \parallel B_1P$. Since ℓ meets B_1P, then by Theorem 14.22, ℓ must meet m_Q. Therefore, m_Q must intersect ℓ at one of its points $B_2, B_3, \ldots, B_n$, say B_R. By Axiom A-2, m_Q must be identical to m_R, otherwise there will be two lines through B_R that are parallel to B_1P. Therefore, if Q is not on B_1P, then Q must be on one of the parallel lines $m_2, m_3, \ldots, m_n$.

Now we can count the number of points in the plane. Each of the lines B_1P, m_2, $m_3, \ldots, m_n$ has n points, and by Theorem 14.23, no two of these lines share a common point. Therefore, since there are n distinct lines, each with n points, there are n^2 points in the affine plane of order n. $\square$

Notice that by following the construction given in the previous proof, we can count the groupings of parallel lines. This produces the following theorem.

Theorem 14.29. *In an affine plane of order n, there are $n - 1$ lines parallel to (and distinct from) any given line.*

By Theorem 14.29, it is clear that each parallel class consists of n lines. When the definition of parallel is expanded to allow a line to be parallel to itself, then it is easy to show that parallelism forms an equivalence relation (reflexive, symmetric and transitive) on the set of all lines in an affine plane. (See Exercise 14.4.13.) The equivalence classes formed by this relation are the parallel classes, hence the name. Before stating this result, we first count the number of lines in a finite affine plane.

Theorem 14.30. *In a finite affine plane of order n, there are $n^2 + n$ lines.*

Proof.   Let P be a point in the plane. By Corollary 14.27, P is on $n+1$ distinct lines, say $\ell_1, \ell_2, \ldots, \ell_{n+1}$. Each of these lines forms a parallel class consisting of n distinct lines. These parallel classes are mutually disjoint, meaning no line can be in more than one class. (Why? See Exercise 14.4.14.) Therefore, there are at least $(n + 1) \cdot n$ distinct lines in the plane.

Claim: There are no other lines. Why? Let m be a line in the plane. If P is on m, then $m = \ell_i$ for some i, and has, therefore, been counted. If P is not on m, then by Axiom A-2 there is a unique line k through P which is parallel to m. Therefore, $k = \ell_j$ for some j. Thus, m is in the parallel class of ℓ_j and has been counted.

Thus there are exactly $n^2 + n$ lines in the plane. $\square$

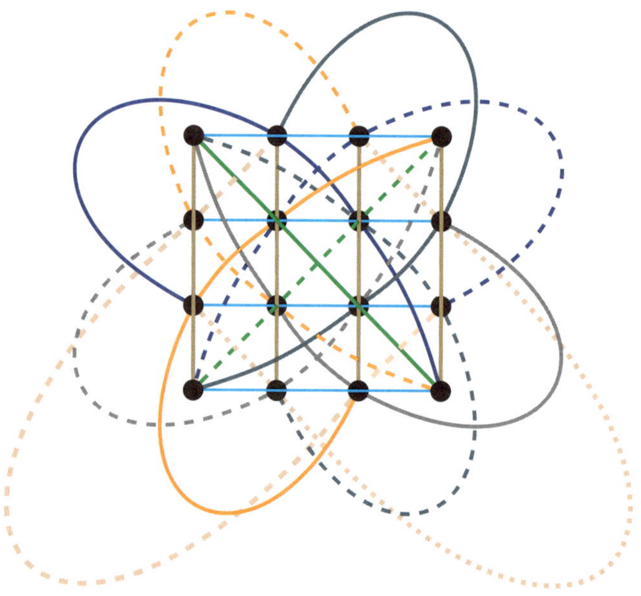

Figure 14.26. Affine plane of order 4

The proof can be modified slightly to establish the previously discussed combinatorial result about the parallel classes.

Theorem 14.31. *In an affine plane of order n, there are n + 1 parallel classes, each containing n lines.*

One more example of a model for the affine plane axioms is shown in Figure 14.26, an affine plane of order 4. After labeling the points in alphabetical order, starting in the lower left corner and proceeding in a similar manner to π_3 (as shown in Figure 14.22), identify the lines and the parallel classes, then verify the axioms and the theorems for this larger plane.

Finally, let's revisit the comment about categorical systems that we made near the end of Section 14.1. When a set of axioms is categorical, any statement about the undefined terms that holds true in one model will be true in all models. We noted that while the set of axioms for Four Point geometry is categorical, this property does not hold for all sets of finite geometry axioms. Clearly, the set of affine plane axioms is not categorical since it admits affine planes of different orders. Likewise, the set of projective plane axioms is not categorical. Even if we were to specify the axioms for an affine plane of order n, this set of axioms is not categorical for all orders. For example, there are seven nonisomorphic models for the affine plane of order 9. On the other hand, the axiom systems for Fano's and Young's geometries are categorical since they both admit a unique model, up to isomorphism.

Exercises 14.4

1. Verify that one of the diagrams for Four Point geometry demonstrated in Figure 14.2 is a model for the affine plane axioms.

2. Give the specific affine plane axiom(s) which is/are not satisfied by each of the proposed models shown in Figure 14.27.

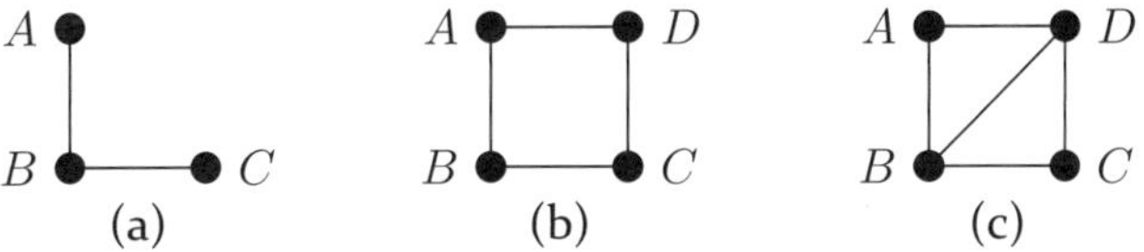

Figure 14.27. Exercise 14.4.2

3. Verify that the model shown in Figure 14.22 satisfies the axioms for an affine plane.

4. Redraw the model shown in Figure 14.22 in a different way, but be sure to maintain the demonstrated incidence structure. [There are infinitely many correct drawings that can be given.]

5. For each part, give an example of a model to verify the independence of the specified affine plane axiom.

   (a) Axiom A-1
   (b) Axiom A-2
   (c) Axiom A-3

6. Is the set of affine plane axioms categorical?

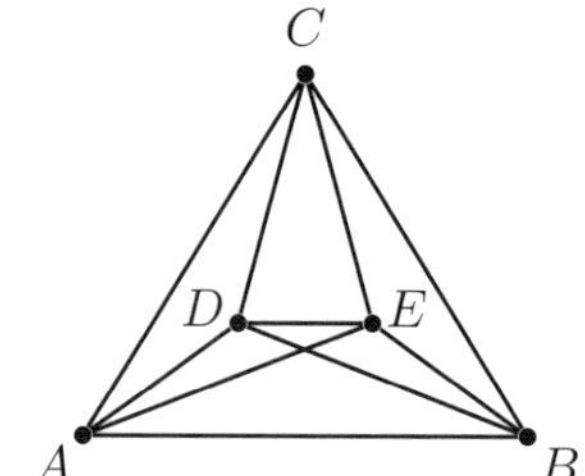

Figure 14.28. Exercise 14.4.7

7. A diagram representing the set of points and lines given in Exercise 14.3.2 is shown
    in Figure 14.28.

    (a) Explain why Axiom A-2 is not valid here.

    (b) How many lines through point A are parallel to line $\{D, E\}$?

    (c) In general, to a point P which is not on a line ℓ, how many lines through P are
         parallel to ℓ in this configuration of points and lines?

    (d) Replace Axiom A-2 with a new axiom (call it A-2*) that is valid for this configu-
         ration of points and lines. Use precise language to create Axiom A-2*.

    (e) What might be a good name for the sets of points and lines satisfying axioms A-1,
         A-2* and A-3?

8. State the dual of Theorem 14.19: *In an affine plane, two distinct lines can meet in at
    most one point.* If the dual is true, then prove it. Otherwise, give a counterexample.

9. State the prove the dual of Theorem 14.20: *There are at least two points on every line
    in any affine plane.*

10. State and prove the dual of Theorem 14.21: *In an affine plane, for any line ℓ there is
     a point not on ℓ.*

11. Show that a finite affine plane does not satisfy the Principle of Duality.

12. Prove Corollary 14.27: In an affine plane of order n, every point is on $n + 1$ lines.

13. Let the relation $\sim$ on the set of lines, $\mathscr{L}$, in an affine plane be given by $\ell \sim m$ iff $\ell \parallel m$
     or $\ell = m$. Prove that $\sim$ is an equivalence relation on the set of lines by completing
     the following three steps to show it is reflexive, symmetric and transitive.

    (a) Show that $\sim$ is reflexive, that is, $\ell \sim \ell$ for all lines ℓ.

    (b) Show that $\sim$ is symmetric, that is, if $\ell \sim m$, then $m \sim \ell$.

    (c) Show that $\sim$ is transitive, that is, if $\ell \sim m$ and $m \sim k$, then $\ell \sim k$.

14. A parallel class is a set consisting of a line together with all the lines parallel to it.
     Let the parallel class of line ℓ be denoted by $[\,\ell\,]$.

    (a) Using Exercise 13, show that $[\,\ell\,] = \{\, m \in \mathscr{L} \mid m \sim \ell \,\}$.

    (b) Show that if $\ell \parallel m$, then $[\,\ell\,] = [\,m\,]$.

    (c) Show that each line in the plane is an element of exactly one parallel class.

15. Modify the technique employed in the proof of Theorem 14.15 in order to provide a different proof of Theorem 14.30: In a finite affine plane of order n, there are $n^2 + n$ lines.

16. Prove Theorem 14.31: In an affine plane of order n, there are $n + 1$ parallel classes, each containing n lines.

17. As we will see in Section 14.6, there is an affine plane of order 5. How many points and lines does it have?

18. **Tic-tac-toe on an Affine Plane.** The standard game of tic-tac-toe is played on a 3×3 board. Since the affine plane of order 3 consists of 9 points that can be arranged in a 3×3 grid, we can play tic-tac-toe on this plane. Each cell in the standard 3×3 board is identified with the corresponding point of π_3. The new twist is that the winning lines are the lines in π_3. Using Figure 14.23 as a guide, we see that there are 12 ways to win when playing on π_3. Eight ways to win are the standard three-in-a-row horizontal, vertical and diagonal winners, but playing on π_3 means that there are four new ways to win. The new winning lines prescribed by π_3 are shown here.

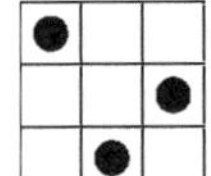 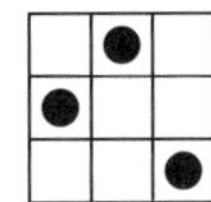 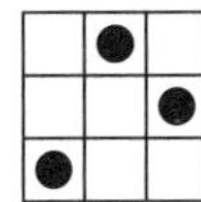 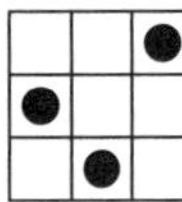

Play this game with one of your classmates for as long as it takes to determine the general outcome of this game.

19. Answer the following questions about the affine plane of order 4, as demonstrated in Figure 14.26.

  (a) How many points are there?
  (b) How many lines?
  (c) How many parallel classes?
  (d) How many lines are in each parallel class?

20. **Tic-tac-toe on π_4.** (Be sure to try Exercise 18 first.) Tic-tac-toe on a standard 4×4 board has 10 winning lines, namely 4 horizontal, 4 vertical and 2 diagonal. By Exercise 19, there are 20 winning lines when playing on the 4×4 arrangement of points on an affine plane of order 4. So, playing on π_4 means there are ten new ways to win. The new winning lines prescribed by π_4 are given below.

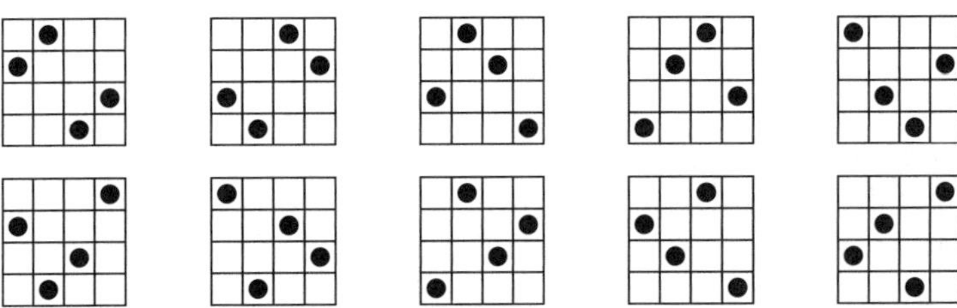

Play this game with one of your classmates for as long as it takes to determine the general outcome of this game. (For an in-depth analysis of the game of tic-tac-toe played on finite planes, see [21].)

14.5 Transforming afine into projective

In this section we explore the relationship between affine and projective planes of the same order. An affine plane of order n has n^2 points and $n^2 + n$ lines, while the projective plane has $n^2 + n + 1$ points and lines. Inspired by the method of perspective drawing employed by Renaissance painters and the points at infinity first introduced by Desargues, we will transform an affine plane of order n into a projective plane of order n by carefully adding one line and $n + 1$ points.

As discussed in Section 14.3.1, long before projective or affine geometry existed, it was artists, architects and cartographers who worked out the mechanics of perspective drawing in order to faithfully represent an object from the three-dimensional world in a two-dimensional way. Anyone who has ever tried to sketch a scene knows how difficult this can be. The wall in the photograph on the left, for example, consists of uniformly sized glass blocks, and yet, those nearer the viewer appear wider. The mortar lines in the wall are parallel, and yet, these lines appear to converge in the distance. The metro corridor displays similar features, showing parallel lines that converge in the distance as well as congruent semicircles that appear smaller in the distance.

The method of perspective drawing developed during this time gave its practitioners techniques to project this visual information onto their sketches. Artists skilled in perspective techniques were able to produce realistic paintings of scenes much more complicated than the ones in these photographs. Thankfully, in the finite geometries of this chapter, we have nothing more complicated than lines with which to contend, and it is within the methods of these Renaissance artists that we find the key to transforming these lines from the affine to the projective. Though it is a bit strange to put these words on the page, Desargues' outrageous idea holds the key: *Parallel lines must converge!* Recall that artists call the point on the canvas at which the image of parallel lines in the three-dimensional world converge the vanishing point, and that there is one vanishing point for every set of parallel lines. Likewise, Desargues wrote of parallel lines meeting at a theoretical infinite distance. This is exactly what we do to extend an affine plane, though there is no notion of distance here.[3] Simply described, in order to transform the affine into the projective, we force the parallel lines of an affine plane to meet at a new point. What artists call a vanishing point, mathematicians call an **ideal point**, or, a **point at infinity**. Unlike Hyperbolic geometry, these ideal points are points in the geometry. By this construction, there is one new point at infinity for each parallel class. Additionally, we need to add the mathematical equivalent of

---

[3]The descriptor "at infinity" makes sense in the real projective plane. See the box at the end of this section.

the horizon line employed by an artist. Fittingly, this line consists entirely of our new points at infinity, and is called the **line at infinity**.

As an example, let's start with our smallest affine plane, π_2, as shown in Figure 14.20. This plane has $n^2 = 4$ points, $n^2 + n = 6$ lines, and $n + 1 = 3$ parallel classes, with each class consisting of $n = 2$ lines. Each point is on $n + 1 = 3$ lines, and each line is on $n = 2$ points. We modify the diagram for π_2 shown in Figure 14.20 by adding one point at infinity to each of the three parallel classes which are illustrated in Figure 14.21. Let's add point P_1 to lines AB and CD, P_2 to lines AC and BD, and P_3 to lines AD and BC, as demonstrated in Figure 14.29, where three different colors are employed to help identify lines ABP_1, CDP_1, ACP_2, BDP_2, ADP_3 and BCP_3. Before we add the line at infinity, notice that the diagram in Figure 14.29 does not satisfy the finite projective plane axioms since, for example, there is no line between points P_1 and P_2. As we see in Figure 14.30, adding the line at infinity, consisting entirely of the new points at infinity, eliminates this problem.

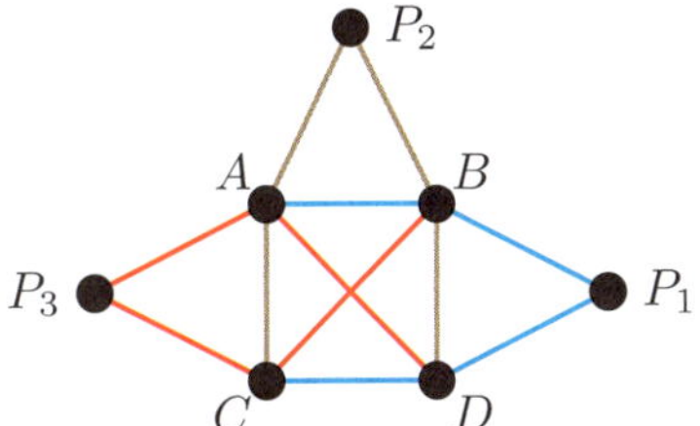

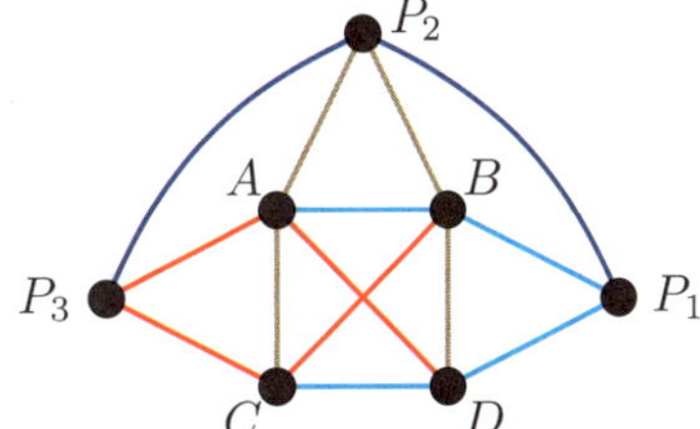

Figure 14.29. Three points at infinity added to π_2

Figure 14.30. Line at infinity $P_1P_2P_3$ added to π_2

With the points configured in this way, the resulting arrangement of points may not be easily recognizable. We can identify the plane we have produced by doing some counting. How many points are there? How many lines? How many points are on each line? How many lines are on each point?[4] If you answered these questions correctly then you should suspect that we have produced the projective plane of order 2. Take a few minutes to verify that this diagram is equivalent to the Fano plane.

Desargues and the Renaissance artists have steered us in the right direction! In general, this process transforms the affine into the projective. Let's see why. Since an affine plane of order n has $n + 1$ parallel classes, we've added exactly $n + 1$ points. This brings our total number of points from n^2 to $n^2 + n + 1$. Since each line of π_n has n points, adding a point at infinity to each line means that we now have $n + 1$ points on each line. Additionally, the line at infinity consists of the $n + 1$ points at infinity. Lastly, adding the line at infinity to the $n^2 + n$ lines of π_n makes $n^2 + n + 1$ lines. Clearly, we have the correct number of points and lines, as well as the correct number of points on a line. Since we started with an affine plane, Axiom P-3 still holds. We leave it as an exercise for the reader to verify that Axioms P-1 and P-2 also hold, and hence, this process transforms an affine plane of order n into a projective plane of order n.

It is interesting to note that, once transformed, there is no difference between an original point and a point at infinity in the projective plane. Likewise, the line at infinity is no different than any of the original lines. This was not the case in the early work

---

[4]Answers in order: 7, 7, 3, 3

of Desargues and Poncelet in the seventeenth and early nineteenth centuries, as these elements maintained a separate nature. The uniformity of points and lines in projective geometry was established in the mid-nineteenth century by von Staudt. Consequently, there was no need to mention either points or lines at infinity in Section 14.3.2.

Transforming the projective into the affine is a simple matter of reversing the process outlined above. The homogeneity of lines in a projective plane, that is, any line has exactly the same properties as any other, means that we may take any line as the line at infinity. So, removing any line and all the points on it from a projective plane of order n forms an affine plane of order n. As an example, try this process with the Fano plane. Once again, we leave it to the reader to verify that the affine plane axioms hold. This line of reasoning results in the following theorem.

Theorem 14.32. *There exists a projective plane of order n if and only if there is an affine plane of order n.*

In the next section, we discuss which orders of finite planes exist. On a final note, in addition to transforming a finite affine plane into a projective plane of the same order, adding ideal points and the line at infinity also transforms the Euclidean plane into an infinite projective plane (see box below).

Model for the real projective plane

To transform the Euclidean plane to a projective plane, called the **real projective plane**, we must grapple with the idea of adding a point to each of infinitely many parallel classes. Furthermore, each of these classes consists of infinitely many lines. How can we visualize this? Surprisingly, our knowledge of Spherical geometry will serve us well.

As illustrated in the diagram, suppose the Euclidean plane sits tangent to a sphere at its north pole, N. A ray from the center of the sphere, C, to any point, A, in the plane will pierce the northern hemisphere at a point, A', creating a one-to-one correspondence between points in the plane and points in the sphere's northern hemisphere. As a result, we will not need the points of the southern hemisphere in our model, though for visualization purposes, we'll continue to draw on our knowledge of the full sphere. Notice that the farther A is from the north pole, the closer A' is to the equator, but no point of the Euclidean plane is identified with a point *on* the equator. With this insight, it would be natural to conclude that the points on the equator will be excluded from our model along with the southern hemisphere, but as we will see, these are the distinguishing points in this model of an infinite projective plane.

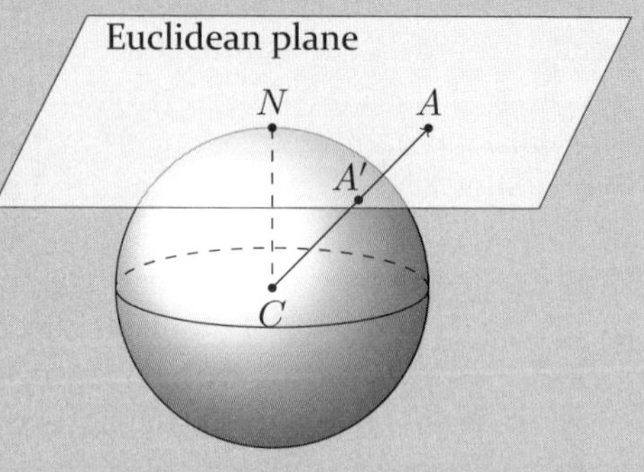

Under this mapping, what happens to a line in the plane? If P and Q are points on a line ℓ in the plane, then the span of rays (or vectors) $\overrightarrow{CP}$ and $\overrightarrow{CQ}$ is itself a plane which intersects the Euclidean plane at line ℓ, and passes through the center of the sphere. Thus, its intersection with the sphere is a great circle, and the image of ℓ is the arc of the great circle that lives strictly in the northern hemisphere.

Since any two great circles intersect at antipodal points, the image of two intersecting lines in the Euclidean plane under this mapping is two arcs of great circles that meet at a point in the northern hemisphere. For example, the image of two lines in the plane through N is two arcs of longitudinal lines in the northern hemisphere meeting at N. Likewise, the image of two lines in the plane through A is two arcs of great circles in the northern hemisphere meeting at A'. So, the image of the point of intersection of two nonparallel lines in the Euclidean plane under this mapping is exactly one point in the northern hemisphere. Lines that are parallel in the Euclidean plane are mapped to arcs of great circles that meet at the equator. Therefore, a point on the equator becomes a point at infinity in this model. Since the images of parallel lines meet at two points on the equator, and we may only add one point at infinity to any set of parallel lines, antipodal equator points must be identified in this model. This is the most difficult element of the model to visualize since, like three-dimensional models of Hyperbolic geometry, this cannot be rigidly embedded in space, that is, we cannot make an actual physical model of this space that we are able to hold in our hands. In this model of the infinite projective plane, there is a one-to-one correspondence between the points of the Euclidean plane and those in the northern hemisphere, and the line at infinity is the equator. We leave it to the reader to verify that the first two projective plane axioms are satisfied by this model.

Exercises 14.5

1. Transform the Fano plane (shown in Figure 14.1) into an affine plane by removing one line from the projective plane.

2. Transform the affine plane of order 3, shown in Figure 14.22, into a projective plane of order 3.

3. Transform the projective plane of order 3, shown in Figure 14.19, into an affine plane by removing one line from the projective plane.

4. Transform the affine plane of order 4 as shown in Figure 14.26 into a projective plane.

5. Explain why Axioms P-1 and P-2 hold for the set of points and set of lines that result from adding $n + 1$ points at infinity and one line at infinity to a finite affine plane of order n. When showing that Axiom P-1 holds, consider three cases, depending on whether none, one or both of the given points are ideal points. When showing that Axiom P-2 holds, consider two cases, depending on whether one of the given lines is the line at infinity.

6. Explain why Axioms A-1, A-2 and A-3 will hold for the set of points and set of lines that result from removing one line and the $n + 1$ point on it from a projective plane of order n.

7. Prove Theorem 14.32: There exists a projective plane of order n if and only if there is an affine plane of order n.

8. Explain why Axioms P-1 and P-2 hold for the set of points and set of lines that result from adding the points at the equator and the arc that contains them to the upper hemisphere model for the infinite projective plane (see box). When showing that Axiom P-1 holds, consider three cases, depending on whether none, one or both of the given points are ideal points. When showing that Axiom P-2 holds, consider two cases, depending on whether one of the given lines is the line at infinity.

14.6 Open problem in finite geometry

One of the great unsolved problems in finite geometry is the determination of the orders for which an affine and projective plane exists. We have seen models for affine planes of order 2, 3 and 4, and projective planes of order 2 and 3. Do we know any others? With the constructions in Section 14.5, we know there is a projective plane of order n if and only if there is an affine plane of order n. Therefore, there must be a projective plane of order 4. Do models exist for all other orders?

In 1906, Veblen and Bussey showed algebraically that planes exist whenever n is a prime power, that is, $n = p^k$, where p is prime and k is a positive integer. Listing the planes of small order known to exist, this means there are affine and projective planes of order $2, 3, 4, 5, 7, 8, 9, 11, 13, 16, 17, 19, 23, 25, \ldots$. Since Euclid proved that there are infinitely many primes, there are clearly infinitely many finite planes. What about the infinitely many composite, nonprime power orders that do not appear in the list? It's here where we venture into predominantly unknown territory. To this date, there is no known finite affine or projective plane of nonprime power order, and it has long been conjectured that prime powers are the only orders for which planes exist. Moreover, there are some composite orders that are known not to exist.

First, let's start with the two smallest composite orders missing from the list, 6 and 10. The question of the existence of an affine plane of order 6 is related to the 36 Officer Problem. Posed by Euler in 1782, the problem asks the following: Can 36 officers, of six different ranks from six different regiments, be arranged in a 6 by 6 square such that each row and column has each rank and regiment exactly once? [**42**] As noted before in this text, the ease with which a question is posed can be inversely proportional to the difficulty in finding a solution. A solution to Euler's question remained unpublished until 1901 when a French colonel, Gaston Tarry (1843–1913), computed (by hand!) all possible arrangements [**115**]. It took Tarry one year to do what would now take a computer under an hour, eventually finding that there was no such arrangement. This is exactly what Euler had predicted, but could not prove.

Euler approached the problem in the following way. He used Latin letters (A,B,C, D,E,F) to represent the ranks and Greek letters $(\alpha, \beta, \gamma, \delta, \epsilon, \zeta)$ to represent the regiments. Thus, each officer has a unique representation as an ordered pair of a Latin letter and a Greek letter, for example (C, β). Euler then decomposed the problem into two sub-problems, the first to arrange the regiments and the second to arrange the ranks. To arrange either, he made a 6×6 square in which each letter appeared exactly once in each row and each column. This object has come to be known as a **Latin square**, regardless of the alphabet employed. For example, the first two matrices in Figure 14.31 are examples of 4×4 Latin squares.

$$\begin{pmatrix} A & B & C & D \\ B & A & D & C \\ C & D & A & B \\ D & C & B & A \end{pmatrix} \text{ and } \begin{pmatrix} \alpha & \beta & \gamma & \delta \\ \delta & \gamma & \beta & \alpha \\ \beta & \alpha & \delta & \gamma \\ \gamma & \delta & \alpha & \beta \end{pmatrix} \text{ form } \begin{pmatrix} A\alpha & B\beta & C\gamma & D\delta \\ B\delta & A\gamma & D\beta & C\alpha \\ C\beta & D\alpha & A\delta & B\gamma \\ D\gamma & C\delta & B\alpha & A\beta \end{pmatrix}$$

Figure 14.31. Overlapping two Latin squares produces the Graeco-Latin square

To demonstrate the advantage of decomposing the problem into two different arrangements of symbols, let's consider a smaller problem. If Euler had posed a 16 Officer Problem, a solution is easily constructed by overlapping the first two Latin squares in Figure 14.31 so that every possible ordered pair of Latin and Greek letters appears exactly once. We say that the two Latin squares are **orthogonal** if this occurs. Overlapping the two given matrices produces the rightmost 4×4 matrix in Figure 14.31, called a **Graeco-Latin square**. The reader should verify that every ordered pair of letters appears exactly once in this Graeco-Latin square. To convince yourself that this process does not work with every pair of 4×4 Latin squares, try overlapping the first matrix with the Latin square

$$\begin{pmatrix} \alpha & \beta & \gamma & \delta \\ \delta & \alpha & \beta & \gamma \\ \gamma & \delta & \alpha & \beta \\ \beta & \gamma & \delta & \alpha \end{pmatrix}.$$

(Also, verify that this *is* a Latin square.)

To solve the 36 Officer Problem, Euler needed to find a pair of orthogonal 6×6 Latin squares. He never did, and he not only conjectured that it could not be done for the 6×6 case, but also that it could not be done for any $n \times n$ case where $n = 4k + 2$ and k is any nonnegative integer. In essence, he conjectured that there is no pair of orthogonal $n \times n$ Latin squares for $n = 2, 6, 10, 14, 18, 22, 26, \ldots$, although he had no proof for all but the trivial 2×2 case. Tarry, an amateur mathematician, must have had a great deal of time on his hands given that his solution boiled down to brute force, checking thousands of matrices.

Euler translated the 36 Officer Problem into a question about Latin squares, but the reader may be wondering how this problem is related to finite planes. It wasn't until 1938 that R.C. Bose (1901–1987) established the connection between orthogonal Latin squares and finite affine planes [**13**]. He proved that an affine plane of order n exists if and only if there is a set of $n - 1$ Latin squares that are mutually (pairwise) orthogonal. In particular, an affine plane of order 6 exists iff there are 5 orthogonal 6×6 Latin squares. Since Tarry had shown that it's impossible to find *two* orthogonal 6×6 Latin squares, this implies that there is no affine (or projective) plane of order 6.

Further progress on the possible orders of finite planes was not made until the middle of the twentieth century. In 1949, Richard Bruck (1914–1991) and Herbert J. Ryser (1923–1985) proved that if n is not a prime power, $n = 4k + 1$ or $n = 4k + 2$ for positive integer k, and n cannot be written as the sum of two squares, then there is no projective plane of order n [**16**]. This has come to be known as the Bruck-Ryser-Chowla Theorem due to its generalization in 1950 with the help of Sarvadaman Chowla (1907–1995). This theorem resolves an infinite number of cases, but leaves just as many unsolved. To list just a few of the cases of low order resolved by the theorem, there are no such planes of order $n = 6, 14, 21$ and 22. For instance, $6 \neq p^k$ for any prime, $6 = 4 \cdot 1 + 2$, and 6 can only be written as a sum of positive integers in three ways, $1 + 5 = 2 + 4 = 3 + 3$, none of which is a sum of squares. While this particular case had already been solved by Tarry, the Bruck-Ryser-Chowla result is certainly more elegant, as well as infinitely more general. Despite its generality, some cases of low order remaining unanswered were $10, 12, 15, 18$, and 20. To demonstrate why such orders are not caught in the net of this theorem, for instance, $10 \neq p^k$ for any prime, and $10 = 4 \cdot 2 + 2$, but $10 = 1^2 + 3^2$.

Ryser's student, Clement Lam, was part of the team of mathematicians and computer scientists at Concordia University who proved the nonexistence of the plane of order 10 in 1989. His award-winning expository article ([**83**]) tells the story of a solution that took nearly a decade, and required thousands of hours of computer time. (See [**25**] for an account in the popular press.) If you are wondering why the case of the nonexistence of a projective plane of order 10 was so difficult, the number of candidates to be checked is a number that is 1449 digits long! The number of candidates grows so quickly that $n = 12$ is the next composite number which remains unresolved to this day! As such, this unsolved mathematics problem demonstrates the limits of our current computing technology. To summarize, while it is conjectured that affine and projective planes of prime-power orders are the only ones to exist, this remains an open problem. Perhaps the next generation of mathematicians and computer scientists will produce a new breakthrough.

Lastly, there's still this matter of Euler's conjecture. Bose resolved the nearly two centuries old question with two other mathematicians in 1959 [**14**]. Not only did they give a pair of 10×10 orthogonal Latin squares, they proved that there is a pair of orthogonal $n \times n$ Latin squares for every n except 2 and 6. So, every n^2 Officer Problem is solvable except when $n = 2$ or 6. After showing Euler to be wrong, they were dubbed "*Euler's spoilers.*" The story of the solution was front-page news in the *New York Times* on April 26, 1959 [**93**].

15

Isometries

Figure 15.1. Reflection of the Jefferson memorial, Washington, D.C.

This chapter takes us back to Euclidean and Hyperbolic geometry but from a different perspective. You may recall that in the "proof" of SAS, Euclid lifts a triangle and "applies," or superposes, it atop another to explain the congruence of the triangles. We rejected Euclid's unstated assumption that a figure can be moved about the plane without affecting any of its properties, and supported this in Section 5.3 by showing that angles and lengths can be distorted in Taxicab geometry when moving a triangle within the plane. With the exception of Euclid's superposition technique, the idea of motion in the plane has been absent in our investigations. Here, we address this topic as we consider mapping, or transforming, one geometric figure, such as a line, circle or triangle, to another through a function satisfying some basic properties. The study of these mappings is at the intersection of geometry and modern algebra.

In the previous chapter, we learned that projective geometry developed in the early nineteenth century during the quest to identify properties of a figure which were unchanged by any number of perspective transformations. The properties impervious to such transformations are called invariants. Since a perspective transformation is just one of many ways to transform one set of points to another, mathematicians began to study invariants of other types of transformations. Adding further complexity to these investigations is the fact that with the nineteenth century came a shattering of the belief in a "mono-geometric" Euclidean world; there were now other geometries to consider. In the latter half of the nineteenth century, a twenty-three year old Felix Klein saw a way to unify the study of these questions. Klein realized that another recent development in mathematics, algebraic group theory, could be used as a new lens through which to view all geometries. For his inaugural professorial address at the University of Erlangen in 1872, Klein used the concept of a group to classify planar geometries by studying invariants under transformations. After its publication, this idea became known as the *Erlangen Programm*, and it is widely regarded as one of the most influential papers of the nineteenth century. In this chapter, we take a look at the result of Klein's bold idea to intertwine geometry and modern algebra as we study a subgroup of transformations known as *rigid motions*, or *isometries*, for the Euclidean and hyperbolic planes. Naturally, we start with the formal definition of an algebraic group.

Definition 15.1. A **group** is a nonempty set of elements, G, together with a binary operation, $\circ$, satisfying the following four properties:

(1) The set of elements is closed under the operation, that is, $a \circ b \in G$ for all $a, b \in G$.

(2) The set contains an identity element, that is, there is an element, e, such that $a \circ e = e \circ a = a$ for all elements $a \in G$.

(3) Each element in the set has an inverse, that is, for each $a \in G$, there exists $b \in G$ such that $a \circ b = b \circ a = e$.

(4) The operation is associative, that is, for all a, b, and $c \in G$, $a \circ (b \circ c) = (a \circ b) \circ c$.

As an example, the set of integers under addition forms a group where the identity is 0, and the additive inverse of an element a is $-a$. Other examples of groups include the set of even integers under addition, or the set of nonzero rationals under multiplication. The set of integers under multiplication is not a group since the third condition is not satisfied. The set G is not restricted to subsets of real numbers. For example, the set of polynomial functions under addition forms a group where the identity is the constant polynomial, $i(x) = 0$, and the additive inverse of polynomial $p(x)$ is $-p(x)$. In this chapter, the set G will consist of a different set of functions under the operation of composition. We will focus on how these functions transform basic geometric figures. To do this, we must start by carefully defining these mapping functions and their properties.

15.1 Rigid motions or isometries

Functions are maps that assign one output to each input. For example, the real-valued function $f(x) = x^2$ assigns the output x^2 to any input x in the set of real numbers. The particular types of functions of interest to us in this chapter are those that preserve distance, that is, the distance between any two inputs equals the distance between their corresponding outputs. Such functions are called **isometries**, from the Greek *iso*, meaning same, and *metria*, meaning measure. For function $f(x)$ given above, since $f(1) = 1$ and $f(2) = 4$, then f is not an isometry because the distance between these inputs, $|1-2| = 1$, does not equal the distance between their outputs, $|f(1)-f(2)| = 3$. As another example, consider $g(x) = -x$ as a function of the real numbers. For any inputs a and b, the distance between the outputs is

$$|g(b) - g(a)| = |-b - (-a)| = |-b + a| = |a - b| = |b - a|,$$

which is equivalent to the distance between the inputs. Since g preserves distance, it possesses the distinguishing feature of the type of function under consideration in this chapter. However, while g is an example of a real-valued function, we will be working in planar geometry, and thus, the inputs and outputs of our functions will be points of the plane.

Definition 15.2. A function from a plane to itself is called a **transformation**, or **mapping**. If A and B are points, and f is a transformation with $f(A) = B$, then we say that f transforms, or maps, A to B. If $f(A) = A$, then we say that f fixes A, or, that A is a **fixed point** of f. The **identity**, i, is a transformation that fixes all points, that is, $i(A) = A$ for all A.

Given two transformations, f and g, we can create a new transformation using composition, denoted $f \circ g$, where $(f \circ g)(A) = f(g(A))$. This means that we work from right to left when encountering composition notation, applying g first, then f. In general, $f \circ g \neq g \circ f$. For example, suppose f is the sock function and g is the shoe function. When f acts on a bare foot, the output is a foot with a sock on it. So, $g \circ f$ acting on a bare foot produces the foot covered by a sock then covered by a shoe, presumably, as desired. On the other hand (or foot, perhaps), $f \circ g$ acting on a bare foot results in the embarrassing and impractical outcome of a sock covering a shoe which covers a foot.

Definition 15.3. An **isometry**, or **rigid motion**, is a mapping (or transformation), f, of a plane onto itself that preserves distances, that is, given any two points A and B in the plane, we have

$$d(A, B) = d(f(A), f(B)).$$

Notice that the word "onto" in the definition specifies that an isometry of a plane must be **onto**, or **surjective**. This means that the range of the map f must be all points of the plane, that is, for any point D on the plane, there is another point C so that $f(C) = D$. [As demonstrated at the beginning of this section, Euclidean analytic geometry can be a useful tool for the reader to build examples and counterexamples to test the properties of functions. For example, take a moment to convince yourself that, in the Euclidean plane where points are given by ordered pairs, $f(A) = f((a_1, a_2)) = (2a_1, 2a_2) = B$ is an onto map but not an isometry since it does not preserve distance. As another example, show that $g(A) = g((a_1, a_2)) = (1 + a_1, 2 + a_2) = B$ is an isometry.]

Since an isometry preserves distances, it must be **one-to-one**, or **injective**; that is, given an isometry, f, and points A and B, if $f(A) = f(B)$, then $A = B$. We leave the proof of this fact as an exercise for the reader. [As an example, show that g as given above is a one-to-one function.] Functions that are both injective and surjective are called **bijective**, and every bijective function has an inverse. Thus, every isometry is a bijection, and consequently, has an inverse. We leave the fact that the set of isometries of a plane forms a group under composition as an exercise. The identity element in this group is the identity transformation, and the unique inverse of an isometry f is denoted by f^{-1}.

Isometries do more than just preserve distances, they also preserve our two main characters, the line and the circle, as well as angles and triangles. We start by showing any isometry is line-preserving. For the remaining results in this section, a reference to Neutral geometry means a plane that satisfies the axioms of Neutral geometry. We also assume that these planes have a distance function.

Theorem 15.4. *In Neutral geometry, every isometry transforms straight lines onto straight lines.*

Proof. Let f be an isometry and let $\ell = \overleftrightarrow{AB}$ be a line. To show that $f(\ell)$ is a line, we need to show two things. First, if C is another point on line ℓ, we must show $C' = f(C)$ lies on the unique line joining A' and B', then we must show there is a point D on ℓ such that $f(D) = D'$. Recall, in Neutral geometry, C is between A and B iff $d(A, B) = d(A, C) + d(C, B)$. For the first part, if C is between A and B, we have

$$d(A', C') + d(C', B') = d(A, C) + d(C, B) = d(A, B) = d(A', B').$$

Hence C' lies on line $A'B'$ between A' and B' as desired. A similar argument applies if C lies outside of the segment AB.

For the second part, let D' be any point of $A'B'$ that lies between A' and B'. Since $d(A', D') + d(D', B') = d(A', B') = d(A, B)$, there is a unique point D on ℓ such that $d(A, D) = d(A', D')$ and $d(B, D) = d(B', D')$. Then, $f(D)$ is the same distance from A' as D', and $f(D)$ is the same distance from B' as D'. So, $D' = f(D)$. A similar argument applies if D' lies on $\overleftrightarrow{AB}$ but outside of the segment $A'B'$.

Hence, f maps every point of line $\overleftrightarrow{AB}$ to a point on $A'B'$, and every point of line $\overleftrightarrow{A'B'}$ is the image of some point on $\overleftrightarrow{AB}$. Thus, f transforms the line $\overleftrightarrow{AB}$ onto the line $\overleftrightarrow{A'B'}$. $\qquad\square$

Lemma 15.5. *In Neutral geometry, if an isometry fixes two points, then it fixes the line joining them.*

Proof. Let f be an isometry and let A and B be two points of the plane such that $f(A) = A$ and $f(B) = B$. By Theorem 15.4, f maps line $\overleftrightarrow{AB}$ to itself.

Moreover, if C is on line $\overleftrightarrow{AB}$ and $f(C) = C'$, then $d(A, C) = d(A, C')$ and $d(B, C) = d(B, C')$. Since there are only two points on $\overleftrightarrow{AB}$ at a given distance from A, then there is only one point that is both $d(A, C)$ from A and $d(B, C)$ from B. Hence, $C = C'$ and the isometry fixes the entire line. $\qquad\square$

We leave the proofs of the following three theorems to the reader.

Theorem 15.6. *In Neutral geometry, the image of a circle under an isometry is a circle with equal radius.*

Theorem 15.7. *In Neutral geometry, the image of a triangle under an isometry is a congruent triangle.*

Theorem 15.8. *In Neutral geometry, the image of an angle under an isometry is a congruent angle.*

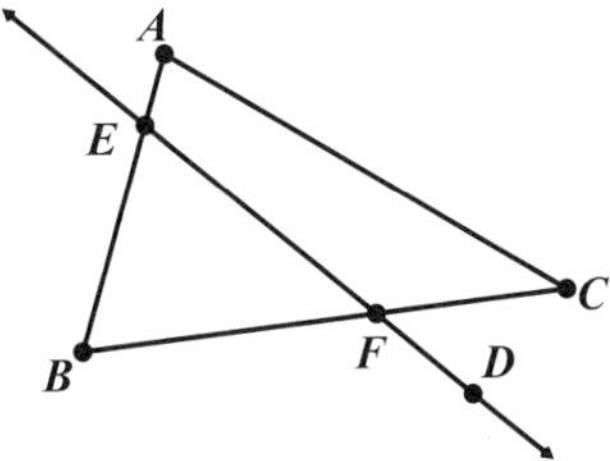

Figure 15.2. Fixed Point Theorem

The following theorem gives a shocking result: an isometry that fixes three non-collinear points fixes every point in the entire plane.

Theorem 15.9. *In Neutral geometry, if an isometry fixes three noncollinear points, then it is the identity.*

Proof. Suppose that f is an isometry such that $f(A) = A$, $f(B) = B$ and $f(C) = C$, where A, B and C are noncollinear. From Lemma 15.5, we know that f fixes lines $\overleftrightarrow{AB}$, $\overleftrightarrow{BC}$ and $\overleftrightarrow{AC}$. Let D be any point in the plane that is not on these three lines. Let E be a point on segment $\overline{AB}$ that lies strictly between A and B. Draw line $\overleftrightarrow{DE}$, as illustrated in Figure 15.2. By Pasch's Theorem, since line $\overleftrightarrow{DE}$ enters triangle $\triangle ABC$ at E, it must intersect either $\overline{BC}$ or $\overline{AC}$. WLOG, assume that it intersects $\overline{BC}$ at point F. Then, since f fixes both E and F, it fixes line $\overleftrightarrow{EF}$, and hence, it must fix D. $\qquad\square$

An immediate and important corollary to the previous theorem is the fact that when isometries agree at three noncollinear points, they must agree everywhere.

Corollary 15.10. *If two isometries agree at three noncollinear points, then they agree everywhere.*

Proof. Let A, B, and C be three noncollinear points, and let f and g be isometries such that $f(A) = g(A)$, $f(B) = g(B)$ and $f(C) = g(C)$. Consider the isometry $g^{-1} \circ f$. Notice that, for example, $(g^{-1} \circ f)(A) = g^{-1}(f(A)) = g^{-1}(g(A)) = A$. Since $(g^{-1} \circ f)(A) = A$, $(g^{-1} \circ f)(B) = B$ and $(g^{-1} \circ f)(C) = C$, then $g^{-1} \circ f = i$ by Theorem 15.9. Composing on the left by g on both sides of the equation gives

$$g \circ (g^{-1} \circ f) = g \circ i$$

$$(g \circ g^{-1}) \circ f = g \quad \text{(group identity and associative properties)}$$

$$i \circ f = g \quad \text{(inverse property)}$$

$$f = g \quad \text{(identity property)}. \qquad\square$$

With this corollary, we see that any isometry in Neutral geometry is uniquely determined by the image of three noncollinear points. In other words, an isometry is completely determined by what it does to the vertices of any triangle! Furthermore, for one of the main results in the next section, we prove that if $\triangle ABC \cong \triangle DEF$, then there is a unique isometry that transforms A to D, B to E, and C to F. That is, there is a unique isometry that transforms $\triangle ABC$ to $\triangle DEF$.

In the remaining sections we consider the different possible types of isometries in both Euclidean geometry as well as Hyperbolic geometry. We start by introducing *reflections*, the basic building blocks for isometries in Neutral geometry.

Exercises 15.1

1. Prove that an isometry must be a one-to-one mapping. That is, given an isometry, f, and points A and B, prove that if $f(A) = f(B)$, then $A = B$.

2. Prove that the set of isometries of a plane forms a group under composition.

3. Prove that, in Neutral geometry, the image of a circle under an isometry is a circle of congruent radius.

4. Prove that, in Neutral geometry, the image of a triangle under an isometry is a congruent triangle.

5. Prove that, in Neutral geometry, the image of an angle under an isometry is a congruent angle.

15.2 Reflections

Informally, the idea of a *reflection* is to "flip", or mirror, the entire plane across a given line. The photograph at the beginning of this chapter is a real-world example of a reflection.

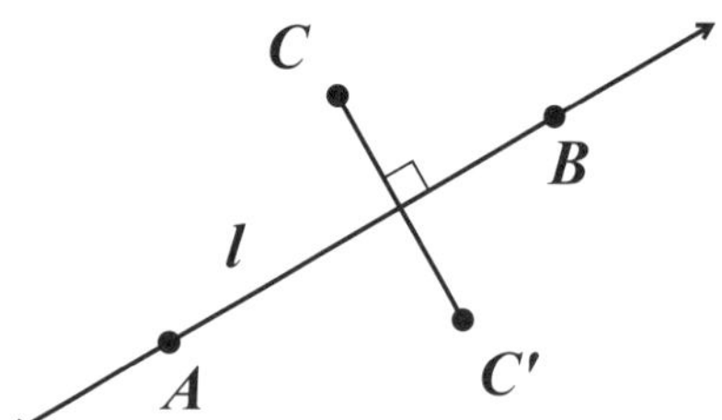

Figure 15.3. Reflection across ℓ

Definition 15.11. Given a line $\ell = \overleftrightarrow{AB}$, the **reflection** across ℓ is denoted by $F_\ell = F_{AB}$, and if C is any point in the plane that does not lie on ℓ, then

$$F_{AB}(C) = C'$$

where C' is the unique point such that ℓ is the perpendicular bisector of CC'. The **line of reflection**, ℓ, is fixed by this motion.

We leave the proof of the following theorem as an exercise.

Theorem 15.12. *In Neutral geometry, a reflection is an isometry.*

Since there are infinitely many lines in the plane, there are infinitely many reflections. Remembering that we are interested in groups of functions under composition, it is natural at this point to ask the following: Does the set of reflections form a group under composition? We start by noting that composition is always associative. It is also clear that a reflection is an **involution**, that is, it is its own inverse since $F_{AB}^{-1} = F_{AB}$.[1] However, it is not possible to express the identity transformation as a reflection since there is no line across which every point of the plane remains fixed. Thus, the set of reflections does not form a group under composition. Additionally, the set of reflections is not closed under composition. To help see why the composition of two reflections cannot be another reflection, we introduce the concepts of *preserving* and *reversing orientation*.

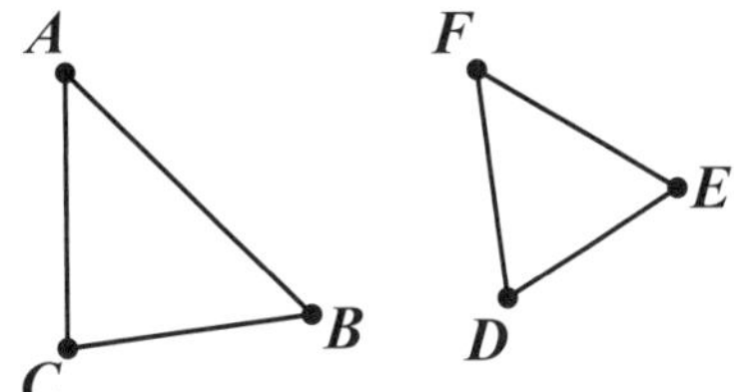

Figure 15.4. Orientation of a triangle

Definition 15.13. A triangle has **clockwise orientation** if when one traverses the vertices in alphabetical order, one travels the circumference of the triangle in a clockwise direction. Otherwise, the triangle has **counterclockwise orientation**.

In Figure 15.4, triangle $\triangle ABC$ has clockwise orientation and triangle $\triangle DEF$ has counterclockwise orientation. Since labeling the vertices of a triangle is arbitrary, we are not interested in the orientation per se, but rather, whether or not it changes under the action of a given isometry.

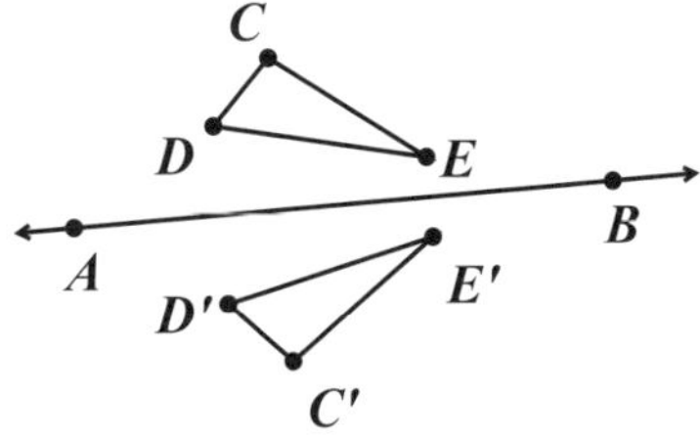

Figure 15.5. Reflections reverse orientation

---

[1] Interestingly, the term *involution* is the only new term introduced by Desargues that remains in the mathematical lexicon. (See page 342 for a discussion of Desarguesian nomenclature.)

Definition 15.14. An isometry, f, is called **orientation preserving** if both triangles $\triangle ABC$ and $\triangle A'B'C'$ have the same orientation (where $A' = f(A), B' = f(B)$ and $C' = f(C)$). If the two triangles have opposite orientations, one clockwise and the other counterclockwise, then f is **orientation reversing**.

Reflections reverse orientation, as demonstrated in Figure 15.5. Since reversing orientation twice results in the original orientation, the composition of two reflections is orientation preserving. Therefore, a composition of two reflections cannot be another reflection.

We have introduced reflections as the first example of an isometry because they are the building blocks for all isometries in Neutral geometry. The following theorem makes this plain: all possible isometries of the plane are the result of at most three reflections.

Theorem 15.15. *Every isometry of Neutral geometry is the composition of three or fewer reflections.*

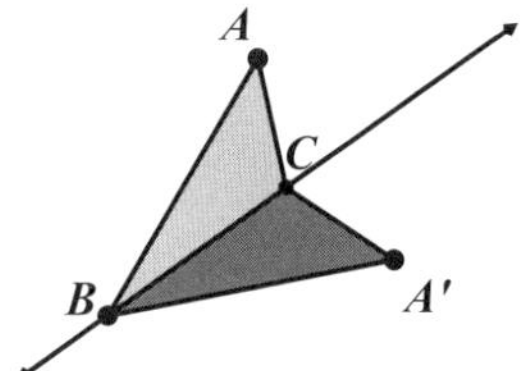

Figure 15.6. One reflection

Proof. Let f be an isometry, $\triangle ABC$ be any triangle in the plane, and $A' = f(A), B' = f(B)$, and $C' = f(C)$. By Theorem 15.7, we have $\triangle ABC \cong \triangle A'B'C'$. If we can find another isometry that also maps A to A', B to B', and C to C', then by Corollary 15.10, the two isometries must agree everywhere. Thus, our goal is to show that we can transform A to A', B to B', and C to C' in three or fewer reflections.

Case 1. Suppose that f fixes all three vertices, A, B, and C. By Theorem 15.9, we have $f = i$. In this case, f can be written as a reflection composed with itself, that is $f = F_\ell \circ F_\ell$, where ℓ is any line in the plane.

Case 2. Suppose that f fixes two of the three vertices. WLOG, we will assume that f fixes B and C. By Lemma 15.5, f must fix BC, as illustrated in Figure 15.6. Since $\triangle ABC \cong \triangle A'BC$, we have $BA = BA'$ and $CA = CA'$. By Theorem 3.5, B and C must both lie on the perpendicular bisector of AA'. Thus, $F_{BC}(A) = A'$, and f is a reflection, specifically, $f = F_{BC}$.

Case 3. Suppose that f fixes one of the three vertices. WLOG, we will assume that f fixes C. Let D be the midpoint of the segment BB', as demonstrated in Figure 15.7. Since $CB = CB'$, C must lie on the perpendicular bisector of BB'. The reflection F_{CD} will take B to B'. Let $A'' = F_{CD}(A)$. The image of $\triangle ABC$ under F_{CD} is then $\triangle A''B'C$

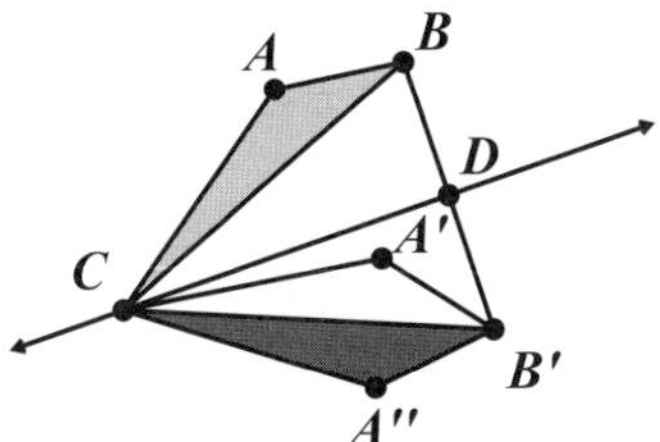

Figure 15.7. Two reflections

which shares at least two vertices with $\triangle A'B'C$. If $A'' = A'$, we are done. If not, then by Case 2, there is a single reflection that will transform A'' to A'. Therefore, f is the composition of at most two reflections.

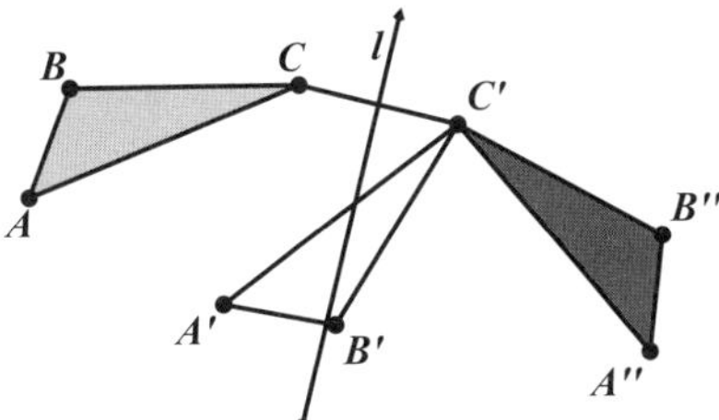

Figure 15.8. Three reflections

Case 4. Suppose that f does not fix any of the three vertices. Let ℓ be the perpendicular bisector of CC'. Then $F_\ell(C) = C'$. Let $A'' = F_\ell(A)$ and $B'' = F_\ell(B)$, as demonstrated in Figure 15.8. The image of $\triangle ABC$ under F_ℓ is then $\triangle A''B''C'$ which shares at least one vertex with $\triangle A'B'C'$. By Case 3, we need at most two reflections to transform $\triangle A''B''C'$ to $\triangle A'B'C'$. Thus, we need at most three reflections to transform $\triangle ABC$ to $\triangle A'B'C'$. $\square$

Theorem 15.15 goes a long way in helping us understand the structure of isometries in Neutral geometry. Equipped with reflections, we now venture off to meet the remaining three types of isometries in the Euclidean plane: *translations, rotations* and *glide reflections.*

Exercises 15.2

1. Prove Theorem 15.12: In Neutral geometry, a reflection is an isometry. [*Hints: Note that there are two cases to consider. The case where C and D are on the same side of ℓ is illustrated in Figure* 15.9. *For the case where C and D are on opposite sides of ℓ, let the intersection of CD with ℓ be E. Join C'E and D'E, then prove that C'ED' forms a straight line.*]

2. A figure in the plane has **reflectional** or **mirror symmetry** if there is a line ℓ such that F_ℓ maps the figure to itself. In this case, line ℓ is called a **line of symmetry** for the figure.

   (a) Find the lines of symmetry for each of the following regular polygons.

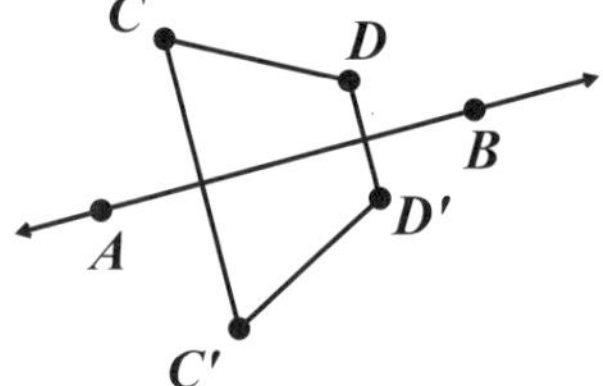

Figure 15.9. Exercise 15.2.1: C and D are on the same side of ℓ

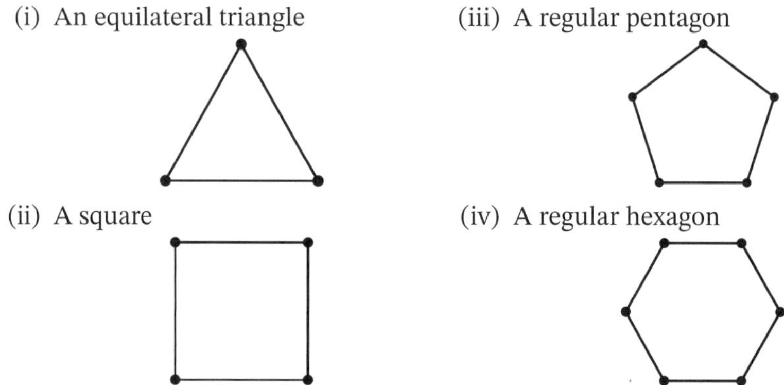

(i) An equilateral triangle

(ii) A square

(iii) A regular pentagon

(iv) A regular hexagon

(b) In general, how many lines of symmetry does a regular n-gon have?

15.3 Isometries of the Euclidean plane

For this section, when we refer to *the plane*, we mean the *Euclidean plane*. There are three types of isometries left to define: *translations, rotations* and *glide reflections*. After initially defining them, we will see that each of them can be thought of as the composition of either two or three reflections.

15.3.1 Translations - Part 1. A *translation* of the plane can be viewed as sliding the entire plane by a given length in a given direction, as the following definition makes clear.

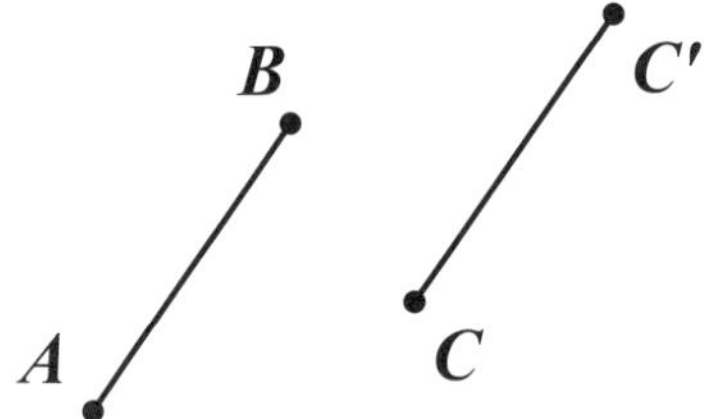

Figure 15.10. Translation by AB

Definition 15.16. Given points A and B, the **translation** that maps A to B is denoted by T_{AB}, and if C is any point in the plane, then

$$T_{AB}(C) = C'$$

where C' is the unique point such that $AB = CC'$, and the rays $\overrightarrow{AB}$ and $\overrightarrow{CC'}$ are parallel and point in the same direction. We may also write $T_{\overrightarrow{AB}}$ to denote this translation.

Students of calculus or physics may recall that a **vector** in the plane is a quantity that has both direction and magnitude, without a specified position. For this reason, we also say that a translation slides the entire plane by a given vector. The vector with starting point A and ending point B is denoted by $\overrightarrow{AB}$. As we use the same notation for vectors and rays, we rely on context to provide the proper interpretation. Since a vector has no position, it is equivalent to any vector in the plane with the same length and in the same direction. Consequently, the name of a translation is not unique. In the above definition, for example, we have $T_{AB} = T_{CC'}$ since vectors $\overrightarrow{AB}$ and $\overrightarrow{CC'}$, having the same length and direction, are equivalent. Hence, they produce the same translation of the plane since translations are determined by these two vector properties. This is why it is sometimes clearer to add vector arrows, instead writing $T_{\overrightarrow{AB}} = T_{\overrightarrow{CC'}}$. Since vector notation is more cumbersome, we only include it when necessary.

As a first step, we'd like to show that translations are isometries. To do this, we need the following lemma relating a translation with a parallelogram.

Lemma 15.17. *Let T_{AB} be a translation. If C and D are any two points in the plane with $C' = T_{AB}(C)$ and $D' = T_{AB}(D)$, and, if C' is not collinear with C and D, then $CC'D'D$ is a parallelogram.*

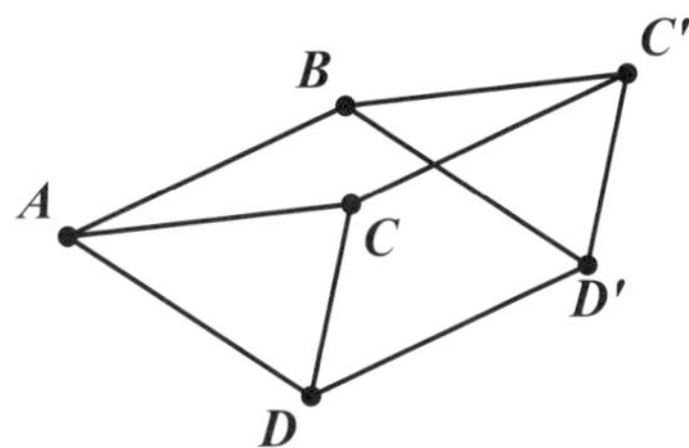

Figure 15.11. Translations are isometries

Proof. Let T_{AB} be a translation, and let C and D be any two points in the plane with $C' = T_{AB}(C)$ and $D' = T_{AB}(D)$ where C' is not collinear with C and D. By definition, $CC' = AB = DD'$, $\overrightarrow{CC'} \parallel \overrightarrow{AB}$ and $\overrightarrow{DD'} \parallel \overrightarrow{AB}$. By Euclid I.30, we have $\overrightarrow{CC'} \parallel \overrightarrow{DD'}$. Consequently, by Euclid I.33, we have $CD = C'D'$ and $CD \parallel C'D'$. Thus, $CC'D'C$ is a parallelogram.   □

Since C and D are any two points in the plane, and we proved that $CD = C'D'$, we have nearly proven that a translation is an isometry. The one extra condition of Lemma 15.17 is that C' is not collinear with C and D. If we can show that $CD = C'D'$ in the case where C', C and D are collinear, then we have the following theorem. We leave it to the reader to show that distance is preserved in this case.

Theorem 15.18. *Every translation is an isometry.*

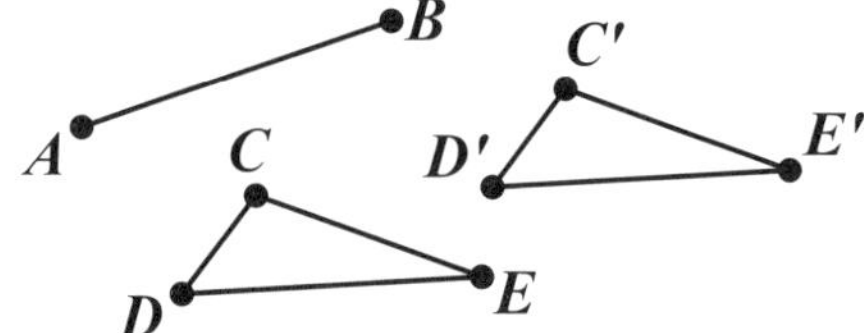

Figure 15.12. Translations preserve orientation

In the previous section, we found that the set of reflections does not form a group since it is not closed under composition. How about the set of translations? As illustrated in Figure 15.12, notice that, unlike reflections, a translation preserves orientation. In fact, the composition of any two translations is always a translation. In order to see this, we need to note that given any translation, we can pick any point in the plane as the starting point for our translation vector. (Alternatively, we could pick any point as its ending point.) Thus, we only need to prove the following theorem, and we leave the proof as an exercise for the reader.

Theorem 15.19. *If A, B and C are any three points, then*

$$T_{BC} \circ T_{AB} = T_{AC}.$$

From this theorem, we have that $T_{AB} \circ T_{BA} = T_{AA} = i$. Thus,

$$T_{AB}^{-1} = T_{BA}.$$

The above discussion leads us to the following observation, the proof of which we leave as an exercise.

Theorem 15.20. *The set of translations forms a group under composition.*

Since it forms a group, the set of translations is self-contained. As we will see, this makes translations unique among the isometries of the Euclidean plane.

The closure property of a group means that the composition of two or more translations must be another translation. By Corollary 15.10, we know that any isometry of the plane is uniquely determined by the image of three noncollinear points. So, we can identify the translation that results from a composition of translations by tracking the image of a triangle. For example, if T_1, T_2 and T_3 are translations such that the composition $T_3 \circ T_2 \circ T_1$ results in translation T_4, then we can easily identify T_4 after finding $(T_3 \circ T_2 \circ T_1)(\triangle ABC)$. To do this, we find the intermediate images of $\triangle ABC$, progressing towards the final resulting image of the given composition as follows: beginning with $T_1(\triangle ABC)$, we then find $T_2(T_1(\triangle ABC))$, and finally, $T_3(T_2(T_1(\triangle ABC)))$. Once we have the final image, we can specify T_4. While the three points of a triangle are sufficient, we find that it is far easier to work with the four points of a square, as demonstrated by the following example, due to its inherent symmetry and the fact that the image of a square under any isometry is another square.

Example 15.21. Consider the composition $T_{DB} \circ T_{AC}$, where points A, B, C and D are given by $\square ABCD$, as shown in Figure 15.13. To identify the composition $T_{DB} \circ T_{AC}$ as a single translation, we first determine the intermediate image $T_{AC}(\square ABCD) =$

$\square A'B'C'D'$. Here, we mean each point x is mapped to x'. For example, since we translate by $\overrightarrow{AC}$, we have $A' = T_{AC}(A) = C$. Next, we find the final image of the resulting composition, $T_{DB}(T_{AC}(\square ABCD)) = T_{DB}(\square A'B'C'D') = \square A''B''C''D''$. Here, for example, since we translate by $\overrightarrow{DB}$, we have $D'' = T_{DB}(D') = B'$. To identify the overall effect of the composition as a single translation, notice that the image of each vertex of $\square ABCD$ under the given composition has moved in the direction of vector $\overrightarrow{AB}$ by twice the length of AB. Therefore, the given composition can be written as the single isometry $T_{2\overrightarrow{AB}}$.

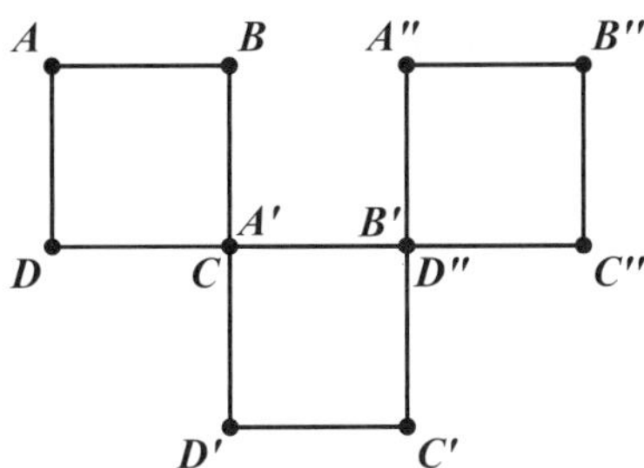

Figure 15.13. Example 15.21: Composition of translations of $\square ABCD$

Exercises 15.3.1

1. Finish the proof of Theorem 15.18 to show that every translation is an isometry by considering the case when C, C' and D (hence D') are all collinear.

2. Prove Theorem 15.19: Given any three points, A, B and C, prove that $T_{BC} \circ T_{AB} = T_{AC}$.

3. Given any three points, A, B and C, prove that

$$T_{BC} \circ T_{AB} = T_{AB} \circ T_{BC}.$$

   Consider two cases: one where A, B and C are collinear, and the other where they are not.

4. Prove Theorem 15.20: The set of translations forms a group under composition.

5. Consider square $\square ABCD$ given in Figure 15.14. For each part, determine the image of the square under the given composition in order to write the resulting composition as a single isometry. Use the method demonstrated in Example 15.21 as a guide. Be sure to include all intermediate images of the square, as well as the final image, for each composition.

   (a) $T_{CD} \circ T_{AB}$ (c) $T_{DB} \circ T_{BC}$
   (b) $T_{AC} \circ T_{BD}$ (d) $T_{BD} \circ T_{CB} \circ T_{AC}$

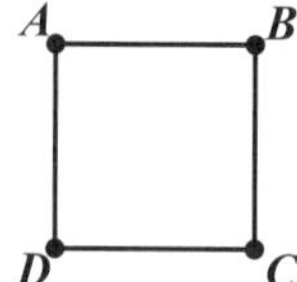

Figure 15.14. Exercise 15.3.1.5

15.3.2 Rotations - Part 1. Our third type of isometry is a *rotation*. This mapping rotates the entire plane around a chosen point by a given angle. (This type of transformation is denoted with the letter R, which explains why reflections do not enjoy their natural letter designation.)

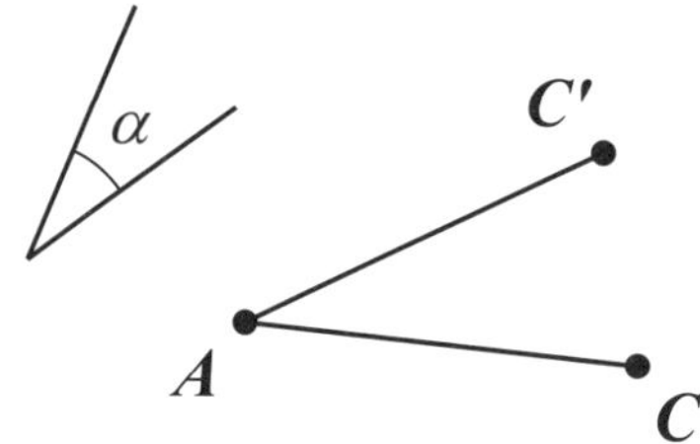

Figure 15.15. Rotation around A by α

Definition 15.22. Given a point A and an angle α, the **rotation** about A by α is denoted by $R_{A,\alpha}$, and if C is any point in the plane other than A, then

$$R_{A,\alpha}(C) = C'$$

where C' is the unique point such that $AC = AC'$ and $\angle CAC' = \alpha$. Positive angles rotate counterclockwise and negative angles rotate clockwise. The point A is the **center of rotation** and is fixed by this motion.

Note that while the center of a rotation is unique, the angle of rotation is not since we can add either an integer multiple of 2π radians, or $360°$, depending on how we measure our angles. In general, we tend to limit our range of angle measure to $-\pi < \alpha < 2\pi$, or $-180° < \alpha < 360°$, since they are easily recognizable. We leave the proof of the following theorem as an exercise.

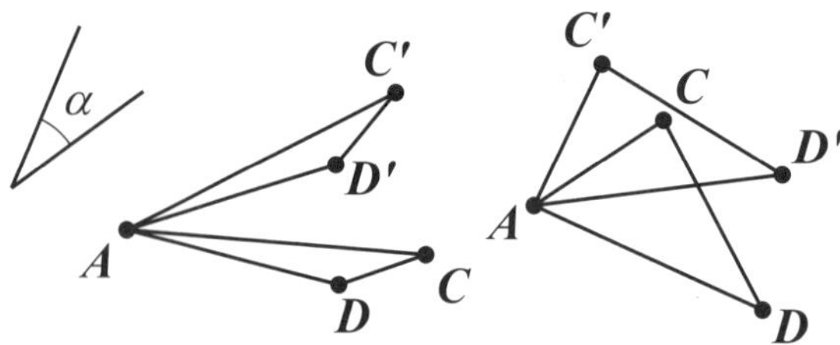

Figure 15.16. Rotations are isometries: two cases

Theorem 15.23. *Every rotation is an isometry.*

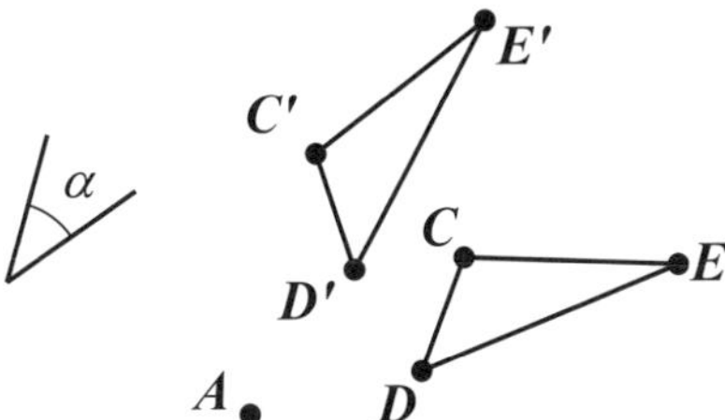

Figure 15.17. Rotations preserve orientation

As we have done with the other types of isometries, we would like to determine whether the set of rotations forms a group. Notice that, like translations, rotations preserve orientation, as illustrated in Figure 15.17. It is clear that $R_{A,\alpha}^{-1} = R_{A,-\alpha}$ and that we can write the identity transformation as a rotation about any point by an angle of zero, that is, $R_{A,0}$. The set of rotations is not, however, closed under composition. (Consider what happens when composing two rotations about the same point versus two rotations about different points.) More specifically, the composition of two rotations *may* be another rotation, but it may also be a translation. Without closure under composition, the set of rotations does not form a group. In order to see why two rotations can result in either a rotation or a translation, we need to determine how reflections, translations and rotations are related.

Exercises 15.3.2
1. Prove Theorem 15.23: Every rotation is an isometry. Note that, as illustrated in Figure 15.16, there are two cases to consider: one where $\triangle AC'D'$ overlaps $\triangle ACD$, and the other where it does not overlap .

2. Consider $\triangle ABC$ shown in Figure 15.18 with corresponding interior angles α, β and γ. Prove that $R_{C,2\gamma} \circ R_{B,2\beta} \circ R_{A,2\alpha} = i$. [*Hint: Find three points that are fixed.*]

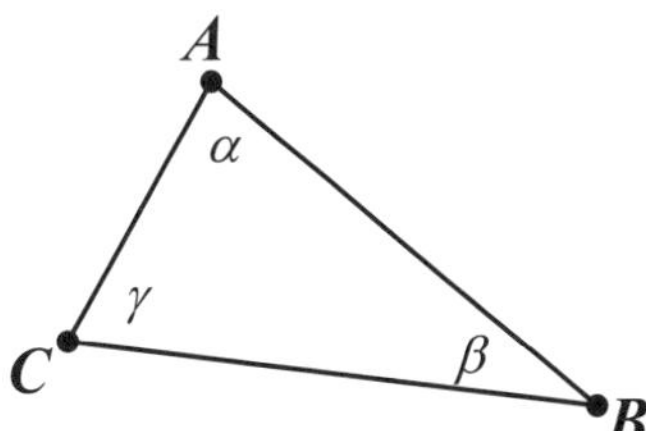

Figure 15.18. Exercise 15.3.2.2

3. A figure in the plane has **rotational symmetry** if there is a point O and an angle α such that $R_{O,\alpha}$ maps the figure to itself.

   (a) A regular polygon has rotational symmetry about its center, O. For each of the following regular n-gons, find all possible angles, α, such that $R_{O,\alpha}$ maps the polygon to itself.

(i) An equilateral triangle   (iii) A regular pentagon

(ii) A square   (iv) A regular hexagon

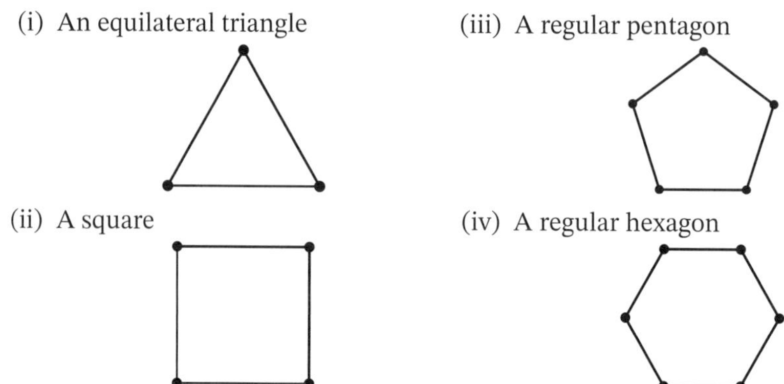

(b) In general, what are the angles of rotation, α, for a regular n-gon with center O such that $R_{O,\alpha}$ maps the polygon to itself?

4. Let A be any point in the plane. Show that the set of rotations about A forms a group.

15.3.3 Reflections as building blocks. We have discussed three isometries of the plane in Euclidean geometry, namely reflections, translations and rotations. In Section 15.2, we noted that the composition of two reflections is not a reflection since every reflection is orientation reversing, but a composition of two reflections is orientation preserving. Translations and rotations also preserve orientation. In this section, we show that a composition of two reflections is either a translation or a rotation. To do this we rely on the fact that when two isometries agree at three noncollinear points, they must agree everywhere.

Theorem 15.24. *Let F_ℓ and F_m be distinct reflections.*

- *If ℓ and m are parallel, then $F_m \circ F_\ell$ is the translation T_{AB}, where ray $\overrightarrow{AB}$ intersects ℓ then m, meeting each perpendicularly, and segment AB has length twice the distance between the lines.*

- *If ℓ and m intersect at a point A, then $F_m \circ F_\ell$ is the rotation $R_{A,2\alpha}$, where α is the counterclockwise angle from ℓ to m at A.*

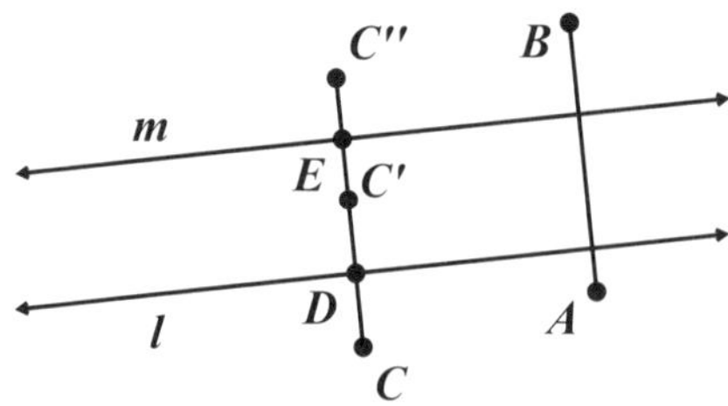

Figure 15.19. Reflecting across two parallel lines

Proof. Let F_ℓ and F_m be distinct reflections.

Case 1. Assume that ℓ and m are parallel. Let $\overrightarrow{AB}$ be a ray that intersects ℓ then m, is perpendicular to both lines, and whose segment AB has length twice the distance between the lines. Choose a point C on the opposite side of ℓ as m, whose distance to ℓ is less than the distance from m to ℓ, as illustrated in Figure 15.19. Let $C' = F_\ell(C)$ and $C'' = F_m(C') = (F_m \circ F_\ell)(C)$. By Proclus' Axiom and Euclid I.29, the line CC' will intersect m at a right angle. Thus, C, C' and C'' are collinear. Let D be the intersection of CC' with ℓ, and E be the intersection of $C'C''$ with m. By definition of reflection, $CD = DC'$ and $C'E = EC''$. Thus, CC'' has twice the length of DE, which is the distance between ℓ and m. Furthermore, ray $\overrightarrow{CC''}$ intersects ℓ then m, meeting both lines perpendicularly. By Euclid I.27, $\overrightarrow{CC''}$ is parallel to $\overrightarrow{AB}$. Since $\overrightarrow{CC''}$ is in the same direction as $\overrightarrow{AB}$ by construction, we have $C'' = T_{AB}(C)$. Clearly, we can find two other points close to C (and on the same side of ℓ as C) that form a triangle. After reproducing the same argument with these two points, we have three noncollinear points on which $F_m \circ F_\ell$ and T_{AB} agree. Therefore, by Corollary 15.10, these isometries must agree everywhere. Thus, $F_m \circ F_\ell$ is a translation.

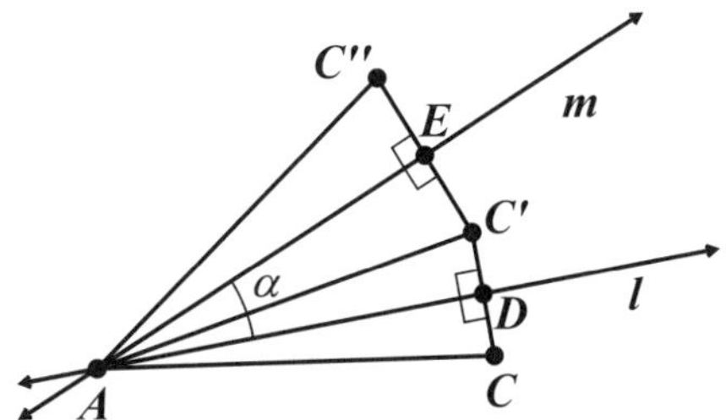

Figure 15.20. Reflecting across two intersecting lines

Case 2. Assume that ℓ and m intersect at a point A. Let α be the counterclockwise angle from ℓ to m at A. Choose a point C outside of α, but close enough to ℓ so that $C' = F_\ell(C)$ lies inside of α, as illustrated in Figure 15.20. Let $C'' = F_m(C') = (F_m \circ F_\ell)(C)$. Let D be the intersection of CC' with ℓ, and E be the intersection of $C'C''$ with m. By definition of reflection, $CD = DC'$ and $C'E = EC''$. We also have that angles $\angle ADC, \angle ADC', \angle AEC'$ and $\angle AEC''$ are all right angles. Thus, by SAS, we have $\triangle ADC \cong \triangle ADC'$ and $\triangle AEC' \cong \triangle AEC''$. Therefore, $AC = AC' = AC''$. Moreover, $\angle CAD = \angle C'AD$ and $\angle C'AE = \angle C''AE$. Thus, $\angle CAC'' = 2\angle DAE = 2\alpha$. By definition, $C'' = R_{A,2\alpha}(C)$. Once again, we can find two other points close to C (and on the same side of ℓ as C) that form a triangle. After reproducing the same argument with these two points, we have three noncollinear points on which $F_m \circ F_\ell$ and $R_{A,2\alpha}$ agree. Therefore, by Corollary 15.10, these isometries must agree everywhere. Thus, $F_m \circ F_\ell$ is a rotation. $\qquad\square$

While Theorem 15.24 starts with two distinct reflections and determines the result of their composition, it also gives us a useful way to decompose either a translation or a rotation into two carefully chosen reflections. The choice of these two reflections is not, however, unique. This is summarized by the following corollary.

Corollary 15.25.

(i) *For any translation T, there exist reflections F_ℓ and F_m such that $T = F_m \circ F_\ell$.*

(ii) *For any rotation R, there exist reflections F_ℓ and F_m such that $R = F_m \circ F_\ell$.*

As a result of this corollary, reflections can be thought of as the building blocks for all rotations and translations. For now, these tools will allow us to prove that the composition of two rotations is either a rotation or a translation.

Exercises 15.3.3

1. Consider square $\square ABCD$ given in Figure 15.21. For each part, use Theorem 15.24 to determine the image of the square under the given composition in order to write the resulting composition as a single isometry. Use the method demonstrated in Example 15.21 as a guide. Be sure to include all intermediate images of the square, as well as the final image, for each composition.

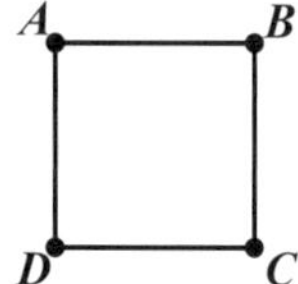

Figure 15.21. Exercise 15.3.3.1

(a) $F_{BC} \circ F_{AD}$

(b) $F_{AD} \circ F_{BC}$

(c) $F_{AC} \circ F_{BD}$

(d) $F_{AC} \circ F_{AD}$

(e) $F_{AD} \circ F_{AC}$

(f) $F_{DC} \circ F_{BC}$

(g) $F_{AD} \circ F_{AB}$

15.3.4 Translations & rotations - Part 2. Equipped with our new tools, we are now ready to prove that the composition of two rotations is either a rotation or a translation, and that the composition of a translation and a rotation, in either order, is a rotation.

Theorem 15.26. *Let $R_{A,\alpha}$ and $R_{B,\beta}$ be two rotations with $0 \le \alpha, \beta < 2\pi$. The composition $R_{B,\beta} \circ R_{A,\alpha}$ is either a rotation or a translation.*

Proof. If $A = B$, then $R_{B,\beta} \circ R_{A,\alpha} = R_{A,\alpha+\beta}$, and we are done.

Assume $A \ne B$. Let $\ell = \overleftrightarrow{AB}$. Construct a line m through A such that the directed angle from m to ℓ at A is $\alpha/2$. Construct a line n through B such that the directed angle from ℓ to n at B is $\beta/2$. By Theorem 15.24, we have $R_{A,\alpha} = F_\ell \circ F_m$ and $R_{B,\beta} = F_n \circ F_\ell$. Thus,

$$R_{B,\beta} \circ R_{A,\alpha} = (F_n \circ F_\ell) \circ (F_\ell \circ F_m) = F_n \circ (F_\ell \circ F_\ell) \circ F_m = F_n \circ i \circ F_m = F_n \circ F_m.$$

Case 1. If $(\alpha + \beta)/2 = \pi$, then by Euclid I.28, lines n and m are parallel, as illustrated in Figure 15.22. Thus, $F_n \circ F_m$ is a translation by Theorem 15.24.

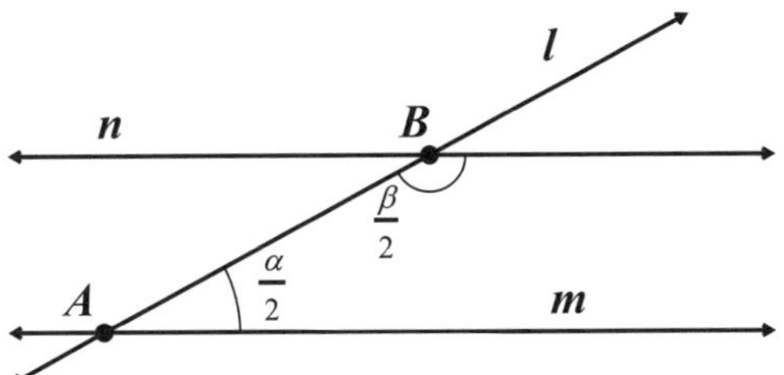

Figure 15.22. $\alpha + \beta = 2\pi$

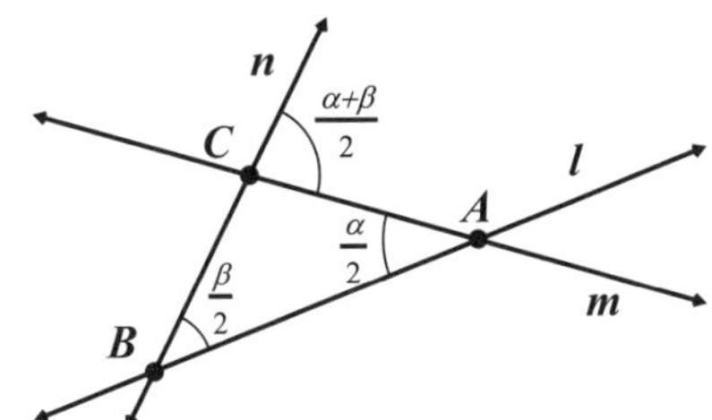

Figure 15.23. $\alpha + \beta < 2\pi$

Case 2. If $(\alpha + \beta)/2 < \pi$, then m and n must intersect at a point C. Moreover, by Euclid I.32, the angle from m to n at C is $(\alpha + \beta)/2$, as illustrated in Figure 15.23. Thus, $F_n \circ F_m = R_{C,\alpha+\beta}$ by Theorem 15.24.

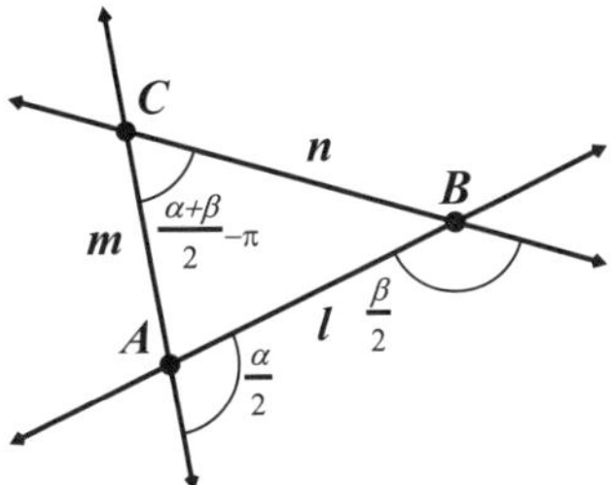

Figure 15.24. $\alpha + \beta > 2\pi$

Case 3. If $(\alpha + \beta)/2 > \pi$, then once again m and n must intersect at a point C, as illustrated in Figure 15.24. We leave it as an exercise to show that the angle from m to n at C is $(\alpha + \beta)/2 - \pi$. Thus, $F_n \circ F_m = R_{C,\alpha+\beta}$ by Theorem 15.24. $\qquad\square$

Theorem 15.27. *Let $R_{A,\alpha}$ be a nonzero rotation and let T be a translation. The compositions $R_{A,\alpha} \circ T$ and $T \circ R_{A,\alpha}$ are both rotations with angle α.*

Proof. Let $R_{A,\alpha}$ be a nonzero rotation and let T be a translation where $T(B) = A$. We will show that $R_{A,\alpha} \circ T$ is a rotation with angle α. Since $B = T^{-1}(A)$, we have $T = T_{BA}$. Let C be the midpoint of BA. Construct two lines, ℓ and m, that are perpendicular to $\overleftrightarrow{AB}$ at C and A, respectively, as illustrated in Figure 15.25. By Theorem 15.24, $T_{BA} = F_m \circ F_\ell$.

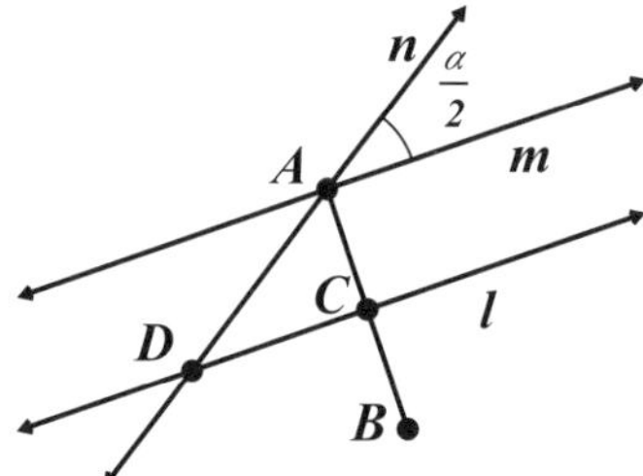

Figure 15.25. $R_{A,\alpha} \circ T$

Construct a line n through A such that the counterclockwise angle from m to n is $\alpha/2$. By Theorem 15.24, $R_{A,\alpha} = F_n \circ F_m$. Hence,

$$R_{A,\alpha} \circ T = (F_n \circ F_m) \circ (F_m \circ F_\ell) = F_n \circ F_\ell.$$

By Proclus' Axiom, n must intersect ℓ at a point D, and by Euclid I.29, we have angle $\angle ADC = \alpha/2$. Thus, by Theorem 15.24, $F_n \circ F_\ell = R_{D,\alpha}$.

We leave it to the reader to show that $T \circ R_{A,\alpha}$ is also a rotation with angle α. □

We summarize the results of this section and Section 15.3.1 with the following table that shows the type of isometry resulting from all possible combinations of the composition of two translation (T) or rotation (R) isometries. In the table, the type of isometry produced by $f \circ g$ is found in the row with f at the left, and the column with g at the top. For example, since $T \circ R$ results in a rotation, there is an R in the last column of the second row. When working in group theory, this type of table is usually called a **multiplication table** since $f \circ g$ is typically abbreviated as fg and referred to as a product.

$\circ$	T	R
T	T	R
R	R	T or R

With this analysis completed, we are now in a position to answer the following question: *What is the smallest group that contains the set of rotations?* We leave it to the reader to explain why the answer is the set containing all rotations and translations. Note that since the set of translations forms a group (Theorem 15.20), we can now see that it also forms what is called a **subgroup** of this larger group containing the rotations and translations.

Exercises 15.3.4

1. In the proof of Theorem 15.26, for Case 3 show that the angle from m to n at C is $(\alpha + \beta)/2 - \pi$ as illustrated in Figure 15.24.

2. Prove that $T \circ R_{A,\alpha}$ is a rotation with angle α.

3. Prove that the smallest group that contains the set of rotations is the set containing all rotations and translations.

4. Consider square $\square ABCD$ given in Figure 15.26. For each part, determine the image of the square under the given composition in order to write the resulting composition as a single isometry. Use the method demonstrated in Example 15.21 as a guide, though for this problem it can be helpful to decompose each rotation or translation into a product of two carefully chosen reflections. For example, $T_{AB} = F_{EH} \circ F_{AD}$, where E and H are the midpoints of AB and CD, respectively. Be sure to include all intermediate images of the square, as well as the final image, for each composition.

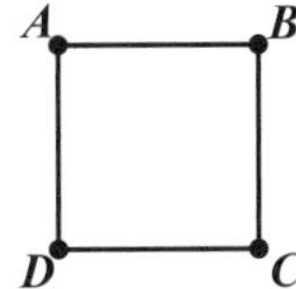

Figure 15.26. Exercise 15.3.4.4

(a) $R_{C,\frac{\pi}{2}} \circ R_{B,\frac{\pi}{2}}$

(b) $R_{A,\frac{\pi}{2}} \circ R_{C,\frac{\pi}{2}}$

(c) $R_{C,\frac{\pi}{2}} \circ R_{A,\frac{\pi}{2}}$

(d) $R_{C,\frac{\pi}{2}} \circ R_{B,\frac{\pi}{2}} \circ R_{A,\frac{\pi}{2}}$

(e) $R_{D,\frac{\pi}{2}} \circ R_{C,\frac{\pi}{2}} \circ R_{B,\frac{\pi}{2}} \circ R_{A,\frac{\pi}{2}}$

(f) $R_{B,\frac{\pi}{2}} \circ T_{AB}$

(g) $T_{AB} \circ R_{B,\frac{\pi}{2}}$

(h) $R_{A,\pi} \circ R_{D,\pi}$

(i) $R_{D,\pi} \circ R_{A,\pi}$

15.3.5 Glide reflections. At the end of the previous section, we determined the smallest group containing the set of rotations, and noted that the set of translations forms its own group. With this group question answered for two of the three types of isometries, the obvious question to ask is: *What is the smallest group that contains the set of reflections?* Since the composition of two reflections is either a translation or a rotation, clearly this group must contain translations and rotations in order to satisfy the group closure property. So, this group contains all three types of isometries, but is that enough? The answer is no. Recall that reflections are orientation reversing while both translations and rotations are orientation preserving. The composition of three reflections is, necessarily, orientation reversing, but the result is not always a reflection. For example, as shown in Figure 15.8 for the proof of Case 4 of Theorem 15.15, $\triangle A'B'C'$ is the image of $\triangle ABC$ under isometry f which is clearly orientation reversing but not a reflection as demonstrated by line ℓ. This means that there is another type of orientation reversing isometry, and it must be a member of the smallest group containing the set of reflections. This fourth isometry is called a *glide reflection*, which is, essentially, a reflection composed with a translation.

Definition 15.28. Given a segment AB, the **glide reflection** along AB is denoted by G_{AB}, and is defined as

$$G_{AB} = T_{AB} \circ F_{AB}.$$

As the composition of two isometries, a glide reflection is an isometry since the set of isometries forms a group. (See Exercise 15.1.2.) Additionally, this composition

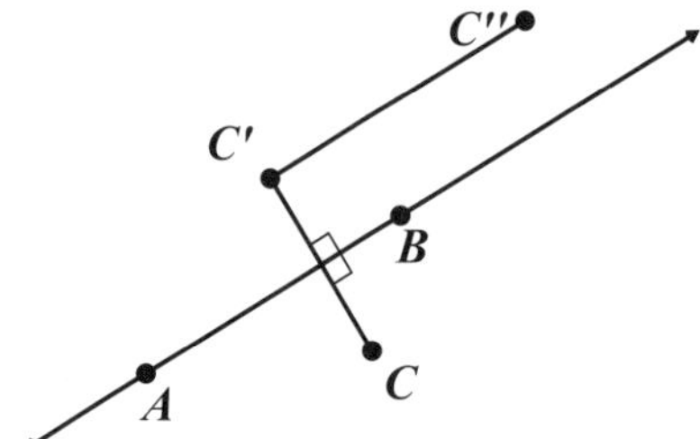

Figure 15.27. $C'' = G_{AB}(C)$

is commutative, and we leave it to the reader to show that $F_{AB} \circ T_{AB} = T_{AB} \circ F_{AB}$. By Corollary 15.25, since every translation is the composition of two reflections, every glide reflection is the composition of three reflections. As we will see, the composition of three reflections can also produce a reflection. Using Theorem 15.24 in combination with the associativity of composition, we can think of the composition of any three reflections as a reflection followed by a translation, or a reflection followed by a rotation. In either case, we will show that the resulting isometry is either a reflection or a glide reflection. Thus, since every isometry of the plane is the composition of three or fewer reflections by Theorem 15.15, glide reflections are the last of the remaining isometries of the Euclidean plane. We start by showing that the composition of a reflection and a translation, in either order, is a reflection or a glide reflection.

Theorem 15.29. *Let T_{AB} be a translation and F_ℓ be a reflection. Then $F_\ell \circ T_{AB}$ and $T_{AB} \circ F_\ell$ are both reflections or both glide reflections.*

Proof. We will prove that $F_\ell \circ T_{AB}$ is either a reflection or a glide reflection and leave the proof that $T_{AB} \circ F_\ell$ is the same type of isometry to the reader.

Case 1. Suppose AB lies on ℓ. Then $F_\ell = F_{AB}$, and combining this with Exercise 15.3.5.1 gives $F_\ell \circ T_{AB} = T_{AB} \circ F_\ell = T_{AB} \circ F_{AB}$, which, by definition, is the glide reflection G_{AB}.

Case 2. Suppose $\overleftrightarrow{AB}$ is parallel to ℓ. Then we can find points C and D on ℓ such that $AB = CD$ and $T_{AB} = T_{CD}$. Combining this with Exercise 15.3.5.1 gives $F_\ell \circ T_{AB} = F_\ell \circ T_{CD} = T_{CD} \circ F_\ell$, which, by definition, is the glide reflection G_{CD}.

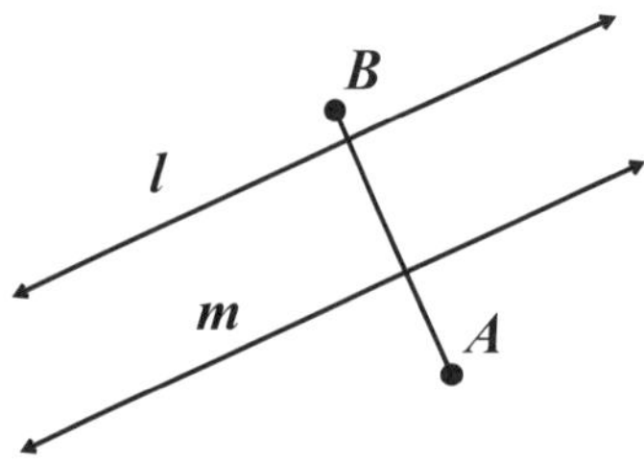

Figure 15.28. $F_\ell \circ T_{AB}$: Case 3

Case 3. Suppose $\overleftrightarrow{AB} \perp \ell$. Construct line m parallel to ℓ whose distance from ℓ is half the length of AB, and on the same side of ℓ as A so that $F_\ell \circ F_m = T_{AB}$. Then, by Theorem 15.24,

$$F_\ell \circ T_{AB} = F_\ell \circ (F_\ell \circ F_m) = F_m,$$

which is a reflection across a line that is parallel to ℓ.

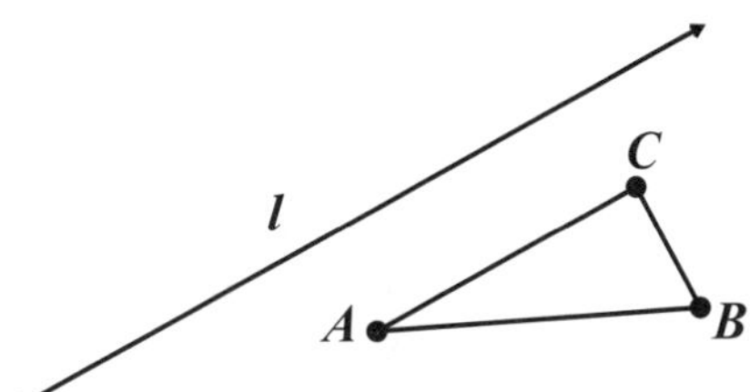

Figure 15.29. $F_\ell \circ T_{AB}$: Case 4

Case 4. Suppose that $\overleftrightarrow{AB}$ is neither parallel nor perpendicular to ℓ. Construct line m parallel to ℓ through A. Construct a line n that is perpendicular to ℓ through B. Let C be the intersection of m and n. Then $\overleftrightarrow{CB} \perp \ell$ and $\overleftrightarrow{AC} \parallel \ell$. By Theorem 15.19, $F_\ell \circ T_{AB} = F_\ell \circ (T_{CB} \circ T_{AC})$. Since $\ell \perp \overleftrightarrow{CB}$, by Case 3, we have $F_\ell \circ T_{CB} = F_k$ where k is parallel to ℓ. Thus,

$$F_\ell \circ T_{AB} = (F_\ell \circ T_{CB}) \circ T_{AC} = F_k \circ T_{AC}$$

where k is parallel to ℓ. By Euclid I.30, $\overleftrightarrow{AC} \parallel k$, and thus, by Case 2, we have $F_k \circ T_{AC} = G_{DE}$, where D and E lie on k, and $T_{AC} = T_{DE}$. $\qquad\square$

Our next theorem shows that the composition of a rotation and a reflection, in either order, is also a reflection or a glide reflection.

Theorem 15.30. *If $R_{A,\alpha}$ is a rotation and F_ℓ a reflection, then $F_\ell \circ R_{A,\alpha}$ and $R_{A,\alpha} \circ F_\ell$ are both reflections or both glide reflections.*

Proof. We will prove that $F_\ell \circ R_{A,\alpha}$ is either a reflection or a glide reflection and leave the proof that $R_{A,\alpha} \circ F_\ell$ is the same type of isometry to the reader.

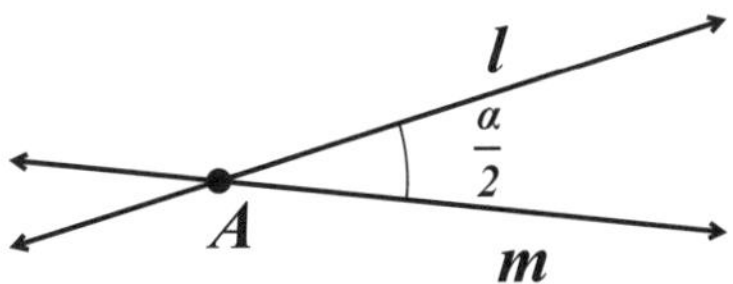

Figure 15.30. $F_\ell \circ R_{A,\alpha}$: Case 1

Case 1. Suppose that the center of rotation, A, lies on the reflection line, ℓ. Construct line m through A such that the angle from m to ℓ is $\alpha/2$, as illustrated in Figure 15.30. By Theorem 15.24, we have $F_\ell \circ F_m = R_{A,\alpha}$. Thus,

$$F_\ell \circ R_{A,\alpha} = F_\ell \circ (F_\ell \circ F_m) = F_m,$$

which is a reflection.

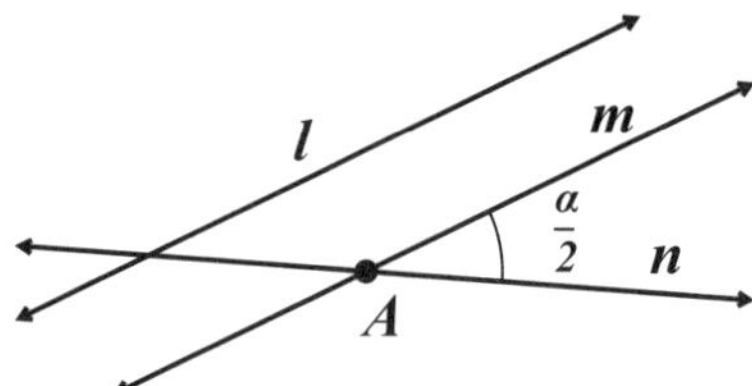

Figure 15.31. $F_\ell \circ R_{A,\alpha}$: Case 2

Case 2. Suppose that A does not lie on ℓ. Construct line m through A parallel to ℓ. Construct line n through A such that the angle from n to m is $\alpha/2$, as illustrated in Figure 15.31. By Theorem 15.24, we have $F_m \circ F_n = R_{A,\alpha}$. Since $\ell \parallel m$, Theorem 15.24 also tells us that $F_\ell \circ F_m$ is a translation. Thus,

$$F_\ell \circ R_{A,\alpha} = F_\ell \circ (F_m \circ F_n) = T \circ F_n.$$

By Theorem 15.29, we know that the composition of a translation and a reflection (in either order) is either a reflection or a glide reflection. In this case, since our translation cannot be perpendicular to n (all three lines would have been parallel), it must be a glide reflection. $\square$

Combining Theorems 15.29 and 15.30, we have the following theorem.

Theorem 15.31. *In Euclidean geometry, the composition of three reflections is either a reflection or a glide reflection.*

We leave it to the reader to explain why the composition of an even number of reflections is a translation or a rotation, and the composition of an odd number of reflections is either a reflection or a glide reflection. Armed with this knowledge, we combine and summarize the results of this and the previous section in a larger multiplication table.

$\circ$	T	R	F	G
T	T	R	F or G	F or G
R	R	T or R	F or G	F or G
F	F or G	F or G	T or R	T or R
G	F or G	F or G	T or R	T or R

Consequently, we can now answer the question posed at the very start of this section: *What is the smallest group that contains the set of reflections?* We leave it to the reader to explain why the answer is the set containing all reflections, translations, rotations and glide reflections.

Since we will be changing our scope of consideration of isometries to the hyperbolic plane after this section, the final question we'd like to answer is whether we have found all of the possible types of isometries of the Euclidean plane. Could there be a distance-preserving motion that we've failed to consider? Fortunately, this is not the case. By Theorem 15.15, every isometry is the composition of three or fewer reflections, and hence, must be one of the four types of isometries that we have studied. Thus, by combining Theorems 15.15, 15.24 and 15.31, we have the following theorem.

Theorem 15.32. *Every isometry in the Euclidean plane is either a reflection, translation, rotation or glide reflection.*

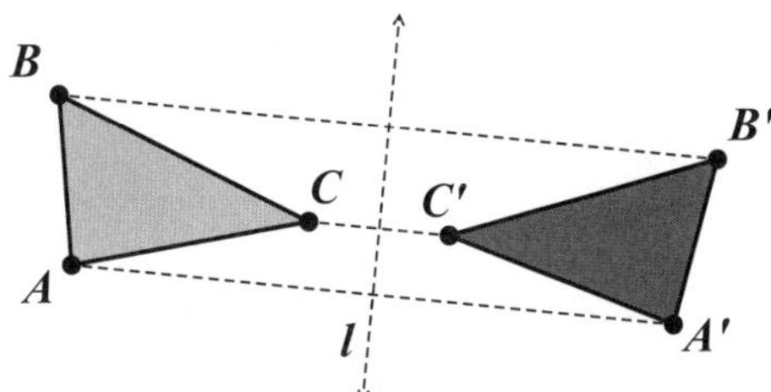

Figure 15.32. Identifying a reflection

We have done it! We have classified all of the isometries of the Euclidean plane.

As a final note, suppose we reverse the process, that is, suppose we are given a triangle $\triangle ABC$ and its image $\triangle A'B'C'$ under some isometry. How can we identify the isometry? Here, the proof of Theorem 15.15 gives us some insight. For example, the isometry is a reflection if the perpendicular bisector of either AA', BB' or CC', as shown in Figure 15.32, is the line of reflection, ℓ.

As demonstrated in Figure 15.33, the isometry is a translation if it is given by $T_{AA'} = T_{BB'} = T_{CC'}$. If, on the other hand, the isometry is a rotation, then the center of rotation is the intersection point, D, of the perpendicular bisectors of AA', BB' and CC' and the angle of rotation is given by $\angle ADA' = \angle BDB' = \angle CDC'$, as shown in Figure 15.34. We leave it to the reader to explain how to identify a glide reflection.

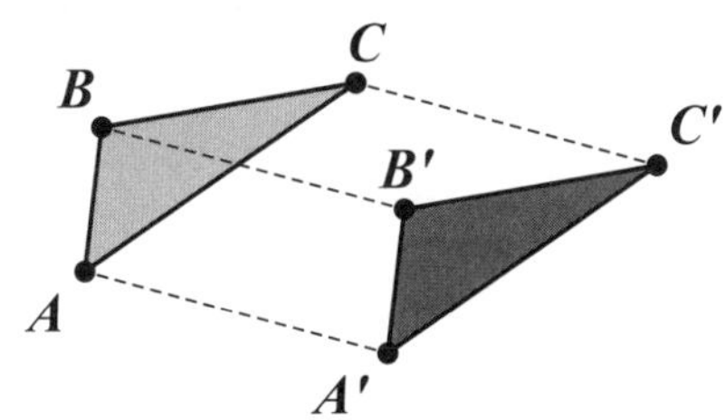

Figure 15.33. Identifying a translation

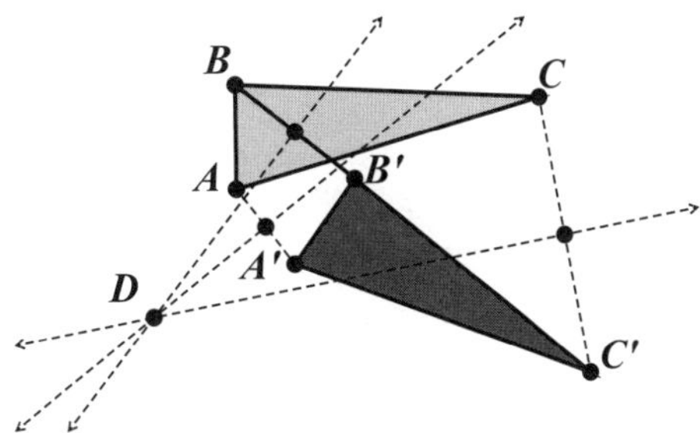

Figure 15.34. Identifying a rotation

We leave the isometries of the Euclidean plane to venture into the isometries of Hyperbolic geometry. In order to determine what a reflection looks like in the Poincaré Half-plane model, we must first discuss another type of transformation of the Euclidean plane, an *inversion*.

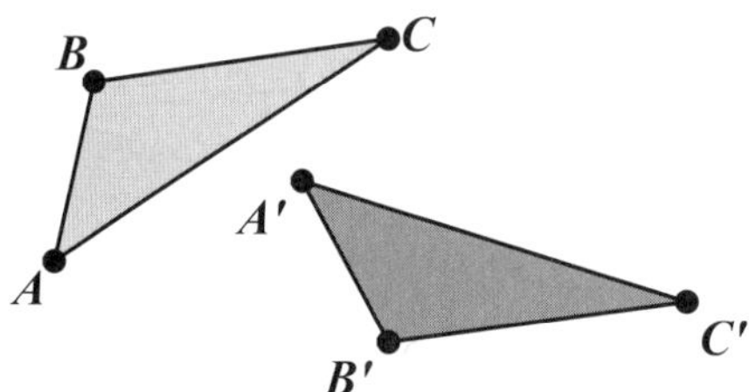

Figure 15.35. Identifying a glide reflection

Exercises 15.3.5

1. Prove that $F_{AB} \circ T_{AB} = T_{AB} \circ F_{AB}$.

2. Finish the proof of Theorem 15.29 by showing that $T_{AB} \circ F_\ell$ is the same type of isometry as $F_\ell \circ T_{AB}$ (either a reflection or a glide reflection) in each of the four cases.

3. Finish the proof of Theorem 15.30 by showing that $R_{A,\alpha} \circ F_\ell$ is the same type of isometry as $F_\ell \circ R_{A,\alpha}$ (a reflection or a glide reflection) in each of the two cases.

4. Explain why the composition of an even number of reflections is a translation or a rotation, and the composition of an odd number of reflections is either a reflection or a glide reflection.

5. In the multiplication table given on page 396, explain why $G \circ T$ and $G \circ R$ each produce either F or G.

6. In the multiplication table given on page 396, explain why $T \circ G$ and $R \circ G$ each produce either F or G.

7. In the multiplication table given on page 396, explain why $G \circ F$, $F \circ G$ and $G \circ G$ each produce either T or R.

8. Explain why the smallest group that contains the set of reflections is the set containing all reflections, translations, rotations and glide reflections.

9. Explain how to identify a glide reflection given a triangle $\triangle ABC$ and its image $\triangle A'B'C'$. [*Hint: Figure 15.35 may be helpful.*]

10. Consider square $\square ABCD$ given in Figure 15.36. For each part, determine the image of the square under the given composition in order to write the resulting composition as a single isometry. Use the method demonstrated in Example 15.21 as a guide, though for this problem it may be helpful to decompose the components of each composition into products of carefully chosen reflections or translations in order to identify the resulting isometry. For example, $T_{AB} = F_{EH} \circ F_{AD}$, where E and H are the midpoints of AB and CD, respectively. Be sure to include all intermediate images of the square, as well as the final image, for each composition.

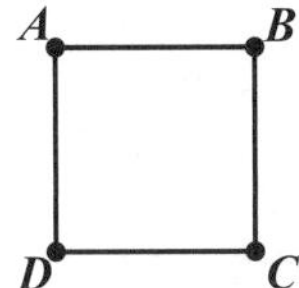

Figure 15.36. Exercise 15.3.5.10

(a) $T_{AB} \circ F_{AD}$

(b) $F_{AD} \circ T_{AB}$

(c) $F_{AD} \circ T_{AC}$

(d) $T_{AC} \circ F_{AD}$

(e) $R_{A,\frac{\pi}{2}} \circ F_{AB}$

(f) $F_{BC} \circ R_{A,\frac{\pi}{2}}$

(g) $G_{BC} \circ G_{AB}$

(h) $F_{AB} \circ F_{AD} \circ F_{BC}$

(i) $F_{AD} \circ F_{AB} \circ F_{BC}$

15.4 Inversions in the Euclidean plane

The isometries of the Poincaré Half-plane model come in a few familiar flavors, but before we can understand one of these isometries, we must introduce a new motion in the Euclidean plane called an *inversion*. While this important type of transformation is not an isometry of the Euclidean plane, it is precisely the mapping needed to comprehend hyperbolic reflections in the Half-plane model.

Definition 15.33. Given a point A and a radius r, the **inversion** in the circle with center A and radius r, is denoted $I_{A,r}$, and if C is any point in the plane other than A, then

$$I_{A,r}(C) = C',$$

where C' lies on the ray $\overrightarrow{AC}$ and satisfies the equation

$$AC \cdot AC' = r^2,$$

as illustrated in Figure 15.37. The point A is the **center of inversion**, r is the **radius of inversion**, and the circle with center A and radius r is the **circle of inversion**.

Technically, $I_{A,r}(A)$ is undefined, making this a mapping of a **punctured** Euclidean plane, that is, the plane without the point A. If $I_{A,r}(C) = C'$, then since $AC \cdot AC' = r^2$, we must have $I_{A,r}(C') = C$ by definition. Hence, an inversion is an involution, that is, it is its own inverse. We note that if C lies on the circle of inversion, then $I_{A,r}(C) = C$ since $AC = AC' = r$. Thus, an inversion fixes its circle of inversion. If C lies inside the circle of inversion, then $AC < r$ and the constraint $AC \cdot AC' = r^2$ guarantees that $AC' > r$. Thus, the image of a point inside the circle lies outside of the circle. Similarly, the image of a point outside the circle must lie inside the circle, and A cannot be the image of any point.

Given a point, C, that doesn't lie on the circle of inversion, using Euclidean geometry, how can we find its image under inversion? We consider two cases: the first where C lies inside the circle of inversion, and the second where it lies outside.

Theorem 15.34. *Given a center A, a radius r, and a point C, where C neither lies on the circle of inversion nor is A, to find $C' = I_{A,r}(C)$.*

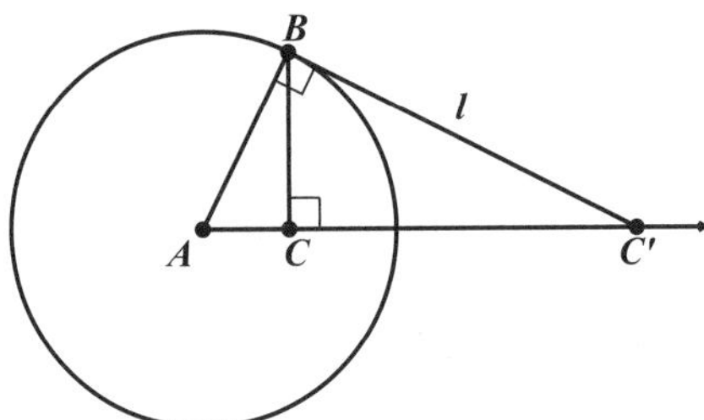

Figure 15.37. Finding C' when C lies inside the circle of inversion

Proof.

Case 1. Suppose C lies inside the circle centered at A. Join AC and extend it to form ray $\overrightarrow{AC}$. Using Euclid I.11, construct the perpendicular to AC at C. Let B be its intersection with the circle. Join AB. Using Euclid I.11, construct the perpendicular ℓ to AB at B. By Euclid's fifth postulate, ℓ and $\overrightarrow{AC}$ will intersect on the opposite side of $\overleftrightarrow{BC}$ as A. Let C' be this intersection, as illustrated in Figure 15.37. We leave it to the reader to prove that $AC \cdot AC' = AB^2$.

Case 2. Suppose C lies outside the circle centered at A. Join AC. Using Euclid III.17, construct a tangent line from C to the circle of inversion. Let B be the point of tangency. Using Euclid I.12, construct a perpendicular from B to AC, labeling the point of intersection C'. We leave it to the reader to illustrate this case and once again prove that $AC \cdot AC' = AB^2$. □

Recall that isometries map lines to lines and circles to circles. This is not quite the case for inversions. Inversions are much more playful with our main characters. Sometimes they map lines to lines and circles to circles, but they can also map lines to circles and circles to lines. To prove this, we need the following lemma.

Lemma 15.35. *Suppose $I_{A,r}$ is an inversion, and A, B and C are distinct, noncollinear points. If $B' = I_{A,r}(B)$ and $C' = I_{A,r}(C)$, then $\triangle ABC \sim \triangle AC'B'$.*

Proof. Consider triangles $\triangle ABC$ and $\triangle AC'B'$. By definition, these triangles share $\angle A$. Moreover, since $BA \cdot B'A = CA \cdot C'A = r^2$, we have

$$\frac{B'A}{C'A} = \frac{CA}{BA}.$$

By Euclid VI.6, we have $\triangle ABC \sim \triangle AC'B'$. □

Notice that the proof of Lemma 15.35 does not rely upon any particular configuration of the points. We leave it as an exercise for the reader to illustrate the different possible cases. We are now ready to show that inversions map lines and circles to lines and circles. As per the comment that the center of an inversion cannot be the image of any point, when we write *onto* in the following theorem, we refer to the punctured plane that does not include the point A.

Theorem 15.36. *Given center A and radius r, the inversion $I_{A,r}$ maps*

- *a line that passes through A onto itself,*
- *a line that does not pass through A onto a circle that passes through A,*
- *a circle that passes through A onto a line that does not pass through A, and*
- *a circle that does not pass through A onto another circle that also does not pass through A.*

Proof.

Case 1. Assume that ℓ is a line that passes through A, and let C be a point on ℓ other than A. By definition, C' lies on the ray $\overrightarrow{AC}$, and hence the inversion $I_{A,r}$ maps ℓ onto itself.

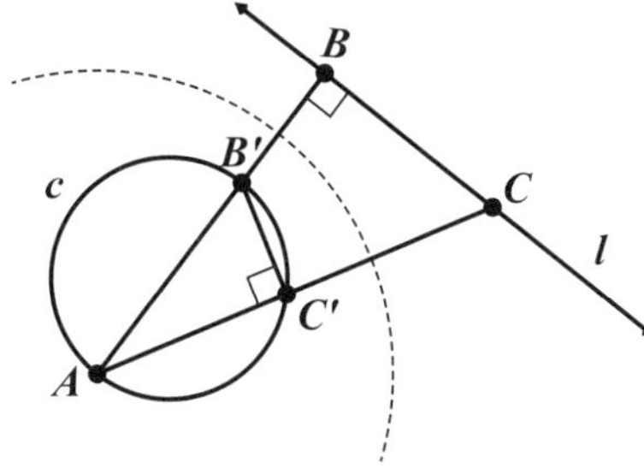

Figure 15.38. Case 2: *B* lies outside the circle of inversion

Case 2. Assume that ℓ is a line that does not pass through A. Using Euclid I.12, construct a perpendicular from A to ℓ, and let B be the point of intersection. Let C be a point on ℓ other than B. Let B' and C' be the images of B and C, respectively, under the inversion $I_{A,r}$. By Lemma 15.35, we have $\triangle ABC \sim \triangle AC'B'$, and therefore, $\angle AC'B'$ is a right angle. By the converse of Thales' theorem (see Exercise 11.2.4), the circumcenter of triangle $\triangle AC'B'$ is the midpoint of AB'. Hence, C' lies on the circle with diameter AB'. Since this holds for an arbitrary point C on ℓ, the image of ℓ is the circle c with diameter AB' (minus the point A). Notice that this proof does not rely upon any particular configuration of the points. With the circle of inversion represented by the dashed arc, Figure 15.38 illustrates the case where B lies outside the circle of inversion, while Figure 15.39 illustrates the case where B lies inside this circle, but C lies outside.

Case 3. Assume that c is a circle that passes through A. The proof of this case relies on revisiting Case 2 from a different perspective. Let AB' be a diameter of this circle. Let B be the image of B' under the inversion $I_{A,r}$. Using Euclid I.11, construct the perpendicular to AB through B and call it ℓ. (Notice that we we can still refer to Figures 15.38 and 15.39.) By Case 2, the image of ℓ under the inversion $I_{A,r}$ is c (minus the point A). Since inversions are involutions, this implies that the image of circle c (minus the point A) under $I_{A,r}$ must be ℓ.

Case 4. Assume that c is a circle that does not pass through A. If c has center A and radius s, then the image of c under the inversion $I_{A,r}$ will be the circle with center A

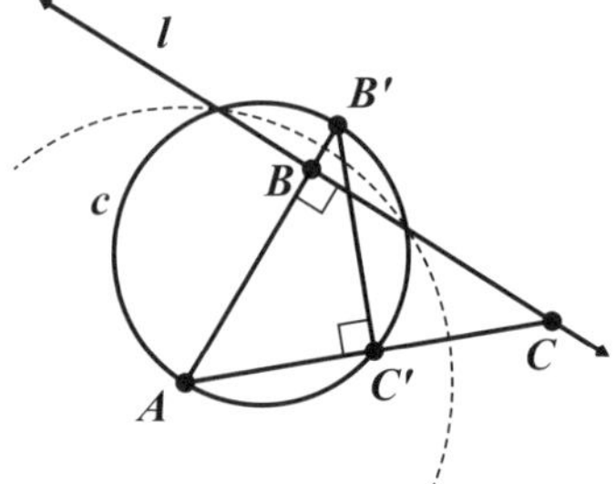

Figure 15.39. Case 2: B lies inside the circle of inversion

and radius r^2/s. (Why? If P is a point on circle c, then by definition, P' lies on ray $\overrightarrow{AP}$ and $AP \cdot AP' = r^2$.)

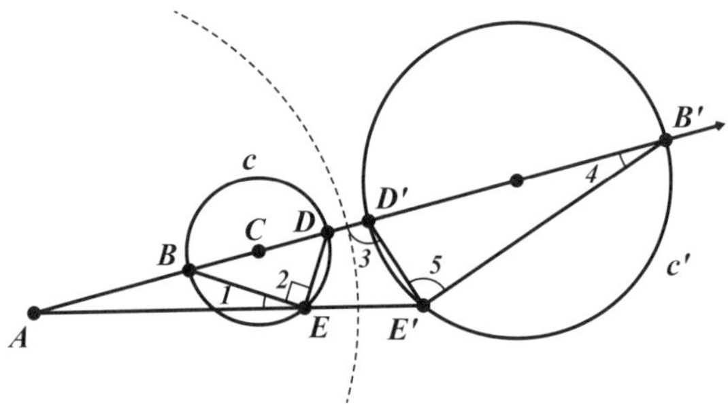

Figure 15.40. Case 4: center of inversion, A, does not lie on BD

Suppose that c has center C, where $C \neq A$. Join AC, and let BD be the diameter of c that lies on $\overleftrightarrow{AC}$. There are several subcases to consider. We start with the case where c lies inside the circle of inversion, but A does not lie on BD. WLOG, we assume that B lies between A and C. Let $B' = I_{A,r}(B)$ and $D' = I_{A,r}(D)$. Pick an arbitrary point E on circle c, that is not on the diameter BD, and let $E' = I_{A,r}(E)$, as illustrated by Figure 15.40. By Thale's theorem, angle $\angle 2$ is a right angle. By Lemma 15.35, we have that $\triangle ABE \sim \triangle AE'B'$ and $\triangle ADE \sim \triangle AE'D'$. Therefore, $\angle 1 = \angle 4$ and $\angle 1 + \angle 2 = \angle 3$. By Euclid I.32, we have that $\angle 3 = \angle 4 + \angle 5$. Substitution gives $\angle 1 + \angle 2 = \angle 4 + \angle 5 = \angle 1 + \angle 5$. Subtracting equals from equals, we have $\angle 5 = \angle 2$, which is a right angle. Thus, by the converse of Thales' theorem, E' lies on the circle with diameter $B'D'$. Since this holds for an arbitrary E on c, the image of c is the circle c' with diameter $B'D'$.

The case where c lies outside the circle of inversion, but does not contain it, is very similar to this one. We leave it to the reader to prove the remaining cases. In Figure 15.41, we illustrate the case where where c lies inside the circle of inversion, but A lies on BD. The case where the circle of inversion lies entirely inside c is very similar. For the final case, consider what happens when c and the circle of inversion intersect, as illustrated in Figure 15.45. $\square$

Based on the proof of Theorem 15.36, we have the following corollary.

Corollary 15.37. *Let $I_{A,r}$ be an inversion, ℓ be a line, and c be a circle with center C.*

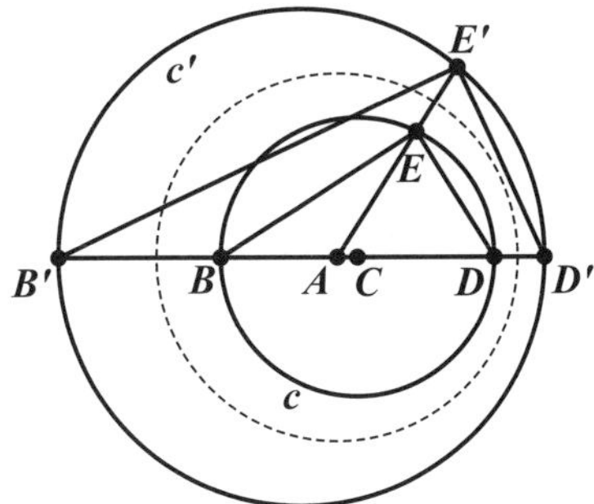

Figure 15.41. Case 4: center of inversion, A, lies on BD

- *If the image of c under inversion $I_{A,r}$ is the circle c' with center D, then A, C and D are collinear.*
- *If the image of c under inversion $I_{A,r}$ is the line ℓ, then $\overleftrightarrow{AC} \perp \ell$.*
- *If the image of ℓ under inversion $I_{A,r}$ is the circle c with center C, then C lies on the perpendicular line from A to ℓ.*

While it is clear that inversions do not preserve distances, they are conformal mappings, that is, they preserve angles. To understand what this means, it will help to recall how we define nonlinear angles. When two circles intersect at a point, an **angle**

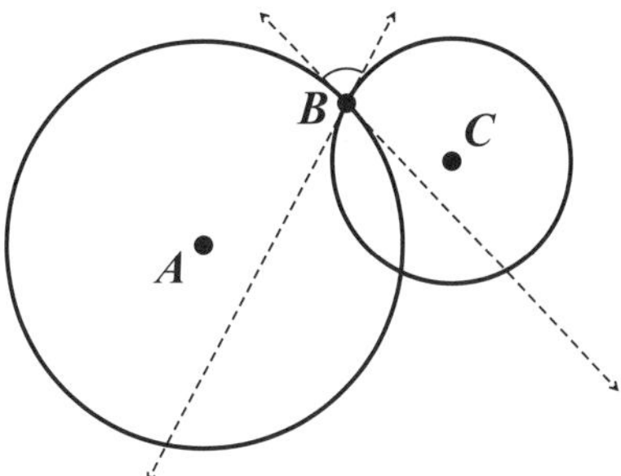

Figure 15.42. An angle of intersection between two circles at B

of intersection of the circles is taken from the angles formed by their tangent lines at that point, as illustrated in Figure 15.42. Similarly, when a line intersects a circle at a point, an angle of intersection between the line and the circle is taken as one of the supplementary angles formed by the line and the tangent line to the circle at the point of intersection, as illustrated in Figure 15.43.

We state the following theorem without proof. (For a proof, see [**80**] or [**110**].)

Theorem 15.38. *Inversions preserve angles.*

Before heading to the Half-plane model for Hyperbolic geometry, we need one more inversion theorem. Recall that two intersecting circles are *orthogonal* when their radii at a point of intersection are perpendicular. Using Euclid III.16, this implies that the radius of one circle is tangent to the other circle, and thus, the angle of intersection is a right angle.

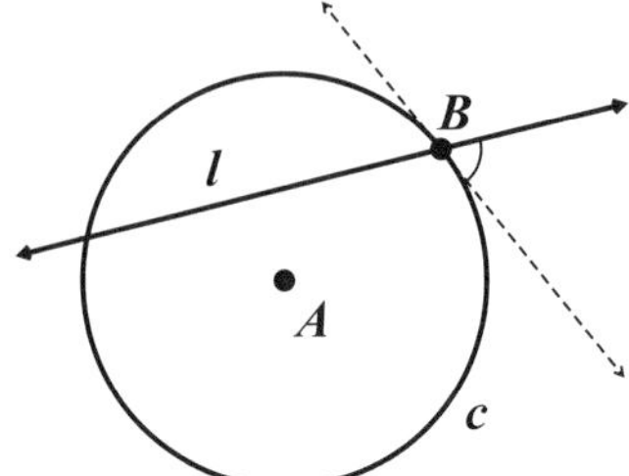

Figure 15.43. An angle of intersection between line ℓ and circle c at point B

Theorem 15.39. *Let $I_{A,r}$ be an inversion, and c be a circle that intersects the circle of inversion.*

- *If c and the circle of inversion are orthogonal, then $I_{A,r}$ maps c onto itself.*
- *If c contains both a point D and its image $I_{A,r}(D) = E$, then c and the circle of inversion are orthogonal.*

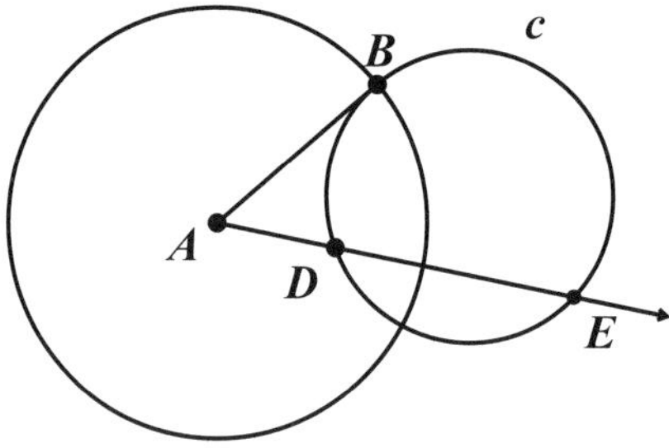

Figure 15.44. Circle c is orthogonal to the circle of inversion

Proof. Suppose that c and the circle of inversion intersect at a point B.

Case 1. Assume that circle c and the circle of inversion are orthogonal. By Euclid III.16, AB is tangent to circle c. Let D be a point on c that does not also lie on the circle of inversion. The ray $\overrightarrow{AD}$ will intersect c at a second point E, as illustrated by Figure 15.44. By Euclid III.36 applied to circle c, we have $AB^2 = AD \cdot AE$. Thus, by definition, $E = I_{A,r}(D)$. Since this holds for an arbitrary point D of circle c, we have shown that $I_{A,r}$ maps c onto itself.

Case 2. Assume that c contains a point D and its image under inversion, E. By definition of inversion, E lies on $\overrightarrow{AD}$ and $AB^2 = AD \cdot AE$. Hence, by Euclid III.37 applied to circle c, AB is tangent to c. Therefore, by Euclid III.18, the circle of inversion and c are orthogonal. $\square$

With this last result, we are now ready to explore the isometries of the hyperbolic plane.

Exercises 15.4

1. Finish the proof of Case 1 of Theorem 15.34.

2. Illustrate and finish the proof of Case 2 of Theorem 15.34.

3. Consider a center A, a radius r, and a point C, where C lies inside the circle of inversion but is not A. Let B be the intersection of the ray $\overrightarrow{AC}$ with the circle of inversion. Prove that $CB \neq C'B'$, and hence the inversion $I_{A,r}$ is not an isometry.

4. Give a diagram to illustrate each of the following cases of Lemma 15.35.

   (a) Assume B and C both lie inside the circle of inversion.
   (b) Assume B and C both lie outside the circle of inversion.
   (c) WLOG, assume B lies inside the circle of inversion and C is outside.

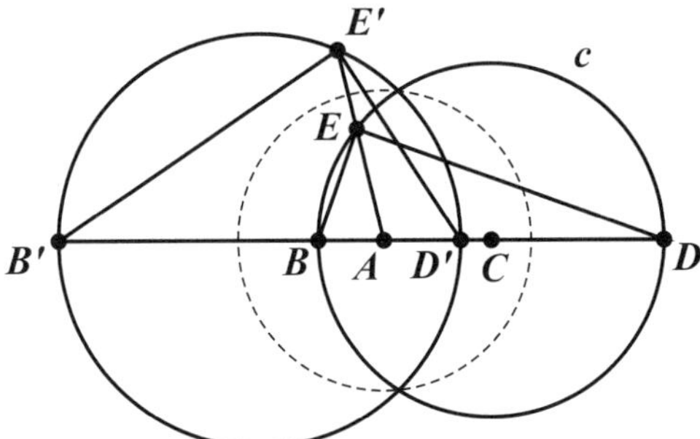

Figure 15.45. Exercise 15.4.5: Case 4 of Theorem 15.36: circle of inversion intersects c

5. Prove the following subcases of Case 4 in Theorem 15.36.

   (a) The case where c lies inside the circle of inversion, but A lies on BD, as illustrated in Figure 15.41.
   (b) The case where the c and the circle of inversion intersect, as illustrated in Figure 15.45.

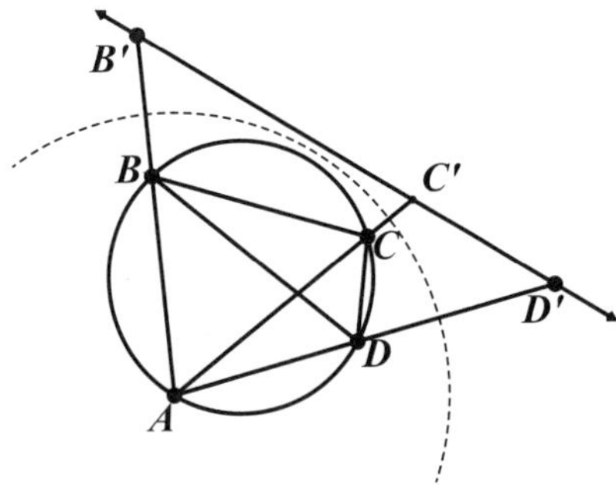

Figure 15.46. Exercise 15.4.6: A proof of Ptolemy's Theorem using inversion

6. In Chapter 10, Ptolemy's Theorem states that for any given cyclic quadrilateral $ABCD$, we have

$$AC \cdot BD = AD \cdot BC + AB \cdot CD.$$

In this exercise, we give an alternative proof of this theorem using inversion.

Proof. Let $ABCD$ be a cyclic quadrilateral in circle c. Consider the inversion $I_{A,r}$ where r is greater than the diameter of c. Let B', C' and D' be the images under the inversion $I_{A,r}$ of B, C and D, respectively.

(a) Using similar triangles, show that

$$B'C' = \frac{AB' \cdot BC}{AC}.$$

(b) Using the definition of inversion, explain why

$$B'C' = \frac{r^2 \cdot BC}{AB \cdot AC}.$$

(c) Using a similar argument, show that

$$B'D' = \frac{r^2 \cdot BD}{AB \cdot AD} \quad \text{and} \quad C'D' = \frac{r^2 \cdot CD}{AC \cdot AD}.$$

(d) Explain why $B'D' = B'C' + C'D'$.

(e) Substitute and simplify to show that this implies

$$AC \cdot BD = AD \cdot BC + AB \cdot CD. \qquad \square$$

15.5 Isometries of the hyperbolic plane

Many of the overall concepts and results of Section 15.3 still apply in the hyperbolic plane. For starters, we defined reflections using only Neutral geometry. To distinguish **hyperbolic reflections** from their Euclidean counterparts, we use the notation HF_ℓ, rather than F_ℓ. Once again, if C is any point in the plane that does not lie on ℓ, then $HF_\ell(C) = C'$, where C' is the unique point such that ℓ is the perpendicular bisector of CC'. Happily, we can also define rotations exactly as we did in Section 15.3.2. That is, given a point A and an angle α, the **hyperbolic rotation** about A by α is denoted $HR_{A,\alpha}$, and if C is any point in the plane other than A, then

$$HR_{A,\alpha}(C) = C'$$

where C' is the unique point such that $AC = AC'$ and $\angle C'AC = \alpha$. Moreover, if ℓ and m are two distinct lines that intersect at a point A, then once again $HF_m \circ HF_\ell$ is the rotation $HR_{A,2\alpha}$, where α is the counterclockwise angle from ℓ to m at A. (This is because the proof of Case 2 of Theorem 15.24 holds in Neutral geometry.) As before, we can represent the identity transformation as a rotation about any point A, by an angle of zero, that is, $HR_{A,0}$.

Translations, on the other hand, are more difficult to extend. The definition given in Section 15.3.1 is no longer well-defined; given points A and B and a point C not on the line $\overleftrightarrow{AB}$, there are infinitely many lines through C that are parallel to $\overleftrightarrow{AB}$. Thus, ray $\overrightarrow{CC'}$ is no longer unique. Additionally, while parallelograms are abundant, they no longer have the property that opposite sides are congruent. Looking to the Euclidean case for inspiration, we note that the composition of two reflections is a translation when the lines of reflection are parallel. Thus, it is tempting to simply define a translation as the composition of two reflections HF_ℓ and HF_m, where ℓ and m are parallel lines. Unfortunately, this definition is a bit too broad. In particular, we would have to distinguish between the case when our lines of reflection are ultraparallel versus when they are sensed parallel. Instead, we avoid any notion of parallel by taking our cue from the proof of Theorem 15.24. Here, given AB with midpoint C, we have $T_{AB} = F_m \circ F_\ell$, where ℓ and m are perpendicular to AB, and pass through A and C, respectively. We take this as our definition of a *hyperbolic translation*.

Definition 15.40. Consider two distinct points A and B. Let C be the midpoint of AB. The **hyperbolic translation** that maps A to B is denoted by HT_{AB} and is defined as

$$HT_{AB} = HF_m \circ HF_\ell,$$

where ℓ and m are perpendicular to AB, and pass through A and C, respectively. When $A = B$, the trivial hyperbolic translation HT_{AA} maps every point to itself.

Since HT is the composition of two isometries, a hyperbolic translation is clearly an isometry. Note that lines ℓ and m in our definition must be ultraparallel by Theorem 13.7 since they share a common perpendicular. Furthermore, by Theorem 13.34, we know that ultraparallel lines have a unique common perpendicular. Thus we have the following lemma.

Lemma 15.41. *If ℓ and m are ultraparallel lines, then $HF_m \circ HF_\ell$ is a hyperbolic translation.*

Proof. Assume ℓ and m are ultraparallel lines. By Theorem 13.34, ℓ and m have a unique common perpendicular $\overleftrightarrow{EF}$, where E lies on ℓ and F lies on m, as illustrated in Figure 15.47. Let G lie on $\overleftrightarrow{EF}$ such that F is the midpoint of EG. Then by definition, we have $HF_m \circ HF_\ell = HT_{EG}$. $\qquad\qquad\square$

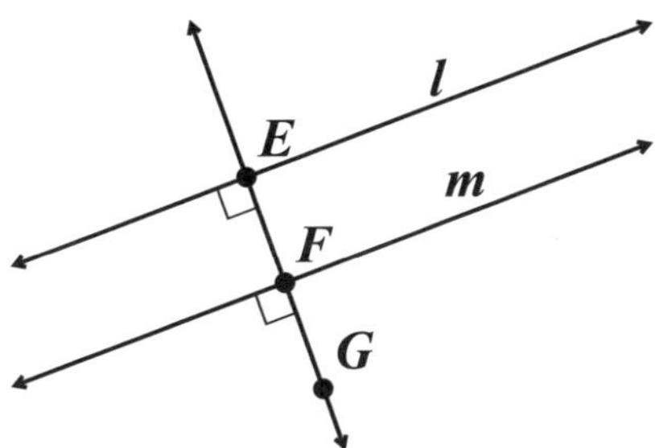

Figure 15.47. $HF_m \circ HF_\ell$ is a hyperbolic translation

Thus, a hyperbolic isometry is a nontrivial hyperbolic translation if and only if it is the composition of reflection across two ultraparallel lines. We have the following definition in the case where the lines of reflection are, instead, sensed parallel.

Definition 15.42. If ℓ and m are sensed parallel, then we say that $HF_m \circ HF_\ell$ is a **parallel displacement** or **horolation**.

We summarize our previous discussion with the following theorem.

Theorem 15.43. *A hyperbolic isometry is either a hyperbolic rotation, a hyperbolic translation, or a parallel displacement if and only if it can be written as the composition of two hyperbolic reflections.*

This leaves us with one final isometry to consider. Recall, that in Section 15.3.5, given AB, the *glide reflection* along AB is denoted by G_{AB}, and is defined as

$$G_{AB} = T_{AB} \circ F_{AB}.$$

Using our updated definition of a hyperbolic translation, we will keep this as our definition of a **hyperbolic glide reflection**, that is

$$HG_{AB} = HT_{AB} \circ HF_{AB}.$$

As we showed in the Euclidean case, once again, we would need to prove that this definition is broad enough to encompass the remaining isometries of the hyperbolic plane. Fortunately, the analogous result to Theorem 15.31 still holds. We state the following theorem without proof. (For a proof, see [**80**].)

Theorem 15.44. *In Hyperbolic geometry, the composition of three hyperbolic reflections is either a hyperbolic reflection or a hyperbolic glide reflection.*

Thus, we have the following classification of Hyperbolic isometries.

Theorem 15.45. *Every isometry of the hyperbolic plane is either a hyperbolic reflection, a hyperbolic rotation, a hyperbolic translation, a parallel displacement, or a hyperbolic glide reflection.*

Given our familiarity with Euclidean geometry in the plane, we have no difficulty envisioning the motion of the four types of isometries in Section 15.3. With some practice, we can visualize the image of any point under under a given reflection, translation, rotation or glide reflection. Visualizing the isometries in Hyperbolic geometry, on the other hand, is much more difficult as the different models of the hyperbolic plane are more foreign. What do reflections, for example, look like in the Poincaré Half-plane model? We answer this question in the following section.

15.5.1 Reflections in the Poincaré Half-Plane model.

As introduced in Chapter 12, the Poincaré Half-plane model is a representation of the hyperbolic plane consisting of the Euclidean upper half-plane with the x-axis boundary excluded. A *point* is any point of the Euclidean upper half-plane, and *lines* in the model are either vertical rays starting at the x-axis, or semicircles centered on the x-axis. Our goal in this section is to determine what a hyperbolic reflection looks like in this model. We will see that there are two types of hyperbolic reflections, depending on whether the hyperbolic line is represented by a vertical ray or a semicircle. We first consider the case where the hyperbolic line is represented by a vertical ray.

Theorem 15.46. *If the hyperbolic line, ℓ, is represented by a vertical ray, then the hyperbolic reflection across ℓ is the same as the Euclidean reflection across ℓ, namely $HF_\ell = F_\ell$.*

Proof. Given hyperbolic line ℓ with equation $x = c$, consider a point $A = (x_1, x_2)$, not on ℓ. Let $A' = F_\ell(A)$ be the image of A under the Euclidean reflection across ℓ. Then $A' = (x_2, y_1)$, where $x_2 = 2c - x_1$. The hyperbolic line, m, joining A and A' is the semicircle centered at $C = (c, 0)$ with radius $r = \sqrt{(x_1 - c)^2 + y_1^2}$. The hyperbolic lines ℓ and m are perpendicular and intersect at the point $B = (c, r)$. We leave it as an exercise for the reader to show that $d_H(A, B) = d_H(A', B)$, and thus, ℓ is the perpendicular bisector for segment AA'. Therefore, $A' = HF_\ell(A)$, that is, A' is the image of A under the hyperbolic reflection across ℓ. $\qquad\square$

The other possibility is that ℓ is represented by a semicircle of radius k and center C on the x-axis. In this case, the hyperbolic reflection across ℓ is the Euclidean inversion $I_{C,k}$. Before showing this, we prove the more general result that the Euclidean inversion $I_{C,k}$, where $C = (c, 0)$, preserves the arc length of a path γ in the Half-plane model. Thus, Euclidean inversions are hyperbolic isometries in Poincaré's model.

Theorem 15.47. *The Euclidean inversion $I_{C,k}$, where $C = (c, 0)$, preserves the arc length of a path γ in the Poincaré Half-plane model.*

Proof. Any path γ can be parameterized as $x(t) = c + r(t)\cos t$ and $y(t) = r(t)\sin t$ for some $r(t)$ where $\alpha \leq t \leq \beta$. [Note that these are merely modified polar coordinates centered at $C = (c, 0)$ rather than $(0, 0)$. Additionally, the parameterization process may require decomposing the path into pieces to ensure that $r(t)$ is a function. In other words, the parameterized path $r(t)$ must have a single value for any given t, that is, rays originating from C may only meet $r(t)$ once. If path γ fails this condition, then we can split the path into as many pieces, $\gamma_i = r_i(t)$, as necessary so that each piece of the parameterization, $r_i(t)$, is a function.] From Section 12.3, we have

$$\left(\frac{dx}{dt}\right)^2 + \left(\frac{dy}{dt}\right)^2 = (r'(t))^2 + (r(t))^2.$$

Thus,

$$s_H(\gamma) = \int_\alpha^\beta \frac{1}{y}\sqrt{\left(\frac{dx}{dt}\right)^2 + \left(\frac{dy}{dt}\right)^2}\, dt$$

$$= \int_\alpha^\beta \frac{1}{r(t)\sin t}\sqrt{(r'(t))^2 + (r(t))^2}\, dt.$$

We note that if $\gamma' = I_{C,k}(\gamma)$, then γ' has parameterization: $x(t) = c + \dfrac{k^2}{r(t)}\cos t$ and $y(t) = \dfrac{k^2}{r(t)}\sin t$ for $\alpha \leq t \leq \beta$. (Why?) We leave it to the reader to show that this parameterization gives

$$\left(\frac{dx}{dt}\right)^2 + \left(\frac{dy}{dt}\right)^2 = \frac{k^4}{(r(t))^2}\left(\frac{(r'(t))^2}{(r(t))^2} + 1\right).$$

Therefore,

$$
\begin{aligned}
s_H(\gamma') &= \int_\alpha^\beta \frac{1}{y}\sqrt{\left(\frac{dx}{dt}\right)^2 + \left(\frac{dy}{dt}\right)^2}\, dt \\
&= \int_\alpha^\beta \frac{r(t)}{k^2 \sin t}\sqrt{\frac{k^4}{(r(t))^2}\left(\frac{(r'(t))^2}{(r(t))^2} + 1\right)}\, dt \\
&= \int_\alpha^\beta \frac{1}{r(t)\sin t}\sqrt{(r'(t))^2 + (r(t))^2}\, dt \\
&= s_H(\gamma). \qquad\qquad\qquad\qquad\qquad\qquad\qquad\qquad \square
\end{aligned}
$$

Now that we have shown that Euclidean inversions centered on the x-axis are hyperbolic isometries, by Theorem 15.4, we know that an inversion maps a hyperbolic line to another hyperbolic line. To develop our intuition, we investigate further by determining the image of some hyperbolic lines in this model under these inversion mappings.

Theorem 15.48. *Consider a semicircle centered at $C = (c, 0)$ with radius k. The Euclidean inversion $I_{C,k}$ maps*

- *a vertical ray that passes through C onto itself,*
- *a vertical ray that does not pass through C onto a semicircle centered on the x-axis that passes through C,*
- *a semicircle centered on the x-axis that passes through C onto a vertical ray that does not pass through C, and*
- *a semicircle centered on the x-axis that does not pass through C onto another semicircle centered on the x-axis that also does not pass through C.*

Proof. We first note that the image of point A under the inversion $I_{C,k}$ lies on the ray $\overrightarrow{CA}$. Thus $I_{C,k}$ maps the upper half-plane onto itself. We consider the following four cases:

Case 1. Suppose that ℓ is the vertical ray $x = c$. Then by Theorem 15.36, the inversion $I_{C,k}$ maps ℓ onto itself.

Case 2. Suppose that ℓ is the vertical ray $x = a$, where $a \neq c$. Let $A = (a, 0)$. Note that in Euclidean geometry, $CA \perp \ell$. Thus, based on the proof of Theorem 15.36, $I_{C,k}$ maps ℓ onto the semicircle, s, in the upper half-plane with diameter CA' where $A' = I_{C,k}(A)$. Since A' lies on the x-axis, the center of semicircle s also lies on the x-axis.

Case 3. Suppose that ℓ is given by a semicircle s with center $A = (a, 0)$ that passes through C. Let CD be the diameter of s. Then based on the proof of Theorem 15.36, the image of s is a ray starting at $D' = I_{C,k}(D)$. Moreover, the ray is perpendicular to $\overleftrightarrow{AC}$, which is the x-axis. Since D' lies on $\overrightarrow{CA}$, we have $D' = (d', 0)$ for some d'. Thus, the image of s is the vertical ray $x = d'$.

Case 4. Suppose that ℓ is given by a semicircle s with center $A = (a, 0)$ that does not pass through C. Then based on the proof of Theorem 15.36, the image of s is another semicircle s' whose diameter lies on $\overleftrightarrow{AC}$, which is the x-axis. Thus, the center of s' also lies on the x-axis. $\qquad\square$

We are now ready to show that Euclidean inversions act as our second type of reflection in the Half-plane model. Suppose the hyperbolic line ℓ is represented by a semicircle of radius k centered at C on the x-axis. Consider a point A not on ℓ, and let $A' = I_{C,k}(A)$. To justify our claim that $A' = HF_\ell(A)$, we need to show that ℓ is the perpendicular bisector of AA'. We will do this in two steps. Using Euclid I.12, we can construct the unique hyperbolic line, m, through A that is perpendicular to ℓ. Our first step will be to show that A' also lies on m. If we let B be the intersection of m with ℓ, our second step will be to show that $d_H(A, B) = d_H(A', B)$.

Lemma 15.49. *Suppose the hyperbolic line ℓ is represented by a semicircle centered at $C = (c, 0)$ with radius k. Consider a point A not on ℓ, and let $A' = I_{C,k}(A)$. If m is the unique perpendicular hyperbolic line from A to ℓ, then A' also lies on m.*

Proof.

Case 1. Suppose A lies on the vertical ray $x = c$. Then the unique perpendicular from A to ℓ is the vertical ray $x = c$. Since A' lies on the ray $\overrightarrow{CA}$, A' also lies on $x = c$.

Case 2. Suppose A does not lie on $x = c$. In this case, the unique perpendicular from A to ℓ is a semicircle m that is orthogonal to ℓ. Based on the proof of Theorem 15.39, $I_{C,k}$ maps m onto itself. Thus A' lies on m. $\qquad\square$

Theorem 15.50. *If the hyperbolic line, ℓ, is represented by a semicircle centered at $C = (c, 0)$ with radius k, then the hyperbolic reflection across ℓ is the Euclidean inversion in ℓ, that is, $HF_\ell = I_{C,k}$.*

Proof. Consider a point A that does not lie on our hyperbolic line, ℓ. Let $A' = I_{C,k}(A)$. By Lemma 15.49, $\ell \perp AA'$. Let B be the intersection of ℓ and AA'. We note that $I_{C,k}$ fixes B. Since $I_{C,k}$ is an isometry, we have $d_H(A, B) = d_H(A', B') = d_H(A', B)$. Therefore, ℓ is the perpendicular bisector of AA', and we have shown that $HF_\ell(A) = I_{C,k}(A)$ for an arbitrary point, A, in the hyperbolic plane. $\qquad\square$

We end this chapter by visualizing the image of a triangle under hyperbolic reflection and hyperbolic rotation. We can use Theorem 15.34, to find the images of the vertices of the triangles. The image of triangle $\triangle ABC$ under reflection across a hyperbolic line, ℓ, is illustrated in Figure 15.48. The image of triangle $\triangle ABC$ under two

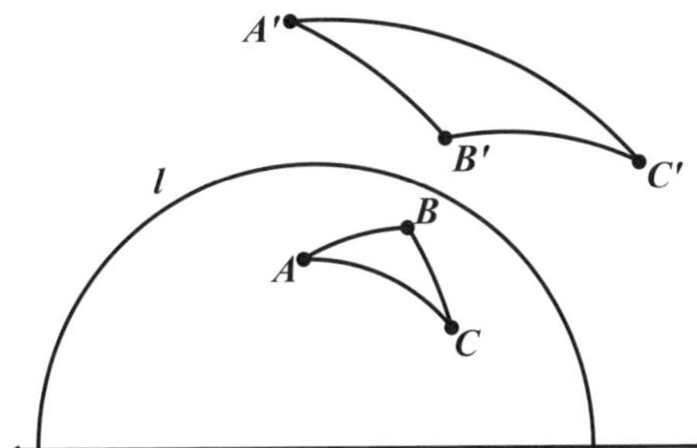

Figure 15.48. The image of $\triangle ABC$ under reflection across hyperbolic line ℓ

reflections, $HF_m \circ HF_\ell$, is illustrated in Figure 15.49. Note that this is equivalent to a hyperbolic rotation centered at D. Since all of these maps are isometries, we know, for example, that $AC = A'C' = A''C''$. These distance-preserving transformations give us another way to experience the inherent distortion of the model as the figure approaches the x-axis boundary. Like the fish in Escher's *Circle Limit III* on page 261, these triangles are all the same size.

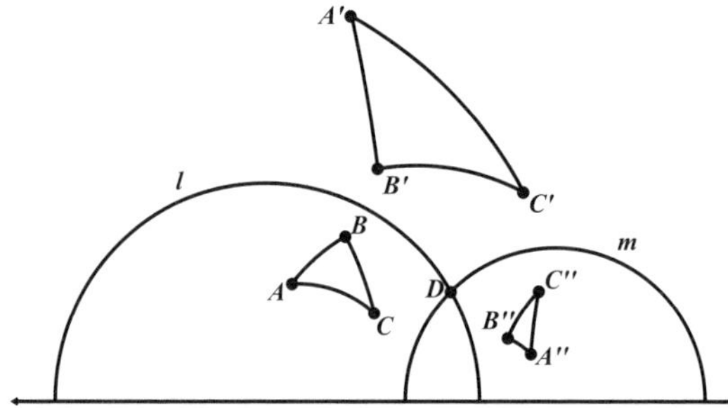

Figure 15.49. The image of $\triangle ABC$ under hyperbolic rotation centered at D

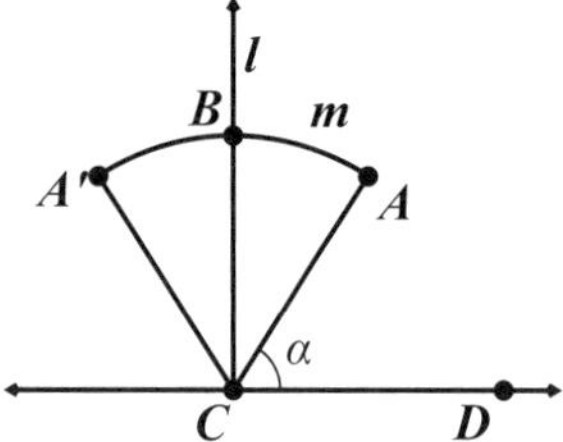

Figure 15.50. Exercise 15.5.1: A hyperbolic reflection across a vertical ray

Exercises 15.5

1. In Theorem 15.46, we wish to show that $d_H(A, B) = d_H(A', B)$. First, parameterize m by letting $x = c + r\cos t$ and $y = r\sin t$. WLOG, assume that $x_1 > c$. Let D be a point on the x-axis to the right of C. Let α be angle $\angle ACD$, as illustrated in Figure 15.50.

   (a) Show that
   $$d_H(A, B) = \ln\left(\frac{1 + \cos\alpha}{\sin\alpha}\right).$$

   (b) Explain why $d_H(A', B) = \ln\left[\csc(\pi - \alpha) - \cot(\pi - \alpha)\right].$

   (c) Finish the proof by establishing the equality of the expressions in the first two parts of this problem. Specifically, prove that
   $$\left[\csc(\pi - \alpha) - \cot(\pi - \alpha)\right] = \left(\frac{1 + \cos\alpha}{\sin\alpha}\right).$$

2. With parameterization $x(t) = c + \dfrac{k^2}{r(t)} \cos t$ and $y(t) = \dfrac{k^2}{r(t)} \sin t$, show that

$$\left(\frac{dx}{dt}\right)^2 + \left(\frac{dy}{dt}\right)^2 = \frac{k^4}{(r(t))^2}\left(\frac{(r'(t))^2}{(r(t))^2} + 1\right).$$

Key for the source of entries in the multiplication table on page 396.

$\circ$	T	R	F	G
T	Ex 15.3.1.3 (Th 15.19)	Ex 15.3.4.2 (Th 15.27)	Ex 15.3.5.2 (Th 15.29)	Ex 15.3.5.6
R	Th 15.27	Th 15.26	Ex 15.3.5.3 (Th 15.30)	Ex 15.3.5.6
F	Th 15.29	Th 15.30	Th 15.24	Ex 15.3.5.7
G	Ex 15.3.5.5	Ex 15.3.5.5	Ex 15.3.5.7	Ex 15.3.5.7

16

Constructibility

Figure 16.1. Detail of Raphael's *The School of Athens* (1510-11)

As we near the end of our story, we take one last look at the basic Euclidean tools used to create our main characters, the compass and unmarked straightedge. It is astounding to realize that these two simple tools give rise to all of Euclidean plane geometry. The ancient Greeks were elegant and skillful builders, creating the complex logical structure of geometry with these elementary tools. Archimedes famously claimed that he could move the world with a lever, but we cannot help but marvel at what he and his fellow geometers were able to do with a mere compass and a straightedge.

Greek geometry was the cutting edge of mathematics for more than a thousand years. Powerful as their tools are, there were constructions that eluded the Greek geometers. We briefly mentioned three such problems in Sections 3.2 (trisecting angles), 7.5 (squaring the circle) and 11.5 (constructing regular n-gons). These are three of the four most famous construction problems in the history of mathematics, the fourth being the doubling of a cube. The fame of each problem is due to the length of time it went unsolved, and the scores of mathematicians who toiled away without success. None of these problems was resolved until the nineteenth century. The significance of each problem lies not in its eventual solution, but in the enormous amount of new mathematics generated in the quest for a solution. The same can be said of the quest to prove the Parallel Postulate. To borrow from Ralph Waldo Emerson, these stories serve as a reminder that mathematics "is a journey, not a destination."

In this section, we discuss each problem and solution. In order to understand the solutions, we must enter briefly into the world of abstract algebra.

16.1 Four famous problems of antiquity

There are three classical construction problems of Greek mathematics and they are commonly referred to as:

(1) Squaring the circle

(2) Doubling the cube

(3) Trisecting the angle

To these three, we will add a fourth of interest to the Greeks:

(4) Constructing a regular n-gon

Let's take a close look at each of these problems.

16.1.1 Squaring the circle. At the end of Book II, Euclid can square any polygon, that is, he can construct a square of equal area to any given polygon. Thus, polygons are quadrable. Given this spectacular success, it is not surprising that the Greeks tried to square plane figures with curved boundaries. In particular, Hippocrates of Chios (ca. 470–ca. 410 BCE) could square a particular **lune**, a crescent-shaped figure bounded by two circular arcs. We give the construction of this type of lune, and leave the proof of its quadrability to the reader.

Construction of Hippocrates' Lune: Consider a semicircle with center A and diameter BC. Construct D on the semicircle such that $AD \perp BC$. Join BD and bisect this segment at a point E. Construct a semicircle with center E and diameter BD. Let F be a point on the arc $\widehat{BD}$ and G be a point on the semicircle BGD.

Claim: The lune $BGDF$ has the same area as $\triangle ABD$, and hence, is quadrable.

Notice that the construction of Hippocrates' lune begins with the side (BD) of a square inscribed in a circle. This is the basis for one type of lune. If we started this process with the side of some other n-gon inscribed in a circle, we would produce a

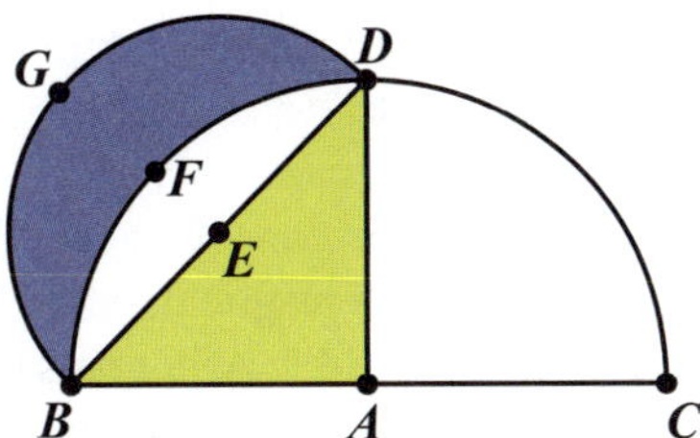

Figure 16.2. Hippocrates' lune

different type of lune. In total, the Greeks demonstrated the quadrability of three different types of lunes. Over fifteen hundred years later, Euler and Finnish mathematician Martin Johan Wallenius (1731–1773) added two others to this total in the eighteenth century. In the twentieth century, two Russian mathematicians, N.G. Tschebotaröw (1894–1947) and his student A.V. Dorodnov (1908–), put an end to the quest for quadrable lunes by showing that these are the *only* types of quadrable lunes, and there can be no others [**106**].

Squaring a nonrectilineal figure surely held great appeal to the Greeks, and while showing the quadrability of three out of the five possible squarable lunes was a fine achievement, the ultimate goal was one of our main characters, the circle. This is our first famous classical question.

Is the circle squarable?

That is, given a circle, is it possible to construct a square of equal area using only a compass and unmarked straightedge?

16.1.2 Doubling the cube. Of the four problems in this chapter, this is the only one that has not been discussed earlier in the book. It is also the only one that does not belong to plane geometry, but its inclusion is mandatory for the sake of completeness. Since our investigations have been in planar geometry, let's start our discussion with the related idea of doubling in the lower dimensions. Given any line segment we can easily construct another of double its length with the use of a compass. We can also double a square, that is, given a square, we can construct another square with twice the area of the original square using only a compass and unmarked straightedge. To do this we simply construct our new square on the diagonal of the given square. (Why does this work? See related Exercise 7.6.5.) Next, we come to three-dimensional volume and present the second famous classical problem.

Given a cube, is it possible to construct another cube with twice its volume using only a compass and unmarked straightedge?

Applying the same technique we used in two dimensions, here constructing a larger cube on the diagonal of the given cube, produces a cube of volume $2\sqrt{2}$ times the original. This problem is a bit frustrating as it seems easy to solve, but often in mathematics, a difficult problem can be easily stated. (Can you think of another deep problem lurking beneath an easily understood question?[1])

---

[1]Goldbach's conjecture is one such example. It states that every even integer greater than two can be expressed as the sum of two primes. For example, $50 = 7 + 43$. While easy to understand, the problem remains unsolved.

There are two different legends as to the origins of this problem. Because the first of these stories takes us to the Greek isle of Delos, this problem is sometimes referred to as *the Delian Problem*. According to Theon of Smyrna (ca. 70–ca. 135) in his writings on Eratosthenes' *Platonicus*, the Athenians consulted the oracle of Apollo at Delos in 430 BCE to ask what they could do to stop a plague that had killed their leader and was devastating their city. The oracle instructed the Athenians to double the existing cubical altar of Apollo. The architects tried to do this by doubling the length of the existing edges of the altar. Legend has it that the enraged god worsened the misery inflicted by the plague. Upon discovering their error, they consulted Plato on the matter. Plato's reply was as follows.

> the god has given this oracle, not because he wanted an altar of double the size, but because he wished in setting this task before them to reproach the Greeks for their neglect of mathematics and their contempt of geometry. [**116**]

The second story comes from a king's request to double the size of his son's tomb. According to the legend, King Minos commissioned a tomb to honor his son Glaucus who had died by falling into a vat of honey. When the architects presented their plans for the royal monument in the form of a cube, the unhappy king found its size lacking. In a letter to King Ptolemy, Eratosthenes quoted King Minos as saying: "Small indeed is the tomb thou hast chosen for a royal burial. Let it be double" [**116**][2].

16.1.3 Trisecting the angle. Of the four problems presented in this chapter, this one is the most misunderstood. In Proposition I.9, Euclid bisects any given angle. By continuing to divide the angle, Euclid can quarter an angle or "eighth" an angle, for example, but he gives no proposition regarding the trisection of an angle. The third problem of antiquity takes up this question, and is as follows:

> Given any angle, is it possible to trisect the angle, dividing it into three equal parts, using only a compass and unmarked straightedge?

It is clear that some angles can be trisected. For example, since we can construct an equilateral triangle, we can construct a 60° angle. Since we can bisect any angle, we can construct both 30° and 15° angles. Thus, we can trisect a 90° angle and a 45° angle. The question, however, is whether there is a construction to trisect *any* given angle.

16.1.4 Constructing a regular *n*-gon. The last construction problem we consider is related to Chapter 11:

> Given a positive integer $n \geq 3$, is it possible to construct a regular n-gon using only a compass and unmarked straightedge?

As we saw in Section 11.5, Gauss answered this question and the answer was no. In this chapter, we will use central angles to see why some specific regular *n*-gons are not constructible.

Now that we have outlined the problems, let's see how they were eventually solved. *Spoiler alert: None of the four constructions is possible, thus they are usually called "The Impossible Constructions."* To understand why, we need to first examine what is possible.

---

[2]Different Ptolemy than the astronomer of Section 10.3.

Exercises 16.1

1. In Hippocrates' quadrature of the lune, let $a = AB$. Prove that the lune $BGDF$ has the same area as the triangle $\triangle ABD$, namely $\frac{1}{2}a^2$.

16.2 Constructible numbers

A positive number is constructible if it is possible to construct a segment whose length is the given number using a compass and unmarked straightedge. To determine exactly which numbers are constructible with the Euclidean tools, we need to start with a **unit length**, that is, a segment that we arbitrarily decide has length 1. Without this starting point, we have no way to know the length of any segment. From this starting point, it is fairly clear that on a given ray we can construct segments of integral lengths 2, 3, 4, and so on. Moreover, since we can bisect a segment we can also construct segments of lengths $\frac{1}{2}, \frac{1}{4}, \frac{1}{8}$, and so on. The Pythagorean Theorem tells us that square roots are constructible. For example, the diagonal of a unit square has length $\sqrt{2}$. Better still, the following construction, known as the **Pythagorean spiral**, generates the sequence of lengths $\sqrt{2}, \sqrt{3}, \sqrt{4}, \sqrt{5}$, and so on.

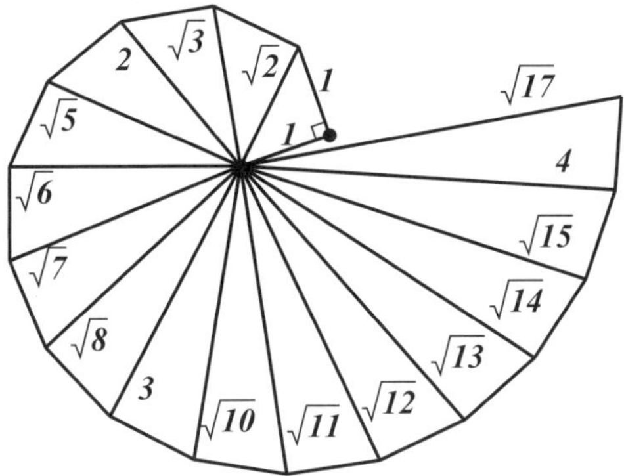

Figure 16.3. Pythagorean spiral

Construction of the Pythagorean spiral: Construct an isosceles right triangle with legs of length 1. Then the hypotenuse is $\sqrt{2}$. Construct a second right triangle on the hypotenuse of the first one by adding a perpendicular segment of unit length. The length of the newly constructed hypotenuse is $\sqrt{1^2 + (\sqrt{2})^2} = \sqrt{3}$. This process can be continued, and at each step we construct a segment of length $\sqrt{n+1}$ from a segment of length $\sqrt{n}$.

So far, we see that from a given unit length, we are able to construct all positive integers and their square roots, and some specific fractions. To find other numbers that are constructible, it is helpful to understand the algebraic operations that we can perform geometrically on constructible lengths with our Euclidean tools. It is in this way that we can grow our set of constructible lengths by using known contructible lengths as building blocks. Of the five operations specified in the following theorem, we give constructions for three operations and leave the proofs to the reader. (There are certainly other ways to construct ab, $\frac{a}{b}$ and $\sqrt{a}$, and we encourage the reader to explore alternative methods.)

Theorem 16.1. *Given a unit length and constructible lengths a and b, we can construct the following:*

(1) $a + b$

(2) $a - b$ *(assuming $a > b$)*

(3) ab

(4) $\frac{a}{b}$

(5) $\sqrt{a}.$

Proof. We leave the constructions of $a + b$ and $a - b$ to the reader.

Construction of ab: Consider a ray starting at a point A. Construct $AB = 1$, and on the same ray, construct $AC = a$. On another ray starting at A, construct $AD = b$. Join BD. Construct a line through C parallel to BD. Let E be the intersection of this line with $\overrightarrow{AD}$.

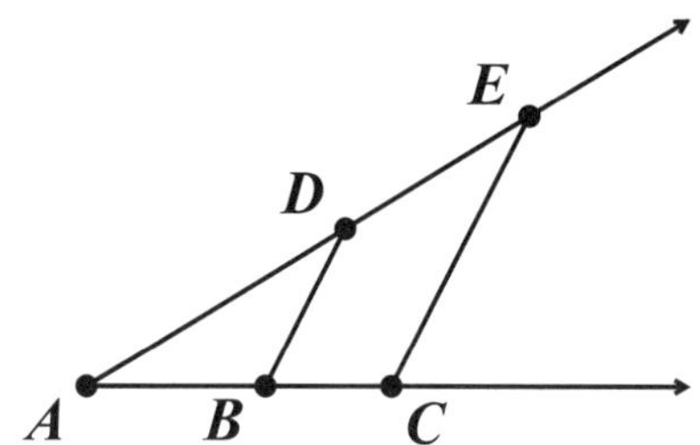

Figure 16.4. Constructions of ab and $\frac{a}{b}$

Claim: $AE = ab.$

Construction of $\frac{a}{b}$: Consider a ray starting at a point A. Construct $AB = 1$, and on the same ray, construct $AC = b$. On another ray starting at A, construct $AE = a$. Join EC. Construct a line through B parallel to EC. Let D be the intersection of this line with $\overrightarrow{AE}$.

Claim: $AD = \frac{a}{b}.$

Construction of $\sqrt{a}$: Consider a ray starting at a point A. Construct $AB = a$, and on the same ray, construct $BC = 1$. Bisect AC at D, and then construct a circle with center, D, and radius, AD. Construct a perpendicular to AC through B, and let E be its intersection with the circle.

Claim: $BE = \sqrt{a}.$ □

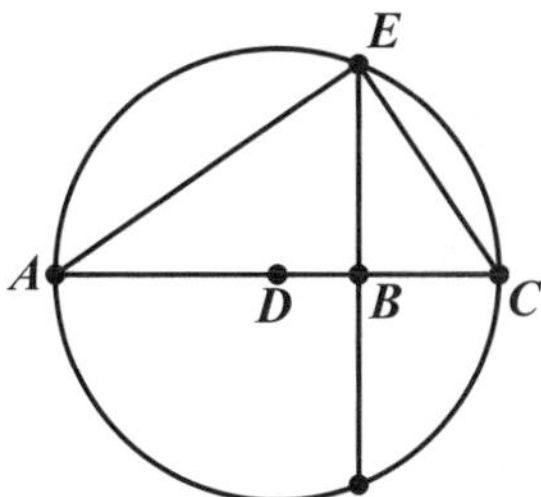

Figure 16.5. Construction of $\sqrt{a}$

With this theorem, we see that the algebraic operations of addition, subtraction, multiplication, division and extraction of square roots have corresponding geometric constructions. More simply, it connects algebra and geometry. As an example, consider $\sqrt{7}$. It is a real number that lies between 2 and 3, and algebraically, it is a solution to the equation $x^2 = 7$. Geometrically, we can produce a segment of length $\sqrt{7}$ by adjoining seven unit lengths in a line, then following the construction outlined for square roots in part 5 of Theorem 16.1. Consider the number $\frac{8+\sqrt{3}}{5}$. We can construct a segment of this length by starting with its component segments of length 8, $\sqrt{3}$ and 5. Next, we construct a segment of length $8 + \sqrt{3}$ using part 1, then the desired length using part 4. Furthermore, for positive integers n, notice that we can construct any integer exponents, a^n, or $(2^n)^{th}$ roots, $\sqrt[2^n]{a}$, through repeated multiplication or extraction of square roots, respectively, of any constructible a. For example, we can construct $(\sqrt{2} + \sqrt[8]{7})^3$.

The important thing to realize is that, starting with a given unit length, the constructions in this theorem provide a recipe for constructing a segment of length equal to any desired positive real number that can be expressed using only integers and a finite number of the algebraic operations of addition, subtraction, multiplication, division and extraction of square roots (as long as there are no square roots of negative numbers). Hence, any positive real number meeting these conditions is of particular interest since it is constructible. We define this type of real number as follows.

Definition 16.2. A real number is called a **surd** if it can be written using only integers and a finite number of the algebraic operations of addition, subtraction, multiplication, division and extraction of square roots.

For example, $\sqrt{4 + 2\sqrt{2}}$ and $\dfrac{2 - \sqrt{3}}{\sqrt[4]{5 + 3\sqrt{2}}}$ are both surds. Theorem 16.1 makes it clear that any positive surd is constructible. Thus, in particular, these examples are constructible. Since the definition also allows for negative numbers, $-\sqrt{4 + 2\sqrt{2}}$ is a surd. While we cannot construct a segment of negative length, we can construct a segment of length $\left| -\sqrt{4 + 2\sqrt{2}} \right|$.

Are all real numbers surds? The answer is no, but the explanation for this answer will come later. While we may be tempted to judge a book by its cover and say that a number like $\cos\left(\dfrac{\pi}{12}\right)$ or $\dfrac{\sqrt[3]{26+15\sqrt{3}}+\sqrt[3]{26-15\sqrt{3}}}{4}$ is not a surd since it contains a cosine or a cube root, two operations that are clearly not one of the five allowable operations, it would be foolish to do so. It just so happens that these are equivalent to the surds $\dfrac{\sqrt{2}+\sqrt{6}}{4}$ and 1, respectively. The number $\sqrt[3]{5}$ is not a surd, but we will need the help of a few theorems from modern algebra at the end of this section to explain why it is not possible to rewrite it, as we were able to do with the two previous numbers. For now, we ask the reader to believe that there are numbers which are not surds, and that $\sqrt[3]{5}$ is one of them.

It turns out that the numbers which can be produced by starting with 1 and using a finite number of the five specified algebraic operations are the *only* numbers that are constructible. We present this result in the following theorem, and in order to prove it, we introduce a few new terms and present a sequence of lemmas.

Theorem 16.3. *A length is constructible if and only if it is a positive surd.*

As established by our discussion above, the backwards direction of this iff is a direct result of Theorem 16.1. Showing the forwards direction, that any constructible length is a positive surd, will take some work, and we will do this without the machinery of extension fields that we would have in a modern algebra course. We will, however, need a systematic way to keep track of exactly what lengths are constructible, and for this, we rely on the Cartesian coordinate system and basic analytic geometry.

In order to show that any length that we can construct with our straightedge and compass can be expressed with integers, addition, subtraction, multiplication, division and square roots, we need to consider all possible ways that lengths can be constructed. As will soon be obvious, this is most easily accomplished by working within a coordinate system. Since we need an origin to define a coordinate system, we assume that $(0,0)$ is a constructible point, and thus, 0 is constructible. We will also assume that we have the point $(1,0)$. (Technically, we can pick any two distinct points in our plane and label them accordingly.) From these two points, we have our given unit length. As always, we have our Euclidean tools. With these building blocks, the first item we are able to construct is the line through these two points, which will serve as the x-axis. Next, we can use our compass to construct the unit circle. The new point on the Cartesian plane that has just been constructed is $(-1,0)$. From this meager beginning, we can use our compass to construct any point (n,m), where n,m are integers. How? If we construct another circle centered at $(1,0)$ that passes through $(0,0)$, then we get a new point on the Cartesian plane, $(2,0)$. Continuing along the x-axis in either direction, and constructing unit circles centered at newly constructed points, will eventually produce any point $(n,0)$. Once we construct the y-axis we can use the same process to produce any point $(0,m)$. One way to do is to start with two circles, one centered at $(1,0)$ that passes through $(-1,0)$, the other centered at $(-1,0)$ that passes through $(1,0)$. Using SSS and the Pythagorean Theorem, we can show that the intersection of these circles produces two new points at $(0,\sqrt{3})$ and $(0,-\sqrt{3})$. Joining either point with the origin will produce the y-axis.

We stop this constructive approach here, but note that the key rules to playing this point construction game is that **new points are constructed using our Euclidean tools, and most importantly, that the coordinates are identifiable**. We could play the pin-the-tail-on-the-donkey game where you drop your pen on any spot on the coordinate plane, but we will not say that a point created in this way has been constructed since we have no way of identifying its coordinates.

Now, suppose that $|x|$ is some constructible number. Using our compass, we can construct the points $(|x|, 0)$ or $(-|x|, 0)$ on the x-axis. In a similar manner to our construction of the y-axis given above, we can construct a right angle from one of these points to construct any of the points $(|x|, |y|)$, $(|x|, -|y|)$, $(-|x|, |y|)$ or $(-|x|, -|y|)$, as long as $|y|$ is constructible. The following definition naturally follows from this discussion.

Definition 16.4. Given an origin $(0, 0)$ and perpendicular x- and y-axes, a point (x, y) is **constructible** iff the absolute value of each coordinate is either 0 or a constructible length.

Specifically, a point (x, y) on the Cartesian plane is constructible if and only if $|x|$ and $|y|$ are both constructible lengths. How many constructible points are there? Given our discussion above, if a and b are surds, then (a, b) is constructible. Since there are infinitely many ways to combine any collection of integers with a finite combination of the allowable operations, there are clearly infinitely many points in addition to the integer coordinate points, (n, m), that are constructible. As our goal is to prove the forwards direction of Theorem 16.3, we need to show that our tools are only capable of producing lengths that are surds. Since this definition identifies constructible lengths with constructible points in a natural way, we would like to show that any point that we can construct with our Euclidean tools has surd coordinates. This means that we need to answer the following two questions:

1. If we start with points $(0, 0)$ and $(1, 0)$, what points (x, y) on the Cartesian plane are constructible with our Euclidean tools?

2. For any constructible (x, y), are both $|x|$ and $|y|$ surds? If the answer is yes, then we have proven Theorem 16.3. If the answer is no, then the theorem is not true.

The advantage to working within the Cartesian coordinate system is that it allows us to utilize our significant analytic geometry skills to answer these questions. Of primary importance, we need to give the algebraic equations that mimic the use of the straightedge and compass. In keeping with Euclid's first postulate, we have the following algebraic definition of a constructible line.

Definition 16.5. A line $ax + by = c$ is **constructible** if it passes through two constructible points.

Essentially, any line is determined by two constructible points and a straightedge. Note that $(1, 1)$ and $(2, 2)$ are both constructible points. Therefore, the line $y = x$ in the Cartesian plane is constructible. This does not, however, imply that all (or any!) other points on this line are constructible. If every point, (x, x), on this line were constructible, then every $|x|$ is a constructible length. This would imply that every positive real number is constructible, an impossibility as the previous spoiler alert has already foretold.

Likewise, let's take a look at the algebraic equivalent of the compass. In keeping with Euclid's third postulate, we have the following definition of a constructible circle.

Definition 16.6. A circle $(x - h)^2 + (y - k)^2 = r^2$ is **constructible** if both the center, (h, k), and the radius, r, are constructible.

As noted above, this does not imply that all (or any!) other points on this circle are constructible. For example, the unit circle, $x^2 + y^2 = 1$, is constructible since both $(0,0)$ and 1 are constructible, but this information alone does not tell us anything about the constructibility of the points on the circle. We certainly know that the "compass points", $(h + r, k)$, $(h, k + r)$, $(h - r, k)$ and $(h, k - r)$, of any circle are constructible, but we do not need the entire circle to tell us that. If constructing a line or a circle does not mean that the points on these objects are constructible, then how can these tools give us new constructible points? The answer can be found in the way we started to build our set of constructible points with those of type (n, m). How did we identify these points? The sweep of a compass or the stroke of a straightedge produces a new constructible point where these constructible objects meet. Consider the two examples we have constructed above, $y = x$ and $x^2 + y^2 = 1$. They give us two new constructible points where they intersect, $\left(\dfrac{\sqrt{2}}{2}, \dfrac{\sqrt{2}}{2}\right)$ and $\left(-\dfrac{\sqrt{2}}{2}, -\dfrac{\sqrt{2}}{2}\right)$. Clearly, these points have surd coordinates.

This example provides the key to answering the first of our questions as it illuminates how we generate new constructible points (hence new constructible lengths) with our Euclidean tools. Since our compass and straightedge only produce constructible lines and constructible circles, we only need to consider all possible points of intersection produced by these constructible objects. How can we possibly check all such constructible points which result from the intersection of two constructible lines, the intersection of a constructible line and a constructible circle, and the intersections of two constructible circles? The simple answer is, a little algebra goes a long way! Let's determine the points of intersection produced by these objects, and in the process, answer the second question for these newly constructed points: Do these points have surd coordinates? We will accomplish this goal with the following sequence of lemmas.

Lemma 16.7 [Lines]. *If a line $ax + by + c = 0$ passes through points (x_1, y_1) and (x_2, y_2), where x_1, x_2, y_1 and y_2 are all surds, then the coefficients a, b and c are also surds.*

Proof. Let a line pass through the points (x_1, y_1) and (x_2, y_2), where x_1, x_2, y_1 and y_2 are all surds. We leave it to the reader to show that the equation of the line is given by $(y_2 - y_1)x + (x_1 - x_2)y + (x_2y_1 - x_1y_2) = 0$, the coefficients of which are clearly surds since they are merely products and differences of surds. $\qquad\square$

Lemma 16.8 [Circles]. *If a circle $x^2 + y^2 + Ax + By + C = 0$ has center, (h, k), and radius, r, where h, k and r are surds, then A, B and C are also surds.*

Proof. A circle with center, (h, k), and radius r, is given by the equation $(x - h)^2 + (y - k)^2 = r^2$. Simplifying, we get $x^2 + y^2 + (-2h)x + (-2k)y + (h^2 + k^2 - r^2) = 0$. Here, $A = -2h$, $B = -2k$, and $C = h^2 + k^2 - r^2$ are surds since they are merely sums, differences, and products of surds. $\qquad\square$

Lemma 16.9 [Line meets Line]. *If the nonparallel lines $ax + by + c = 0$ and $a'x + b'y + c' = 0$ have surd coefficients, then their point of intersection has surd coordinates.*

Proof. We leave it to the reader to show that the intersection of the lines $ax+by+c=0$ and $a'x+b'y+c'=0$ is the point

$$\left(\frac{bc'-b'c}{ab'-a'b},\ \frac{a'c-ac'}{ab'-a'b}\right). \qquad \square$$

Lemma 16.10 [Line meets Circle]. *If the line with equation $ax+by+c=0$ intersects the circle with equation $x^2+y^2+a'x+b'y+c'=0$, and both have surd coefficients, then any point intersection has surd coordinates.*

Proof. We leave it to the reader to show that any intersection point of the line, $ax+by+c=0$, and the circle, $x^2+y^2+a'x+b'y+c'=0$, is of the form (x,y) where

$$x=\frac{-B\pm\sqrt{B^2-4AC}}{2A},$$

with $A=a^2+b^2$, $B=2ac+a'b^2-abb'$, and $C=c^2-bb'c+c'b^2$, and

$$y=\frac{-ax-c}{b}. \qquad \square$$

Lemma 16.11 [Circle meets Circle]. *If two intersecting circles, with equations $x^2+y^2+ax+by+c=0$ and $x^2+y^2+a'x+b'y+c'=0$, both have surd coefficients, then any point of intersection has surd coordinates.*

Proof. We leave it to the reader to show the intersection of the circles, $x^2+y^2+ax+by+c=0$ and $x^2+y^2+a'x+b'y+c'=0$, is of the form (x,y) where

$$x=\frac{-B\pm\sqrt{B^2-4AC}}{2A},$$

with $A=(a'')^2+(b'')^2$, $B=2a''c''+a'(b'')^2-a''b''b'$, $C=(c'')^2-b''b'c''+c'(b'')^2$, $a''=a-a'$, $b''=b-b'$, and $c''=c-c'$, and

$$y=\frac{-a''x-c''}{b''}. \qquad \square$$

These five lemmas show that if we start with points whose coordinates are surds, and radii which are surds, then the resulting constructible lines and circles have surd coefficients, and any resulting intersections also have surd coordinates. Therefore, since we started with just $(0,0)$ and $(1,0)$, any point on the Cartesian plane that results from the use of our Euclidean tools must have surd coordinates. One last thing to check is the length of a line segment joining any two constructible points. Notice that any line segment is constructible when its endpoints are constructible. Since we are trying to show that any constructible length is a surd, we must take a moment to examine the length of such a line segment.

Lemma 16.12 [Length of Segment]. *If a segment joins the points (x_1,y_1) and (x_2,y_2), where x_1,x_2,y_1 and y_2 are all surds, then the length of the segment is also a surd.*

Proof. The length of the segment joining the points (x_1,y_1) and (x_2,y_2) is

$$\sqrt{(x_1-x_2)^2+(y_1-y_2)^2}. \qquad \square$$

Thus, any constructible line segment has a surd length, that is, it can be expressed using only integers and a finite number of the five allowable algebraic operations.

In total, we have shown that any new lengths or points that can be generated with our Euclidean tools must be surds or have surd coordinates. Thus, we cannot escape the set of surds using only a compass and unmarked straightedge, and we have proven Theorem 16.3: *A length is constructible if and only if it is a positive surd.*

In order to discuss the solutions of the four classical construction problems, we must briefly classify the surds within the larger context of the real numbers. Before doing that, take a moment to appreciate the ease with which we were able to answer the questions posed earlier in this section once we brought our considerable analytic geometry skills to bear upon the problem.

16.2.1 Algebraic numbers. Just as the set of positive integers can be partitioned into disjoint sets consisting of the evens and odds, the real numbers can also be partitioned. One way to do this is to separate the reals into the negatives and nonnegatives, while another is to split the reals into the *rational* and *irrational* numbers. Recall that a real number is **rational** if it can be expressed as the ratio of two integers. The modern reader may know that $\sqrt{2}$ cannot be written as a ratio of integers as it is a nonterminating, nonrepeating decimal, while the Pythagoreans proved that $\sqrt{2}$ is not rational without the help of a decimal system. There are other ways to partition the set of real numbers, and we are particularly interested in classifying the set of surds within the set of of real numbers. In order to do this, we introduce *algebraic* and *transcendental* numbers, two other disjoint sets which partition the real numbers.

Definition 16.13. A real number is **algebraic** if it is the solution to a polynomial equation of some positive integer degree, n,

$$a_n x^n + a_{n-1} x^{n-1} + \cdots + a_2 x^2 + a_1 x + a_0 = 0,$$

where all of the coefficients, $a_n, a_{n-1}, \ldots, a_2, a_1$, and a_0, are integers. A real number that is not algebraic is called **transcendental**.

For example, both $2/3$ and $\sqrt{2}$ are algebraic as they satisfy the equations $3x - 2 = 0$ and $x^2 - 2 = 0$, respectively. Joseph Liouville (1809–1882) gave the first example of a transcendental number in 1844, namely

$$0.11000100000000000000000010\ldots = \sum_{k=1}^{\infty} 10^{-k!}.$$

Charles Hermite (1822–1901) proved that e is transcendental in 1873.[3] Based on Hermite's ideas, Lindemann was able to show that π is transcendental in 1882.[4] Surds are decidedly algebraic in nature, and we state the following theorem relating surds and algebraic numbers without proof.

Theorem 16.14. *If x is a surd, then x is algebraic.*

The proof of this result lies squarely in the field of algebra and can be found in undergraduate abstract algebra books. (See [55].) While a formal proof is well outside the

---

[3] For a proof that e is transcendental, see [72].
[4] For a proof that π is transcendental, see [91].

scope of this book, we can nevertheless develop an understanding for why a particular surd is algebraic. Consider, for example, the surd $\frac{\sqrt{2}+\sqrt{6}}{4}$. We will produce a polynomial equation with integer coefficients that our surd satisfies as follows. Let $x = \frac{\sqrt{2}+\sqrt{6}}{4}$; then $4x = \sqrt{2} + \sqrt{6}$. Squaring both sides and simplifying, we have $4x^2 - 2 = \sqrt{3}$. Squaring once again and simplifying, we have $16x^4 - 16x^2 + 1 = 0$. We leave it as an exercise for the reader to produce a polynomial for the surd $\sqrt{4 + 2\sqrt{2}}$. In general, for any surd x, we can square as many times as is necessary to remove all square roots, and what remains will be a polynomial. This polynomial will necessarily have rational coefficients since surds are merely integers combined with the operations of addition, subtraction, multiplication, division and extraction of square roots. To produce integer coefficients is a simple matter of multiplying the resulting polynomial by the least common denominator of its finitely many rational coefficients.

The converse, however, is not true as there are solutions to polynomial equations with integer coefficients that are not surds. For example, $\sqrt[3]{5}$ is algebraic since it is a solution of the equation $x^3 - 5 = 0$, but we have requested your trust in that it is not a surd. Thus, as a set, the surds sit within the set of algebraic numbers. Likewise, since every rational number is a surd, then the set of rational numbers is nested inside the set of surds. The Venn diagram in Figure 16.6 illustrates these relationships.

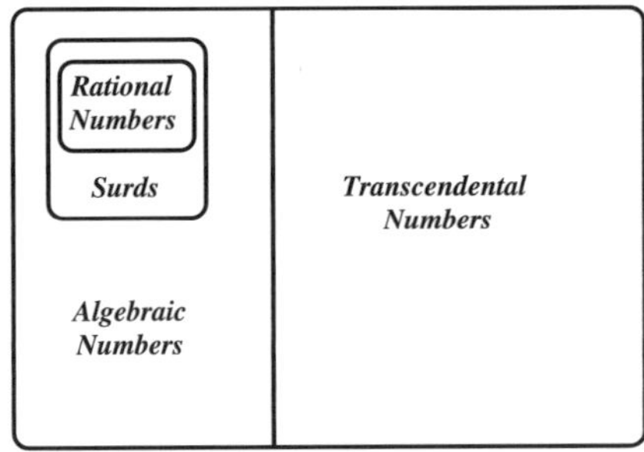

Figure 16.6. Venn diagram for the set of real numbers

While Theorems 16.3 and 16.14 make it easy to see that a transcendental number cannot be constructible, it offers no help in separating the algebraic numbers from the surds. Clearly, the real roots of linear and quadratic equations with integer coefficients are constructible, but it is more difficult to determine constructibility for roots of higher-order polynomials. Consider the cubic equation $x^3 - 6x^2 + 9x - 2 = 0$. This equation has three roots, 2, $2 + \sqrt{3}$ and $2 - \sqrt{3}$, all of which are constructible, but this is not always the case for cubic equations. The equation $x^3 = 5$ has $\sqrt[3]{5}$ as a solution, which is not a surd, and therefore, not constructible. To make good on our promise to explain why $\sqrt[3]{5}$ is not a surd, and to allow us to show that a few other algebraic numbers are not surds in the next section, we need the following two theorems. The second is a common result from high-school algebra. For a proof of the first, see Edwin Moise's *Elementary Geometry from an Advanced Standpoint*.

Theorem 16.15. *A cubic equation,*

$$ax^3 + bx^2 + cx + d = 0,$$

with integer coefficients has a root that is a surd iff it has a rational root.

Theorem 16.16 [Rational Root Theorem]. *Let*

$$a_n x^n + a_{n-1} x^{n-1} + \cdots + a_2 x^2 + a_1 x + a_0 = 0,$$

where each a_i is an integer. If $x = \dfrac{p}{q}$ is a rational solution to this equation written in lowest terms, then p divides a_0 and q divides a_n.

The Rational Root Theorem is typically taught in high-school algebra as a way of finding a first root to a cubic equation. For example, by the Rational Root Theorem, if $3x^3 - 8x^2 - 5x + 6 = 0$ has a rational root then it must be of form $\dfrac{\text{divisor of } 6}{\text{divisor of } 3}$. Thus, the only possible rational roots of this equation are $\pm 1, \pm \frac{1}{3}, \pm 2, \pm \frac{2}{3}, \pm 3$ or ± 6. It is easily checked that this equation has three rational roots, $-1, 3$ and $\frac{2}{3}$. As another example, $3x^3 - 8x^2 - x - 6 = 0$ has the same list of possible rational roots, and it is easily checked that 3 is its only rational root.

With these theorems, we can finally make good on our promise of showing that $\sqrt[3]{5}$ is not a surd. Consider the equation $x^3 - 5 = 0$, of which $\sqrt[3]{5}$ is a root. If $\sqrt[3]{5}$ were a surd, then by Theorem 16.15, the equation $x^3 - 5 = 0$ must have a rational root. However, by Theorem 16.16, the only possible rational roots for this equation are $\{\pm 1, \pm 5\}$, none of which is a root. Thus, $\sqrt[3]{5}$ is not a surd.

We are now ready to demonstrate the impossibility of each of the four famous construction problems of antiquity.

Exercises 16.2

1. In the construction of ab, prove that $AE = ab$.

2. In the construction of $\frac{a}{b}$, prove that $AD = \frac{a}{b}$.

3. In the construction of $\sqrt{a}$, prove that $BE = \sqrt{a}$.

4. Given a unit length, explain how to construct a segment of length $2\frac{3}{8}$.

5. Show that the equation of the line passing through the points (x_1, y_1) and (x_2, y_2) is given by $(y_2 - y_1)x + (x_1 - x_2)y + (x_2 y_1 - x_1 y_2) = 0$.

6. Show that the intersection of the lines $ax + by + c = 0$ and $a'x + b'y + c' = 0$ is the point
$$\left(\frac{bc' - b'c}{ab' - a'b}, \; \frac{a'c - ac'}{ab' - a'b} \right).$$

7. Show that the intersections of the line $ax + by + c = 0$ and the circle $x^2 + y^2 + a'x + b'y + c' = 0$ are the points (x, y) where
$$x = \frac{-B \pm \sqrt{B^2 - 4AC}}{2A},$$

with $A = a^2 + b^2$, $B = 2ac + a'b^2 - abb'$, and $C = c^2 - bb'c + c'b^2$, and

$$y = \frac{-ax - c}{b}.$$

8. Show the intersections of the circles $x^2 + y^2 + ax + by + c = 0$ and $x^2 + y^2 + a'x + b'y + c' = 0$ are the points (x, y) where

$$x = \frac{-B \pm \sqrt{B^2 - 4AC}}{2A},$$

with $A = (a'')^2 + (b'')^2$, $B = 2a''\,c'' + a'\,(b'')^2 - a''\,b''\,b'$, $C = (c'')^2 - b''\,b'\,c'' + c'\,(b'')^2$, $a'' = a - a'$, $b'' = b - b'$, and $c'' = c - c'$, and

$$y = \frac{-a''x - c''}{b''}.$$

9. Show that $\sqrt{4 + 2\sqrt{2}}$ is algebraic.

16.3 Four counterexamples

16.3.1 Squaring the circle is impossible.

Consider a circle of radius 1, and assume that we can construct a square with the same area as the circle. Since the area of a circle of radius 1 is π units squared, we have constructed a segment of length $\sqrt{\pi}$. By Theorem 16.1, we can construct a segment of length $\sqrt{\pi} \cdot \sqrt{\pi} = \pi$. But π is transcendental, hence not algebraic, and consequently, not a surd. Therefore, π is not constructible and we have a contradiction.

Conclusion: A circle of unit radius cannot be squared.

16.3.2 Doubling the cube is impossible.

A cube of side 1 has volume 1 unit cubed. Doubling this requires a cube of volume 2 units cubed, which itself requires the construction of a side of length $\sqrt[3]{2}$. To see that a segment of this length is not constructible, consider the equation $x^3 - 2 = 0$, of which $\sqrt[3]{2}$ is a root. If $\sqrt[3]{2}$ were a surd, then by Theorem 16.15, the equation $x^3 - 2 = 0$ must have a rational root. However, by Theorem 16.16, the only possible rational roots for this equation are $\{\pm 1, \pm 2\}$, none of which is a root. Thus, $\sqrt[3]{2}$ is not constructible.

Conclusion: A cube with side of unit length cannot be doubled.

Therefore, as regards the legend of the origin of this problem: The Greek gods were cruel!

16.3.3 Trisecting a general angle is impossible.

In order to show the impossibility of trisecting any given angle, we only need to find one angle that cannot be trisected. For our counterexample, we prove that the 60° angle cannot be trisected. We leave the proof of the following lemma to the reader.

Lemma 16.17. *Angle α is constructible if and only if the length $\cos \alpha$ is constructible.*

We recall the following two identities from trigonometry:

$$\cos(\alpha + \beta) = \cos \alpha \cos \beta - \sin \alpha \sin \beta$$

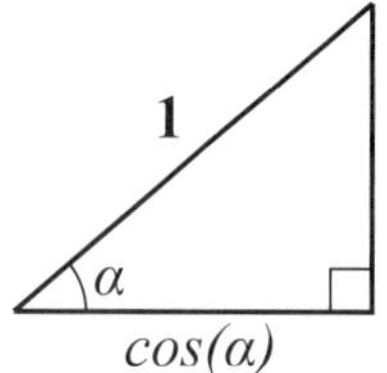

Figure 16.7. Cosine of α

and

$$\sin(\alpha + \beta) = \sin\alpha\cos\beta + \cos\alpha\sin\beta.$$

These allow us to prove the following double angle formulas:

$$\cos(2\alpha) = 2\cos^2\alpha - 1 \quad \text{and} \quad \sin(2\alpha) = 2\sin\alpha\cos\alpha.$$

We leave it to the reader to prove that

$$\cos(3\alpha) = 4\cos^3\alpha - 3\cos\alpha. \tag{16.1}$$

Letting $\alpha = 20°$, we have

$$\frac{1}{2} = 4\cos^3(20°) - 3\cos(20°).$$

If we set $x = \cos 20°$, then by Theorem 16.15, if x is a surd then the equation

$$\frac{1}{2} = 4x^3 - 3x$$

has a rational root. This equation can be rewritten as $8x^3 - 6x - 1 = 0$. We leave it to the reader to check that none of the possible rational roots, namely $\{\pm 1, \pm\frac{1}{2}, \pm\frac{1}{4}, \pm\frac{1}{8}\}$, is a root. Thus, $\cos 20°$ is not a surd, hence not constructible. This further implies that a $20°$ angle is not constructible by Lemma 16.17. Therefore, we cannot trisect a $60°$ angle.

Conclusion: An angle of $60°$ cannot be trisected.

French mathematician Pierre Laurent Wantzel (1814–1848) proved that the $60°$ angle cannot be trisected in 1837 in his paper "On the means of ascertaining whether a problem in geometry can be solved with ruler and compass" [124]. In the same article, he also proved that it is not possible to double the cube. Thus Wantzel resolved two of the three classical construction problems of Greek mathematics in one seven page paper.

Every year amateur mathematicians claim to have trisected a general angle. They are, of course, always proved wrong! Any successful trisection requires the use of additional tools beyond the Euclidean straightedge and compass. Perhaps it is worthwhile to note here that the Euclidean straightedge is *unmarked*. Were we to change our tools to include a marked ruler, the answer to this question would be different. We will explore this further at the end of this chapter.

16.3.4 Constructing all regular *n*-gons is Impossible. While there are many from which to choose, and though we only need one counterexample, we will prove that it is impossible to construct a regular nonagon (9-gon) *and* a regular heptagon (7-gon). For the nonagon, we use our previous result on the nonconstructibility of the 20° angle, and for the heptagon, we use our results on surds.

From our discussion on constructing regular *n*-gons in Chapter 11, we recall the following lemma.

Lemma 11.13. *A regular n-gon is constructible if and only if its central angle $\frac{360°}{n}$ is constructible.*

If we could construct a regular 9-gon, then we could construct its central angle, a 40° angle. Bisecting this, we would be able to construct a 20° angle, which we know is impossible. Thus, we cannot construct a regular 9-gon.

Conclusion: A regular 9-gon is not constructible.

What does the result about central angles imply for Gauss's regular 17-gon, as discussed in Section 11.5? Well, its central angle, $\frac{360°}{17}$, *must* be constructible. Likewise, $\cos\left(\frac{360°}{17}\right)$ must also be constructible. This means that $\cos\left(\frac{360°}{17}\right)$ is a surd, that is, it can be written using only addition, subtraction, multiplication, division and square roots. With a bit of effort, in his *Disquisitiones Arithmeticae* of 1801, Gauss showed that $\cos\left(\frac{360°}{17}\right)$ equals

$$-\frac{1}{16} + \frac{1}{16}\sqrt{17} + \frac{1}{16}\sqrt{34 - 2\sqrt{17}} + \frac{1}{8}\sqrt{17 + 3\sqrt{17} - \sqrt{34 - 2\sqrt{17}} - 2\sqrt{34 + 2\sqrt{17}}},$$

which is clearly a surd! [57]

The smallest regular polygon that is not constructible is the regular heptagon (7-gon). Let $\alpha = \frac{360°}{7}$. Using a few standard trigonometric identities, we leave it to the reader to prove another identity similar to Equation (16.1):

$$\cos 4\alpha = 8\cos^4\alpha - 8\cos^2\alpha + 1. \tag{16.2}$$

Since $\cos\theta = \cos(360° - \theta)$ for any θ, and $3\alpha + 4\alpha = 360°$ for our chosen α, we have $\cos 3\alpha = \cos 4\alpha$. Combining this with Equations (16.1) and (16.2), then substituting $x = \cos\alpha$ gives

$$4x^3 - 3x = 8x^4 - 8x^2 + 1.$$

So, x satisfies the equation $8x^4 - 4x^3 - 8x^2 + 3x + 1 = 0$. Factoring produces $(x - 1)(8x^3 + 4x^2 - 4x - 1)$, and since $x \neq 1$, x must be a root of $8x^3 + 4x^2 - 4x - 1 = 0$. By Theorem 16.15, if x is a surd then $8x^3 + 4x^2 - 4x - 1 = 0$ has a rational root. We leave it to the reader to check that none of the possible rational roots for this equation, namely $\{\pm 1, \pm\frac{1}{2}, \pm\frac{1}{4}, \pm\frac{1}{8}\}$, satisfies the equation. Therefore, x is not a surd, and hence, is not constructible.

Conclusion: A regular 7-gon is not constructible.

What if we change the rules by changing our tools?

Throughout this book, our construction tools have been the Euclidean tools, a compass and an unmarked straightedge. As evidenced in this chapter, these tools present a limit to the admissible constructions. The only way to enhance our ability to construct is to add a new tool. So, if we would like the ability to trisect any angle, let's say, then what tool do we need? The Greeks provided an answer to this question by adding a **marked straightedge** to their toolbox, where the straightedge has two marks to indicate a given distance. With it, we can, for example, slide the marked straightedge so that the marks lie on two given lines (or a line and a circle) while it passes through some other given point. The Greeks called a construction using this tool a **neusis construction**. Archimedes is credited with the neusis trisection that we give in Exercise 16.3.7. In Exercise 16.3.8, we share a different neusis trisection attributed to Nicomedes (ca. 240 BCE). We can also use this tool to double any cube, and we give such a construction in Exercise 16.3.10.

In addition to trisecting angles and doubling cubes, this new tool allows us to construct more, but not all, regular polygons. In Chapter 11, we noted that the following twelve regular n-gons, where $n \leq 25$, are constructible with the Euclidean tools: $3, 4, 5, 6, 8, 10, 12, 15, 16, 17, 20, 24$. Given our new ability to trisect any angle, it is clear that we can now construct the regular 9- and 18-gons. Furthermore, since we can construct cube roots, we can also construct the regular 7- and 14-gons. In fact, using a marked straightedge, we can construct the regular $7, 9, 11, 13, 14, 18, 19, 21$ and 22-gons. The 23-gon is the first regular polygon that is known to be impossible to construct with these tools.[a] Currently, it is unknown whether there is a neusis construction for the 25-gon (the next regular n-gon in our list) [**8**].

The regular heptagon is the polygon of lowest order that is not constructible with Euclidean tools since it is impossible to produce the required angle, $\pi/7$. So, how does a marked straightedge allow us to construct this angle? The following original neusis construction was published in 1975 by Crockett Johnson, an author and illustrator of comic strips and children's books. Starting with segment BC, construct a square on BC with diagonal CD. Construct the circle with center C and radius CD. Construct the perpendicular bisector to BC. Using a marked straightedge, construct a point A on the perpendicular bisector so that if AB intersects the circle at a point, F, then $AF = BC$. Claim: $\angle BAC = \dfrac{\pi}{7}$. For Johnson's trigonometric proof, see [**75**].

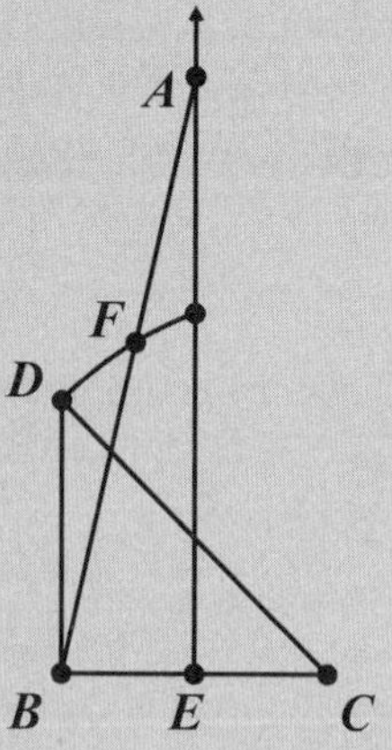

Johnson's motivation for constructing the heptagon was purely artistic, but like the Renaissance painters in search of perspective techniques, his quest for a solution led him to some interesting mathematics. In explanation, Johnson ends his article by

> noting, "The writer, a painter and not a mathematician, made use of the fascinatingly co-operative internal geometry of the polygon in constructions for a series of abstract paintings, the drawing for one of which led to the discovery of the $\sqrt{2}$ line and the neusis construction." One of his abstract paintings based on the heptagon is shown in Figure 16.8. To understand the relationship between the large triangle in the painting and the regular heptagon, see Exercise 7.2.5 and its related diagram given in Figure 7.12.
>
> ---
> [a]The neusis construction of the regular 29-gon is also impossible.

Figure 16.8. Crockett Johnson's *Heptagon from Its Seven Sides* (1973)

Exercises 16.3

1. Prove Lemma 16.17, that is, an angle α is constructible if and only if the length $\cos\alpha$ is constructible.

2. Prove the trigonometric identity in Equation (16.1):
$$\cos(3\alpha) = 4\cos^3\alpha - 3\cos\alpha.$$

3. Prove that $8x^3 - 6x - 1 = 0$ has no rational roots.

4. Prove the trigonometric identity in Equation (16.2):
$$\cos 4\alpha = 8\cos^4\alpha - 8\cos^2\alpha + 1.$$

5. Prove that $8x^3 + 4x^2 - 4x - 1 = 0$ has no rational roots.

6. Is it possible to construct a $1°$ angle?

7. Archimedes is credited with the following trisection of a given acute angle $\angle ABC$ using a neusis construction, that is, a construction produced with a compass and *marked* straightedge.

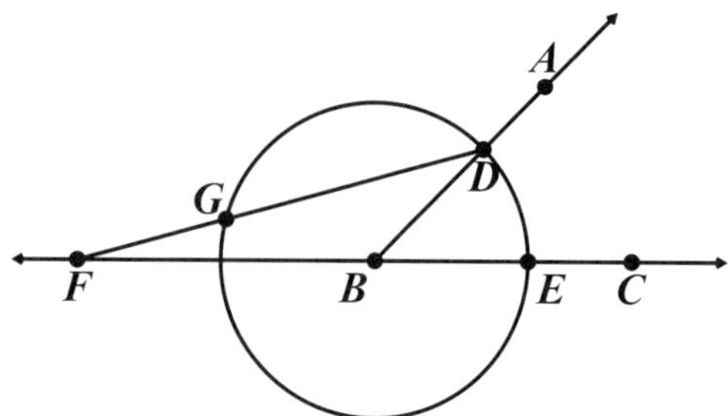

Figure 16.9. Exercise 16.3.7: Archimedes' trisection of $\angle ABC$ using *marked* straightedge

(a) Draw a circle with center B of any radius and let D and E be its intersections with rays $\overrightarrow{BA}$ and $\overrightarrow{BC}$, respectively.

(b) Using a marked ruler, find a point F that lies outside of the circle on the ray $\overrightarrow{CB}$ such that the segment DF intersects the circle at a point G where $FG = BD$, as illustrated in Figure 16.9.

Prove that in Archimedes' trisection, $\angle DFC = \frac{1}{3}\angle ABC$. [*Hint: Join GB.*]

8. Nicomedes (ca. 240 BCE) is credited with the following trisection of a given acute angle $\angle ABC$ using a neusis construction, that is, a construction produced with a compass and *marked* straightedge.

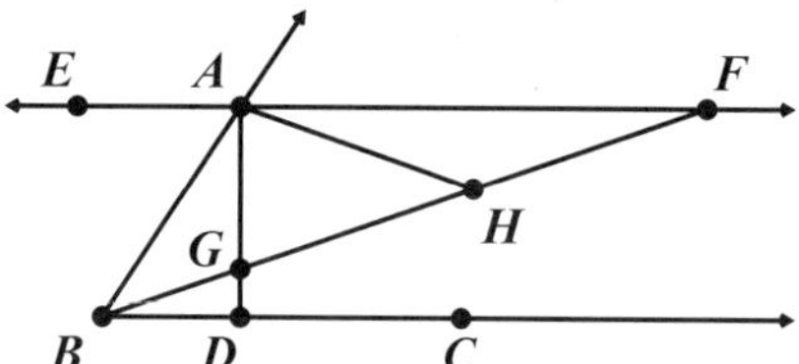

Figure 16.10. Exercise 16.3.8: Nicomedes' trisection of $\angle ABC$ using marked straightedge

(a) Construct AD perpendicular to ray $\overrightarrow{BC}$ so that D lies on $\overrightarrow{BC}$.

(b) Construct $\overleftrightarrow{AE}$ parallel to $\overleftrightarrow{BC}$.

(c) Using a marked straightedge, construct F on $\overleftrightarrow{AE}$ so that F lies inside of angle $\angle ABC$, and if FB intersects AD at G, then $GF = 2 \cdot AB$.

Prove that in Nicomedes' construction, $\angle FBC = \frac{1}{3}\angle ABC$. [*Hint: Let J be the midpoint of AF. After joining HJ, show that AH = FH.*]

9. Prove that if any acute angle can be trisected, then any obtuse angle can be trisected.

10. Consider the following neusis construction for doubling a cube. Let AB have length a. Using Euclid I.11, construct perpendicular CB to AB at B. Construct D on the opposite side of BC as A so that angle $\angle CBD = 30°$, as illustrated in Figure 16.11. Using a marked straightedge, construct E on $\overrightarrow{BD}$ such that, if AE intersects $\overrightarrow{BC}$ at F,

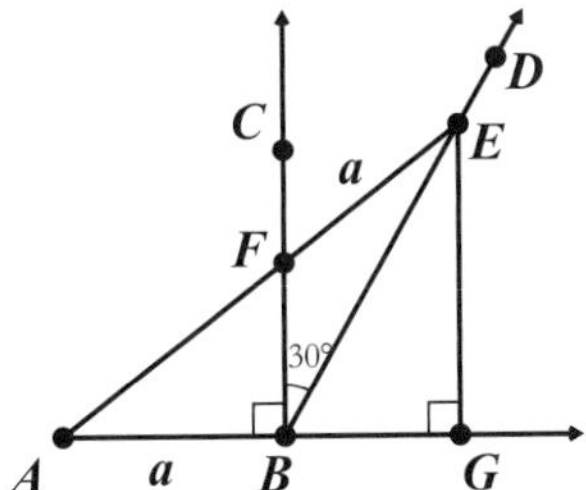

Figure 16.11. Exercise 16.3.10: Constructing $a\sqrt[3]{2}$ using marked straightedge

then $FE = AB$. Using the Pythagorean Theorem and the similarity of triangles $\triangle ABF$ and $\triangle AGE$, prove that $(AF)^3 = 2a^3$.[5]

16.4 The limits of geometry

As is the case at the end of every book, it has come time to bid farewell to our two main characters. Before we do, let's review our travels with them. Since much of the history of mathematics *is* the story of geometry, our quest to understand these characters has taken us from ancient Greece to the twenty-first century. As we laid our assumptions bare, we journeyed far from the comfortable world of the Euclidean plane, exploring the new and wonderful worlds of Spherical, Taxicab, Hyperbolic, and finally, Affine and Projective geometries. Immersing ourselves in other geometric worlds is akin to experiencing life in another country. The initial culture shock gives way to an enriching experience, transforming us into 'geometrically global citizens' with an understanding for what constitutes the essence of a geometry.

We started our journey with Euclid and found that the two-thousand year quest to prove his fifth postulate from the other four had an unattainable goal. Nineteenth-century mathematicians realized that accepting this postulate leads to one type of geometry, Euclidean, while not accepting it leads to an equally valid, but different geometry, Hyperbolic. Euclid was thus vindicated in his choice to make the Parallel Postulate an axiom, though we also learned that his set of axioms was not without its flaws. Taxicab and Spherical geometries revealed gaps in Euclid's reasoning, thus identifying essential axioms absent from his list. While Hilbert provided a solid foundation by filling these gaps, Gödel made us painfully aware of the limits inherent in any set of axioms for Euclidean geometry, or for that matter, any logical system.

We discovered other limits of geometry in this chapter, specifically, limitations of the Euclidean tools. Euclid's first three postulates provide our elementary construction tools, the compass and unmarked straightedge, which in turn, give rise to our two main characters, the line and the circle. While it is amazing to consider the wealth of constructions made possible by these humble devices, it is perhaps surprising that, in order to understand the limitations of these tools, we must journey beyond the realm of geometry. The classic construction problems of antiquity described in this chapter

---

[5]This construction is given in [**67**], p. 270.

cannot be solved with the techniques and tools of Euclidean geometry. It was only after the development of the field of abstract algebra in the early nineteenth century that these impossibility proofs *could* be given.

It bears noting that the resolution to these construction problems is of a completely different nature than that of the Parallel Postulate. In the case of the classic constructions, it is impossible to solve these problems with unmarked straightedge and compass alone. In the case of the Parallel Postulate, the statement can neither be proven nor disproven. While these can both be considered roadblocks in our journey, they are actually different types of obstructions. The former is a proof that something is impossible while the latter shows that something is impossible to prove.

Since the impossible constructions have taken us beyond the world of geometry, we have reached an appropriate end to our story. To be sure, there is certainly much more to learn about our main characters. So while this book has ended, we hope that from time to time you will check back in with the *line* and the *circle* to see what they have been doing. The field of geometry in the twentieth century offers many avenues of further exploration with these characters, sequels if you will, including differential geometry, fractal geometry, topology and algebraic geometry, to name a few.   □

Euclid's Definitions and Axioms

A.1 Definitions

(1) A *point* is that which has no part.

(2) A *line* is breadthless length.

(3) The extremities of a line are points.

(4) A *straight line* is a line which lies evenly with the points on itself.

(5) A *surface* is that which has length and breadth only.

(6) The extremities of a surface are lines.

(7) A *plane surface* is a surface which lies evenly with the straight lines on itself.

(8) A *plane angle* is the inclination to one another of two lines in a plane which meet one another and do not lie in a straight line.

(9) And when the lines containing the angle are straight, the angle is called *rectilineal*.

(10) When a straight line set up on a straight line makes the adjacent angles equal to one another, each of the equal angles is *right*, and the straight line standing on the other is called a *perpendicular* to that on which it stands.

(11) An *obtuse angle* is an angle greater than a right angle.

(12) An *acute angle* is an angle less than a right angle.

(13) A *boundary* is that which is an extremity of anything.

(14) A *figure* is that which is contained by any boundary or boundaries.

(15) A *circle* is a plane figure contained by one line such that all the straight lines falling upon it from one point among those lying within the figure are equal to one another;

(16) And the point is called the *centre* of the circle.

(17) A *diameter* of the circle is any straight line drawn through the centre and terminated in both directions by the circumference of the circle, and such a straight line also bisects the circle.

(18) A *semicircle* is the figure contained by the diameter and the circumference cut off by it. And the centre of the semicircle is the same as that of the circle.

(19) *Rectilineal figures* are those which are contained by straight lines, *trilateral* figures being those contained by three, *quadrilateral* those contained by four, and *multilateral* those contained by more than four straight lines.

(20) Of trilateral figures, an *equilateral triangle* is that which has its three sides equal, an *isosceles triangle* that which has two of its sides alone equal, and a *scalene triangle* that which has its three sides unequal.

(21) Further, of trilateral figures, a *right-angled triangle* is that which has a right angle, an *obtuse-angled triangle* that which has an obtuse angle, and an *acute-angled triangle* that which has its three angles acute.

(22) Of quadrilateral figures, a *square* is that which is both equilateral and right-angled; an *oblong* that which is right-angled but not equilateral; a *rhombus* that which is equilateral but not right-angled; and a *rhomboid* that which has its opposite sides and angles equal to one another but is neither equilateral nor right-angled. And let quadrilaterals other than these be called *trapezia*.

(23) *Parallel* straight lines are straight lines which, being in the same plane and being produced indefinitely in both directions, do not meet one another in either direction.

A.2 Postulates

(1) To draw a straight line from any point to any point.

(2) To produce a finite straight line continuously in a straight line.

(3) To describe a circle with any center and distance.

(4) That all right angles are equal to one another.

(5) That, if a straight line falling on two straight lines make the interior angles on the same side less than two right angles, the two straight lines, if produced indefinitely, meet on that side on which are the angles less than the two right angles.

A.3 Common notions

(1) Things which are equal to the same thing are also equal to one another.

(2) If equals be added to equals, the wholes are equal.

(3) If equals be subtracted from equals, the remainders are equal.

(4) Things which coincide with one another are equal to one another.

(5) The whole is greater than the part.

Euclid's Propositions

B.1 Book I

Proposition I.1. *On a given finite straight line to construct an equilateral triangle.*

Proposition I.2. *To place at a given point [as an extremity]* [1] *a straight line equal to a given straight line.*

Proposition I.3. *Given two unequal straight lines, to cut off from the greater a straight line equal to the less.*

Proposition I.4 [SAS]. *If two triangles have the two sides equal to two sides respectively, and have the angles contained by the equal straight lines equal, they will also have the base equal to the base, the triangle will be equal to the triangle, and the remaining angles will be equal to the remaining angles respectively, namely those which the equal sides subtend.*

Proposition I.5. *In isosceles triangles the angles at the base are equal to one another, and, if the equal straight lines be produced further, the angles under the base will be equal to one another.*

Proposition I.6. *If in a triangle two angles be equal to one another, the sides which subtend the equal angles will also be equal to one another.*

Proposition I.7. *Given two straight lines constructed on a straight line [from its extremities] and meeting in a point, there cannot be constructed on the same straight line [from its extremities], and on the same side of it, two other straight lines meeting in another point and equal to the former two respectively, namely each to that which has the same extremity with it.*

Proposition I.8 [SSS]. *If two triangles have the two sides equal to two sides respectively, and have also the base equal to the base, they will also have the angles equal which are contained by the equal straight lines.*

---

[1] The square brackets within the statement of a proposition indicate material added by Heath to clarify the Greek text.

Proposition I.9. *To bisect a given rectilineal angle.*

Proposition I.10. *To bisect a given finite straight line.*

Proposition I.11. *To draw a straight line at right angles to a given straight line from a given point on it.*

Proposition I.12. *To a given infinite straight line, from a given point which is not on it, to draw a perpendicular straight line.*

Proposition I.13. *If a straight line set up on a straight line make angles, it will make either two right angles or angles equal to two right angles.*

Proposition I.14. *If with any straight line, and at a point on it, two straight lines not lying on the same side make the adjacent angles equal to two right angles, the two straight lines will be in a straight line with one another.*

Proposition I.15 [Vertical Angle Theorem]. *If two straight lines cut one another, they make the vertical angles equal to one another.*

Proposition I.16 [Exterior Angle Theorem]. *In any triangle, if one of the sides be produced, the exterior angle is greater than either of the interior and opposite angles.*

Proposition I.17. *In any triangle two angles taken together in any manner are less than two right angles.*

Proposition I.18. *In any triangle the greater side subtends the greater angle.*

Proposition I.19. *In any triangle the greater angle is subtended by the greater side.*

Proposition I.20. *In any triangle two sides taken together in any manner are greater than the remaining one.*

Proposition I.21. *If on one of the sides of a triangle, from its extremities, there be constructed two straight lines meeting within the triangle, the straight lines so constructed will be less than the remaining two sides of the triangle, but will contain a greater angle.*

Proposition I.22. *Out of three straight lines, which are equal to three given straight lines, to construct a triangle: thus it is necessary that two of the straight lines taken together in any manner should be greater than the remaining one. [I.20]*

Proposition I.23. *On a given straight line and at a point on it to construct a rectilineal angle equal to a given rectilineal angle.*

Proposition I.24. *If two triangles have the two sides equal to two sides respectively, but have the one of the angles contained by the equal straight lines greater than the other, they will also have the base greater than the base.*

Proposition I.25. *If two triangles have the two sides equal to two sides respectively, but have the base greater than the base, they will also have the one of the angles contained by the equal straight lines greater than the other.*

Proposition I.26 [ASA], [AAS]. *If two triangles have the two angles equal to two angles respectively, and one side equal to one side, namely, either the side adjoining the equal angles, or that subtending one of the equal angles, they will also have the remaining sides equal to the remaining sides and the remaining angle to the remaining angle.*

Proposition I.27. *If a straight line falling on two straight lines make the alternate angles equal to one another, the straight lines will be parallel to one another.*

Proposition I.28. *If a straight line falling on two straight lines make the exterior angle equal to the interior and opposite angle on the same side, or the interior angles on the same side equal to two right angles, the straight lines will be parallel to one another.*

Proposition I.29. *A straight line falling on parallel straight lines makes the alternate angles equal to one another, the exterior angle equal to the interior and opposite angle, and the interior angles on the same side equal to two right angles.*

Proposition I.30. *Straight lines parallel to the same straight line are also parallel to one another.*

Proposition I.31. *Through a given point to draw a straight line parallel to a given straight line.*

Proposition I.32. *In any triangle, if one of the sides be produced, the exterior angle is equal to the two interior and opposite angles, and the three interior angles of the triangle are equal to two right angles.*

Proposition I.33. *The straight lines joining equal and parallel straight lines (at the extremities which are) in the same directions (respectively) are themselves also equal and parallel.*

Proposition I.34. *In parallelogrammic areas the opposite sides and angles are equal to one another, and the diameter bisects the areas.*

Proposition I.35. *Parallelograms which are on the same base and in the same parallels are equal to one another.*

Proposition I.36. *Parallelograms which are on equal bases and in the same parallels are equal to one another.*

Proposition I.37. *Triangles which are on the same base and in the same parallels are equal to one another.*

Proposition I.38. *Triangles which are on equal bases and in the same parallels are equal to one another.*

Proposition I.39. *Equal triangles which are on the same base and on the same side are also in the same parallels.*

Proposition I.40. *Equal triangles which are on equal bases and on the same side are also in the same parallels.*

Proposition I.41. *If a parallelogram have the same base with a triangle and be in the same parallels, the parallelogram is double of the triangle.*

Proposition I.42. *To construct, in a given rectilineal angle, a parallelogram equal to a given triangle.*

Proposition I.43. *In any parallelogram the complements of the parallelograms about the diameter are equal to one another.*

Proposition I.44. *To a given straight line to apply, in a given rectilineal angle, a parallelogram equal to a given triangle.*

Proposition I.45. *To construct, in a given rectilineal angle, a parallelogram equal to a given rectilineal figure.*

Proposition I.46. *On a given straight line to describe a square.*

Proposition I.47 [The Pythagorean Theorem]. *In right-angled triangles the square on the side subtending the right angle is equal to the squares on the sides containing the right angle.*

Proposition I.48 [The Pythagorean Theorem's Converse]. *If in a triangle the square on one of the sides be equal to the squares on the remaining two sides of the triangle, the angle contained by the remaining two sides of the triangle is right.*

B.2 Book II

Proposition II.1. *If there be two straight lines, and one of them be cut into any number of segments whatever, the rectangle contained by the two straight lines is equal to the rectangles contained by the uncut straight line and each of the segments.*

Proposition II.2. *If a straight line be cut at random, the rectangles contained by the whole and both of the segments are equal to the square on the whole.*

Proposition II.3. *If a straight line be cut at random, the rectangle contained by the whole and one of the segments is equal to the rectangle contained by the segments and the square on the aforesaid segment.*

Proposition II.4. *If a straight line be cut at random, the square on the whole is equal to the squares on the segments and twice the rectangle contained by the segments.*

Proposition II.5. *If a straight line be cut into equal and unequal segments, the rectangle contained by the unequal segments of the whole together with the square on the straight line between the points of section is equal to the square on the half.*

Proposition II.6. *If a straight line be bisected and a straight line be added to it in a straight line, the rectangle contained by the whole with the added straight line and the added straight line together with the square on the half is equal to the square on the straight line made up of the half and the added straight line.*

Proposition II.7. *If a straight line be cut at random, the square on the whole and that on one of the segments both together are equal to twice the rectangle contained by the whole and the said segment and the square on the remaining segment.*

Proposition II.8. *If a straight line be cut at random, four times the rectangle contained by the whole and one of the segments together with the square on the remaining segment is equal to the square described on the whole and the aforesaid segment as on one straight line.*

Proposition II.9. *If a straight line be cut into equal and unequal segments, the squares on the unequal segments of the whole are double of the square on the half and of the square on the straight line between the points of section.*

Proposition II.10. *If a straight line be bisected, and a straight line be added to it in a straight line, the square on the whole with the added straight line and the square on the* *added straight line both together are double of the square on the half and of the square described on the straight line made up of the half and the added straight line as on one straight line.*

Proposition II.11. *To cut a given straight line so that the rectangle contained by the whole and one of the segments is equal to the square on the remaining segment.*

Proposition II.12 [Law of Cosines]. *In obtuse-angled triangles the square on the side subtending the obtuse angle is greater than the squares on the sides containing the obtuse angle by twice the rectangle contained by one of the sides about the obtuse angle, namely that on which the perpendicular falls, and the straight line cut off outside by the perpendicular towards the obtuse angle.*

Proposition II.13 [Law of Cosines]. *In acute-angled triangles the square on the side subtending the acute angle is less than the squares on the sides containing the acute angle by twice the rectangle contained by one of the sides about the acute angle, namely that on which the perpendicular falls, and the straight line cut off within by the perpendicular towards the acute angle.*

Proposition II.14. *To construct a square equal to a given rectilineal figure.*

B.3 Book III

Proposition III.1. *To find the centre of a given circle.*

Proposition III.3. *If in a circle a straight line through the centre bisect a straight line not through the centre, it also cuts it at right angles; and if it cut it at right angles, it also bisects it.*

Proposition III.16. *The straight line drawn at right angles to the diameter of a circle from its extremity will fall outside the circle, and into the space between the straight line and the circumference another straight line cannot be interposed.*

Proposition III.17. *From a given point to draw a straight line touching a given circle.*

Proposition III.18. *If a straight line touch a circle, and a straight line be joined from the centre to the point of contact, the straight line so joined will be perpendicular to the tangent.*

Proposition III.19. *If a straight line touch a circle, and from the point of contact a straight line be drawn at right angles to the tangent, the centre of the circle will be on the straight line so drawn.*

Proposition III.20 [Star Trek Theorem]. *In a circle the angle at the centre is double of the angle at the circumference, when the angles have the same circumference as base.*

Proposition III.21. *In a circle the angles in the same segment are equal to one another.*

Proposition III.22. *The opposite angles of quadrilaterals in circles are equal to two right angles.*

Proposition III.26. *In equal circles equal angles stand on equal circumferences, whether they stand at the centres or at the circumferences.*

Proposition III.27. *In equal circles angles standing on equal circumferences are equal to one another, whether they stand at the centres or at the circumferences.*

Proposition III.28. *In equal circles equal straight lines cut off equal circumferences, the greater equal to the greater and the less to the less.*

Proposition III.29. *In equal circles equal circumferences are subtended by equal straight lines.*

Proposition III.30. *To bisect a given circumference.*

Proposition III.31 [Thales' Theorem]. *In a circle the angle in the semicircle is right, that in a greater segment less than a right angle, and that in a less segment greater than a right angle.*

Proposition III.32. *Given a circle with chord BD and a tangent line EF which intersects the circle at B, if A is any point not on $\overset{\frown}{BD}$, lying on the opposite side of $\overleftrightarrow{BD}$ as F, then the inscribed angle $\angle BAD$ is equal to the angle $\angle DBF$.*

Proposition III.35 [Intersecting Chords Theorem]. *If in a circle two straight lines cut one another, the rectangle contained by the segments of the one is equal to the rectangle contained by the segments of the other.*

Proposition III.36. *Given a point D that lies outside of a given circle, if BD is tangent to the circle at B and ACD is a line that intersects the circle at two points, producing the chord AC, then $BD^2 = AD \cdot CD$.*

Proposition III.37. *Given a point D that lies outside of a given circle, if BD intersects the circle at B and ACD is a line that intersects the circle at two points, producing the chord AC, then if $BD^2 = AD \cdot CD$, BD is tangent to the circle.*

B.4 Book IV

Proposition IV.2. *In a given circle to inscribe a triangle equiangular with a given triangle.*

Proposition IV.4. *In a given triangle to inscribe a circle.*

Proposition IV.5. *About a given triangle to circumscribe a circle.*

Proposition IV.6. *In a given circle to inscribe a square.*

Proposition IV.7. *About a given circle to circumscribe a square.*

Proposition IV.8. *In a given square to inscribe a circle.*

Proposition IV.9. *About a given square to circumscribe a circle.*

Proposition IV.10. *To construct an isosceles triangle having each of the angles at the base double of the remaining one.*

Proposition IV.11. *In a given circle to inscribe an equilateral and equiangular pentagon.*

Proposition IV.12. *About a given circle to circumscribe an equilateral and equiangular pentagon*

Proposition IV.13. *In a given pentagon, which is equilateral and equiangular, to inscribe a circle.*

Proposition IV.14. *About a given pentagon, which is equilateral and equiangular, to circumscribe a circle.*

Proposition IV.15. *In a given circle to inscribe an equilateral and equiangular hexagon.*

Proposition IV.16. *In a given circle to inscribe a fifteen-angled figure which shall be both equilateral and equiangular.*

B.5 Book VI

Proposition VI.2. *If a straight line be drawn parallel to one of the sides of a triangle, it will cut the sides of the triangle proportionally; and, if the sides of the triangle be cut proportionally, the line joining the points of section will be parallel to the remaining side of the triangle.*

Proposition VI.4 [$\widehat{\text{AAA}}$]. *Equiangular triangles are similar.*

Proposition VI.5 [$\widehat{\text{SSS}}$]. *If two triangles have their sides proportional, the triangles will be equiangular and will have those angles equal which the corresponding sides subtend.*

Proposition VI.6 [$\widehat{\text{SAS}}$]. *If two triangles have one angle equal to one angle and the sides about the equal angles proportional, the triangles will be equiangular and will have those angles equal which the corresponding sides subtend.*

Proposition VI.8. *If in a right-angled triangle a perpendicular be drawn from the right angle to the base, the triangles adjoining the perpendicular are similar both to the whole and to one another.*

Proposition VI.19. *Similar triangles are to one another in the duplicate ratio of the corresponding sides.*

Proposition VI.20. *Similar polygons are divided into similar triangles, and into triangles equal in multitude and in the same ratio as the wholes, and the polygon has to the polygon a ratio duplicate of that which the corresponding side has to the corresponding side.*

Proposition VI.31 [Generalized Pythagorean Theorem]. *In right-angled triangles the figure on the side subtending the right angle is equal to the similar and similarly described figures on the sides containing the right angle.*

Proposition VI.33. *In equal circles angles have the same ratio as the circumferences on which they stand, whether they stand at the centres or at the circumferences.*

Appendix C

Visual Guide to Euclid's Propositions

C.1 Book I

I.1 Given 

construct 

I.2 Given 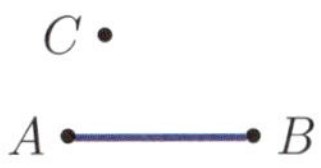

construct $CD = AB$ 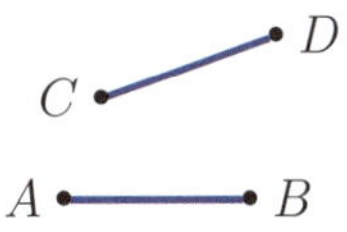

I.3 Given 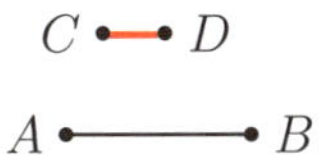

construct E such that $AE = CD$ 

I.4 SAS 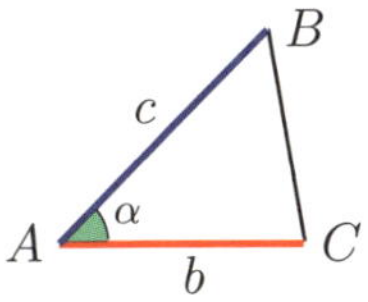

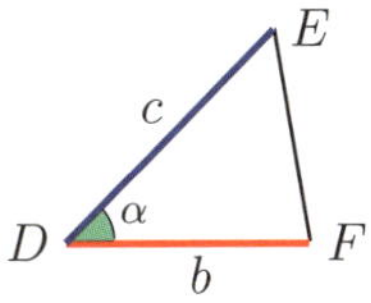

$\Rightarrow \triangle ABC \cong \triangle DEF$

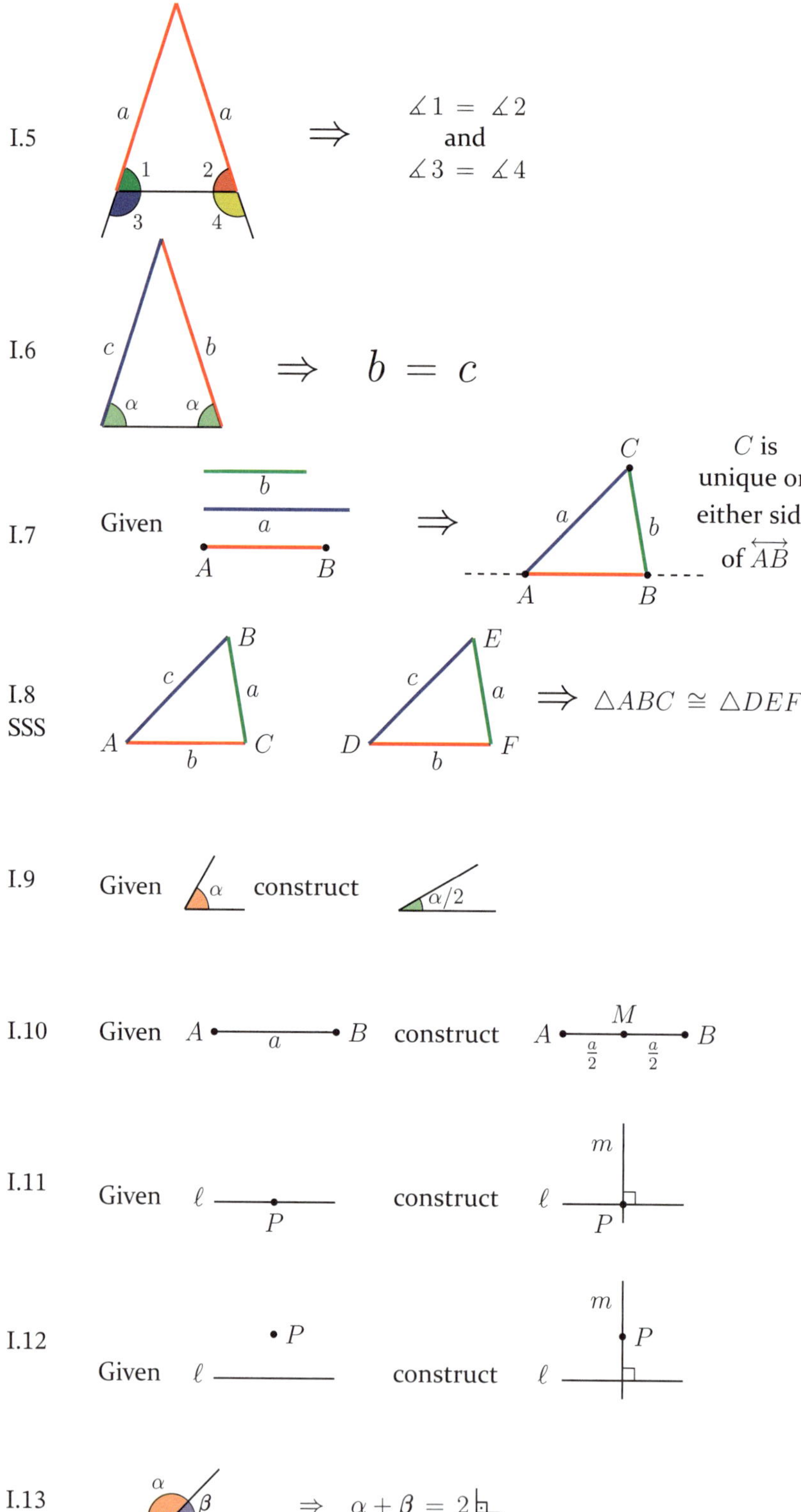
I.5
a   a
1   2
3   4
∠1 = ∠2
and
∠3 = ∠4
I.6
c   b
α   α
⇒ b = c
I.7
Given
b
a
A   B
⇒
C
a   b
A   B
C is unique on either side of AB
I.8
SSS
B
c   a
A   C
b
E
c   a
D   F
b
⇒ △ABC ≅ △DEF
I.9
Given α construct α/2
I.10
Given A a B construct A M B
a/2 a/2
I.11
Given ℓ P construct m ℓ P
I.12
P
Given ℓ construct m P ℓ
I.13
α β ⇒ α + β = 2⌐

I.14 $\qquad$ $\alpha + \beta = 2\,\llcorner\quad \Rightarrow\quad \overleftrightarrow{ABC}$

I.15 $\qquad$ $\Rightarrow\quad \alpha = \beta$

I.16 $\qquad$ $\Rightarrow\quad \begin{aligned}\gamma &> \alpha\\ \gamma &> \beta\end{aligned}$

I.17 $\qquad$ $\Rightarrow\quad \begin{aligned}\alpha + \beta &< 2\,\llcorner\\ \alpha + \gamma &< 2\,\llcorner\\ \beta + \gamma &< 2\,\llcorner\end{aligned}$

I.18 $\qquad$ $b > a\quad \Rightarrow\quad \beta > \alpha$

I.19 $\qquad$ $\beta > \alpha\quad \Rightarrow\quad b > a$

I.20 $\qquad$ $\Rightarrow\quad \begin{aligned}a + b &> c\\ a + c &> b\\ b + c &> a\end{aligned}$

I.21 $\qquad$ $\Rightarrow\quad \begin{aligned}b + c &> r + s\\ \delta &> \alpha\end{aligned}$

I.22 Given where $\begin{aligned}a + b &> c\\ a + c &> b\\ b + c &> a\end{aligned}$ construct

I.23 Given construct

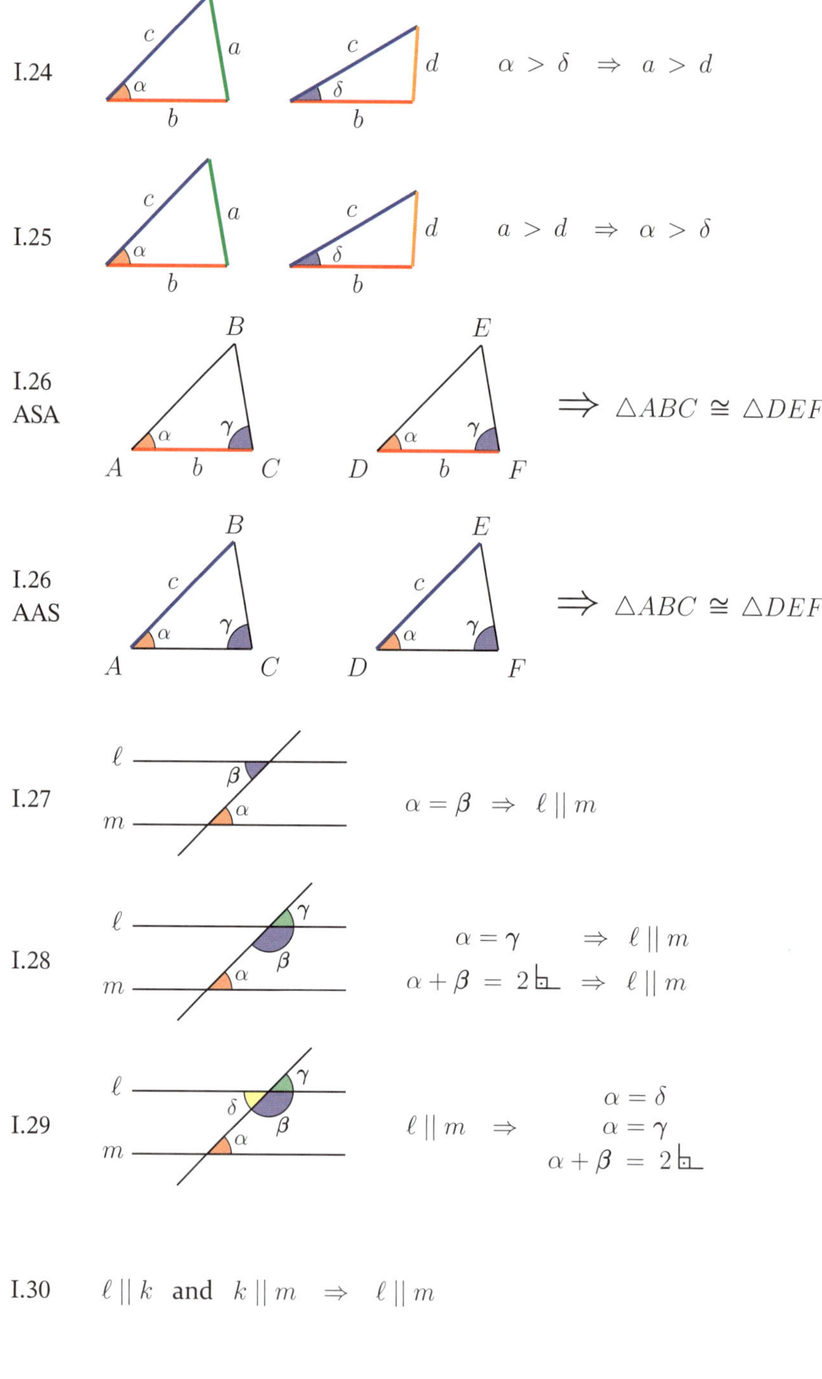

I.30 $\ell \parallel k$ and $k \parallel m$ ⇒ $\ell \parallel m$

I.31   Given   construct $m \parallel \ell$ through A

I.32 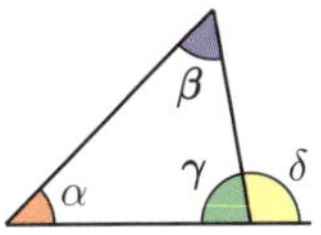 $\Rightarrow$ $\alpha + \beta + \gamma = 2\,\llcorner$
$\alpha + \beta = \delta$

I.33 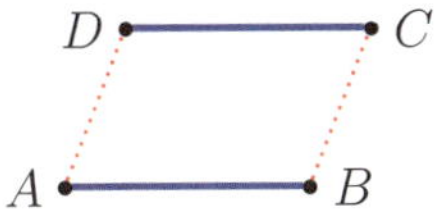 $AB \parallel CD$
$AB = CD$ $\Rightarrow$ $AD \parallel BC$
$AD = BC$

I.34 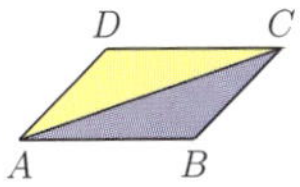 $AB \parallel CD$
$BC \parallel AD$ $\Rightarrow$ $AB = CD$
$BC = AD$
$\angle A \cong \angle C$
$\angle B \cong \angle D$ and $\quad = \quad$

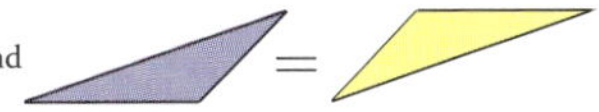

I.35 $\Rightarrow$ $\quad = \quad$

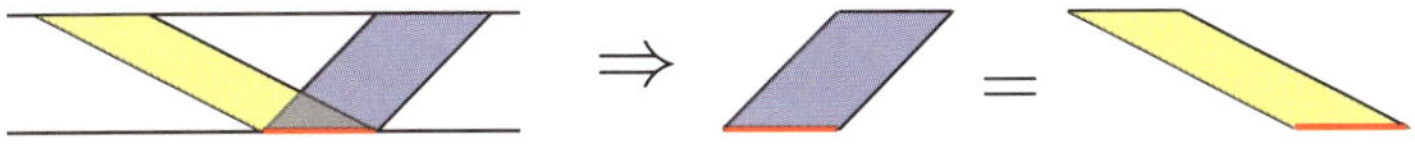

I.36 $\Rightarrow$ $\quad = \quad$

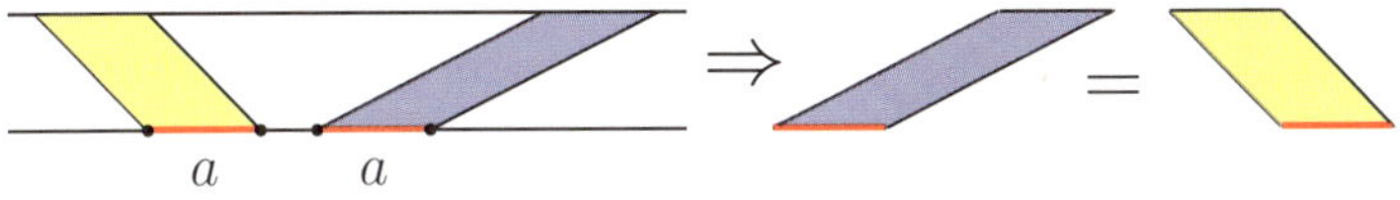

I.37 $\Rightarrow$ $\quad = \quad$

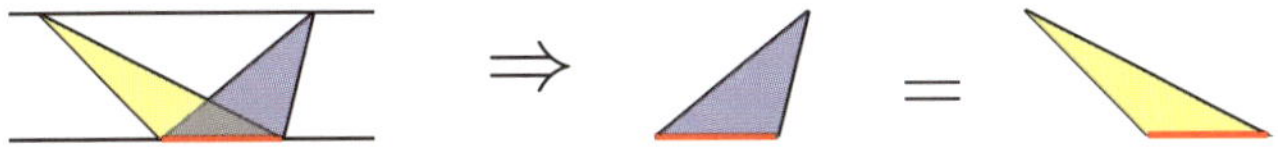

I.38 $\Rightarrow$ $\quad = \quad$

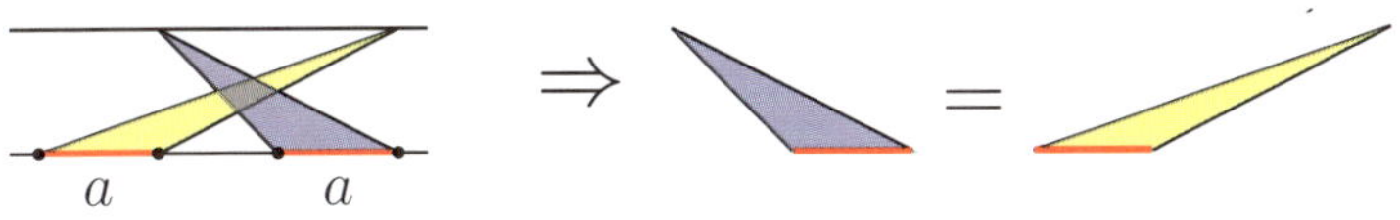

I.39 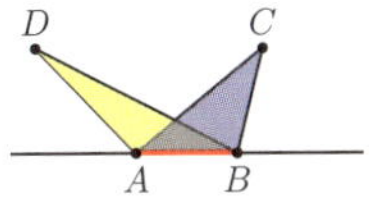 and 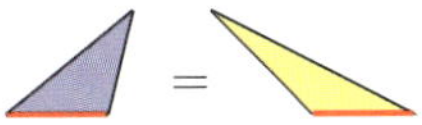 $\quad = \quad$ $\Rightarrow$ $AB \parallel CD$

I.40 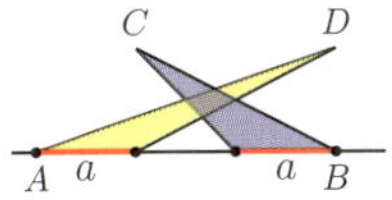 and 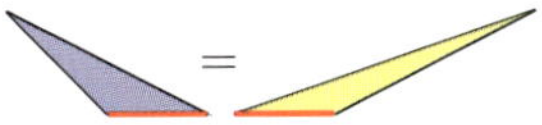 $\quad = \quad$ $\Rightarrow$ $AB \parallel CD$

I.41 $\Rightarrow$ $\quad = 2$

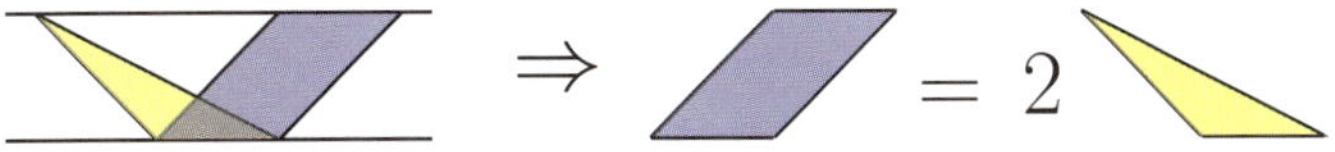

I.42 Given 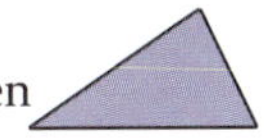 and α construct $\alpha \quad = $

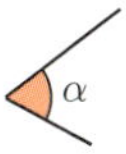

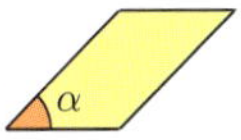

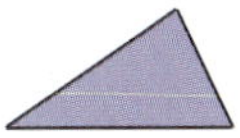

I.43 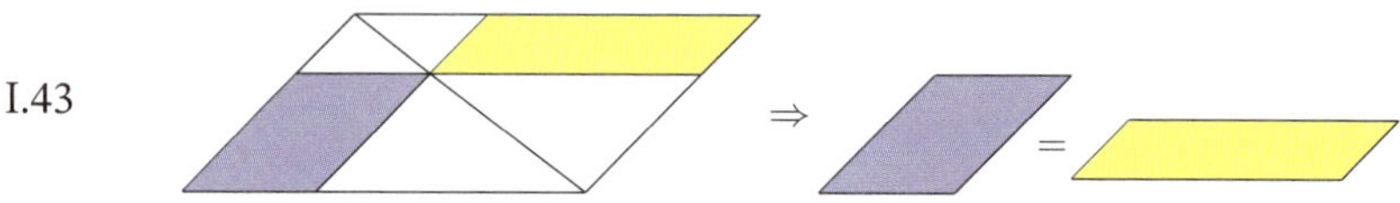

I.44 Given ... and ... construct 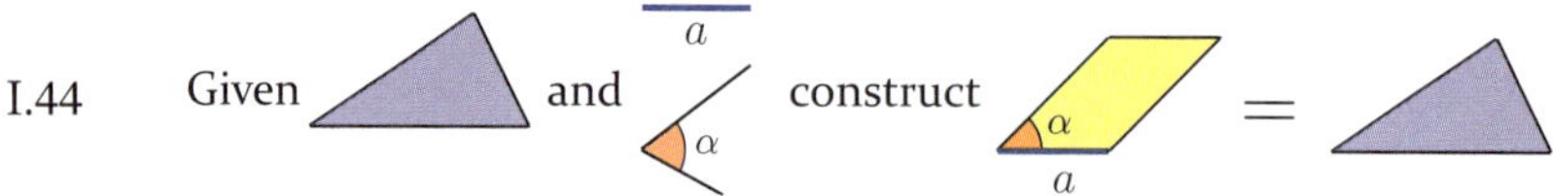

I.45 Given ... and ... construct 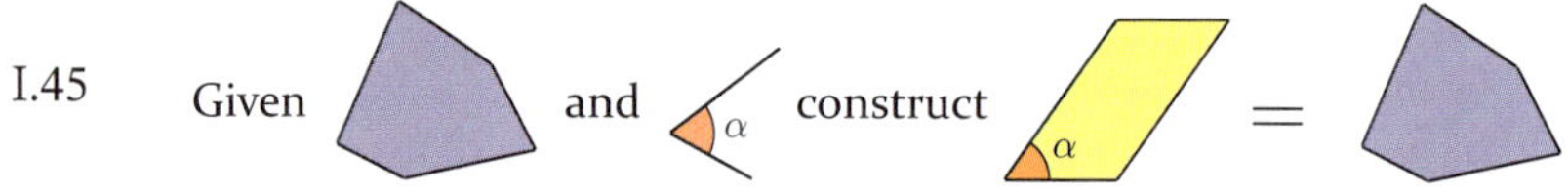

I.46 Given ... construct 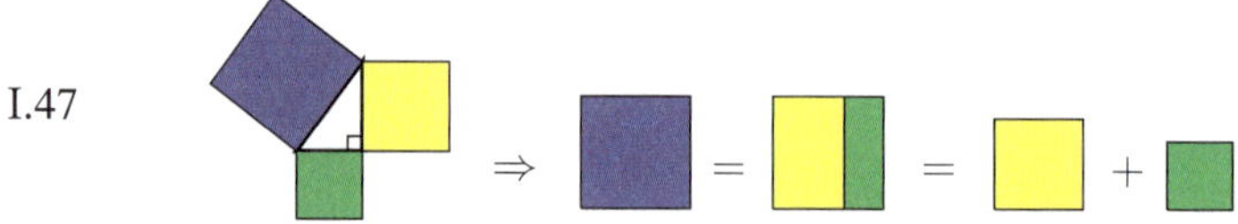

I.47 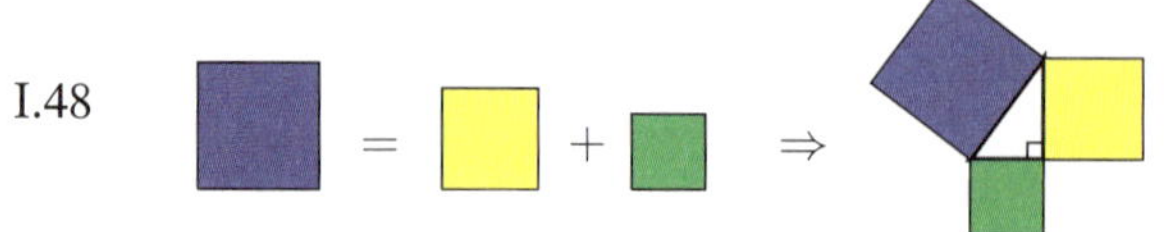

I.48

C.2 Book II

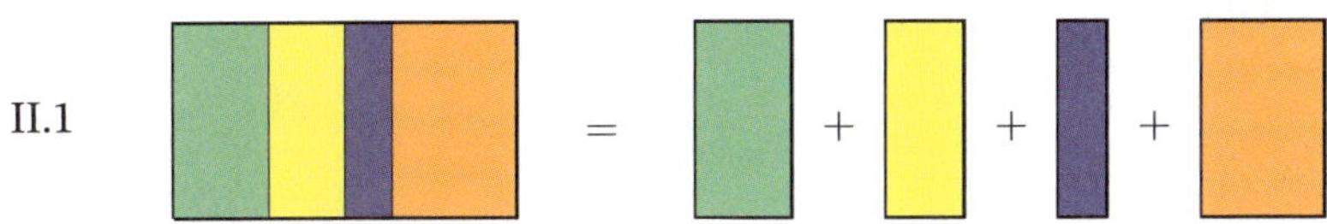

II.1

II.2

II.3

II.4

II.11   Given   construct   such that   =

C.3 Book III

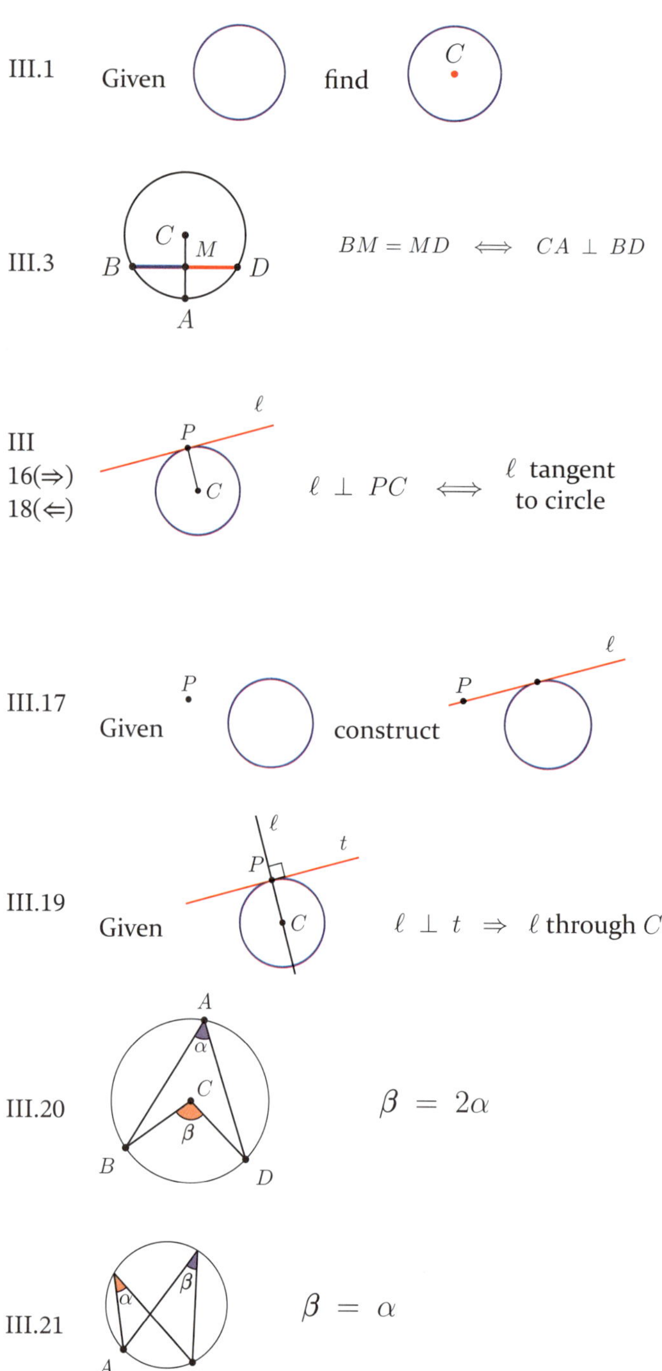

III.1　Given　find

III.3　$BM = MD \iff CA \perp BD$

III 16($\Rightarrow$) 18($\Leftarrow$)　$\ell \perp PC \iff \ell$ tangent to circle

III.17　Given　construct

III.19　Given　$\ell \perp t \Rightarrow \ell$ through C

III.20　$\beta = 2\alpha$

III.21　$\beta = \alpha$

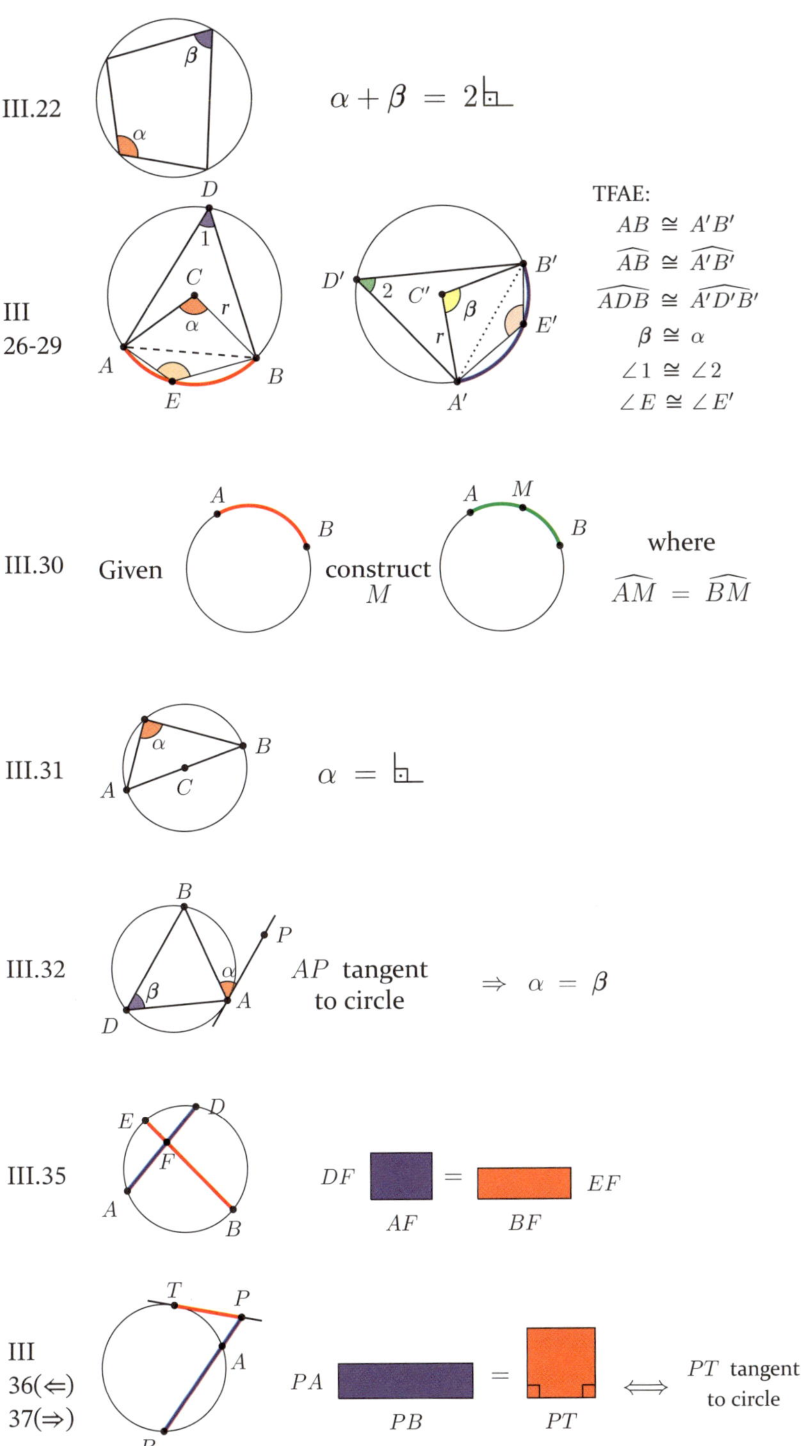

III.22
$\alpha + \beta = 2\,\llcorner$

III 26-29
TFAE:
$AB \cong A'B'$
$\widehat{AB} \cong \widehat{A'B'}$
$\widehat{ADB} \cong \widehat{A'D'B'}$
$\beta \cong \alpha$
$\angle 1 \cong \angle 2$
$\angle E \cong \angle E'$

III.30   Given   construct M   where $\widehat{AM} = \widehat{BM}$

III.31 $\alpha = \llcorner$

III.32 AP tangent to circle $\Rightarrow \alpha = \beta$

III.35 $DF \cdot AF = EF \cdot BF$

III 36($\Leftarrow$) 37($\Rightarrow$) $PA \cdot PB = PT \Longleftrightarrow PT$ tangent to circle

C.4 Book IV

IV.2: Inscribe similar triangle in given circle
IV.5: Circumscribe circle about given triangle

IV.6: Inscribe square in given circle
IV.9: Circumscribe circle about given square

IV.11: Inscribe regular pentagon in given circle
IV.14: Circumscribe circle about given regular pentagon

IV.15: Inscribe regular hexagon in given circle
IV.15 Cor: Circumscribe circle about given regular hexagon

IV.16: Inscribe regular 15-gon in given circle
IV.16 Cor: Circumscribe circle about given regular 15-gon

IV.3: Circumscribe similar triangle about given circle
IV.4: Inscribe circle in given triangle

IV.7: Circumscribe square about given circle
IV.8: Inscribe circle in a given square

IV.12: Circumscribe regular pentagon about given circle
IV.13: Inscribe circle in given regular pentagon

IV.15 Cor: Circumscribe regular hexagon about given circle
IV.15 Cor: Inscribe circle in given regular hexagon

IV.16 Cor: Circumscribe regular 15-gon about given circle
IV.16 Cor: Inscribe circle in given regular 15-gon

Appendix **D**

Euclid's Proofs

D.1 Book I

The following proofs are not included within the chapters.

Proposition I.38. *Triangles which are on equal bases and in the same parallels are equal to one another.*

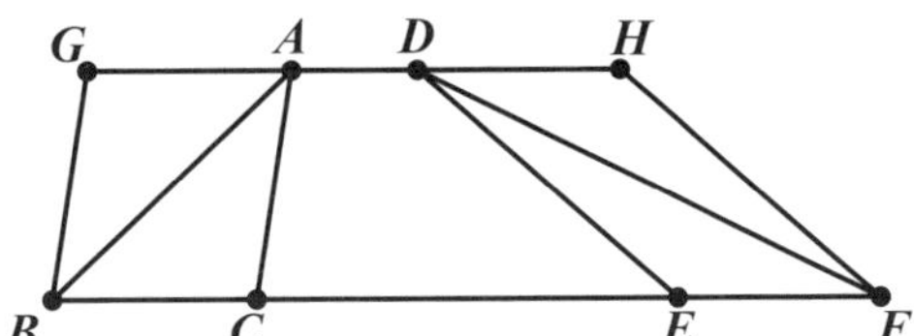

Figure D.1. Proposition I.38

Proof. Let ABC, DEF be triangles on equal bases BC, EF and in the same parallels BF, AD; I say that the triangle ABC is equal to the triangle DEF.

For let AD be produced in both directions to G, H; through B let BG be drawn parallel to CA, [I.31] and through F let FH be drawn parallel to DE.

Then each of the figures $GBCA, DEFH$ is a parallelogram; and $GBCA$ is equal to $DEFH$;

for they are on equal bases BC, EF and in the same parallels BF, GH. [I.36]

Moreover the triangle ABC is half of the parallelogram $GBCA$; for the diameter AB bisects it. [I.34]

And the triangle FED is half of the parallelogram $DEFH$; for the diameter DF bisects it. [I.34]

[But the halves of equal things are equal to one another.]

Therefore the triangle ABC is equal to the triangle DEF.

Therefore etc. Q.E.D. $\square$

Proposition I.40. *Equal triangles which are on equal bases and on the same side are also in the same parallels.*

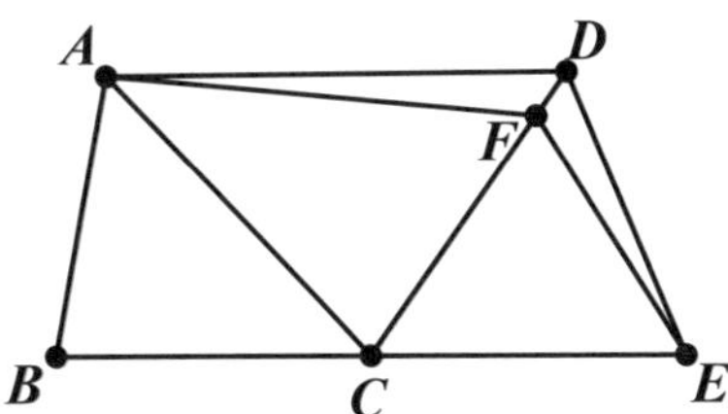

Figure D.2. Proposition I.40

Proof.[1]   Let ABC, CDE be equal triangles on equal bases BC, CE and on the same side.
I say that they are also in the same parallels.
For let AD be joined; I say that AD is parallel to BE.
For, if not, let AF be drawn through A parallel to BE [I.31], and let FE be joined.
Therefore the triangle ABC is equal to the triangle FCE; for they are on equal bases BC, CE and in the same parallels BE, AF. [I.38]
But the triangle ABC is equal to the triangle DCE; therefore the triangle DCE is also equal to the triangle FCE, [C.N. 1] the greater to the less: which is impossible. Therefore AF is not parallel to BE.
Similarly we can prove that neither is any other straight line except AD; therefore AD is parallel to BE.
Therefore etc. Q.E.D.   □

Proposition I.41. *If a parallelogram have the same base with a triangle and be in the same parallels, the parallelogram is double of the triangle.*

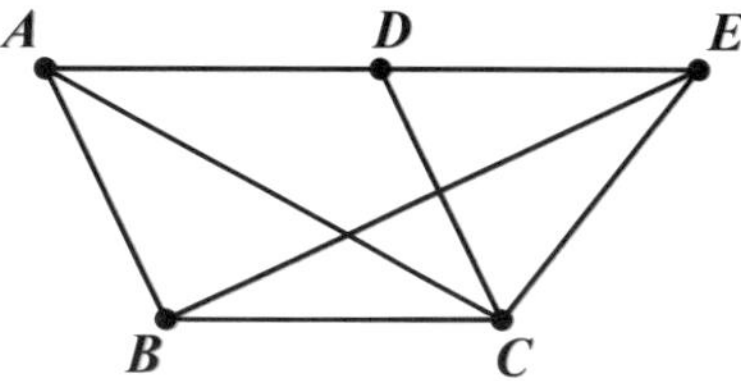

Figure D.3. Proposition I.41

Proof.   For let the parallelogram $ABCD$ have the same base BC with the triangle EBC, and let it be in the same parallels BC, AE;
I say that the parallelogram $ABCD$ is double of the triangle BEC.
For let AC be joined.

---

[1]Point F must be the intersection of the ray $\overrightarrow{CD}$ with the line through A that is parallel to BE. Also, while Euclid's notation and the figure imply that the triangles share a common point C, the proof does not require this to be the case.

Then the triangle *ABC* is equal to the triangle *EBC*; for it is on the same base *BC* with it and in the same parallels *BC*, *AE*. [I.37]

But the parallelogram *ABCD* is double of the triangle *ABC*; for the diameter *AC* bisects it; [I.34] so that the parallelogram *ABCD* is also double of the triangle *EBC*.

Therefore etc. Q.E.D. □

Hilbert's Axioms for Plane Euclidean Geometry

Hilbert's axioms are reprinted by permission of Open Court Publishing Company, a division of Carus Publishing Company, Chicago, IL, from *Foundations of Geometry* by D. Hilbert (trans. L. Unger), ©1971 by Open Court Publishing Company.

Undefined Terms. *point, line, plane, lie (or lie on), between, congruent*

I. Axioms of Incidence

I.1 For every two points A, B there exists a line a that contains each of the points A, B.[1]

I.2 For every two points A, B there exists no more than one line that contains each of the points A, B.

I.3 There exist at least two points on a line. There exist at least three points that do not lie on a line.

I.4 For any three points A, B, C that do not lie on the same line there exists a plane α that contains each of the points A, B, C. For every plane, there exists a point which it contains.

II. Axioms of Order

II.1 If a point B lies between a point A and a point C, then the points A, B, C are three distinct points of a line, and B also lies between C and A.

II.2 For two points A and C, there always exists at least one point B on the line AC such that C lies between A and B.

II.3 Of any three points on a line there exists no more than one that lies between the other two.

---

[1] Two or more points "are always to be understood as distinct points."

II.4 Let A, B, C be three points that do not lie on a line and let a be a line in the plane ABC which does not meet any of the points A, B, C. If the line a passes through a point of the segment AB, it also passes through a point of the segment AC, or through a point of the segment BC.

III. Axioms of Congruence

III.1 If A, B are two points on a line a, and if A' is a point on the same or on another line a' then it is always possible to find a point B' on a given side of the line a' through A' such that the segment AB is congruent or equal to the segment $A'B'$. In symbols $AB \equiv A'B'$.

III.2 If a segment $A'B'$ and a segment $A''B''$ are congruent to the same segment AB, then the segment $A'B'$ is also congruent to the segment $A''B''$, or briefly, if two segments are congruent to a third one they are congruent to each other.

III.3 On a line a let AB and BC be two segments which except for B have no point in common. Furthermore, on the same or on another line a' let $A'B'$ and $B'C'$ be two segments which except for B' also have no point in common. In that case, if $AB \equiv A'B'$ and $BC \equiv B'C'$ then $AC \equiv A'C'$.

III.4 Let $\angle(h, k)$ be an angle in a plane α and a' a line in a plane α' and let a definite side of a' in α' be given. Let h' be a ray on the line a' that emanates from the point O'. Then there exists in the plane α' one and only one ray k' such that the angle $\angle(h, k)$ is congruent or equal to the angle $\angle(h', k')$ and at the same time all interior points of the angle $\angle(h', k')$ lie on the given side of a'. Symbolically $\angle(h, k) \equiv \angle(h', k')$. Every angle is congruent to itself, i.e., $\angle(h, k) \equiv \angle(h, k)$ is always true.

III.5 If for two triangles ABC and $A'B'C'$ the congruences $AB \equiv A'B'$, $AC \equiv A'C'$, $\angle BAC \equiv \angle B'A'C'$ hold, then the congruence $\angle ABC \equiv \angle A'B'C'$ is also satisfied.[2]

IV. Axiom of Parallels

IV.1 Let a be any line and A a point not on it. Then there is at most one line in the plane, determined by a and A, that passes through A and does not intersect a.

V. Axioms of Continuity

V.1 (Axiom of measure or Archimedes' Axiom) If AB and CD are any segments then there exists a number n such that n segments CD constructed contiguously from A, along the ray from A to B, will pass beyond the point B.

V.2 (Axiom of line completeness) An extension of a set of points on a line with its order and congruence relations that would preserve the relations existing among the original elements as well as the fundamental properties of line order and congruence that follows from Axioms I-III and from V.1 is impossible.

---

[2]"Here, and in what follows, the vertices of a triangle shall always be supposed not to lie on the same line."

Credits, Permissions and Acknowledgements

Figure 1.1: *Circles in a Circle* by Vasily Kandinsky (1923) courtesy of the Philadelphia Museum of Art, Object Number 1950-134-104, The Louise and Walter Arensberg Collection, 1950.

Excerpt from "The Stretched String" in Section 1.2: Excerpt of the essay from Inner Issues of *The Mathematical Experience* by Philip J. Davis, Reuben Hersh and Elena Anne Marchisotto, is reprinted by permission from Springer Nature ©2012 [**30**].

Excerpt from *Black Elk Speaks: The Complete Edition* in Section 1.2: Excerpt from the book [**90**] by John G. Neihardt is reprinted by permission of the University of Nebraska Press ©2014 by the Board of Regents of the University of Nebraska.

Figure 1.2: National Museum of the American Indian, Washington, D.C., photograph in the Carol M. Highsmith Archive, Library of Congress, Prints and Photographs Division.

Figure 2.1: CALVIN AND HOBBES ©1991 Watterson. Reprinted with permission of UNIVERSAL Uclick. All rights reserved.

Euclid's *Elements*: Common notions, postulates, selected definitions, propositions and proofs from Books I-VI of Sir Thomas Heath's translation of the *Elements* [**40**] appearing in Chapters 2, 3, 7-11, and Appendices A, B and D, are reprinted with the permission of Dover Publications.

Figure 3.1: Proposition I.1 of Oliver Byrne's *The Elements of Euclid* [**19**] courtesy of the University of Toronto Libraries.

Figure 4.1: Photograph courtesy of the NASA Earth Observatory, *earthobservatory.nasa.gov*

Figure 6.1: Hilbert photograph from the Archives of the Mathematisches Forschungsinstitut Oberwolfach Archives. Gödel photograph from the Kurt Gödel Papers, the Shelby White and Leon Levy Archives Center, Institute for Advanced Study, Princeton, NJ, on deposit at Princeton University. Photographer unknown.

Hilbert's axioms: As found in Section 6.3 and Appendix E, Hilbert's axioms are reprinted by permission of Open Court Publishing Company, a division of Carus Publishing Company, Chicago, IL, from *Foundations of Geometry* [**74**] by D. Hilbert (trans. L. Unger), ©1971 by Open Court Publishing Company.

Figure 7.35: Photograph of Babylonian tablet Plimpton 322 by C. Proust courtesy of the Rare Book & Manuscript Library at Columbia University. Used with permission. *isaw.nyu.edu/exhibitions/before-pythagoras*

Figure 10.1: Olympic rings used with permission from the United States Olympic Committee.

Figure 11.35: Close-up of 17-point star on statue of Gauss from Benutzer:Brunswyk via Wikimedia Commons, licensed under the Creative Commons Attribution-Share Alike 3.0 Unported license

Figure 11.36: Homage to Carl Friedrich Gauss by Eugen Jost. Plate 23, p. 74, from BEAUTIFUL GEOMETRY by Eli Maor and Eugen Jost [86]. Copyright ©2014 by Princeton University Press. Reprinted by permission. Artist contact: eugenjost@bluewin.ch

Figure 12.1: M.C. Escher's *Circle Limit III*, ©2013 The M.C. Escher Company-The Netherlands. All rights reserved. *www.mcescher.com*

Figure 12.3: Mercator projection, derived from NASA Earth Observatory Blue Marble series composite photograph, by Daniel R. Strebe, December 16, 2011, via Wikimedia Commons, lines enhanced by Elyn Rykken, licensed under the Creative Commons Attribution-Share Alike 3.0 Unported license.

Figure 12.4: Mollweide projection, derived from NASA Earth Observatory Blue Marble series composite photograph, by Daniel R. Strebe, August 15, 2011, via Wikimedia Commons, licensed under the Creative Commons Attribution-Share Alike 3.0 Unported license.

Figure 12.8: M.C. Escher's *Circle Limit IV*, ©2013 The M.C. Escher Company-The Netherlands. All rights reserved. *www.mcescher.com*

Figure 14.6: Albrecht Dürer's illustration, Wikimedia Commons.

Figure 14.7: Detail of Hans Holbein the Younger's *The Ambassadors* (1533), Wikimedia Commons.

Figure 14.8: Leonardo da Vinci's *The Last Supper* (1498), Wikimedia Commons, lines added by Maureen T. Carroll.

Section 14.5: Photograph of Waterloo tube station, London, courtesy of Daniel Carroll. All rights reserved.

Figure 15.1: Photograph of Thomas Jefferson memorial, Washington, DC, courtesy of Kathy Wahl. All rights reserved.

Figure 16.1: Detail of Raphael's *The School of Athens* (1510-11), Wikimedia Commons.

Figure 16.8: Crockett Johnson's *Heptagon from Its Seven Sides (1973)*, negative number 2008-2545, appears courtesy of the Division of Medicine & Science, National Museum of American History, the Smithsonian Institution.

Every effort has been made to contact copyright holders and obtain permissions for the use of copyright material. Notification of any corrections for future reprints would be greatly appreciated.

Bibliography

[1] Edwin Abbott Abbott, *Flatland, A romance of many dimensions*, Dover Publications, New York, 1992.

[2] Norbert A'Campo and Athanase Papadopoulos, *Notes on non-Euclidean geometry*, Strasbourg master class on geometry, IRMA Lect. Math. Theor. Phys., vol. 18, Eur. Math. Soc., Zürich, 2012, pp. 1–182, DOI 10.4171/105-1/1. MR2931886

[3] Archimedes, *The works of Archimedes*, Dover Publications, Inc., Mineola, NY, 2002. Reprint of the 1897 edition and the 1912 supplement; Edited by T. L. Heath. MR2000800

[4] The Archimedes Palimpsest, Walters Art Museum, *archimedespalimpsest.org*

[5] M.N. Aref and William Wernick, *Problems and solutions in Euclidean geometry*, Dover Publication, New York, 1968.

[6] Marcia Ascher, *Ethnomathematics: A multicultural view of mathematical ideas*, Brooks/Cole Publishing Co., Pacific Grove, CA, 1991. MR1095781

[7] Leon Bankoff, *The metamorphosis of the butterfly problem*, Math. Mag. **60** (1987), no. 4, 195–210, DOI 10.2307/2689339. MR960422

[8] Arthur Baragar, *Constructions using a compass and twice-notched straightedge*, Amer. Math. Monthly **109** (2002), no. 2, 151–164, DOI 10.2307/2695327. MR1903152

[9] Elliot Benjamin and C. Snyder, *On the construction of the regular hendecagon by marked ruler and compass*, Math. Proc. Cambridge Philos. Soc. **156** (2014), no. 3, 409–424, DOI 10.1017/S0305004113000753. MR3181633

[10] M. K. Bennett, *Affine and projective geometry*, A Wiley-Interscience Publication, John Wiley & Sons, Inc., New York, 1995. MR1344447

[11] Garrett Birkhoff and M. K. Bennett, *Felix Klein and his "Erlanger Programm"*, History and philosophy of modern mathematics (Minneapolis, MN, 1985), Minnesota Stud. Philos. Sci., XI, Univ. Minnesota Press, Minneapolis, MN, 1988, pp. 145–176. MR945470

[12] R. D. Bohannan, *The nine-points circle*, Ann. of Math. **1** (1884), no. 5, 112, DOI 10.2307/1967589. MR1502022

[13] R. C. Bose, *On the application of the properties of Galois fields to the problem of construction of hyper-Graeco-Latin squares*, Sankhya **3** (1938), 323–338.

[14] R. C. Bose and S. S. Shrikhande, *On the falsity of Euler's conjecture about the non-existence of two orthogonal Latin squares of order 4t + 2*, Proc. Nat. Acad. Sci. U.S.A. **45** (1959), 734–737. MR0104590

[15] Carl B. Boyer, *A history of mathematics*, Reprint of the 1968 original. Princeton University Press, Princeton, NJ, 1985. MR784225

[16] R. H. Bruck and H. J. Ryser, *The nonexistence of certain finite projective planes*, Canadian J. Math. **1** (1949), 88–93. MR0027520

[17] Burton, David M, *The History of mathematics: An introduction*, 6th ed., McGraw Hill Higher Education, New York, 2007.

[18] W. H. Bussey, *Geometric constructions without the classical restriction to ruler and compasses*, Amer. Math. Monthly **43** (1936), no. 5, 265–280, DOI 10.2307/2301710. MR1523652

[19] Oliver Byrne, *The first six books of the Elements of Euclid in which coloured diagrams and symbols are used instead of letters for the greater ease of learners*, Facsimile of the first edition of 1847; The accompanying commentary volume contains two essays by Werner Oechslin. Taschen GmbH, Cologne, 2010. MR2807282

[20] Florian Cajori, *A history of mathematical notations*, Including Vol. I. Notations in elementary mathematics; Vol. II. Notations mainly in higher mathematics; Reprint of the 1928 and 1929 originals. Dover Publications, Inc., New York, 1993. MR3363427

[21] Maureen T. Carroll and Steven T. Dougherty, *Tic-tac-toe on a finite plane*, Math. Mag. **77** (2004), no. 4, 260–274, DOI 10.2307/3219284. MR2087313

[22] Stephanie Cawthorne and Judy Green, *Harold and the purple heptagon*, Math Horizons, Vol. 17, No.1 (Sept. 2009), 5-9.

[23] Judith N. Cederberg, *A course in modern geometries*, Undergraduate Texts in Mathematics, Springer-Verlag, New York, 1989. MR1013997

[24] Cicero, Marcus Tullius, *Tusculan disputations*, trans. by C.D. Yonge, Harper & Brothers, New York, 1877.

[25] B. A. Cipra, *Computer search solves an old math problem*, Science **242** (1988), 1507-1508.

[26] Calvin C. Clawson, *Mathematical mysteries: The beauty and magic of numbers*, Plenum Press, New York, 1996. MR1450868

[27] J. L. Coolidge, *The Rise and Fall of Projective Geometry*, Amer. Math. Monthly **41** (1934), no. 4, 217–228, DOI 10.2307/2302023. MR1523063

[28] H. S. M. Coxeter and S. L. Greitzer, *Geometry revisited*, The Mathematical Association of America, Washington, DC, 1967.

[29] H. S. M. Coxeter, *Projective geometry*, Revised reprint of the second (1974) edition. Springer-Verlag, New York, 1994. MR1335229

[30] Philip J. Davis, Reuben Hersh, and Elena Anne Marchisotto, *The mathematical experience, study edition*, Modern Birkhäuser Classics, Birkhäuser/Springer, New York, 2012. MR2893927

[31] Khoshchard DeLong, *A profile of mathematical logic*, Addison-Wesley Publishing Co., Reading, Mass.-London-Don Mills, Ont., 1970. MR0258584

[32] R. Descartes, *Oeuvres de Descartes*, Vol. II, Eds. C. Adam and P. Tannery, Librairie Philosophique J. Vrin, Paris, 1988.

[33] A. V. Dorodnov, *On circular lunes quadrable with the use of ruler and compass* (Russian), Doklady Akad. Nauk SSSR (N. S.) **58** (1947), 965–968. MR0022822

[34] Heinrich Dörrie, *100 great problems of elementary mathematics: Their history and solution*, Translated by David Antin, Dover Publications, Inc., New York, 1965. MR0183598

[35] *The dot and the line: A romance in lower mathematics*, dirs: Chuck Jones and Maurice Noble, Metro-Goldwyn-Mayer, 1965. Animated short film.

[36] Douglas Cardinal Architect, *www.djcarchitect.com*

[37] William Dunham, *Journey through genius: The great theorems of mathematics*, Penguin Books, New York, 1991. MR1147417

[38] G. Waldo Dunnington, *Carl Friedrich Gauss: Titan of science. A study of his life and work*, Exposition Press, New York, 1955. MR0072814

[39] Euclid, *Euclid's Elements*, All thirteen books complete in one volume; The Thomas L. Heath translation; Edited by Dana Densmore. Green Lion Press, Santa Fe, NM, 2002. MR1932864

[40] Euclid, *The thirteen books of Euclid's Elements translated from the text of Heiberg. Vol. I: Introduction and Books I, II. Vol. II: Books III–IX. Vol. III: Books X–XIII and Appendix*, Translated with introduction and commentary by Thomas L. Heath; 2nd ed. Dover Publications, Inc., New York, 1956. MR0075873

[41] Leonhard Euler, *Elements of algebra*, Translated from the German by John Hewlett; Reprint of the 1840 edition; With an introduction by C. Truesdell. Springer-Verlag, New York, 1984. MR766740

[42] Leonhard Euler, *Recherches sur une nouvelle espace de quarees magiques*, Verh. Zeeuwsch Genootsch. Wetensch. Vlissengen **9** (1782), 85-239. Reprinted in L. Euler, *Opera Omnia*, ser. 1, vol. **7**, Tuebner, Berlin-Leipzig, 1923, 291-392.

[43] Howard Eves, *An Introduction to the history of mathematics*, Third edition, Holt, Rinehart and Winston, New York-Montreal, Que.-London, 1969. MR0241230

[44] Howard Eves, *A survey of geometry*, Revised edition, Allyn and Bacon, Inc., Boston, Mass., 1972. MR0322653

[45] Howard Eves, *Foundations and fundamental concepts of mathematics*, 3rd ed., With a foreword by Bruce E. Meserve. The Prindle, Weber & Schmidt Series in Advanced Mathematics, PWS-KENT Publishing Co., Boston, MA, 1990. MR1157634

[46] G. Fano, *Sui postulati fondamentali della geometria Proiettiva*, Giornale di Matematiche **30** (1892), 106–132.

[47] Robert Fano, *In loving memory of my father Gino Fano*, The Fano Conference, Univ. Torino, Turin, 2004, pp. 1–4. MR2112563

[48] John Fauvel and Jeremy Gray (eds.), *The history of mathematics: a reader*, Reprint of the 1987 edition. Open University Set Book, Macmillan Press, Ltd., Basingstoke; distributed by Sheridan House, Inc., Dobbs Ferry, NY, 1988. MR942906

[49] *Final report of the National Committee of Fifteen on geometry syllabus*, The Mathematics Teacher, **5** (1912), 46-131.

[50] Archibald H. Finlay, *Zig-zag paths*, Mathmatical Gazette Vol. 43, No. 345 (Oct. 1959), 199.

[51] *Flatland: The Film,* dir: Ladd Ehlinger Jr., Flatland Productions, 2007. Film.

[52] *Flatland: The Movie,* dirs: Dano Johnson and Jeffrey Travis, Flat World Productions, 2007. Film.

[53] *Flatland 2: Sphereland,* dir: Dano Johnson, Sphere World Productions, 2012. Film.

[54] Hans Freudenthal, The Main Trends in the Foundations of Geometry in the 19th Century, chapter in *Logic, methodology and philosophy of science,* Ed. by E. Nagel, P. Suppes and A. Tarski, Stanford University Press, Stanford, 1962.

[55] Joseph A. Gallian, *Contemporary abstract algebra,* 8th ed., Brooks/Cole, Cengage Learning, United States, 2013.

[56] James, A. Garfield, *New-England Journal of Education,* Vol. 3, No.14, (April 1, 1876), 161.

[57] Carl Friedrich Gauss, *Werke,* Vol. I, Königlichen Gesellschaft Der Wissenschaften, Göttingen, 1900.

[58] Carl Friedrich Gauss, *Werke,* Vol. VIII, Königlichen Gesellschaft Der Wissenschaften, Göttingen, 1900.

[59] Andrew M. Gleason, *Angle trisection, the heptagon, and the triskaidecagon,* Amer. Math. Monthly **95** (1988), no. 3, 185–194, DOI 10.2307/2323624. MR935432

[60] Louise Golland, *Karl Menger and taxicab geometry,* Math. Mag. **63** (1990), no. 5, 326–327, DOI 10.2307/2690903. MR1081277

[61] Judith V. Grabiner, *Why did Lagrange "prove" the parallel postulate?,* Amer. Math. Monthly **116** (2009), no. 1, 3–18, DOI 10.4169/193009709X469779. MR2478750

[62] Jeremy J. Gray, *János Bolyai, non-Euclidean geometry, and the nature of space,* With a foreword by Benjamin Weiss, a facsimile of Bolyai's *Appendix,* and an 1891 English translation by George Bruce Halsted. Burndy Library Publications. New Series, vol. 1, Burndy Library, Cambridge, MA, 2004. MR2091226

[63] Jeremy Gray, *Gauss and non-Euclidean geometry,* Math. Appl. (N. Y.), vol. 581, Springer, New York, 2006, pp. 61–80, DOI 10.1007/0-387-29555-0_2. MR2190839

[64] Jeremy Gray, *Worlds out of nothing*: A course in the history of geometry in the 19th century, A course in the history of geometry in the 19th century; Second edition [of MR2305283]. Springer Undergraduate Mathematics Series, Springer-Verlag London, Ltd., London, 2011. MR2760764

[65] Marvin Jay Greenberg, *Euclidean and non-Euclidean geometries: Development and history,* 4th ed., W. H. Freeman and Co., New York, 2008.

[66] Laura Guggenbuhl and Karl Wilhelm Feuerbach, *Mathematician,* The Scientific Monthly **81** (1955), 71-76.

[67] Robin Hartshorne, *Geometry: Euclid and beyond,* Undergraduate Texts in Mathematics, Springer-Verlag, New York, 2000. MR1761093

[68] Susan M. Hawes and Sid Kolpas, *Oliver Byrne: The Matisse of mathematics,* Convergence **12** (August 2015).

[69] Thomas Hawkins, *The Erlanger programm of Felix Klein: Reflections on its place in the history of mathematics* (English, with French and German summaries), Historia Math. **11** (1984), no. 4, 442–470, DOI 10.1016/0315-0860(84)90028-4. MR773797

[70] David W. Henderson, *Experiencing geometry: In Euclidean, spherical, and hyperbolic spaces,* 2nd ed., Prentice Hall, Upper Saddle River, 2001.

[71] Herodotus, translation by A. D. Godley. *Herodotus,* Book 2, Chap. 109, Harvard University Press, Cambridge, MA, 1990.

[72] I. N. Herstein, *Topics in algebra,* 2nd ed., John Wiley & Sons, New York, 1975.

[73] G. Hessenberg, *Uber die projective geometrie,* Sitzungsberichte der Berliner Mathematischen Gesellschaft 1902-03, Berliner Mathematische Gesellschaft, B. G. Teubner, Berlin, 1903.

[74] David Hilbert, *Foundations of geometry,* Second edition. Translated from the tenth German edition by Leo Unger, Open Court, LaSalle, Ill., 1971. MR0275262

[75] Crockett Johnson, *A construction for a regular heptagon,* Mathematical Gazette Vol. 59, No. 407 (Mar. 1975), 17-21.

[76] Juster, Norton, *The dot and the line: A romance in lower mathematics,* Random House, New York, 1963.

[77] Victor J. Katz, *A history of mathematics: An introduction,* HarperCollins College Publishers, New York, 1993. MR1200894

[78] Victor Katz (ed.), *The mathematics of Egypt, Mesopotamia, China, India, and Islam: A sourcebook,* Princeton University Press, Princeton, NJ, 2007. MR2368469

[79] Deborah A. Kent and David J. Muraki, *A geometric solution of a cubic by Omar Khayyam ... in which colored diagrams are used instead of letters for the greater ease of learners,* Amer. Math. Monthly **123** (2016), no. 2, 149–160, DOI 10.4169/amer.math.monthly.123.2.149. MR3470505

[80] L. C. Kinsey, T. Moore, and E. Prassidis, *Geometry & symmetry,* John Wiley & Sons, Hoboken, 2011.

[81] Morris Kline, *Mathematical thought from ancient to modern times,* Oxford University Press, New York, 1972. MR0472307

[82] Eugene F. Krause, *Taxicab geometry: An adventure in non-Euclidean geometry*, Dover Publications, Inc, 1986.

[83] C. W. H. Lam, *The search for a finite projective plane of order* 10, Amer. Math. Monthly **98** (1991), no. 4, 305–318, DOI 10.2307/2323798. MR1103185

[84] D. N. Lehmer, *An elementary course in synthetic projective geometry*, Ginn and Co., Boston, 1917.

[85] Elisha Scott Loomis, *The Pythagorean proposition*, National Council of Teachers of Mathematics, Washington, 1968.

[86] Eli Maor and Eugen Jost, *Beautiful geometry*, Princeton University Press, Princeton, NJ, 2014. MR3155027

[87] Edwin E. Moise, *Elementary geometry from an advanced standpoint*, 3rd ed., Addison-Wesley Publishing Company, Advanced Book Program, Reading, MA, 1990. MR1243182

[88] Christopher Morley, *Hide and seek*, George H. Doran Co., New York, 1920.

[89] Ernest Nagel and James R. Newman, *Gödel's proof*, Revised edition, Edited and with a new foreword by Douglas R. Hofstadter, New York University Press, New York, 2001. MR1871678

[90] Neihardt, John G., *Black Elk speaks: The complete edition*, University of Nebraska Press, Lincoln, 2014. (First edition 1932.)

[91] Ivan Niven, *The transcendence of* π, Amer. Math. Monthly **46** (1939), 469–471, DOI 10.2307/2302515. MR0000415

[92] Helen McKelvey Oakley, *Three hours for lunch: The life and times of Christopher Morley: A biography*, Watermill Publishers, New York, 1976.

[93] J. A. Osmundsen, Major Mathematical Conjecture Propounded 177 Years Ago Is Disproved: 3 MATHEMATICIANS SOLVE, *New York Times*, Apr. 26, 1959, 1, 42.

[94] Alpay Özdural, *Mathematics and arts: connections between theory and practice in the medieval Islamic world* (English, with English and Turkish summaries), Text in English and Arabic. Historia Math. **27** (2000), no. 2, 171–201, DOI 10.1006/hmat.1999.2274. MR1757142

[95] Dan Pedoe, *Geometry and the visual arts*, Dover Publications, Inc., New York, 1976.

[96] Plato, *The dialogues of Plato translated into English with analyses and introductions by B. Jowett, M.A. in Five Volumes*, 3rd edition revised and corrected, Oxford University Press, London, 1892.

[97] J. V. Poncelet, *Traité des propriétés projectives des figures*, Bachelier, Paris, 1822.

[98] Proclus, *A commentary on the first book of Euclid's Elements*, Translated with introduction and notes by Glenn R. Morrow, Princeton University Press, Princeton, N.J., 1970. MR0269463

[99] Constance Reid, *Hilbert*, Springer-Verlag, New York, 1970.

[100] H. W. Richmond, *A construction for a regular polygon of seventeen sides*, Quarterly Journal of Pure and Applied Mathematics, **26** (1893), 206-207.

[101] Fred S. Roberts, *Applied combinatorics*, Prentice Hall, Inc., Englewood Cliffs, NJ, 1984. MR735619

[102] Saccheri, Girolamo, *Euclides vindicatus*, Edited and translated George Bruce Halsted, Open Court Publishing, Chicago, 1920.

[103] Juan Carlos Salazar, *Fuss' theorem*, Mathematical Gazette **90** (2006), 306-07.

[104] Charles T. Salkind, comp., *The contest problem book I: Annual high school mathematics examinations 1950 - 1960*, Mathematical Association of America, Washington, D.C., 1961.

[105] C. Edward Sandifer, *How Euler did even more*, With a preface by Rob Bradley; With chapters by Bradley and Dominic Klyve. Mathematical Association of America, Washington, DC, 2015. MR3308385

[106] Christoph J. Scriba, *Welche Kreismonde sind elementar quadrierbar? Die 2400 jährige Geschichte eines Problems bis zur endgültigen Lösung in den Jahren 1933/1947* (German), Mitt. Math. Ges. Hamburg **11** (1988), no. 5, 517–539. MR935082

[107] Thomas Q. Sibley, *The geometric viewpoint: A survey of geometries*, Addison-Wesley, Reading, 1998.

[108] D.E. Smith, *A source book in mathematics*, McGraw-Hill Book Co., New York, 1929.

[109] Katye O. Sowell, *Taxicab geometry—a new slant*, Math. Mag. **62** (1989), no. 4, 238–248, DOI 10.2307/2689762. MR1020431

[110] Saul Stahl, *The Poincaré half-plane: A gateway to modern geometry*, Jones and Bartlett Publishers, Boston, MA, 1993. MR1217085

[111] Saul Stahl, *Geometry from Euclid to knots*, Dover Publications, Mineola, 2010.

[112] Anthony Stanley, *An elementary treatise of spherical geometry and trigonometry*, Revised Edition, Durrie and Peck, New Haven, 1854.

[113] Fredrick W. Stevenson, *Projective planes*, W. H. Freeman and Co., San Francisco, Calif., 1972. MR0344995

[114] Mabel Sykes, *A source book of problems for geometry, based upon industrial design and architectural ornament*, Dale Seymour Publications, 1994. Originally published in 1912 by Norwood Press, Norwood, MA.

[115] G. Tarry, *Le problème des 36 officers*, Compte Rendu Ass. Franc. Pour l'avancement des Sciences **2** (1901), 170-203.

[116] Ivor Thomas, *Selections illustrating the history of Greek mathematics*, Harvard University Press, Cambridge, MA, 1957.

[117] I. Todhunter, *Spherical trigonometry: For the use of colleges and schools*, Third Edition, Enlarged, Macmillan and Co, London and New York, 1871.

[118] Michael Toepell, *On the origins of David Hilbert's Grundlagen der geometrie*, Arch. Hist. Exact Sci. **35** (1986), no. 4, 329–344, DOI 10.1007/BF00357305. MR851072

[119] Richard J. Trudeau, *The non-Euclidean revolution*, With an introduction by H. S. M. Coxeter. Birkhäuser Boston, Inc., Boston, MA, 1987. MR888822

[120] Nikolaj Tschebotaröw, *Über quadrierbare Kreisbogenzweiecke. I* (German), Math. Z. **39** (1935), no. 1, 161–175, DOI 10.1007/BF01201352. MR1545496

[121] Glen Van Brummelen, *Heavenly mathematics: The forgotten art of spherical trigonometry*, Princeton University Press, Princeton, NJ, 2013. MR3012466

[122] Oswald Veblen and W. H. Bussey, *Finite projective geometries*, Trans. Amer. Math. Soc. **7** (1906), no. 2, 241–259, DOI 10.2307/1986438. MR1500747

[123] Edward C. Wallace and Stephen F. West, *Roads to geometry*, 3rd ed., Pearson Education, Upper Saddle River, NJ, 2004.

[124] Pierre Wantzel, *Recherches sur les moyens de reconnaître si un problème de Géométrie peut se résoudre avec la règle et le compas*, Journal de mathématiques pures et appliquées 1re série, tome 2 (1837), 366-372.

[125] A. Wilansky, *Algebra and geometry: mathematics or science?*, The Mathematics Teacher **54** (1961), 339-343.

[126] Raymond L. Wilder, *Introduction to the foundations of mathematics*, Second edition, John Wiley & Sons, Inc., New York-London-Sydney, 1965. MR0182552

[127] J. W. Young, *Lectures on fundamental concepts of algebra and geometry*, Macmillan Co., New York, 1911.

Notation Index

Index